D0290637

Biological Magnetic Resonance
Volume 16

Modern Techniques in Protein NMR

A Continuation Order Plan is available for this series. A continuation order will bring delivery of each new volume immediately upon publication. Volumes are billed only upon actual shipment. For further information please contact the publisher.

Biological Magnetic Resonance
Volume 16

Modern Techniques in Protein NMR

Edited by

N. Rama Krishna
University of Alabama at Birmingham
Birmingham, Alabama

and

Lawrence J. Berliner
Ohio State University
Columbus, Ohio

KLUWER ACADEMIC / PLENUM PUBLISHERS
NEW YORK, BOSTON, DORDRECHT, LONDON, MOSCOW

The Library of Congress has catalogued the first volume of this series as follows:

Main entry under title:

Biological magnetic resonance:

Includes bibliographies and indexes.
1. Magnetic resonance. 2. Biology–Technique. I. Berliner, Lawrence, J. II. Reuben, Jacques.
QH324.9.M28B56 574.19'285 78-16035
AACR1

ISBN 0-306-45952-3

233 Spring Street, New York, N.Y. 10013

10 9 8 7 6 5 4 3 2 1

A C.I.P. record for this book is available from the Library of Congress.

Printed in the United States of America

In memory of
A. Murray Berliner

Contributors

G. Marius Clore • Laboratory of Chemical Physics, National Institute of Diabetes and Digestive and Kidney Diseases, National Institutes of Health, Bethesda, Maryland 20892-0520

Bennett T. Farmer II • Macromolecular NMR, Pharmaceutical Research Institute, Bristol-Myers Squibb, Princeton, New Jersey 08543-4000

Kevin H. Gardner • Protein Engineering Network Centres of Excellence and Departments of Medical Genetics and Microbiology, Biochemistry, and Chemistry, University of Toronto, Toronto, Ontario M5S 1A8, Canada

Jennifer J. Gesell • Department of Chemistry, University of Pennsylvania, Philadelphia, Pennsylvania 19104

C. Griesinger • Institut für Organische Chemie, Johann Wolfgang Goethe Universität Frankfurt, D-60439 Frankfurt am Main, Germany

Angela M. Gronenborn • Laboratory of Chemical Physics, National Institute of Diabetes and Digestive and Kidney Diseases, National Institutes of Health, Bethesda, Maryland 20892-0520

M. Hennig • Institut für Organische Chemie, Johann Wolfgang Goethe Universität Frankfurt, D-60439 Frankfurt am Main, Germany

Yasmin Karimi-Nejad • Department of NMR Spectroscopy, Bijvoet Center for Biomolecular Research, Utrecht University, 3584 CH Utrecht, The Netherlands

Lewis E. Kay • Protein Engineering Network Centres of Excellence and Departments of Medical Genetics and Microbiology, Biochemistry, and Chemistry, University of Toronto, Toronto, Ontario M5S 1A8, Canada

Ēriks Kupče • Varian NMR Instruments, Walton-on-Thames, Surrey KT12 2QF, England

Francesca M. Marassi • Department of Chemistry, University of Pennsylvania, Philadelphia, Pennsylvania 19104

J. P. Marino • Center for Advanced Research in Biotechnology, Rockville, Maryland 20850

Hiroshi Matsuo • Department of Biological Chemistry and Molecular Pharmacology, Harvard Medical School, Boston, Massachusetts 02115

Stanley J. Opella • Department of Chemistry, University of Pennsylvania, Philadelphia, Pennsylvania 19104

B. Reif • Institut für Organische Chemie, Johann Wolfgang Goethe Universität Frankfurt, D-60439 Frankfurt am Main, Germany

C. Richter • Institut für Organische Chemie, Johann Wolfgang Goethe Universität Frankfurt, D-60439 Frankfurt am Main, Germany

H. Schwalbe • Institut für Organische Chemie, Johann Wolfgang Goethe Universität Frankfurt, D-60439 Frankfurt am Main, Germany

Marco Tessari • Department of Biophysical Chemistry NSR Center, University of Nȳmegen Tuernoviveld 1, 6525 ED Nȳmegen, The Netherlands.

Ronald A. Venters • Duke University Medical Center, Durham, North Carolina 27710

Geerten W. Vuister • Department of Biophysical Chemistry NSR Center, University of Nȳmegen Tuernoviveld 1, 6525 ED Nȳmegen, The Netherlands.

Gerhard Wagner • Department of Biological Chemistry and Molecular Pharmacology, Harvard Medical School, Boston, Massachusetts 02115

Brian Whitehead • Department of Biophysical Chemistry NSR Center, University of Nȳmegen Tuernoviveld 1, 6525 ED Nȳmegen, The Netherlands.

Preface

Volume 16 marks the beginning of a special topic series devoted to modern techniques in protein NMR, under the Biological Magnetic Resonance series. This volume is being followed by Volume 17 with the subtitle *Structure Computation and Dynamics in Protein NMR*. Volumes 16 and 17 present some of the recent, significant advances in biomolecular NMR field with emphasis on developments during the last five years. We are honored to have brought together in these volumes some of the world's foremost experts who have provided broad leadership in advancing this field. Volume 16 contains advances in two broad categories: the first, Large Proteins, Complexes, and Membrane Proteins, and second, Pulse Methods. Volume 17, which will follow covers major advances in Computational Methods, and Structure and Dynamics.

In the opening chapter of Volume 16, Marius Clore and Angela Gronenborn give a brief review of NMR strategies including the use of long range restraints in the structure determination of large proteins and protein complexes. In the next two chapters, Lewis Kay and Ron Venters and their collaborators describe state-of-the-art advances in the study of perdeuterated large proteins. They are followed by Stanley Opella and co-workers who present recent developments in the study of membrane proteins. (A related topic dealing with magnetic field induced residual dipolar couplings in proteins will appear in the section on Structure and Dynamics in Volume 17).

The second section of Volume 16 is on pulse methods, and starts with a survey by Gerhard Wagner's research group on spin decoupling schemes in protein NMR including adiabatic pulses. Then Geerten Vuister and collaborators present some new pulse sequences for measuring coupling constants. In the final chapter, Christian Griesinger and his colleagues give a critical summary of new advances in determining torsion angle constraints in biomolecules, including cross-correlated relaxation measurements.

We are extremely proud of this compilation of excellent contributions describing significant advances in the biomolecular NMR field. Whether the field has already reached a state of maturity with only a few new advances (as occasionally suggested), or it is still rapidly evolving with exciting new developments around every corner, is a question that we will leave it up to the reader to ponder. As always, we welcome suggestions, comments, and criticisms for future volumes.

N. Rama Krishna
Lawrence J. Berliner

Contents

Section I. Large Proteins, Complexes, and Membrane Proteins

Chapter 1

Determining Structures of Large Proteins and Protein Complexes by NMR

G. Marius Clore and Angela M. Gronenborn

Chapter 2

Multidimensional ^{2}H-Based NMR Methods for Resonance Assignment, Structure Determination, and the Study of Protein Dynamics

Kevin H. Gardner and Lewis E. Kay

Chapter 3

NMR of Perdeuterated Large Proteins

Bennett T. Farmer II and Ronald A. Venters

Chapter 4

Recent Developments in Multidimensional NMR Methods for Structural Studies of Membrane Proteins

Francesca M. Marassi, Jennifer J. Gesell, and Stanley J. Opella

Section II. Pulse Methods

Chapter 5

Homonuclear Decoupling in Proteins

Ēriks Kupče, Hiroshi Matsuo, and Gerhard Wagner

Chapter 6

Pulse Sequences for Measuring Coupling Constants

Geerten W. Vuister, Marco Tessari, Yasmin Karimi-Nejad, and Brian Whitehead

Chapter 7

Methods for the Determination of Torsion Angle Restraints in Biomacromolecules

C. Griesinger, M. Hennig, J. P. Marino, B. Reif, C. Richter, and H. Schwalbe

Biological Magnetic Resonance
Volume 16

Modern Techniques in Protein NMR

I

Large Proteins, Complexes, and Membrane Proteins

1

Determining Structures of Large Proteins and Protein Complexes by NMR

G. Marius Clore and Angela M. Gronenborn

1. INTRODUCTION

It is axiomatic that a detailed understanding of the function of a macromolecule necessitates knowledge of its three-dimensional structure. At the present time, the two main techniques that can provide a complete description of the structures of macromolecules at the atomic level are X-ray crystallography in the solid state (single crystals) and nuclear magnetic resonance spectroscopy in solution. One of the rate-limiting factors in solving an X-ray structure is obtaining not only suitable crystals that diffract to sufficient resolution, but also appropriate heavy atom derivatives to determine the phases of the reflections accurately. Despite significant advances in crystallization methods and the advent of new developments such as multiple anomalous dispersion to facilitate phase determination, the number of protein X-ray structures solved to date is several orders of magnitude smaller than the number of available protein sequences. Unlike crystallography, NMR measure-

G. Marius Clore and Angela M. Gronenborn • Laboratory of Chemical Physics, National Institute of Diabetes and Digestive and Kidney Diseases, National Institutes of Health, Bethesda, Maryland 20892-0520.

Biological Magnetic Resonance, Volume 16: Modern Techniques in Protein NMR, edited by Krishna and Berliner. Kluwer Academic / Plenum Publishers, 1999.

ments are carried out in solution under potentially physiological conditions, and are therefore not hampered by the ability or inability of a protein to crystallize.

The size of macromolecular structures that can be solved by NMR has been dramatically increased over the last few years (Clore and Gronenborn, 1991a). The development of a wide range of 2D NMR experiments in the early 1980s culminated in the determination of the structures of a number of small proteins (Wüthrich, 1986; Clore and Gronenborn, 1987). Under exceptional circumstances, 2D NMR techniques can be applied successfully to the structure determination of proteins of up to 100 residues (Dyson *et al.*, 1990; Forman-Kay *et al.*, 1991). Beyond 100 residues, however, 2D NMR methods fail, principally because of spectral complexity that cannot be resolved in two dimensions. In the late 1980s and early 1990s, a series of major advances took place in which the spectral resolution was increased by extending the dimensionality to three and four dimensions (see Clore and Gronenborn, 1991a, for a review). In addition, by combining such multidimensional experiments with heteronuclear NMR, problems associated with large linewidths can be circumvented by making use of heteronuclear couplings that are large relative to the linewidths. Concomitant with the spectroscopic advances, improvements have also taken place in the accuracy with which macromolecular structures can be determined. Thus, it is now possible to determine the structures of proteins in the range of 15–35 kDa at a resolution comparable to crystal structures of ~2.5-Å resolution. The theoretical upper limit of applicability is around 60–70 kDa, and the largest single-chain proteins solved to date are ~30 kDa comprising ~260 residues (Garrett *et al.*, 1997; Martin *et al.*, 1997). Improvements in refinement techniques include direct refinement against accurate coupling constants (Garrett *et al.*, 1994) and ^{13}C and ^{1}H shifts (Kuszewski *et al.*, 1995a,b, 1996a), as well as the use of a conformational data-base potential (Kuszewski *et al.*, 1996b, 1997). More recently, new methods have been developed to obtain structural restraints that characterize long-range order *a priori* (Tjandra *et al.*, 1997a,b). These include making use of the dependence of heteronuclear relaxation on the rotational diffusion anisotropy of nonspherical molecules and of the field dependence of one-bond heteronuclear couplings arising from magnetic susceptibility anisotropy. In this review, we summarize some of the recent developments in multidimensional NMR of biological macromolecules.

2. BASIC PRINCIPLES OF MULTIDIMENSIONAL NMR

All 2D-NMR experiments comprise the same basic scheme (Ernst *et al.*, 1987): a preparation period, an evolution period (t_1) during which time the spins are labeled according to their chemical shift, a mixing period during which time the spins are correlated with each other, and finally a detection period (t_2). A number of experiments are recorded with successively incremented values of the evolution

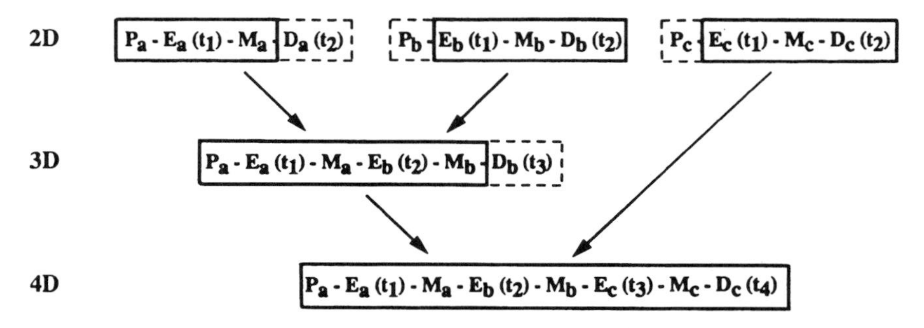

Figure 1. General representation of pulse sequences used in multidimensional NMR illustrating the relationship between the basic schemes used to record 2D, 3D, and 4D NMR spectra. Note construction of 3D and 4D experiments from the appropriate linear combination of 2D ones. P, preparation; E, evolution; M, mixing; D, detection. In 3D and 4D NMR, the evolution periods are incremented independently.

period t_1 to generate a data matrix $s(t_1,t_2)$. Two-dimensional Fourier transformation of $s(t_1,t_2)$ then yields the desired 2D frequency spectrum $S(\omega_1,\omega_2)$. In most homonuclear 2D experiments, the diagonal corresponds to the 1D spectrum and the symmetrically placed cross-peaks on either side of the diagonal indicate the existence of an interaction between two spins. The nature of the interaction depends on the type of experiment. Thus, in a correlation (COSY) experiment the cross-peaks arise from through-bond scalar correlations, whereas in a nuclear Overhauser enhancement (NOE) experiment they arise from through-space correlations.

The extension from 2D NMR to 3D and 4D NMR is straightforward and illustrated schematically in Fig. 1 (Oschkinat *et al.*, 1988). Thus, a 3D experiment is constructed from two 2D experiments by leaving out the detection period of the first 2D experiment and the preparation pulse of the second. This results in a pulse train comprising two independently incremented evolution periods t_1 and t_2, two corresponding mixing periods M_1 and M_2, and a detection period t_3. Similarly, a 4D experiment is obtained by combining three 2D experiments in an analogous fashion. The real challenge of 3D and 4D NMR is twofold: first, to ascertain which 2D experiments should be combined to best advantage, and second, to design the pulse sequences in such a way that undesired artifacts, which may severely interfere with the interpretation of the spectra, are removed.

3. THE NUCLEAR OVERHAUSER EFFECT

The main source of geometric information used in protein structure determination lies in the nuclear Overhauser effect (NOE), which can be used to identify protons separated by less than 5 Å. This distance limit arises from the fact that the NOE (at short mixing times) is proportional to the inverse sixth power of the distance between the protons. Hence, the NOE intensity falls off very rapidly with increasing distance between proton pairs. Despite the short-range nature of the observed interactions, the short approximate interproton distance restraints derived from NOE measurements can be highly conformationally restrictive, particularly when they involve residues that are far apart in the sequence but close together in space.

4. GENERAL STRATEGY FOR THE STRUCTURE DETERMINATION OF PROTEINS AND PROTEIN COMPLEXES BY NMR

The power of NMR over other spectroscopic techniques results from the fact that every NMR-active nucleus gives rise to an individual resonance in the spectrum which can be resolved by higher dimensional (i.e., 2D, 3D, and 4D) techniques. With this in mind, the principles of structure determination by NMR can be summarized very simply by the scheme shown in Fig. 2. The first step is to obtain sequential resonance assignments using a combination of through-bond and through-space correlations; the second step is to obtain stereospecific assignments at chiral centers and torsion-angle restraints using three-bond scalar couplings combined with intraresidue and sequential interresidue NOE data; the third step is to identify through-space connectivities between protons separated by less than 5 Å; and the final step involves calculating 3D structures on the basis of the experimental NMR restraints using one or more of a number of algorithms such as distance geometry and/or simulated annealing (Clore and Gronenborn, 1989). It is not essential to assign all of the NOEs initially. Indeed, many may be ambiguous and several possibilities may exist for their assignments. In such cases, the NOE restraints can be dealt with as an ambiguous $(\Sigma r^{-6})^{-1/6}$ sum restraint such that the restraint is satisfied if at least one of the potential proton pairs is close. Once a low-resolution structure has been calculated from a subset of the NOE data that can be interpreted unambiguously, it is then possible to employ iterative methods to resolve the vast majority of ambiguities. Consider for example an NOE cross-peak that could be attributable to a through-space interaction between either protons A and B or between protons A and C. Once a low-resolution structure is available, it is usually possible to discriminate between these two possibilities. Thus, if protons A and C are significantly greater than 5 Å apart while protons A and B are less than 5 Å apart, it is clear that the cross-peak must arise from an NOE between protons A and B.

4.1. Sample Requirements for NMR Spectroscopy

In the study of macromolecules, concentrations of about 1 mM are typically employed in a sample volume of 0.3–0.5 ml. A key requirement is that the

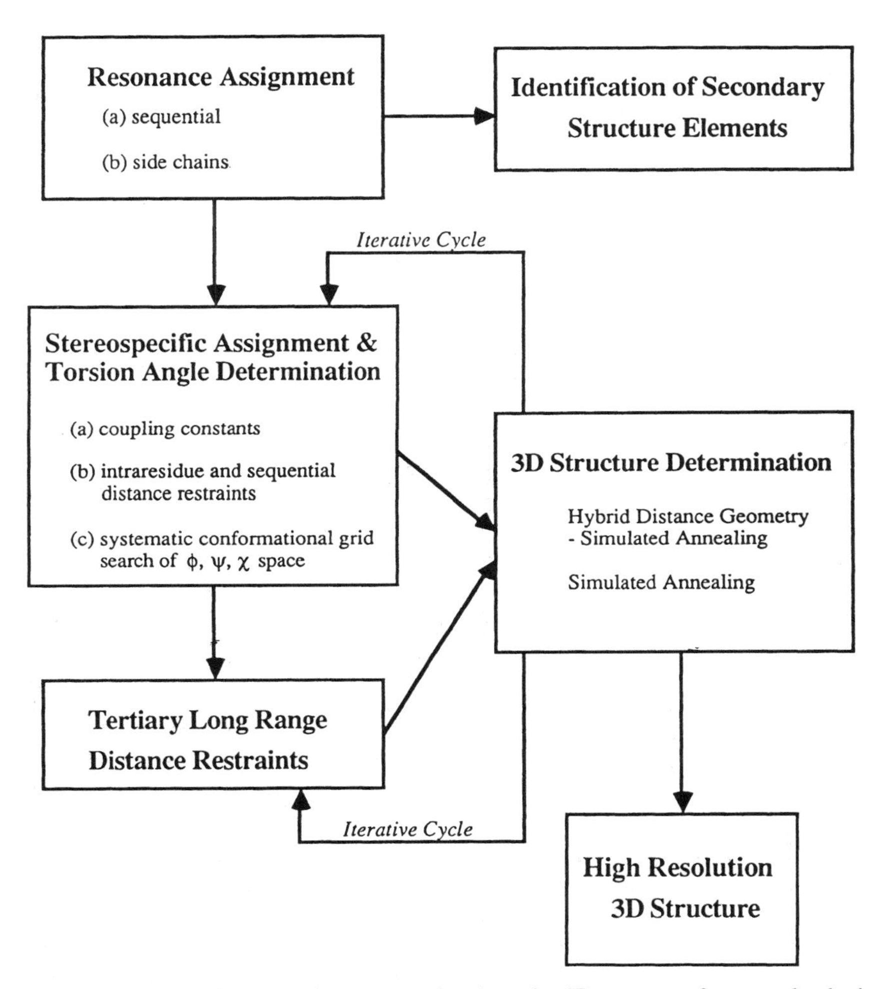

Figure 2. Summary of the general strategy employed to solve 3D structures of macromolecules by NMR.

macromolecule under study should be soluble, should not aggregate, and should be stable for many weeks at room temperature. For ^1H homonuclear work, it is also important to ensure that the buffer employed does not contain any protons. In general, two samples are required, one in D_2O for the observation of nonexchangeable protons only, and the other in 95% H_2O/5% D_2O to permit the observation of exchangeable protons.

Although it is possible to use ^1H homonuclear methods to solve structures of proteins of up to about 100 residues in certain very favorable cases (Dyson *et al.*, 1990; Forman-Kay *et al.*, 1991), it is generally the case that extensive resonance overlap makes this task very time consuming and complex. Hence, providing a protein can be overexpressed in a bacterial system, it is now desirable, even for proteins as small as 30 residues, to make use of the full panoply of multidimensional heteronuclear NMR experiments (Clore and Gronenborn, 1991a,b; Bax and Grzesiek, 1993). This necessitates the use of uniform ^{15}N and ^{13}C labeling, which can be achieved by growing the bacteria on minimal medium containing $^{15}NH_4Cl$ and $^{13}C_6$-glucose as the sole nitrogen and carbon sources, respectively. In general, the following samples are required: ^{15}N-labeled sample in 95% H_2O/5% D_2O, $^{15}N/^{13}C$-labeled sample in 95% H_2O/5% D_2O, and $^{15}N/^{13}C$-labeled sample in D_2O. For very large systems, deuteration and specific labeling is advantageous (Grzesiek *et al.*, 1995a; Venters *et al.*, 1995; Metzler *et al.*, 1996; Gardner *et al.*, 1997; Garrett *et al.*, 1997). For example, a particularly useful strategy for aromatic residues is one of reverse labeling in which the aromatics are at natural isotopic abundance, while the other residues are $^{13}C/^{15}N$ labeled (Vuister *et al.*, 1994). This can be achieved by adding the aromatic amino acids at natural isotopic abundance to the minimal medium in addition to $^{15}NH_4Cl$ and $^{13}C_6$-glucose. A similar approach can also be used for various aliphatic amino acids. Deuteration of nonexchangeable protons is achieved by growing the bacteria in D_2O as opposed to H_2O medium.

4.2. Sequential Assignment

Conventional sequential resonance assignment relies on 2D homonuclear ^1H–^1H correlation experiments to identify amino acid spin systems coupled with 2D ^1H–^1H NOE experiments to identify sequential connectivities along the backbone of the type $C^{\alpha}H(i)$–$NH(i + 1,2,3,4)$, $NH(i)$–$NH(i \pm 1)$, and $C^{\alpha}H(i)$–$C^{\beta}H(i + 3)$ (Wüthrich, 1986; Clore and Gronenborn, 1987). This methodology has been successfully applied to proteins of less than 100 residues, albeit with considerable effort (Dyson *et al.*, 1990; Forman-Kay *et al.*, 1991). For larger proteins, the spectral complexity is such that 2D experiments no longer suffice, and it is essential to increase the spectral resolution by increasing the dimensionality of the spectra (Clore and Gronenborn, 1991a). In some cases it is still possible to apply the same sequential assignment strategy by making use of 3D heteronuclear (^{15}N or ^{13}C) edited experiments to increase the spectral resolution (Driscoll *et al.*, 1990). Frequently, however, numerous ambiguities still remain and it is advisable to adopt a sequential assignment strategy based solely on well-defined heteronuclear scalar couplings along the polypeptide chain (Ikura *et al.*, 1990; Clore and Gronenborn, 1991b; Bax and Grzesiek, 1993), as shown in Fig. 3. The double- and triple-resonance experiments that we currently use together with the correlations that they demonstrate, are summarized in Table 1. With the advent of

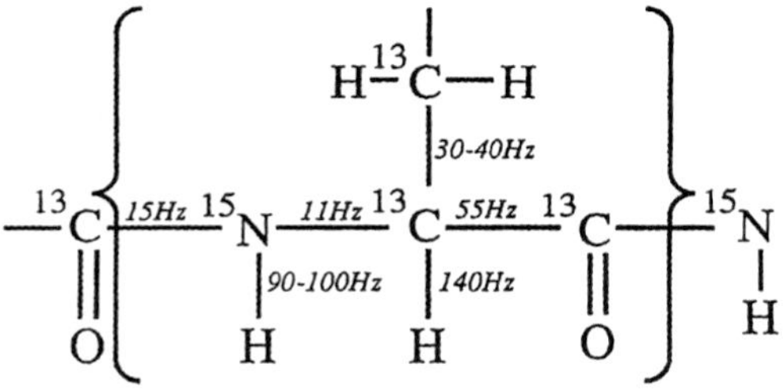

Figure 3. Summary of the one-bond heteronuclear couplings along the polypeptide chain utilized in 3D and 4D NMR experiments.

Table 1
Summary of Correlations Observed in the 3D Double- and Triple-Resonance Experiments Used for Sequential and Sidechain Assignments[a]

Experiment	Correlation	J Coupling[a]
^{15}N-edited HOHAHA	$C^{\alpha}H(i)-^{15}N(i)-NH(i)$	$^{3}J_{HN\alpha}$
	$C^{\beta}H(i)-^{15}N(i)-NH(i)$	$^{3}J_{HN\alpha}$ and $^{3}J_{\alpha\beta}$
HNHA	$C^{\alpha}H(i)-^{15}N(i)-NH(i)$	$^{3}J_{HN\alpha}$
H(CA)NH	$C^{\alpha}H(i)-^{15}N(i)-NH(i)$	$^{1}J_{NC\alpha}$
	$C^{\alpha}H(i-1)-^{15}N(i)-NH(i)$	$^{2}J_{NC\alpha}$
HNCA	$^{13}C^{\alpha}(i)-^{15}N(i)-NH(i)$	$^{1}J_{NC\alpha}$
	$^{13}C^{\alpha}(i-1)-^{15}N(i)-NH(i)$	$^{2}J_{NC\alpha}$
HN(CO)CA	$^{13}C^{\alpha}(i-1)-^{15}N(i)-NH(i)$	$^{1}J_{NCO}$ and $^{1}J_{C\alpha CO}$
HNCO	$^{13}CO(i-1)-^{15}N(i)-NH(i)$	$^{1}J_{NCO}$
HCACO	$C^{\alpha}H(i)-^{13}C^{\alpha}(i)-^{13}CO(i)$	$^{1}J_{C\alpha CO}$
HCA(CO)N	$C^{\alpha}H(i)-^{13}C^{\alpha}(i)-^{15}N(i+1)$	$^{1}J_{C\alpha CO}$ and $^{1}J_{NCO}$
CBCA(CO)NH	$^{13}C^{\beta}(i-1)/^{13}C^{\alpha}(i-1)-^{15}N(i)-NH(i)$	$^{1}J_{C\alpha CO}$, $^{1}J_{NCO}$, and $^{1}J_{CC}$
CBCANH or HNCACB	$^{13}C^{\beta}(i)/^{13}C^{\alpha}(i)-^{15}N(i)-NH(i)$	$^{1}J_{NC\alpha}$ and $^{1}J_{CC}$
	$^{13}C^{\beta}(i-1)/^{13}C^{\alpha}(i-1)-^{15}N(i)-NH(i)$	$^{2}J_{NC\alpha}$ and $^{1}J_{CC}$
HBHA(CO)NH	$C^{\beta}H(i-1)/C^{\alpha}H(i-1)-^{15}N(i)-NH(i)$	$^{1}J_{C\alpha CO}$, $^{1}J_{NCO}$, and $^{1}J_{CC}$
HBHA(CBCA)NH	$C^{\beta}H(i)/C^{\alpha}H(i)-^{15}N(i)-NH(i)$	$^{1}J_{NC\alpha}$ and $^{1}J_{CC}$
	$C^{\beta}H(i-1)/C^{\alpha}H(i-1)-^{15}N(i)-NH(i)$	$^{2}J_{NC\alpha}$ and $^{1}J_{CC}$
C(CO)NH	$^{13}C^{j}(i-1)-^{15}N(i)-NH(i)$	$^{1}J_{C\alpha CO}$, $^{1}J_{NCO}$, and $^{1}J_{CC}$
H(CCO)NH	$H^{j}(i-1)-^{15}N(i)-NH(i)$	$^{1}J_{C\alpha CO}$, $^{1}J_{NCO}$, and $^{1}J_{CC}$
HCCH-COSY	$H^{j}-^{13}C^{j}-^{13}C^{j\pm1}-H^{j\pm1}$	$^{1}J_{CC}$
HCCH-TOCSY	$H^{j}-^{13}C^{j}....^{13}C^{j\pm n}-H^{j\pm n}$	$^{1}J_{CC}$

[a]In addition to the couplings indicated, all of the experiments make use of the $^{1}J_{CH}$ (~140 Hz) and/or $^{1}J_{NH}$ (~95 Hz) couplings. The values of the couplings employed are as follows: $^{3}J_{HN\alpha}$ ~3–10 Hz, $^{1}J_{CC}$ ~35 Hz, $^{1}J_{C\alpha CO}$ ~55 Hz, $^{1}J_{NCO}$ ~15 Hz, $^{1}J_{NC\alpha}$ ~11Hz, $^{2}J_{NC\alpha}$ ~7 Hz. Details of the experiments and original references are provided in the reviews by Clore and Gronenborn (1991a,b), Bax and Grzesiek (1993), and Gronenborn and Clore (1995).

pulsed field gradients either to eliminate undesired coherence transfer pathways (BaxandPochapsky, 1992) or to select particular coherence pathways coupled with sensitivity enhancement (Kay *et al.*, 1992), it is now possible to employ only two- to four-step phase cycles without any loss in sensitivity (other than that related to the reduction in measurement time) such that each 3D experiment can be recorded in as little as 7 h. In most cases, however, signal-to-noise requirements necessitate a measuring time of 1–3 days depending on the experiment. For proteins of greater than ~25 kDa, the assignment of the backbone and sidechain carbons is facilitated by making use of a sample in which the nonexchangeable protons are deuterated, thereby dramatically reducing the linewidths (Grzesiek *et al.*, 1995a; Venters *et al.*, 1995; Garrett *et al.*, 1997). Thus, for example, in the case of the 30-kDa N-terminal domain of enzyme I (EIN), the average T_2 for the backbone amides is increased from ~13 ms in the protonated sample to ~28 ms in the perdeuterated sample (Garrett *et al.*, 1997). The resulting spectral improvement is readily ascertained from a comparison of the 2D $^{1}H–^{15}N$ correlation spectra of protonated and perdeuterated EIN (Fig. 4). The dramatic improvements attainable by deuteration

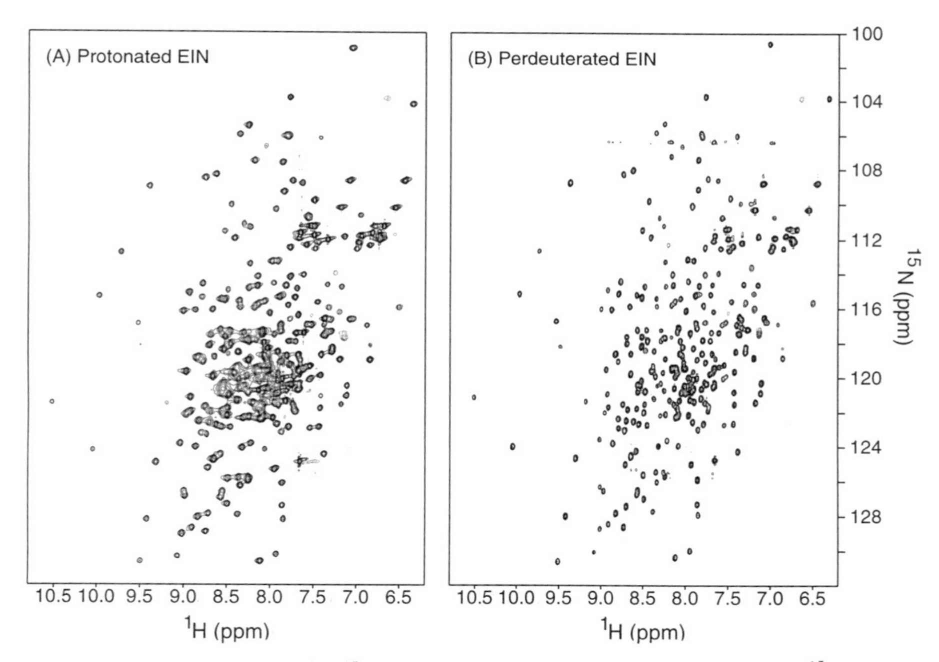

Figure 4. Comparison of the $^{1}H–^{15}N$ HSQC spectrum of protonated and perdeuterated ^{15}N-labeled EIN. From Garrett *et al.* (1997).

in 3D triple-resonance NMR experiments are illustrated in Fig. 5, which shows a comparison of the CBCA(CO)NH and d-CBCA(CO)NH, CBCANH and d-HNCACB, and C(CO)NH and d-C(CO)NH experiments recorded on $^1H–^{15}N/^{13}C$- and $^2H/^{15}N/^{13}C$-labeled EIN.

4.3. Stereospecific Assignments and Torsion-Angle Restraints

Torsion-angle restraints (*see* chapters 6 and 7) can be derived from coupling constant data, as there exist simple geometric relationships between three-bond couplings and torsion angles. In simple systems, the coupling constant can be measured directly from the in-phase or antiphase splitting of a particular resonance in the 1D or 2D spectrum. For larger systems where the linewidths exceed the coupling, it becomes difficult to extract accurate couplings in the manner. An alternative approach involves the use of ECOSY experiments to generate reduced cross-peak multiplets (Griesinger *et al.*, 1986). Although this permits accurate couplings to be obtained, the sensitivity of ECOSY experiments is generally quite low. Furthermore, in multidimensional experiments its utility is restricted by the fact that the couplings have to be measured in the indirectly detected dimensions, and hence are influenced by limited digital resolution. More recently, a series of highly sensitive quantitative J correlation experiments have been developed that circumvent these problems (Bax *et al.*, 1994). These experiments quantitate the loss in magnetization when dephasing caused by coupling is active versus inactive. In some J quantitative correlation experiments, the coupling is obtained from the ratio of cross-peak to diagonal-peak intensities. In others, it is obtained by the ratio of the cross-peaks obtained in two separate experiments (with the coupling active and inactive), recorded in an interleaved manner.

For small proteins, it is often possible to obtain stereospecific assignments of β-methylene protons on the basis of a qualitative interpretation of the homonuclear $^3J_{\alpha\beta}$ coupling constants and the intraresidue NOE data involving the NH, $C^{\alpha}H$, and $C^{\beta}H$ protons (Wagner *et al.*, 1987). A more rigorous approach, which also permits one to obtain ϕ, ψ, and χ_1 restraints, involves the application of a conformational grid search of ϕ,ψ,χ_1 space on the basis of the homonuclear $^3J_{HN\alpha}$ and $^3J_{\alpha\beta}$ coupling constants (which are related to ϕ and χ_1, respectively), and the intraresidue and sequential interresidue NOEs involving the NH, $C^{\alpha}H$, and $C^{\beta}H$ protons (Güntert *et al.*, 1989; Nilges *et al.*, 1990). This information can be supplemented and often supplanted by the measurement of heteronuclear couplings by quantitative J correlation spectroscopy. The most useful couplings in this regard are the $^3J_{C\gamma CO}$ and $^3J_{NC\gamma}$, which are sufficient, when used in combination, to derive the appropriate χ_1 sidechain rotamer. A summary of the heteronuclear quantitative J correlation experiments that we currently employ is provided in Table 2.

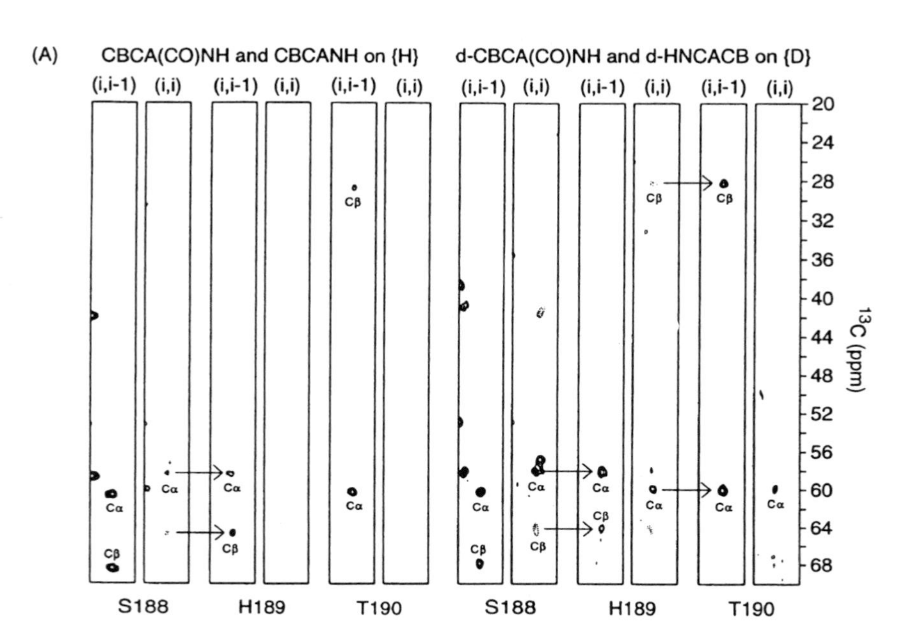
(A)
CBCA(CO)NH and CBCANH on {H}
d-CBCA(CO)NH and d-HNCACB on {D}
(i,i-1) (i,i) (i,i-1) (i,i) (i,i-1) (i,i)
(i,i-1) (i,i) (i,i-1) (i,i) (i,i-1) (i,i)
Cα
Cβ
20
24
28
32
36
40
44
48
52
56
60
64
68
13C (ppm)
S188
H189
T190
S188
H189
T190

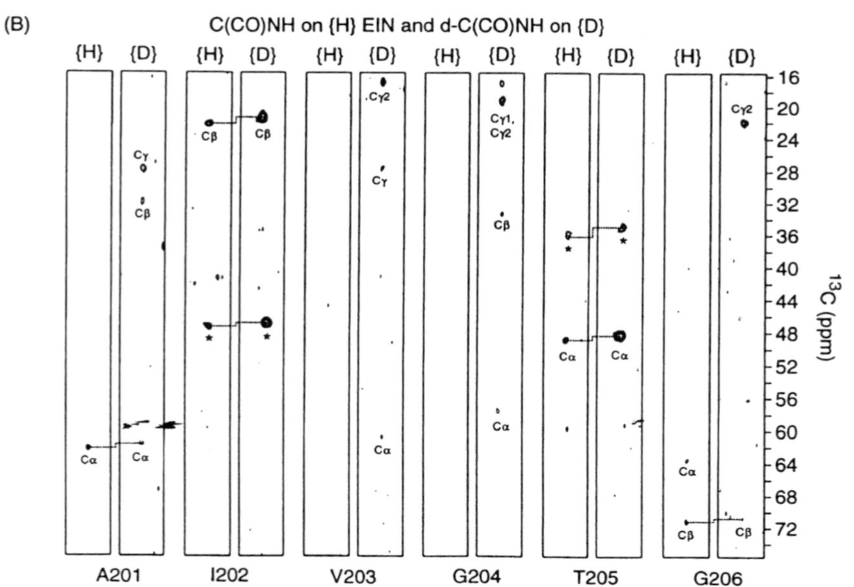
(B)
C(CO)NH on {H} EIN and d-C(CO)NH on {D}
{H} {D} {H} {D} {H} {D} {H} {D} {H} {D} {H} {D}
Cα
Cβ
Cγ
Cγ2
Cγ1, Cγ2
*
16
20
24
28
32
36
40
44
48
52
56
60
64
68
72
13C (ppm)
A201
I202
V203
G204
T205
G206

Figure 5. Selected strips taken from several 3D triple-resonance experiments comparing the results on uniformly protonated and perdeuterated U-^{15}N/^{13}C EIN. (A) CBCANH and CBCA(CO)NH experiments on protonated EIN versus the d-HNCACB and d-CBCA(CO)NH experiments on perdeuterated EIN. (B) C(CO)NH and d-C(CO)NH experiments on protonated and perdeuterated EIN, respectively (peaks labeled with an asterisk in B arise from resonances that have their maximal intensities on an adjacent slice). The CBCANH and d-HNCACB experiments correlate the Cα/Cβ resonances of both the (i – 1) and i residues with the ^{15}N^{1}H resonances of residue i. The CBCA(CO)NH and d-CBCA(CO)NH experiments correlate only the Cα/Cβ resonances of the (i – 1) residue with the ^{15}N^{1}H resonances of residue i; while the C(CO)NH and d-C(CO)NH experiments correlate the sidechain and Cα resonances of residue (i – 1) with the ^{15}N^{1}H resonances of residue i. From Garrett *et al.* (1997).

Table 2
Experiments for Determining Three-Bond Coupling Constants by Quantitative *J* Correlation Spectroscopy[a]

Experiment	Three-bond coupling	Torsion angle
3D HNHA	$^{3}J_{HN\alpha}$	ϕ
3D (HN)CO(CO)NH	$^{3}J_{COCO}$	ϕ
2D ^{13}C-{^{15}N} spin-echo difference CT-HSQC	$^{3}J_{C\gamma N}$	χ_1 of Thr and Val
2D ^{13}C-{^{13}CO} spin-echo difference CT-HSQC	$^{3}J_{C\gamma CO}$	χ_1 of Thr and Val
2D ^{13}CO-{^{13}Cγ(aro)} spin-echo difference ^{1}H–^{15}N HSQC	$^{3}J_{C\gamma(aromatic)CO}$	χ_1 of aromatics
2D ^{15}N-{^{13}Cγ(aro)} spin-echo difference ^{1}H–^{15}N HSQC	$^{3}J_{C\gamma(aromatic)N}$	χ_1 of aromatics
2D ^{15}N–{^{13}Cγ} spin-echo difference ^{1}H–^{15}N HSQC	$^{3}J_{C\gamma(aliphatic)N}$	χ_1 of aliphatics
3D HN(CO)C	$^{3}J_{C\gamma(aliphatic)CO}$	χ_1 of aliphatics
3D HN(CO)HB	$^{3}J_{COH\beta}$	χ_1
3D HNHB	$^{3}J_{NH\beta}$	χ_1
3D HACAHB	$^{3}J_{\alpha\beta}$	χ_1
2D or 3D ^{1}H-detected long-range C–C COSY	$^{3}J_{CC}$	χ_2 of Leu and Ile χ_3 of Met
3D ^{1}H-detected [^{13}C–^{1}H] long-range COSY	$^{3}J_{CH}$	χ_2 of Leu and Ile χ_3 of Met

[a]Details of all experiments are provided in the review by Bax *et al.* (1994) with the exception of the 3D (HN)CO(CO)NH, HN(CO)C, and HACAHB experiments and the 2D ^{13}CO-{^{13}Cγ(aro)}, ^{15}N-{^{13}Cγ(aro)}, and ^{15}N-{^{13}Cγ} spin-echo difference ^{1}H–^{15}N HSQC experiments, which are described in Grzesiek and Bax (1997), Hu and Bax (1997a); Grzesiek *et al.* (1995b), Hu *et al.* (1997) and Hu and Bax (1997b), respectively.

4.4. Assignment of Through-Space Proton–Proton Interactions within a Protein

Although the panoply of 3D heteronuclear experiments is sufficient for the purposes of spectral assignment, yet further increases in resolution are required for the reliable identification of NOE through-space interactions. This can be achieved by extending the dimensionality still further to four dimensions (Kay *et al.*, 1990; Clore *et al.*, 1991a), as illustrated in Fig. 6. Consider a simple 2D spectrum

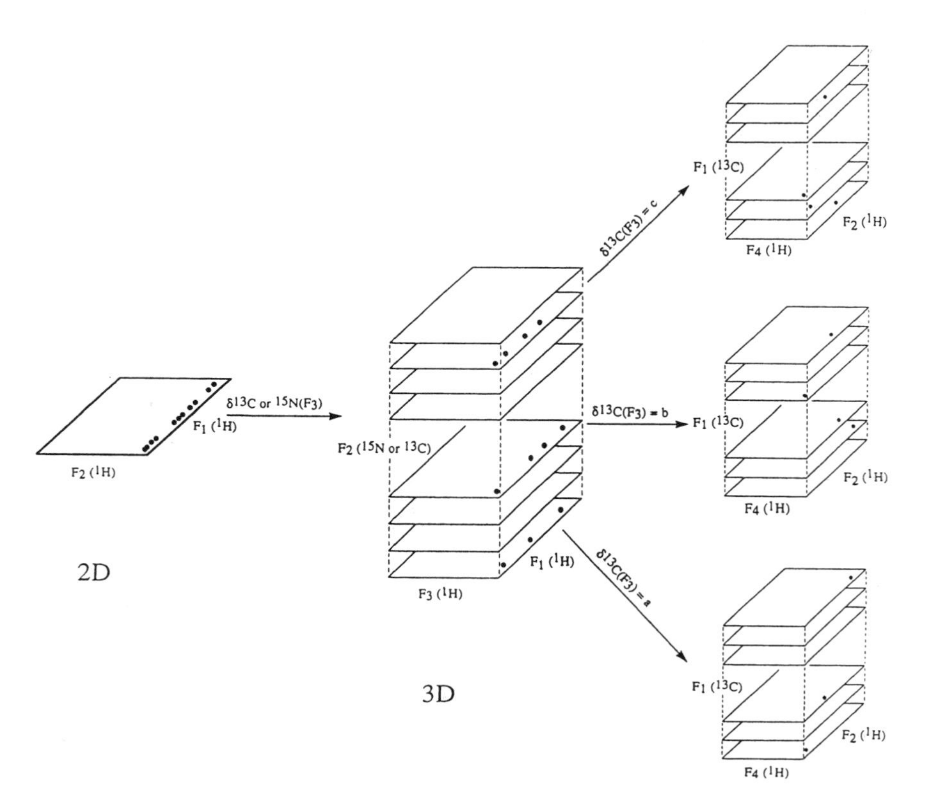

Figure 6. Schematic diagram illustrating the effects of increasing dimensionality on the spectral resolution of an NOE spectrum. In the 2D spectrum, the proton chemical shift of the destination resonances (along the F_2 axis) for all 11 cross-peaks is the same so that one cannot assess the number of destination protons involved. In the 3D spectrum, the cross-peaks appear in three planes, edited according to the shift of the heavy atom (^{15}N or ^{13}C) attached to the destination protons. The identity of the originating protons, however, is only defined by their proton chemical shifts. Finally, in the 4D spectrum, each cross-peak is characterized by four chemical shift coordinates, the proton chemical shifts of the two protons involved, and the chemical shifts of the heavy atoms to which they are attached. From Clore and Gronenborn (1991a).

demonstrating 11 cross-peaks from aliphatic resonances to a single proton resonance position. In the 2D spectrum it is impossible to ascertain whether this destination resonance involves one proton or many. Extending the spectrum to 3D by separating the NOE interactions according to the chemical shift of the heavy atom (^{15}N or ^{13}C) attached to each proton reveals that there are three individual protons involved. The identity of the originating aliphatic protons, however, is only specified by their proton chemical shifts. Because the extent of spectral overlap in the aliphatic region of the spectrum is considerable, additional editing is necessary. This is achieved by adding a further dimension such that each plane of the 3D spectrum now constitutes a cube in the 4D spectrum edited by the ^{13}C shift of the carbon atom attached to each of the originating protons. In this manner, each ^{1}H–^{1}H NOE interaction is specified by four chemical shift coordinates, the two protons giving rise to the NOE and the heavy atoms to which they are attached.

Because the number of NOE interactions present in each 2D plane of a 4D ^{13}C/^{15}N or ^{13}C/^{13}C-edited NOESY spectrum is so small, the inherent resolution in a 4D spectrum is extremely high, despite the low level of digitization (Clore and Gronenborn, 1991a). Indeed, spectra with equivalent resolution can be recorded at magnetic field strengths considerably lower than 600 MHz, although this would obviously lead to a reduction in sensitivity. Further, it can be calculated that 4D spectra with virtual lack of resonance overlap and good sensitivity can be obtained on proteins with as many as 400 residues. Thus, once complete ^{1}H, ^{15}N, and ^{13}C assignments are obtained, analysis of 4D ^{15}N/^{13}C- and ^{13}C/^{13}C-separated NOE spectra should permit the assignment of almost all NOE interactions in a relatively straightforward manner (Clore and Gronenborn, 1991a). The first successful application of these methods to the structure determination of a protein of greater than 15 kDa was achieved in 1991 with the determination of the solution structure of interleukin-1β, a protein of 17 kDa and 153 residues (Clore *et al.*, 1991b). This has now been extended to two 30-kDa single-chain proteins, namely, EIN (259 residues; Garrett *et al.*, 1997) and the serine protease PB92 (269 residues; Martin *et al.*, 1997). The solution structure of EIN is shown in Fig. 7.

4.5. Protein–Ligand and Protein–Protein Complexes

Provided one of the partners in the complex (e.g., a peptide, oligonucleotide, or drug) presents a relatively simple spectrum that can be assigned by 2D methods, the most convenient strategy for dealing with protein–ligand complexes involves one in which the protein is labeled with ^{15}N and ^{13}C and the ligand is unlabeled (i.e., at natural isotopic abundance). It is then possible to use a combination of heteronuclear filtering and editing to design experiments in which correlations involving only protein resonances, only ligand resonances, or only through-space interactions between ligand and protein are observed (Gronenborn and Clore, 1995). These experiments are summarized in Table 3 and have been successfully

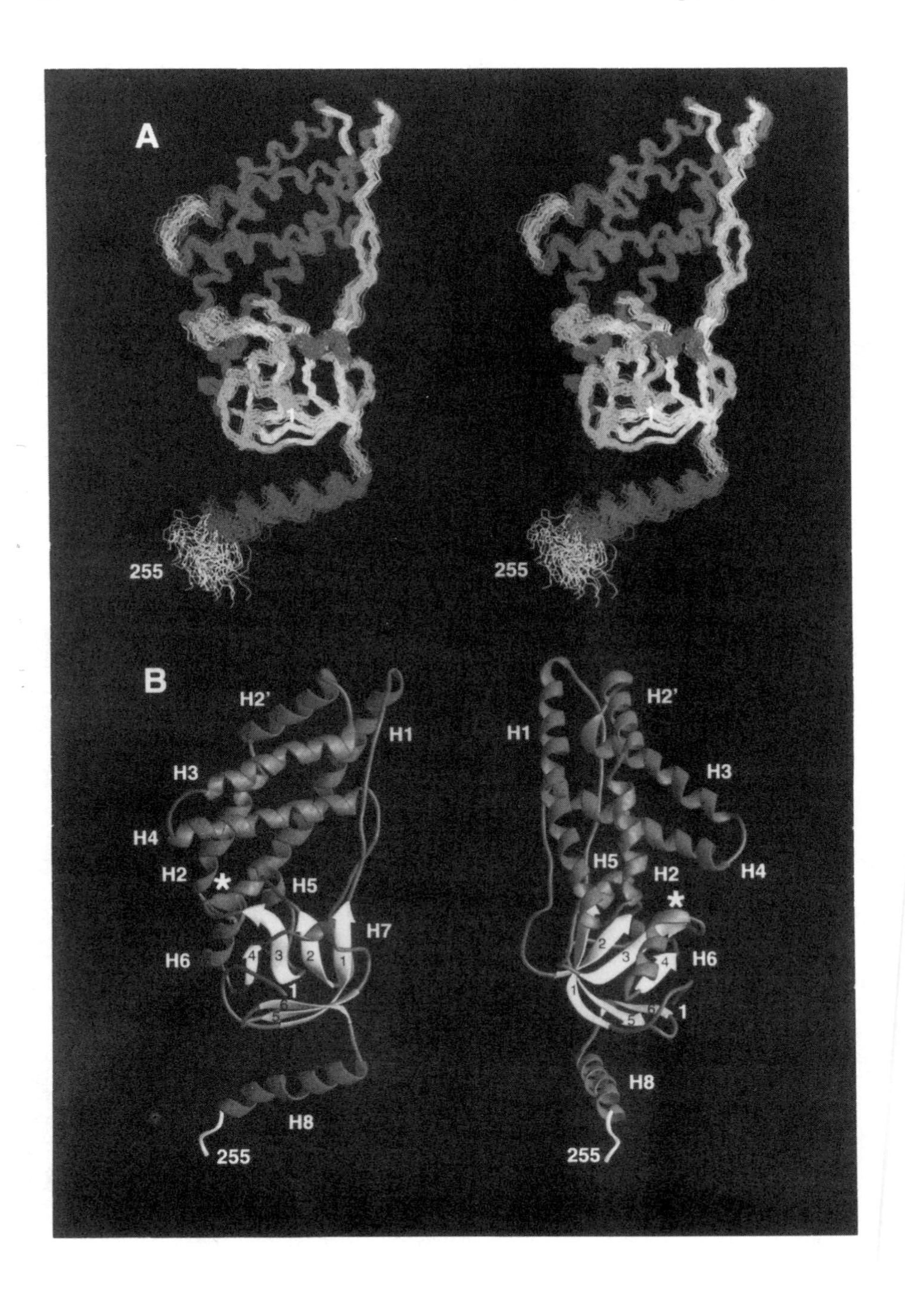
A
1
255
1
255
B
H2'
H1
H3
H4
H2
*
H5
H7
H6
4
3
2
1
1
6
5
H8
255
H2'
H1
H3
H5
H2
H4
*
2
3
4
H6
1
6
5
1
H8
255

Figure 7. (A) Stereoview showing a superposition of the backbone (N, Cα, C) atoms of 50 simulated annealing structures of EIN. (B) Ribbon diagrams illustrating two views of the backbone of EIN. Helices are shown in red, strands in yellow, loops in blue, and the disordered C-terminus in white. The asterisk in (B) indicates the location of the active site histidine (His189). From Garrett *et al.* (1997). A color representation of this figure can be found following page 18.

Table 3

Summary of Heteronuclear-Filtered and -Separated NOE Experiments Used to Study Protein–Ligand Complexes (Including Protein–Nucleic Acid and Protein–Protein Complexes)

Type of contact	Connectivity
U-^{15}N/^{13}C protein + U-^{14}N/^{12}C ligand	
Intra and intermolecular contacts	
3D ^{15}N-separated NOE in H_2O	H(*j*)-^{15}N(*j*)----------H(*i*)
3D ^{13}C-separated NOE in D_2O	H(*j*)-^{13}C(*j*)----------H(*i*)
Intramolecular protein contacts	
4D ^{13}C/^{13}C-separated NOE in D_2O	H(*j*)-^{13}C(*j*)----------H(*i*)-^{13}C(*i*)
4D ^{15}N/^{13}C-separated NOE in H_2O	H(*j*)-^{15}N(*j*)----------H(*i*)-^{13}C(*i*)
3D ^{15}N/^{15}N-separated NOE in H_2O	H(*j*)-^{15}N(*j*)----------H(*i*)-^{15}N(*i*)
Intramolecular ligand contacts	
2D ^{12}C,^{14}N(F_1)/ ^{12}C,^{14}N(F_2)-filtered NOE in H_2O[b]	H(*j*)-^{12}C(*j*)-------H(*i*)-^{12}C(*i*)
	H(*j*)-^{14}N(*j*)-------H(*i*)-^{12}C(*i*)
	H(*j*)-^{12}C(*j*)-------H(*i*)-^{14}N(*i*)
	H(*j*)-^{14}N(*j*)-------H(*i*)-^{14}N(*i*)
2D ^{12}C(F_1)/^{12}C(F_2)-filtered NOE in D_2O[a]	H(*j*)-^{12}C(*j*)-------H(*i*)-^{12}C(*i*)
Intermolecular protein–ligand contacts[c]	
3D ^{15}N-separated(F_1)/^{14}N,^{12}C(F_3)-filtered NOE in H_2O	H(*j*)-^{15}N(*j*)-------H(*i*)-^{12}C(*i*)
	H(*j*)-^{15}N(*j*)-------H(*i*)-^{14}N(*i*)
3D ^{13}C-separated(F_1)/^{12}C(F_3)-filtered NOE in D_2O	H(*j*)-^{13}C(*j*)-------H(*i*)-^{12}C(*i*)
Intramolecular	
3D ^{15}N/^{15}N-separated NOE in H_2O	H(*j*)-^{15}N(*j*)----------H(*i*)-^{15}N(*i*)
4D ^{13}C/^{13}C-separated NOE in D_2O	H(*j*)-^{13}C(*j*)----------H(*i*)-^{13}C(*i*)
4D ^{15}N/^{15}N-separated NOE in H_2O	H(*j*)-^{15}N(*j*)----------H(*i*)-^{15}N(*i*)
Intermolecular	
3D ^{13}C-separated/^{15}N-filtered	H(*j*)-^{13}C(*j*)----------H(*i*)-^{15}N(*i*)
3D ^{15}N-separated/^{13}C-filtered	H(*j*)-^{15}N(*j*)----------H(*i*)-^{13}C(*i*)
4D ^{13}C-separated/^{15}N-separated	H(*j*)-^{13}C(*j*)----------H(*i*)-^{15}N(*i*)

[a] Details of the experiments and original references are cited in the reviews by Clore and Gronenborn (1991a,b), Bax and Grzesiek (1993), and Gronenborn and Clore (1995).

[b] Similar heteronuclear-filtered 2D correlation and Hartmann–Hahn spectra can also be recorded to assign the spin systems of the ligand.

[c] For homomultimeric systems, the multimer needs to be reconstituted from an equimixture of uniformly ^{15}N/^{13}C and ^{14}N/^{12}C or ^{15}N/^{2}H and ^{13}C/^{1}H labeled subunits.

employed in a number of laboratories for a range of systems. Examples of protein–drug complexes are cyclophylin–cyclosporin (Theriault *et al.*, 1993; Spitzfaden *et al.*, 1994) and FK506 binding protein–ascomycin (Meadows *et al.*, 1993). The cyclophylin–cyclosporin studies employed both uniformly ^{13}C, ^{15}N labeled protein as well as ^{13}C labeled drug. Structures of several protein–peptide complexes have also been determined by NMR. The first example was the structure of calmodulin bound to its target peptide from skeletal muscle myosin light-chain kinase (Ikura *et al.*, 1992). Since then a number of other protein–peptide complexes have been determined. These include several SH2– and SH3–peptide complexes (Pascal *et al.*, 1994; Wittekind *et al.*, 1994; Terasawa *et al.*, 1994; Goudreau *et al.*, 1994; Yu *et al.*, 1994; Feng *et al.*, 1994; Xu *et al.*, 1995) and mixed disulfide intermediates of glutaredoxin and glutathione (Bushweller *et al.*, 1994), and of human thioredoxin and its target peptides from the transcription factor NF-κB (Qin *et al.*, 1995) and the redox regulator Ref-1 (Qin *et al.*, 1996). Structures of protein–DNA complexes have been the focus of several laboratories and to date a number have been determined by NMR: These include the transcription factor GATA-1 (Omichinski *et al.*, 1993), the lac repressor headpiece (Chuprina *et al.*, 1993), the antennapedia homeodomain (Billeter *et al.*, 1993), the minimal DNA binding domain of the protooncogene c-*myb* (Ogata *et al.*, 1994), the *trp* repressor (Zhang *et al.*, 1994), the male sex determining factor SRY (Werner *et al.*, 1995), the architectural factor LEF-1 (Love *et al.*, 1995), the chromatin remodeling factor GAGA (Omichinski *et al.*, 1997), and the transcriptional coactivator HMG-I(Y) (Huth *et al.*, 1997).

Oligomeric proteins represent complexes between identical subunits. The first dimer to be solved in our laboratory by NMR was the chemokine interleukin-8 (Clore *et al.*, 1990). Since that time a number of other homodimeric systems have been solved including the Arc repressor (Bonvin *et al.*, 1994), the gene 5 protein from M13 (Folkers *et al.*, 1994), the chemokines hMIP-1β (Lodi *et al.*, 1994), GRO/MGSA (Fairbrother *et al.*, 1994; Kim *et al.*, 1994) and RANTES (Skelton *et al.*, 1995), the Mnt repressor (Burgering *et al.*, 1994), and the C- (Lodi *et al.*, 1995; Eijkelenboom *et al.*, 1995) and N- (Cai *et al.*, 1997) terminal domains of HIV-1 integrase. More recently, the methodology has been extended to a tetramer, namely, the oligomerization domain of the tumor suppressor p53 (Clore *et al.*, 1994, 1995a,b; Lee *et al.*, 1994), and a 45-kDa trimer, namely, the ectodomain of SIV gp41 (Caffrey *et al.*, 1997, 1998). For multimeric proteins, additional labeling schemes can also be used to facilitate the identification of intermolecular NOEs. For example, one subunit can be labeled with $^{15}N/^{2}H$, and the other with $^{13}C/^{1}H$, enabling high-sensitivity 3D and 4D experiments to be recorded in which NOEs are only observed between protons attached to ^{15}N in one subunit and ^{13}C in the other (Caffrey *et al.*, 1997, 1998).

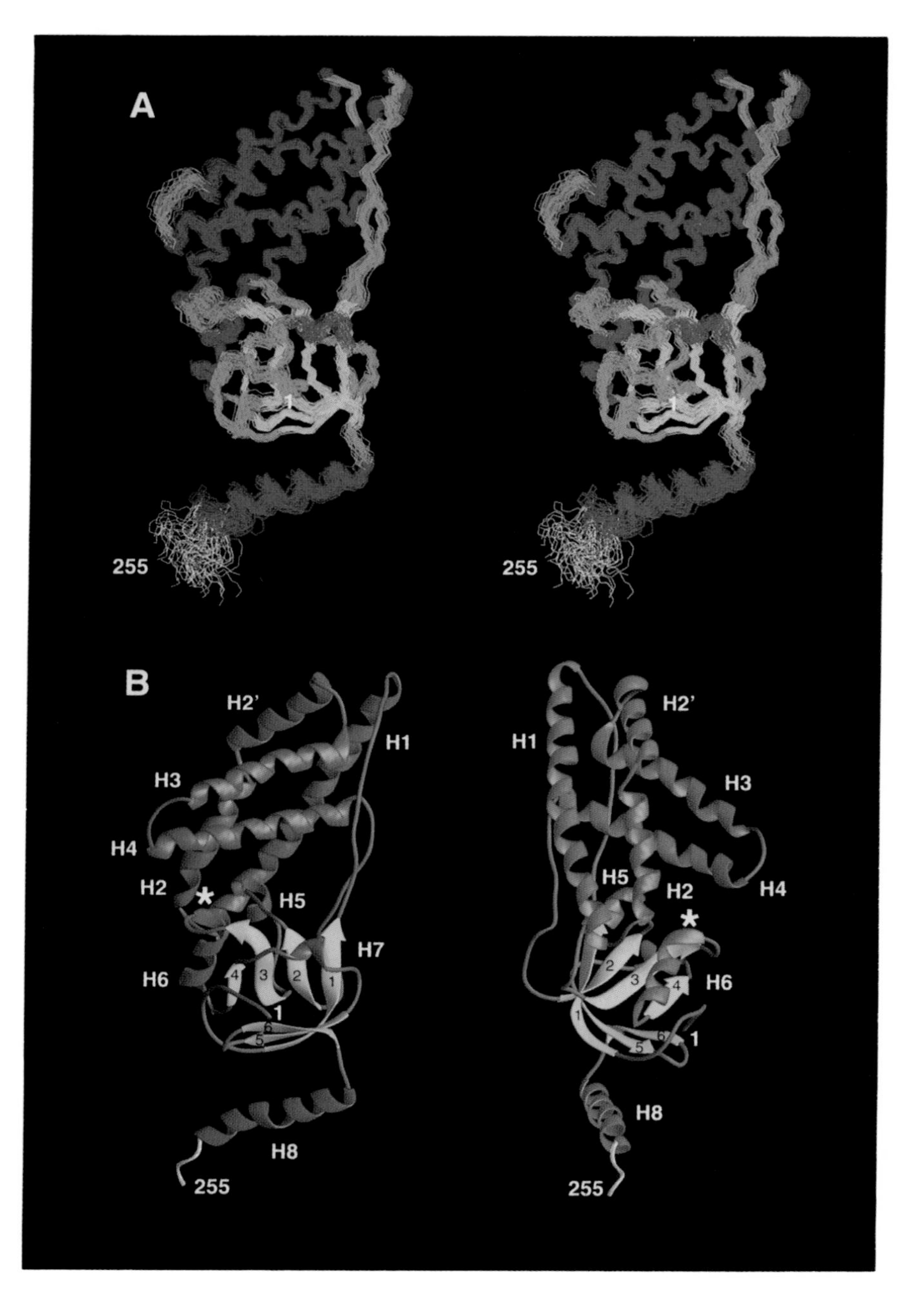

Figure 7. (A) Stereoview showing a superposition of the backbone (N, Cα, C) atoms of 50 simulated annealing structures of EIN. (B) Ribbon diagrams illustrating two views of the backbone of EIN. Helices are shown in red, strands in yellow, loops in blue, and the disordered C-terminus in white. The asterisk in (B) indicates the location of the active site histidine (His189). From Garrett *et al.* (1997).

be supplemented by torsion angle, coupling constant, ^{13}C secondary shift, and ^{1}H shift restraints. The success of the NOE-based approach is related to the fact that short interproton distances between units far apart in a linear array are conformationally highly restrictive (Clore and Gronenborn, 1991a). However, there are numerous cases where restraints that define long-range order can supply invaluable structural information. In particular, they permit the relative positioning of structural elements that do not have many short interproton distance contacts between them. Examples of such situations include modular and multidomain proteins and linear nucleic acids. Two novel approaches have recently been introduced that directly provide restraints that characterize long-range order *a priori*. The first relies on the dependence of heteronuclear (^{15}N or ^{13}C) longitudinal (T_1) and transverse (T_2) relaxation times (specifically T_1/T_2 ratios) on rotational diffusion anisotropy (Tjandra *et al.*, 1997a), and the second on residual dipolar couplings in magnetically oriented macromolecules (Tjandra *et al.*, 1997b). The two methods provide restraints that are related in a simple geometric manner to the orientation of N–H and C–H internuclear vectors relative to the diffusion and molecular magnetic susceptibility tensors, respectively.

For the heteronuclear ^{15}N T_1/T_2 method to be applicable, the molecule must tumble anisotropically (i.e., it must be nonspherical). The minimum ratio of the diffusion anisotropy $D_{\parallel}/D_{\perp}$ for which heteronuclear T_1/T_2 refinement will be useful depends entirely on the accuracy and uncertainties in the measured T_1/T_2 ratios. In practice, the difference between the maximum and minimum T_1/T_2 ratio must exceed the uncertainty in the measured T_1/T_2 values by an order of magnitude. This typically means that $D_{\parallel}/D_{\perp}$ should be greater than ~1.5.

Likewise the applicability of the residual dipolar coupling method (*see* Volume 17, chapter 8) depends on the magnitude of the magnetic susceptibility anisotropy (that is, the degree of alignment of the molecule in the magnetic field). The magnetic susceptibility of most diamagnetic proteins is dominated by aromatic residues, and also contains contributions from the susceptibility anisotropies of the peptide bonds. Because the magnetic susceptibility anisotropy tensors of these individual contributors are generally not collinear, the net value of the magnetic susceptibility anisotropy in diamagnetic proteins is usually small. Much larger magnetic susceptibility anisotropies are obtained if many aromatic groups are stacked on each other such that their magnetic susceptibility contributions are additive, as in the case of nucleic acids. Hence, this particular method is ideally suited to protein–nucleic acid complexes. In practice, the residual dipolar couplings must exceed the uncertainty in their measured values by an order of magnitude, which typically means that the magnetic susceptibility anisotropy should be ~–20 × 10^{-34} m^3/molecule, which is about 20 times greater than that for benzene. In addition, other methods of alignment can potentially be employed. These include exploitation of a protein's anisotropic electrical polarization tensor and optical absorption tensor to obtain

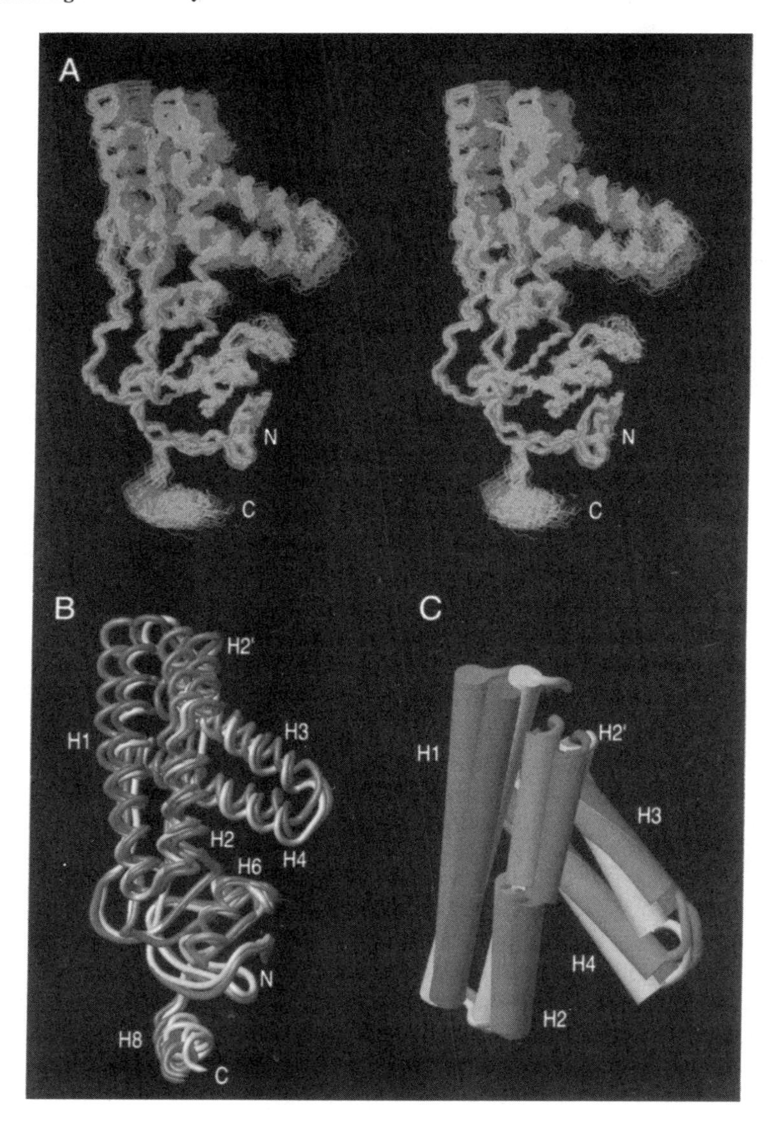

Figure 8. (A) Best-fit superposition of the backbone (N, Cα, C) atoms of the ensemble of simulated annealing structures (30 each) of EIN calculated with (red) and without (blue) ^{15}N T_1/T_2 refinement, best-fitted to the α/β domain (residues 2–20 and 148–230). (B, C) Views showing superpositions, best-fitted to the α/β domain, of the restrained regularized mean structures derived from the ensembles calculated with (red) and without (blue) ^{15}N T_1/T_2 refinement, and of the X-ray structure (yellow). In (B) the backbone of residues 1–249 is displayed as a tubular representation; in (C) the helices of the α domain are shown as cylinders. From Tjandra *et al.* (1997a). A color representation of this figure can be found following page 18.

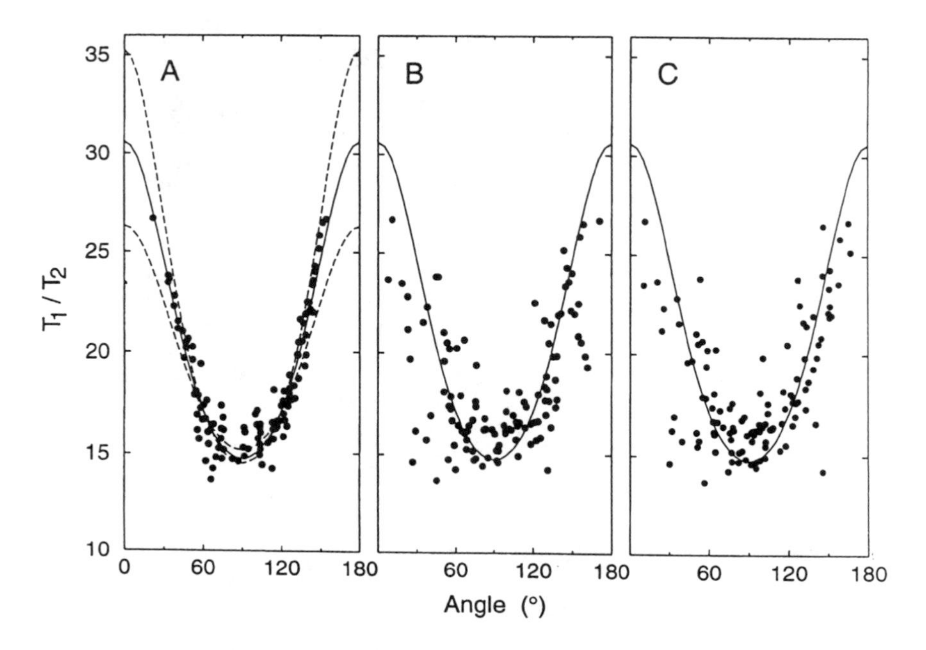

Figure 9. Dependence of the observed ^{15}N T_1/T_2 ratios at 600 MHz on the angle θ between the NH bond vectors and the unique axis of the diffusion tensor for the restrained regularized mean structures of EIN obtained from the ensembles calculated with (A) and without (B) ^{15}N T_1/T_2 refinement, and for the X-ray structure (C). The solid lines represent the theoretical dependence of T_1/T_2 versus θ for a diffusion anisotropy of 2.1 and an effective correlation time of 13.1 ns. The dashed lines in (A) illustrate the effects on the theoretical dependence of T_1/T_2 versus θ of increasing or decreasing the diffusion anisotropy by 15%. From Tjandra *et al.* (1997).

considerable degrees of alignment by means of strong low-frequency electric fields and polarized light, respectively (Tjandra *et al.*, 1997b).

An application of direct refinement against ^{15}N T_1/T_2 ratios is illustrated in Figs. 7 and 8 for the protein EIN (Tjandra *et al.*, 1997b). EIN is elongated in shape with a diffusion anisotropy of ~ 2. As a result the observed T_1/T_2 ratios range from ~14 when the N–H vector is perpendicular to the diffusion axis to ~30 when the N–H vector is parallel to the diffusion axis (Fig. 7). EIN consists of two domains, and of the 2818 NOEs used to determine its structure, only 38 involve interdomain contacts. Refinement against the T_1/T_2 ratios results in a small change in the relative orientations of the two domains (Fig. 8). This is best described in terms of the average angular difference between equivalent helices of the α-domain when best-fitting the structures to the backbone of the α/β domain. This angular difference is 5–7° between the NMR structures refined with and without T_1/T_2 ratios and between the NMR structure refined without T_1/T_2 ratios and the X-ray structure, and ~10° between the the NMR structure refined with T_1/T_2 ratios and the X-ray structure.

7. PERSPECTIVES AND CONCLUDING REMARKS

The recent development of a whole range of highly sensitive multidimensional heteronuclear edited and filtered NMR experiments has revolutionized the field of protein structure determination by NMR. Proteins and protein complexes in the range of 20–50 kDa are now amenable to detailed structural analysis in solution.

Despite these advances, it should always be borne in mind that there are a number of key requirements that have to be satisfied to permit a successful structure determination of larger proteins and protein complexes by NMR. The protein in hand must be soluble and should not aggregate up to concentrations of about 0.5–1 mM; it must be stable at room temperature or slightly higher for considerable periods of time (particularly as it may take several months of measurement time to acquire all of the necessary NMR data); it should not exhibit significant conformational heterogeneity that could result in extensive line broadening; and finally it must be amenable to uniform ^{15}N and ^{13}C labeling. At present, there are still only relatively few examples in the literature of proteins in the range of 15–30 kDa that have been solved by NMR. Likewise, only a handful of protein complexes (with DNA or peptides) and oligomers have been determined to date using these methods. One can anticipate, however, that over the next few years, by the widespread use of multidimensional heteronuclear NMR experiments coupled with semiautomated assignment procedures, many more NMR structures of proteins and protein complexes will become available.

ACKNOWLEDGMENTS. We thank A. Bax for numerous stimulating discussions. The work in the authors' laboratory was supported in part by the AIDS Targeted Antiviral Program of the Office of the Director of the National Institutes of Health.

REFERENCES

Bax, A., and Grzesiek, S., 1993, *Acc. Chem. Res.* **26:**131.

Bax, A., and Pochapsky, S. S., 1992, *J. Magn. Reson.* **99:**638.

Bax, A., Vuister, G. W., Grzesiek, S., Delaglio, F., Wang, A. C., Tschudin, R., and Zhu, G., 1994, *Methods Enzymol.* **239:**79.

Billeter, M., Qian, Y. Q., Otting, G., Müller, M., Gehring, W., and Wüthrich, K., 1993, *J. Mol. Biol.* **234:**1084.

Bonvin, A. M. J. J., Vis, H., Breg, J. N., Burgering, M. J. M., Boelens, R., and Kaptein, R., 1994, *J. Mol. Biol.* **236:**328.

Burgering, M. J. M., Boelens, R., Gilbert, D. E., Breg, J., Knight, K. L., Sauer, R. T., and Kaptein, R., 1994, *Biochemistry* **33:**15036.

Bushweller, J. H., Billeter, M., Holmgren, A., and Wüthrich, K., 1994, *J. Mol. Biol.* **235**:1585.

Caffrey, M., Cai, M., Kaufman, J., Stahl, S. J., Wingfield, P. T., Gronenborn, A.M., and Clore, G.M., 1997, *J. Mol. Biol.* **271:**819

Caffrey, M., Cai, M., Kaufman, J., Stohl, S. J., Wingfield, P. T., Gronenborn, A. M., and Clore, G. M., 1998, *Embo J.* **17:**4572.

Cai, M., Zheng, R., Caffrey, M., Craigie, R., Clore, G. M., and Gronenborn, A. M., 1997, *Nature Struct. Biol.* **4:**567.

Chuprina, V. P., Rullman, J. A. C., Lamerichs, R. M. N. J., van Boom, J. H., Boelens, R., and Kaptein, R., 1993, *J. Mol. Biol.* **234:**446.

Clore, G. M., and Gronenborn, A. M., 1987, *Protein Eng.* **1:**275.

Clore, G. M., and Gronenborn, A. M., 1989, *CRC Crit. Rev. Biochem. Mol. Biol.* **24:**479.

Clore, G. M., and Gronenborn, A. M., 1991a, *Science* **252:**1390.

Clore, G. M, and Gronenborn, A. M., 1991b, *Prog. Nucl. Magn. Reson. Spectrosc.* **23:**43.

Clore, G. M., Appella, E., Yamada, M., Matsushima, K., and Gronenborn, A. M., 1990, *Biochemistry* **29:**1689.

Clore, G. M., Kay, L. E., Bax, A., and Gronenborn, A. M., 1991a, *Biochemistry* **30:**12.

Clore, G. M., Wingfield, P. T., and Gronenborn, A. M., 1991b, *Biochemistry* **30:**2315.

Clore, G. M., Omichinski, J. G., Sakaguchi, K., Zambrano, N., Sakamoto, H., Appella, E., and Gronenborn, A. M., 1994, *Science* **265:**386.

Clore, G. M., Omichinski, J. G., Sakaguchi, K., Zambrano, N., Sakamoto, H., Appella, E., and Gronenborn, A. M., 1995a, *Science* **267:**1515.

Clore, G. M., Ernst, J., Clubb, R. T., Omichinski, J. G., Sakaguchi, K., Appella, E., and Gronenborn, A. M., 1995b, *Nature Struct. Biol.* **2:**321.

Driscoll, P. C., Clore, G. M., Marion, D., Wingfield, P. T., and Gronenborn, A. M., 1990, *Biochemistry* **29:**3542.

Dyson, H. J., Gippert, G. P., Case, D. A., Holmgren, A., and Wright, P. E., 1990, *Biochemistry* **29:**4129.

Eijkelenboom, A. P., Lutzke, R. A., Boelens, R., Pasterk, R. H. A., Kaptein, R., and Hard, K., 1995, *Nature Struct. Biol.* **2:**807.

Ernst, R.R., Bodenhausen, G., and Wokaun, A., 1987, *Principles of Nuclear Magnetic Resonance in One and Two Dimensions,* Oxford University Press (Clarendon), London.

Fairbrother, W. J., Reilly, D., Colby, T. J., Hesselgesser, J., and Horuk, R., 1994, *J. Mol. Biol.* **242:**252.

Feng, S., Chen, J. K., Yu, H., Simon, J. A., and Schreiber, S. L., 1994, *Science* **266:**1241.

Folkers, P. J. M., Nilges, M., Folmer, R. H. A., Konnings, R. N. H., and Hilbers, C. W., 1994, *J. Mol. Biol.* **236:**229.

Forman-Kay, J. D., Clore, G. M., Wingfield, P. T., and Gronenborn, A. M., 1991, *Biochemistry* **30:**2685.

Gardner, K. H., Rosen, M. K., and Kay, L. E., 1997, *Biochemistry* **36:**1389.

Garrett, D. S., Kuszewski, J., Hancock, T. J., Lodi, P. J., Vuister, G. W., Gronenborn, A. M., and Clore, G. M., 1994, *J. Magn. Reson. Ser. B* **104:**99.

Garrett, D. S., Seok, Y.-J., Liao, D.-I., Peterkofsky, A., Gronenborn, A. M., and Clore, G. M., 1997, *Biochemistry* **36**:2517.

Goudreau, N., Cornille, F., Duchesne, M., Tocque, B., Garbay, C., and Roques, B. P., 1994, *Nature Struct. Biol.* **1:**898.

Griesinger, C., Sørensen, O. W., and Ernst, R. R., 1986, *J. Chem. Phys.* **85:**6837.

Gronenborn, A. M., and Clore, G. M., 1995, *CRC Crit. Rev. Biochem. Mol. Biol.* **30:**3515.

Grzesiek, S., and Bax, A., 1997, *J. Biomol. NMR* **9:**207.

Grzesiek, S., Wingfield, P. T., Stahl, S. J., and Bax, A., 1995a, *J. Am. Chem. Soc.* **117:**9594.

Grzesiek, S., Kuboniwa, H., Hinck, A. P., and Bax, A., 1995b, *J. Am. Chem. Soc.* **117:**5312.

Güntert, P., Braun, W., Billeter, M., and Wüthrich, K., 1989, *J. Am. Chem.Soc.* **111:**3397.

Hu, J.-S., and Bax, A., 1997a, *J. Am. Chem. Soc.* **119:**6360.

Hu, J.-S., and Bax, A., 1997b, *J. Biomol. NMR* **439:**323.

Hu, J.-S., Grzesiek, S., and Bax, A., 1997, *J. Am. Chem. Soc.* **119:**1803.

Huth, J. R., Bewley, C. A., Nissen, M. S., Evans, J. N. S., Reeves, R., Gronenborn, A. M., and Clore, G. M., 1997, *Nature Struct. Biol.* **4:**657

Ikura, M., Kay, L. E., and Bax, A., 1990, *Biochemistry* **29:**4659.

Ikura, M., Clore, G. M., Gronenborn, A. M., Zhu, G., Klee, C. B., and Bax, A., 1992, *Science* **256:**632.

Kay, L. E., Clore, G. M., Bax, A., and Gronenborn, A. M., 1990, *Science* **249:**411.

Kay, L. E., Keifer, P., and Saarinen, T., 1992, *J. Am. Chem. Soc.* **114:**10663.

Kim, K.-S., Clark-Lewis, I., and Sykes, B. D., 1994, *J. Biol. Chem.* **269:**32909.

Kuszewski, J., Qin, J., Gronenborn, A. M., and Clore, G. M., 1995a, *J. Magn. Reson. Ser. B* **106:**92.

Kuszewski, J., Gronenborn, A. M., and Clore, G. M., 1995b, *J. Magn. Reson. Ser. B* **107:**293.

Kuszewski, J., Gronenborn, A. M., and Clore, G. M., 1996a, *J. Magn. Reson. Ser. B* **112:**79.

Kuszewski, J., Gronenborn, A. M., and Clore, G. M., 1996b, *Protein Sci.* **5:**1067.

Kuszewski, J., Gronenborn, A. M., and Clore, G. M., 1997, *J. Magn. Reson.* **125:**171.

Lee, W., Harvey, T. S., Yin, Y., Yau, P., Litchfield, D., and Arrowsmith, C. H., 1994, *Nature Struct. Biol.* **1:**877.

Lodi, P. J., Garrett, D. S., Kuszewski, J., Tsang, M. L.-S., Weatherbee, J. A., Leonard, W. J., Gronenborn, A. M., and Clore, G. M., 1994, *Science* **263:**1762.

Lodi, P. J., Ernst, J. A., Kuszewski, J., Hickman, A. B., Engelman, A., Craigie, R., Clore, G. M., and Gronenborn, A. M., 1995, *Biochemistry* **34:**9826.

Love, J. J., Li, X., Case, D. A., Giese, K., Grosschedl, R., and Wright, P. E., 1995, *Nature* **376:**791.

Martin, J. R., Mulder, F. A. A., Karimi-Nejad, Y., van der Zwan, J., Mariani, M., Schipper, D., and Boelens, R., 1997, *Structure* **5:**521.

Meadows, R. P., Nettesheim, D. G., Xu, R. X., Olejniczak, E. T., Petros, A. M., Holzman, T. F., Severin, J., Gubbins, E., Smith, H., and Fesik, S. W., 1993, *Biochemistry* **32:**754.

Metzler, W. J., Wittekind, M., Goldfarb, V., Mueller, L., and Farmer, B. T., II, 1996, *J. Am. Chem. Soc.* **118:**6800.

Nilges, M., Clore, G. M., and Gronenborn, A. M., 1990, *Biopolymers* **29:**813.

Ogata, K., Morikawa, S., Nakamura, H., Sekikawa, A., Inoue, T., Kanai, H., Sarai, A., Ishii, S., and Nishimura, Y., 1994, *Cell* **79:**639.

Omichinski, J. G., Clore, G. M., Schaad, O., Felsenfeld, G., Trainor, C., Appella, E., Stahl, S. J., and Gronenborn, A. M., 1993, *Science* **261:**438.

Omichinski, J. G., Pedone, P. V., Felsenfeld, G., Gronenborn, A. M., and Clore, G. M., 1997, *Nature Struct. Biol.* **4:**122.

Oschkinat, H., Griesinger, C., Kraulis, P. J., Sørensen, O. W., Ernst, R. R., Gronenborn, A. M., and Clore, G. M., 1988, *Nature* **332:**3746.

Pascal, S. M., Singer, A. U., Gish, G., Tamazaki, T., Shoelson, S. E., Pawson, T., Kay, L. E., and Forman-Kay, J. D., 1994, *Cell* **77:**461.

Qin, J., Clore, G. M., Kennedy, W. M. P., Huth, J. R., and Gronenborn, A. M., 1995, *Structure* **3:**289.

Qin, J., Clore, G. M., Kennedy, W. P., Kuszewski, J., and Gronenborn, A. M., 1996, *Structure* **4:**613.

Skelton, N. J., Aspiras, F., Ogez, J., and Scall, T. J., 1995, *Biochemistry* **34:**5329.

Spitzfaden, C., Braun, W., Wider, G., Widmer, H., and Wüthrich, K., 1994, *J. Biomol. NMR* **4:**463.

Terasawa, H., Kohda, D., Hatanaka, H., Tsuchiya, S., Ogura, K., Nagata, K., Ishii, S., Mandiyan, V., Ullrich, A., Schlessinger, J., and Inagaki, F., 1994, *Nature Struct. Biol.* **1:**891.

Theriault, Y., Logan, T. M., Meadows, R., Yu, L., Olejniczak, E. T., Holzman, T., Simmer, R. L., and Fesik, S. W., 1993, *Nature* **361:**88.

Tjandra, N., Garrett, D. S., Gronenborn, A. M., Bax, A., and Clore, G. M., 1997a, *Nature Struct. Biol.* **4:**443.

Tjandra, N., Omichinski, J. G., Gronenborn, A. M., Clore, G. M., and Bax, A., 1997b, *Nature Struct. Biol.* **4:**732.

Venters, R. A., Metzler, W. J., Spicer, L. D., Mueller, L., and Farmer, B. T., II, 1995, *J. Am. Chem. Soc.* **117:**9592.

Vuister, G. W., Kim, S.-J., Wu, C., and Bax, A., 1994, *J. Am. Chem. Soc.* **116:**9206.

Wagner, G., Braun, W., Havel, T. F., Schaumann, T., Go, N., and Wüthrich, K., 1987, *J. Mol. Biol.* **196:**611.

Werner, M. H., Huth, J. R., Gronenborn, A. M., and Clore, G. M., 1995, *Cell* **81:**705.

Wittekind, M., Mapelli, C., Garmer, B. T., Suen, K.-L., Goldfarb, V., Tsao, J., Lavoie, T., Barbacid, M., Meyers, C. A., and Mueller, L., 1994, *Biochemistry* **33:**13531.

Wüthrich, K., 1986, *NMR of Proteins and Nucleic Acids*, Wiley, New York.

Xu, R. X., Word, J. M., David, D. G., Rink, M. J., Willard, D. H., and Gampe, R. T., 1995, *Biochemistry* **34:**2107.

Yu, H., Chen, J. K., Feng, S., Dalgarno, D. C., Brauer, A. W., and Schreiber, S. L., 1994, *Cell* **76:**933.

Zhang, H., Zhao, D., Revington, M., Lee, W., Jia, X., Arrowsmith, C., and Jardetzky, O., 1994, *J. Mol. Biol.* **238:**592.

2

Multidimensional ^{2}H-Based NMR Methods for Resonance Assignment, Structure Determination, and the Study of Protein Dynamics

Kevin H. Gardner and Lewis E. Kay

1. INTRODUCTION

1.1. Background: Deuteration prior to 1993

Over the past 30 years, deuterium labeling methods have played a critical role in solution NMR studies of macromolecules, in many cases improving the quality of spectra by both a reduction in the number of peaks and a concomitant narrowing of linewidths. Deuteration was initially used in a set of elegant experiments by the groups of Crespi and Jardetzky to reduce the complexity of one-dimensional (1D) ^{1}H spectra of proteins (Crespi and Katz, 1969; Crespi *et al.*, 1968; Markley *et al.*, 1968). To this end, highly deuterated proteins were produced by growing algae or

---

Kevin H. Gardner and Lewis E. Kay • Protein Engineering Network Centres of Excellence and Departments of Medical Genetics and Microbiology, Biochemistry, and Chemistry, University of Toronto, Toronto, Ontario M5S 1A8, Canada.

Biological Magnetic Resonance, Volume 16: Modern Techniques in Protein NMR, edited by Krishna and Berliner. Kluwer Academic / Plenum Publishers, 1999.

bacteria in D_2O media supplemented with either uniformly or selectively protonated amino acids. By monitoring the chemical shifts of the few remaining protons, conformational changes that occurred on ligand binding (Markley *et al.*, 1968) or oligomerization (Crespi and Katz, 1969) were investigated. Since these initial experiments, the preparation of proteins with amino acid selective protonation in a deuterated environment (or conversely, selective deuteration in an otherwise protonated molecule) has been regularly used for spectral simplification and residue type assignment (Oda *et al.*, 1992; Anglister, 1990; Arrowsmith *et al.*, 1990a; Brodin *et al.*, 1989).

In the 1980s, random fractional deuteration was employed to improve the quality of homonuclear proton two-dimensional (2D) NMR spectra. Because of the significantly lower gyromagnetic ratio of ^{2}H compared with ^{1}H ($\gamma[^2\text{H}]/\gamma[^1\text{H}]=0.15$), replacement of protons with deuterons removes contributions to proton linewidths from proton–proton dipolar relaxation and ^{1}H–^{1}H scalar couplings. At deuteration levels between 50–75%, the expected decrease in sensitivity related to the limited number of protons is offset to a large extent by a reduction in peak linewidths. Sensitivity gains were initially demonstrated in 1D ^{1}H spectra of the 43-kDa *E. coli* EF-Tu protein (Kalbitzer *et al.*, 1985) and subsequently in 2D homonuclear spectra of *E. coli* thioredoxin used for chemical shift assignment (LeMaster and Richards, 1988). Significant improvements have also been noted in many heteronuclear experiments. This is demonstrated for the case of an ^{15}N–^{1}H HSQC experiment in Fig. 1 where spectra recorded on fully protonated and perdeuterated samples (all nonexchangeable hydrogens substituted with deuterons) of a 30-kDa N-terminal domain of enzyme I of the *E. coli* phosphoenolpyruvate:sugar phosphotransferase system (EIN) are illustrated (Garrett *et al.*, 1997). As observed for the 14-kDa villin 14T protein (Markus *et al.*, 1994), deuterating all nonexchangeable sites to levels of 80–90% decreases amide proton transverse relaxation rates by approximately twofold. The improvement in sensitivity and resolution in the NH-detected dimension is particularly important because many classes of experiments record the amide proton chemical shift during acquisition (Clore and Gronenborn, 1994).

Deuteration can also improve the quality of NOESY experiments. In particular, substitution of aliphatic/aromatic protons with deuterons results in impressive sensitivity gains in NOESY spectra which record NH–NH correlations (LeMaster and Richards, 1988; Torchia *et al.*, 1988a). Initially demonstrated in perdeuterated systems, similar benefits have also been observed for proteins and peptides that are protonated at specific positions in an otherwise highly deuterated background (Pachter *et al.*, 1992; Reisman *et al.*, 1991; Arrowsmith *et al.*, 1990b; Tsang *et al.*, 1990). The sensitivity gains in these NOESY data sets are largely the result of reduced NH linewidths, as described above. However, the decrease in spin diffusion pathways and concomitant increase in selective T_1 values of the diagonal resonances also result in improvements. Moreover, it is possible to employ longer mixing times and, because of reduced spin diffusion rates, relate cross-peak intensities to inter-

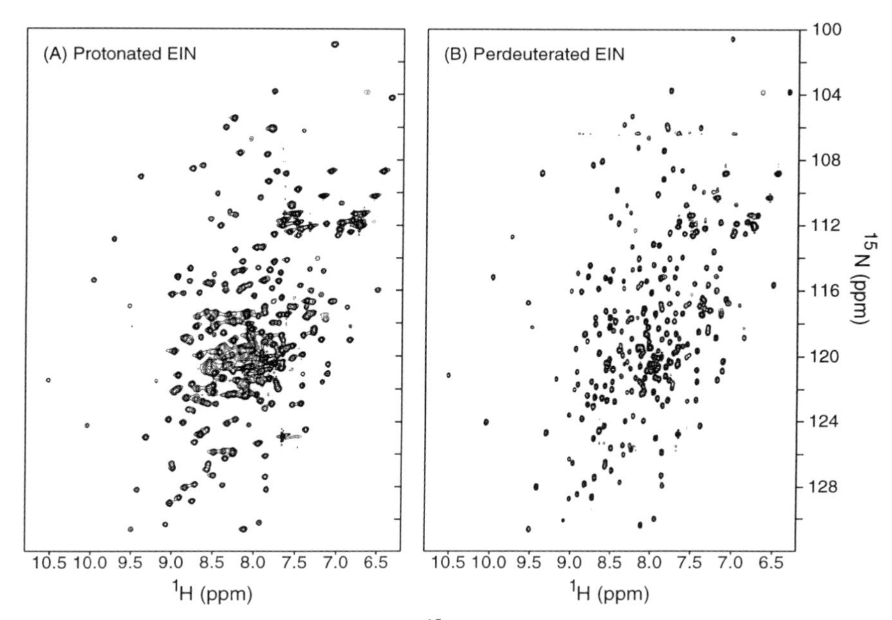

Figure 1. Effect of perdeuteration on the 2D ^{15}N–NH HSQC spectrum of the 30-kDa N-terminal domain of Enzyme I from the *E. coli* phosphoenolpyruvate-sugar phosphotransferase system. Both proteins are uniformly ^{15}N labeled, while the deuterated protein is approximately 90% uniformly deuterated. Reprinted with permission from Garrett *et al.* (1997).

nuclear distances more accurately than is possible from data recorded on highly protonated samples. This is of particular significance for longer range distance restraints. Finally, it is important to note that even in highly deuterated molecules the effects of spin diffusion cannot be neglected. In the case of highly deuterated proteins containing protonated amino acids, NOE spectra have been recorded with long mixing times to specifically promote intraresidue spin diffusion. In this way sidechain protons for several dimeric proteins with molecular masses above 20 kDa have been assigned (Reisman *et al.*, 1993; Arrowsmith *et al.*, 1990a,b).

1.2. Scope of This Review: Deuteration since 1993

As already described, deuteration can result in substantial improvements in NMR spectra used to study the structures of proteins and protein complexes. An alternative powerful strategy is triple resonance, multidimensional NMR spectroscopy of uniformly ^{15}N, ^{13}C labeled proteins (Bax, 1994; Clore and Gronenborn, 1994). In this approach sets of intra- and interresidue chemical shifts are correlated by transferring magnetization from one nucleus to another through one- or two-bond scalar couplings. Data analysis from several of these experiments in combination facilitates the assignment of nitrogen, carbon, and proton chemical shifts of proteins with molecular masses of approximately 25 kDa or less. For systems larger than this, long molecular correlation times result in very efficient relaxation of the participating spins, especially ^{13}C nuclei, attenuating signal intensity and degrading spectral resolution.

A straightforward approach to decrease the relaxation rates of many of the nuclei that are key players in triple-resonance experiments is to substitute carbon-bearing protons with deuterons. At the field strengths currently in use, the major contribution to relaxation of carbon magnetization, for example, derives from one-bond ^{13}C–^{1}H dipolar interactions (Browne *et al.*, 1973; Allerhand *et al.*, 1971). In the case of a ^{13}C–^{1}H spin pair attached to a macromolecule, replacement of the proton with a deuteron attenuates the dipolar interaction by a factor of approximately 15. This results in a substantial lengthening of carbon relaxation times leading to improvements in the sensitivity of experiments relying on scalar-coupling-based transfers of magnetization involving carbon nuclei (Venters *et al.*, 1996; Yamazaki *et al.*, 1994a,b; Grzesiek *et al.*, 1993b; Kushlan and LeMaster, 1993b). The first demonstration of the utility of deuteration in concert with triple-resonance spectroscopy occurred in 1993 when Bax and co-workers described the 4D HN(COCA)NH experiment for correlating sequential amides. Subsequently a suite of triple-resonance experiments for assignment of backbone chemical shifts of deuterated proteins was developed and demonstrated on a 37-kDa protein/DNA complex (Yamazaki *et al.*, 1994a,b) and later extended for use on a 64-kDa version of this system (Shan *et al.*, 1996). The dramatic improvement in both the sensitivity and the resolution of triple-resonance spectra that deuteration provides is illustrated in Fig. 2 by comparing 2D ^{13}C/^{1}H projections of 3D HNCA spectra recorded on a fully protonated ^{15}N, ^{13}C labeled 23-kDa Shc PTB domain/peptide complex (a) and a 75% deuterated version of the same complex (b). The decrease in ^{13}C linewidth allows individual peaks to be resolved in the crowded center of these spectra and the improved signal-to-noise facilitates the observation of weaker cross-peaks.

A large number of triple-resonance pulse sequences have since been optimized for use on deuterated proteins (Section 3.2). The application of many of these methods to high-molecular-mass proteins (40 kDa and beyond) will require significant levels of deuteration. For example, complete assignment of ^{15}N, ^{13}C$^{\alpha}$, ^{13}C$^{\beta}$ and NH chemical shifts was achieved for a 37-kDa trp repressor/DNA ternary complex (Yamazaki *et al.*, 1994a,b) where the protein component was labeled at approximately 70%. However, a 90% deuteration level was required for similar studies on a 64-kDa trp repressor/DNA complex (Shan *et al.*, 1996). Regrettably, the advantages associated with deuteration are not without compromise. The substitution of deuterons for protons depletes the number of protons available for NOE-based interproton distance restraints. In the limiting case of perdeuteration, only exchangeable protons remain, resulting in a drastic reduction in both numbers and

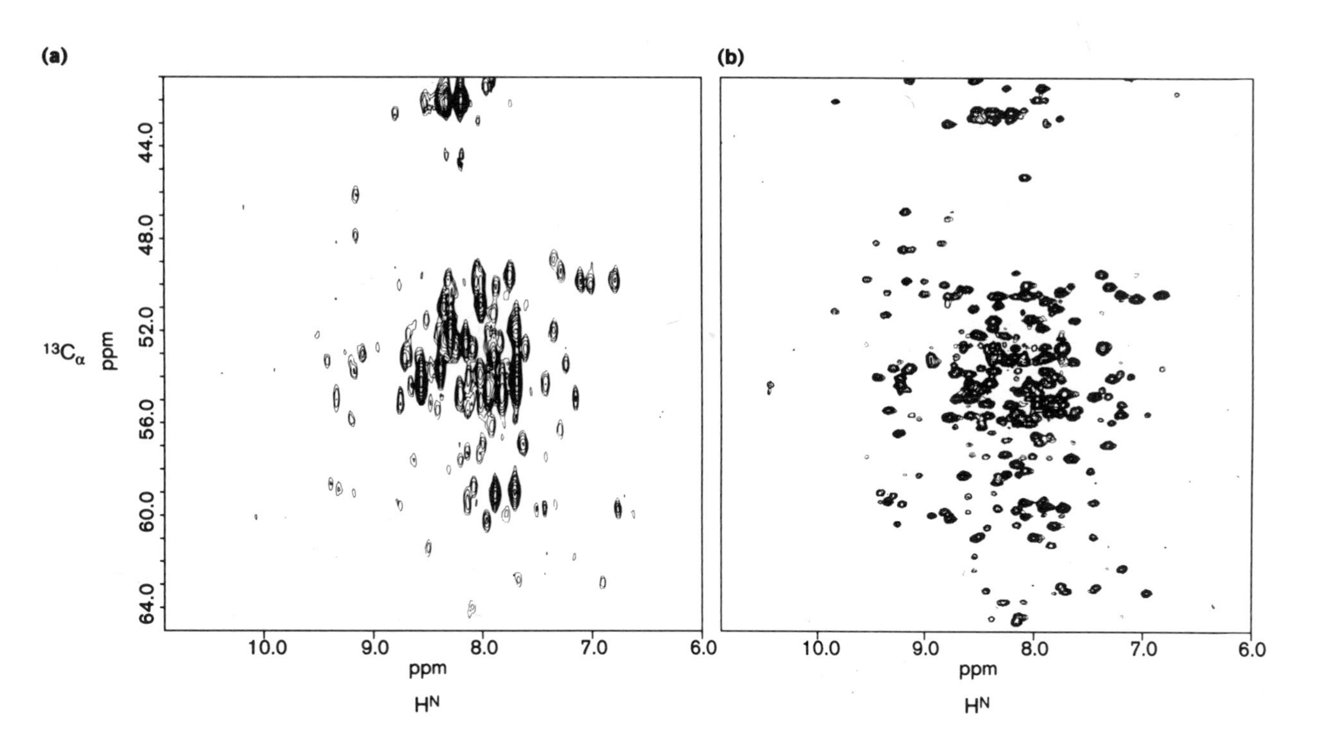

Figure 2. Comparison of 2D ($^{13}C_{\alpha}$, NH) projections from 3D HNCA spectra recorded of a 23-kDa Shc PTB domain/phosphotyrosine peptide complex. (a) ^{15}N, ^{13}C, ^{1}H uniformly labeled Shc PTB domain. (b) ^{15}N, ^{13}C, 75% ^{2}H uniformly labeled Shc PTB domain. Reprinted with permission from Sattler and Fesik (1996).

types of NOEs that can be measured in relation to fully protonated samples. It is not surprising, therefore, that structures determined from NOEs between backbone NH protons only have very low precision and accuracy (Gardner *et al.*, 1997; Smith *et al.*, 1996; Venters *et al.*, 1995b). This has stimulated the development of several different methods for the production of proteins that are selectively protonated at key positions in the molecule, allowing additional distance restraints to be measured, while maintaining a high level of deuteration at other sites so that experiments for sidechain and backbone assignment can enjoy the sensitivity gains that accompany deuteration.

To this point we have introduced the use of deuterium as an approach to improve the efficacy of large classes of triple-resonance experiments that would otherwise suffer significantly from relaxation loses. Outside of the low gyromagnetic ratio of the deuteron, no other properties of ^{2}H are exploited in these experiments. In contrast to the relatively recent use of deuterium in this role, the utility of ^{2}H NMR spectroscopy for the study of molecular dynamics has been recognized for several decades. Most of the work focused primarily on liquid-crystal samples, samples in the solid state, or solution state samples with deuterium labeling restricted to a small number of sites (Vold and Vold, 1991; Johnson *et al.*, 1989; Keniry, 1989; Keniry *et al.*, 1983; Schramm and Oldfield, 1983; Jelinski *et al.*, 1980; Seelig, 1977). Methods have emerged more recently for the measurement of sidechain dynamics properties of uniformly ^{13}C labeled, fractionally deuterated proteins through the indirect measurement of deuterium relaxation times, T_1 and $T_{1\rho}$ (Pervushin *et al.*, 1997; Kay *et al.*, 1996; Muhandiram *et al.*, 1995).

In this review, recent developments in the use of deuterium for studying protein structure and dynamics will be described. This includes the preparation of highly deuterated proteins, optimization of triple-resonance NMR experiments for use on highly deuterated samples, prospects for the structure determination of highly deuterated proteins, and the use of ^{2}H relaxation for the study of protein dynamics. The interested reader should consult previous reviews (LeMaster, 1994, 1990, 1989; Anglister, 1990) for a more thorough coverage of work in these fields prior to 1993.

2. DEUTERIUM LABELING METHODS

Proteins employed in NMR studies can be labeled using a number of different protocols that produce molecules with different patterns of deuterium incorporation. One labeling strategy results in deuterium incorporation throughout a protein in a roughly site-independent manner (uniform or random labeling) whereas a second approach produces a high level of deuteration (or protonation) at a restricted number of sites (site-specific labeling). The optimal labeling pattern for a particular sample is of course determined by the intended application(s). In what follows a number of methods are described for the generation of proteins deuterated on aliphatic/aromatic carbon sites. Although considerable progress has been realized in the past several years in the production of uniformly deuterated (Batey *et al.*, 1996) or site-specifically deuterated (Földesi *et al.*, 1996; Ono *et al.*, 1996; Tolbert and Williamson, 1996; Agback *et al.*, 1994) nucleotides for use in NMR studies of RNA and DNA molecules, the present discussion will focus on protein applications exclusively.

2.1. Uniform Deuteration

As already mentioned, the optimal level of deuteration for a specific application, ranging from the complete substitution of protons with deuterons to moderate levels of fractional replacement, will very much depend on the experiments that are planned. Completely deuterated proteins have been used to eliminate contributions to spectra from one component of a macromolecular complex, as demonstrated in studies of a complex of fully deuterated calmodulin with a protonated peptide (Seeholzer *et al.*, 1986) and deuterated cyclophilin with protonated cyclosporin (Hsu and Armitage, 1992). However, a wider range of applications have made use of samples of (random) fractionally deuterated molecules. For example, random fractionally deuterated proteins were initially used to improve the sensitivity of 2D ^{1}H–^{1}H homonuclear spectra by reducing dipolar relaxation pathways, spin diffusion, and passive scalar couplings (LeMaster and Richards, 1988; Torchia *et al.*, 1988a). Fractional deuteration also significantly improves the sensitivity of many triple-resonance (^{15}N, ^{13}C, ^{1}H) experiments (Farmer and Venters, 1996; Nietlispach *et al.*, 1996; Shan *et al.*, 1996; Farmer and Venters, 1995; Yamazaki *et al.*, 1994a,b; Grzesiek *et al.*, 1993b), as will be discussed in greater detail in Section 3. In addition, the sidechain dynamics of proteins prepared in this manner can be studied using a number of new ^{13}C and ^{2}H spin relaxation experiments (Pervushin *et al.*, 1997; LeMaster and Kushlan, 1996; Muhandiram *et al.*, 1995; Kushlan and LeMaster, 1993a) as addressed in Section 5.

To a first approximation, randomly deuterated proteins can be generated in a straightforward manner. A culture of bacteria containing an overexpression vector for the protein of interest is grown in minimal media with the D_2O/H_2O ratio chosen to reflect the desired level of deuterium incorporation. Proteins generated will be composed of amino acids containing approximately the same deuterium content as in the overexpression media (see below). Cultures of simple eukaryotes such as algae and yeast can also grow in media containing up to 100% D_2O, facilitating the use of these organisms for the biosynthetic production of highly deuterated proteins (Crespi *et al.*, 1968; Katz and Crespi, 1966).

Expression of deuterated proteins in higher eukaryotic cells (plant and animal) is more challenging because these cells will not grow in media containing more than 30–50% D_2O (Katz and Crespi, 1966). This problem can be partially circumvented by culturing these cells in H_2O media supplemented with commercially available ^{2}H-labeled algal amino acid extracts, analogous to previously established methods for expressing uniformly ^{15}N, ^{13}C, ^{1}H labeled proteins in these systems (Hansen *et al.*, 1992). Proteins produced in this manner will be highly deuterated at most carbon positions, although significant levels of potentially nondesired protonation can still occur through $^{2}H/^{1}H$ exchange with solvent protons, most notably at $C\alpha$ sites.

Note that several factors can lead to a nonuniform distribution of deuterium throughout "randomly" fractionated proteins. In this regard, the carbon compounds used as growth substrates for *E. coli* are particularly important as protons from these molecules can be efficiently retained at specific sites within several amino acids. For example, proteins produced in bacteria grown in deuterated media containing ^{1}H-glucose typically have relatively high levels of protonation in sidechains of the aromatic amino acids. This results from the fact that aromatic groups are synthesized from glucose-derived carbohydrates (LeMaster, 1994). In certain cases, specific retention of protons from a protonated carbon source can be exploited to generate useful patterns of site-specific protonation within a highly deuterated background (Section 2.2.2). For applications requiring perdeuterated proteins, bacteria can be grown on deuterated carbon sources [e.g., ^{2}H-glucose or ^{2}H-succinate (LeMaster and Richards, 1988)] or simple protonated carbon sources where the carbon-bound protons are replaced by solvent deuterons prior to or during amino acid biosynthesis [e.g., ^{1}H-sodium acetate (Venters *et al.*, 1995a)].

Nonuniform deuteration can still be significant even when deuterated carbon sources exclusively are employed. Kinetic and thermodynamic isotope effects can alter the activity of metabolic and biosynthetic enzymes to the point of significantly biasing the distribution of deuterium in partially deuterated samples (Martin and Martin, 1990; Galimov, 1985). For example, a protein overexpressed in *E. coli* grown in an 80:20% D_2O:H_2O medium supplemented with deuterated succinate and 75% ^{2}H, DL-alanine was deuterated on average to a level of approximately 75% (LeMaster, 1997). However, specific sites within the protein had significantly lower ^{2}H incorporation, including the Ile $\gamma 1$ methylene positions ($< 50\%$ deuterated). In another case, nonuniform type-specific deuterium incorporation has been observed in the CH_3:CH_2D:CHD_2 isotopomer distribution of methyl groups of an SH2 domain overexpressed in *E. coli* grown in 65:35% D_2O:H_2O medium (Kay *et al.*, 1996).

2.2. Site-Specific Protonation in a Highly Deuterated Environment

2.2.1. Protonated Amino Acid Sidechains in a Deuterated Background

In contrast to the approaches described above for producing proteins with deuterium substituted at an approximately uniform level throughout most of the aliphatic sites, proteins can also be generated where only specific sites are protonated in an otherwise highly deuterated background. Proteins can be labeled in this manner by overexpression in bacterial cultures grown in minimal D_2O media supplemented with protonated small organic molecules such as amino acids, amino acid precursors, or carbon sources (LeMaster, 1994, 1990). Alternatively, bacteria can be cultivated in H_2O-based media supplemented with deuterated algal cell lysates that have been chemically or enzymatically treated to promote site-specific protonation.

Some of the earliest ^{2}H-based NMR studies of proteins made use of deuterated, site-protonated molecules isolated from organisms grown in minimal media with high levels of D_2O and supplemented with fully or partially protonated amino acids (Crespi and Katz, 1969; Crespi *et al.*, 1968; Markley *et al.*, 1968). The resulting simplified 1D ^{1}H NMR spectra of these proteins facilitated the observation of well-separated peaks reporting on sites dispersed throughout the primary sequence of the molecule. Initial studies focused on the incorporation of leucine (Crespi *et al.*, 1968). Subsequently, most of the 20 natural amino acids have been successfully incorporated in this manner, either alone or in combination with other residues (Metzler *et al.*, 1996b; Smith *et al.*, 1996; Oda *et al.*, 1992; Brodin *et al.*, 1989). In general, wild-type strains of bacteria and algae can be used for amino acid specific labeling so long as the growth medium is supplemented with relatively high concentrations (> 30 mg/liter) of protonated compounds. In cases where it is necessary to use smaller amounts of these potentially expensive labeled compounds or to limit nondesirable labeling of amino acids derived from the labeled compound, amino acid-specific auxotrophic strains have been employed (Oda *et al.*, 1992; LeMaster, 1990). Similar approaches have also been used in the context of specific ^{15}N and ^{13}C labeling (Waugh, 1996; McIntosh and Dahlquist, 1990).

As previoulsy mentioned, proteins produced by bacteria in highly deuterated media containing a number of fully protonated amino acids usually retain most of the sidechain protons on these residues. In a typical D_2O-based growth medium, deuterons will replace between 30 and 80% of the Hα protons of the protonated amino acids (Metzler *et al.*, 1996b; Smith *et al.*, 1996; Crespi *et al.*, 1968), while the sidechains remain almost entirely protonated. On one hand, this high level of residual protonation is beneficial for the sensitivity of many proton-detected experiments. However, the presence of these spins can lead to rapid transverse relaxation rates for sidechain ^{13}C nuclei (Metzler *et al.*, 1996b; Smith *et al.*, 1996), reducing the sensitivity of triple-resonance experiments designed to assign sidechain chemical shifts, such as the HCC(CO)NH-TOCSY (Section 3.3). As well, protons located on the sidechain provide effective intraresidue spin diffusion pathways in NOE-based experiments, complicating the quantitation of cross-peak intensities in terms of internuclear distances (Pachter *et al.*, 1992).

2.2.2. Protonated Methyl Groups in a Deuterated Background: Pyruvate

To circumvent the problems associated with the use of fully protonated amino acids in the production of deuterated site-protonated proteins, Rosen and co-workers have developed a method that produces molecules with a more limited number of highly protonated sites (Rosen *et al.*, 1996). The approach involves growing bacteria in a D_2O-based minimal medium with a protonated nonglucose carbon source. As discussed in Section 2.1 for the case of glucose, the use of a protonated carbon source introduces protons at specific positions in various amino acids in a manner dependent on the details of the metabolism of the compound. Rosen *et al.* have taken advantage of the fact that the methyl group of pyruvate is the metabolic precursor of the methyls of Ala, Val, Leu, and Ile ($\gamma 2$ only) to produce proteins that are highly deuterated at all positions with the exceptions of the methyl groups mentioned above. In this approach, proteins are overexpressed in *E. coli* grown in a D_2O-based minimal medium with protonated pyruvate as the sole carbon source. The utility of this approach is demonstrated in Fig. 3, which compares ^{13}C–^{1}H HSQC spectra of ^{15}N, ^{13}C, fully protonated (a) and ^{15}N, ^{13}C, methyl protonated, highly deuterated (b) samples of the C-terminal SH2 domain of bovine phospholipase $C_{\gamma 1}$ (PLCC). The majority of the aliphatic sites in the pyruvate-derived protein are completely deuterated while the methyl groups of Ala, Val, Leu, and Ile ($\gamma 2$) are significantly protonated. However, the methylene groups of several amino acids identified in Fig. 3b are also protonated to a lower extent. It is noteworthy that the methyl peaks in the spectrum of the deuterated sample are highly asymmetric. This is the result of the production of CH_3, CH_2D, and CHD_2 isotopomers coupled with the significant one-bond deuterium isotope shift for both ^{13}C (–0.3 ppm per ^{2}H) and ^{1}H (–0.02 ppm per ^{2}H) (Gardner *et al.*, 1997). The extent of protonation for the pyruvate-derived methyl groups ranges from 40% (Ala) to 60% (Val and Ile) to 80% (Leu), as established on the basis of both mass spectrometry and NMR studies (Rosen *et al.*, 1996). These levels can be further increased by shortening the induction step during growth (Gardner *et al.*, 1996). The moderately high levels of protonation at Ala, Ile ($\gamma 2$), Val, and Leu methyl groups and the significant deuteration at most other sites (most notably, $C\alpha > 95\%$ and $C\beta > 80\%$ deuterated) ensure that high-sensitivity triple-resonance spectra both for backbone assignment and for correlation of sidechain ^{13}C and ^{1}H chemical shifts with backbone ^{15}N–^{1}H spin pairs can be obtained (Gardner *et al.*, 1996).

2.2.3. Protonated Methyl Groups in a Deuterated Background: Specifically Protonated Amino Acids and Amino Acid Precursors

Despite the utility of the pyruvate strategy described above, the presence of methyl isotopomers is limiting, in terms of both resolution and sensitivity. In addition, the level of protein produced in *E. coli* cultures grown in pyruvate-based

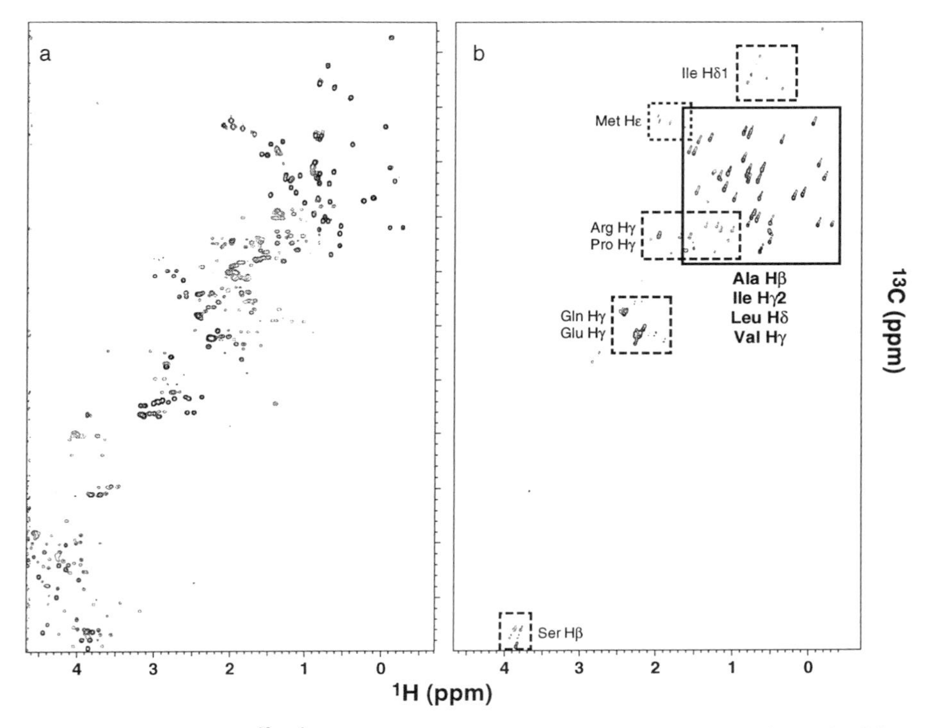

Figure 3. Constant time ^{13}C–^{1}H HSQC spectra of the PLCC SH2 domain prepared from (a) a fully protonated, H_2O medium with a ^{13}C-glucose carbon source and (b) a D_2O-based medium with ^{13}C, ^{1}H pyruvate the sole carbon source. Reprinted from Rosen *et al.* (1996).

media is approximately twofold reduced relative to cultures that use glucose as the carbon source. It is therefore necessary to develop an alternative approach in which proteins are overproduced in bacteria grown in a highly deuterated, glucose medium supplemented with amino acid precursors and amino acids that have useful patterns of protonation and deuteration. Both the precursors and the amino acids can be prepared *in vitro* using a combination of synthetic and enzymatic approaches with the conversion of the amino acid precursors to the appropriate amino acids occurring *in vivo* once the compounds are provided to *E. coli.* Intermediates can be chosen based on several criteria, including the ability to enter a biosynthetic pathway without complications from subsequent $^{1}H/^{2}H$ exchange reactions, the extent of *E. coli* assimilation and the ease of preparation. An example of this approach is illustrated in the production of proteins with high levels of deuteration at all positions with the exception of methyl positions of Val, Leu, and Ile ($\delta 1$ only) (Gardner and Kay, 1997). Proteins are overexpressed in *E. coli* grown in minimal

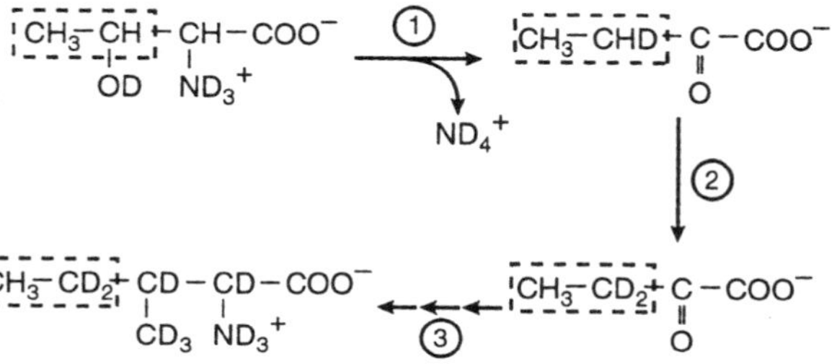

Figure 4. Generation of protein labeled with ^{13}C, ^{2}H (^{1}H-$\delta 1$ methyl) Ile. Step 1: *In vitro* conversion from Thr into [3-^{2}H] 2-ketobutyrate, catalyzed by *E. coli* biosynthetic threonine deaminase (Eisenstein, 1991). Step 2: *In vitro* conversion of [3-^{2}H] 2-ketobutyrate into [3,3-$^{2}H_2$] 2-ketobutyrate by proton/deuterium exchange at C3 using pH* 10.2, 45 °C. Step 3: *In vivo* conversion of [3,3-$^{2}H_2$] 2-ketobutyrate into Ile and eventual incorporation into overexpressed protein, carried out by *E. coli* metabolism. All steps performed in 99.5% D_2O. Reprinted with permission from Gardner and Kay (1997).

^{15}N, ^{13}C, ^{2}H medium containing deuterated glucose as the carbon source and supplemented with [2,3-$^{2}H_2$]-^{15}N, ^{13}C Val (available commercially) and [3,3-$^{2}H_2$]-^{13}C 2-ketobutyrate. Addition of [2,3-$^{2}H_2$]-Val to the growth medium (50 mg/liter, see below) results in the production of proteins in which both Val and Leu are highly deuterated at all nonmethyl positions, and in which only the CH_3 methyl isotopomer is produced. Ile, protonated only at the C$\delta 1$ position, is generated using the set of reactions illustrated in Fig. 4. Commercially available ^{15}N, ^{13}C labeled Thr is the starting compound and is stoichiometrically converted to [3-^{2}H]-^{13}C 2-ketobutyrate in a process catalyzed by *E. coli* biosynthetic threonine deaminase. Note that because this reaction is carried out in D_2O, one of the two methylene hydrogens of 2-ketobutyrate becomes deuterated. Substitution of the remaining methylene proton with a deuteron is achieved via base-catalyzed exchange. The resultant product is sterile filtered and added without any further purification to a D_2O-based minimal medium supplemented with [2,3-$^{2}H_2$]-^{15}N, ^{13}C Val. A ^{13}C–^{1}H HSQC spectrum of the methyl region of the PLCC SH2 domain expressed from *E. coli* grown in this supplemented medium is shown in Fig. 5. Despite the fact that a prototrophic strain has been employed in the overexpression, over 90% of Val, Leu, and Ile in the protein is derived from the added Val and 2-ketobutyrate (Gardner and Kay, 1997). The nonmethyl Ile sidechain positions are highly deuterated and, as is the case with Val and Leu, only the CH_3 isotopomer is produced. In this particular example, fully protonated ^{15}N, ^{13}C Val has been used and therefore Val and Leu retain some protonation at the methine (Val β, Leu γ) positions. The undesired protonation can be eliminated by replacing the ^{15}N, ^{13}C Val added to the growth medium in the production of the SH2 domain with [2,3-$^{2}H_2$]-^{15}N,^{13}C Val, as discussed above. In this context, we have recently obtained a sample of the 40-kDa maltose-binding protein by overexpression in *E. coli* grown in minimal medium (D_2O) using ^{2}H,

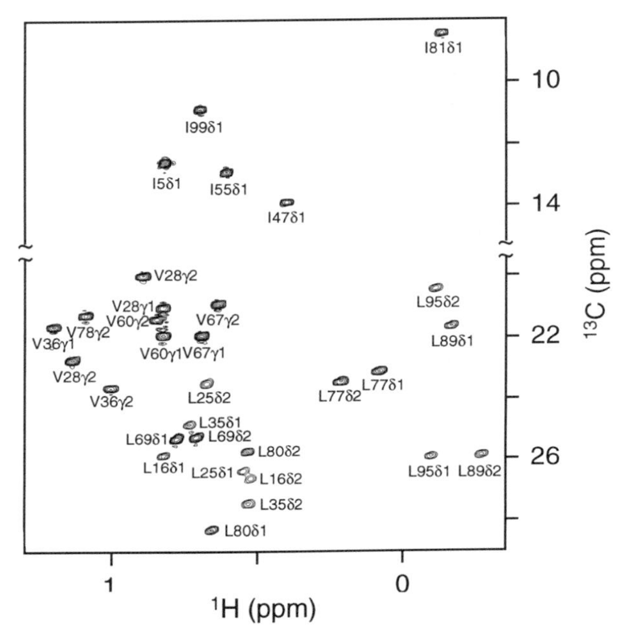

Figure 5. Methyl region of a ^{13}C–^{1}H constant-time HSQC spectrum of a highly deuterated, methyl-protonated PLCC SH2 domain, labeled using D_2O medium supplemented with ^{15}N, ^{13}C Val and [3,3-$^2H_2$]-^{13}C 2-ketobutyrate (Gardner and Kay, 1997). This protein is highly deuterated at all aliphatic sites except for the Val Cβ and Leu Cγ methine positions; these sites can be deuterated by the substitution of [2,3-$^2H_2$]-^{15}N, ^{13}C Val in place of ^{15}N,^{13}C Val. Over 90% of Val, Ile, and Leu derive from the added 2-ketobutyrate or Val. Reprinted with permission from Gardner and Kay (1997).

^{13}C glucose and $^{15}NH_4Cl$ as the sole carbon and nitrogen sources and supplemented with [2,3-$^2H_2$]-^{15}N, ^{13}C Val and [3,3-$^2H_2$]-^{13}C 2-ketobutyrate. ^{13}C–^{1}H HSQC spectra have established the high level of protonation at methyl groups of Val, Leu, and Ile ($\delta 1$) and the very significant extent of deuteration at other carbon positions.

2.2.4. Protonated Cα in a Deuterated Background

While significant emphasis has been placed on producing highly deuterated proteins with protonation at select sidechain positions, methods have also emerged that place protons at specific backbone sites. Recently, Yamazaki *et al.* (1997) have produced a sample of the α subunit of *E. coli* RNA polymerase that is ^{15}N, ^{13}C, ^{2}H, ^{1}Hα-labeled. By acetylating an ^{15}N, ^{13}C, ^{2}H-labeled amino acid mixture and hydrolyzing the resulting esters in H_2O under highly acidic conditions, the deuterons at the Cα positions of most amino acids are efficiently replaced by solvent protons. The sidechain positions, however, remain highly deuterated during this procedure. The resulting mixture of ^{2}H, $^{1}H\alpha$ amino acids was added to *E. coli* growing in an H_2O-based minimal medium immediately prior to protein overexpression. Ten of the amino acid types in RNA polymerase produced in this manner were over 80% protonated at the $C\alpha$ position and most sidechain positions retained high levels of deuteration. One drawback to this methodology in terms of cost effectiveness is that the ester hydrolysis step racemizes the $C\alpha$ position; in principle this could be circumvented by the use of enzyme-based approaches (Homer *et al.*, 1993).

2.3. Practical Aspects of Producing Deuterated Proteins in *E. coli*

Although many of the labeling methods described above can be readily performed in any laboratory that routinely overexpresses proteins in bacterial systems, several aspects deserve special commentary. Figure 6 outlines the approach that we have used to generate both randomly deuterated and deuterated, site-protonated proteins.

At the outset of each growth, we use freshly transformed *E. coli* [typically but not exclusively strain BL21(DE3)] that has been plated onto solid H_2O-based rich medium. This is in contrast to approaches that utilize bacteria that have been previously adapted to growth in deuterated medium by culturing through steps with progressively higher percentages of D_2O (e.g., Venters *et al.*, 1995a) and then stored as glycerol stabs in a high-level D_2O medium. Because the culture is manipulated through multiple liquid medium steps (Fig. 6), we rely on several basic guidelines to ensure robust growth. All of the cultures are maintained at subsaturating cell densities, with optical densities measured at 600 nm (A_{600}) typically below 0.6. When the culture reaches this level the medium is briefly centrifuged and removed and the cells resuspended in an amount of fresh medium to bring the A_{600} to 0.05–0.1. In this manner, the culture is kept essentially in log-phase growth without significant lag periods. Note that the D_2O used during protein production can be recycled by flash chromatography to minimize the cost of this process (Moore, 1979).

It is extremely important that the composition of medium used in each step of the growth process be carefully regulated to ensure a constantly high growth rate. In particular, only one variable of the growth medium is changed (e.g., solvent or carbon source) between subsequent steps. As a result, we have found that we can completely change many variables (e.g., H_2O to D_2O or glucose to pyruvate) over the course of multiple steps without causing the bacterial culture to enter growth lag phases of longer than 1 to 2 h. Note that at least two D_2O growth steps occur prior to induction in each of the paths illustrated in Fig. 6, significantly minimizing the level of residual protonation in overexpressed proteins. With the typical dou-

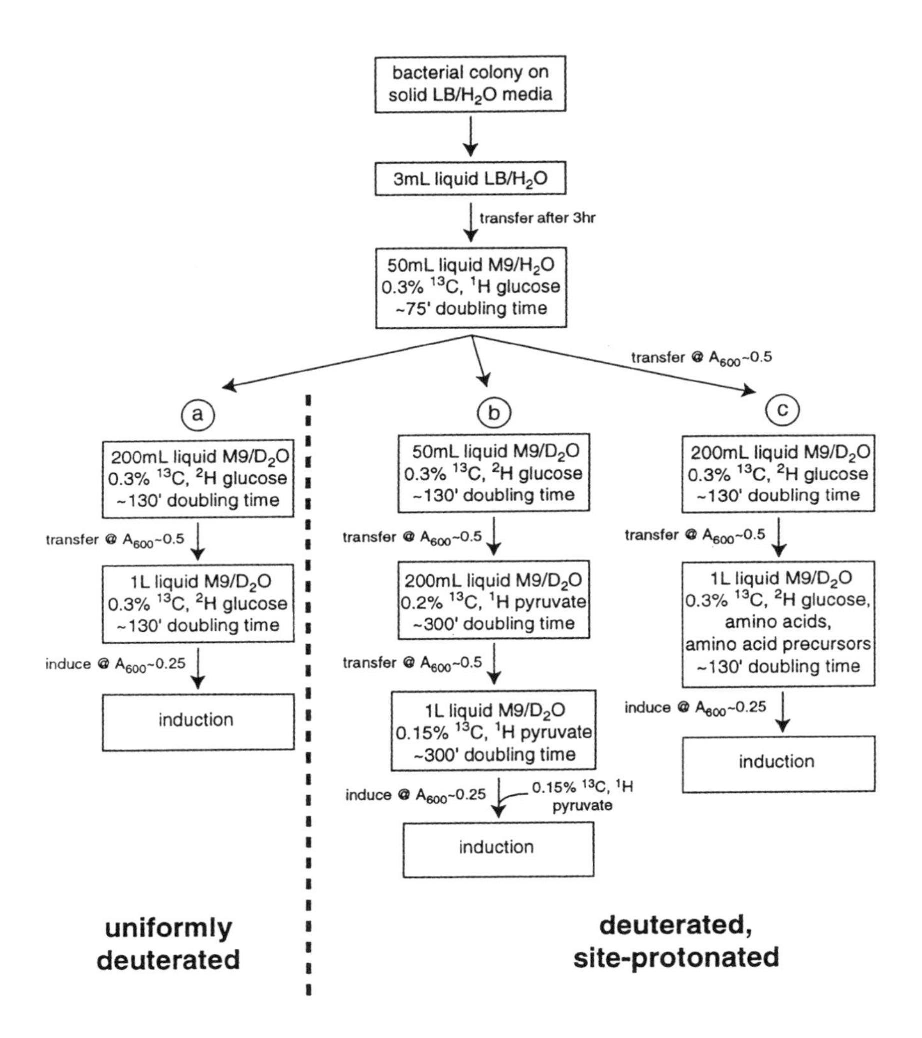

Figure 6. Typical bacterial growth protocols for producing highly deuterated proteins. (a) Uniformly deuterated protein production achieved by growing bacteria in D_2O medium (Section 2.1). (b) Production of deuterated, methyl-protonated protein at Ala, Val, Ile ($\gamma2$), and Leu sites using ^{1}H-pyruvate as a carbon source in a D_2O medium (Section 2.2.2). A total concentration of 0.3% ^{1}H-pyruvate is used in the final induction culture, added incrementally as described in Rosen *et al.* (1996). (c) Production of deuterated, methyl-protonated protein at Val, Ile ($\delta1$), and Leu sites using protonated amino acids or amino acid precursors in a D_2O medium (Section 2.2.3). In all cases, ^{2}H-labeled glucose should be used for steps in D_2O medium to obtain as uniform a level of deuteration as possible. M9 is defined as the medium without the carbon source.

bling times indicated on the figure for bacterial cultures grown at 37 °C, a usual growth requires between 18 and 60 h.

The purification of a highly deuterated protein is essentially unchanged relative to procedures used for the corresponding protonated versions of the molecule. It is important to recognize that proteins synthesized in D_2O will typically retain deuterons at a significant fraction of backbone amide sites even though purification takes place in H_2O-based buffers. Introduction of protons at labile sites can be achieved by a denaturation/renaturation cycle of the protein in an H_2O-containing buffer using chemical denaturants such as guanidine hydrochloride or urea (Constantine *et al.*, 1997; Venters *et al.*, 1996). Alternatively, thermostabile proteins can be incubated at high temperatures for extended periods of time to catalyze NH exchange.

3. TRIPLE-RESONANCE METHODS

3.1. General Comments

Despite the demonstrated utility of triple-resonance (^{15}N, ^{13}C, ^{1}H) methods for structural studies of proteins, a number of significant limitations occur when these techniques are applied to proteins or protein complexes with molecular masses in excess of approximately 20 kDa. The first problem relates to the rapid relaxation rates of many of the nuclei that participate in the magnetization transfer steps that occur during the course of these complex pulse sequences. The situation is particularly acute in the case of ^{13}C spins, where, for example, ^{13}Cα transverse relaxation times are on the order of 16 ms for a protein tumbling with a correlation time of 15 ns (Yamazaki *et al.*, 1994a). The second limiting feature is the lack of resolution in spectra (in particular in carbon dimensions), to a large extent the product of short acquisition times that are necessitated by rapid transverse relaxation rates. As described in Section 1 and illustrated in Fig. 7 for the case of a ^{13}Cα carbon with a single attached hydrogen, deuteration of carbon sites provides a solution to these problems. Figure 7 compares carbon linewidths for a carbon–hydrogen spin pair as a function of molecular correlation time, where relaxation contributions to the ^{13}C spin from chemical shift anisotropy, ^{13}Cα–^{1}Hα or ^{13}Cα–^{2}Hα dipolar interactions are considered. It is clear in the case of a ^{13}C–^{1}H pair that the dominant interaction is dipolar and that this effect can be largely suppressed through deuteration. For example, for a molecular correlation time of 14.5 ns, deuteration is predicted to increase the carbon transverse relaxation time from 16.8 ms (^{13}C–^{1}H) to 257 ms (^{13}C–^{2}H), where only one-bond ^{13}C–^{1}H/^{13}C–^{2}H dipolar effects have been considered. When contributions from additional dipolar interactions involving the adjacent carbonyl, ^{13}Cβ, and ^{15}N spins as well as chemical shift anisotropy are included, the predicted carbon transverse relaxation time decreases from 257 ms to

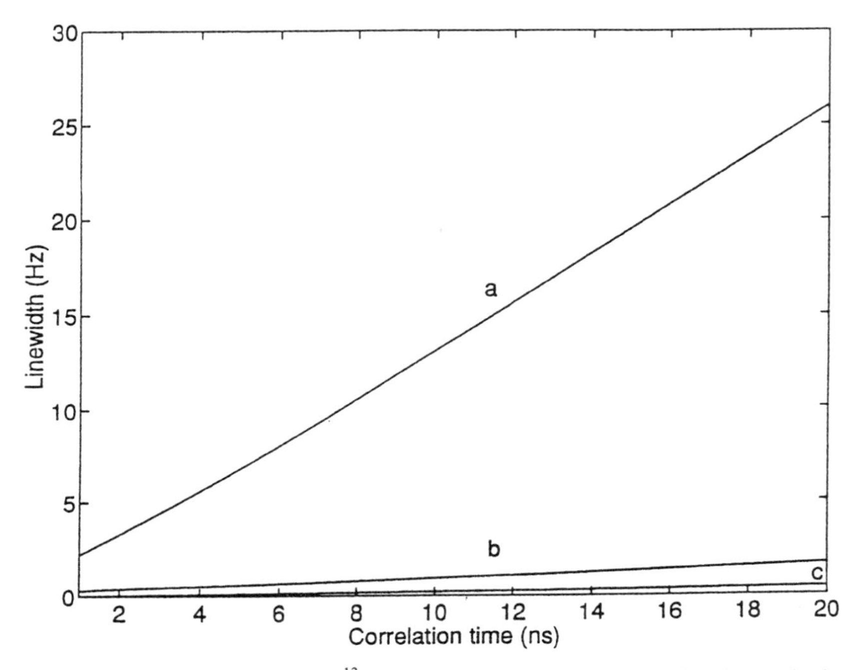

Figure 7. Expected contributions to the ^{13}Cα linewidth from dipolar and CSA relaxation mechanisms assuming an isolated ^{13}Cα–^{1}Hα or ^{13}Cα–^{2}Hα spin pair as part of an isotropically tumbling molecule in solution. (a) Contribution to the ^{13}Cα linewidth from ^{13}Cα–^{1}Hα dipolar interactions (assuming a carbon–proton separation of 1.10 Å); (b) Contribution from ^{13}Cα–^{2}Hα dipolar interactions. (c) Contribution from ^{13}Cα CSA (assuming an axially symmetric ^{13}Cα CSA tensor, $\Delta\sigma$ = 34 ppm). Reprinted with permission from Yamazaki *et al.* 1994a).

approximately 160 ms. Experimentally, transverse relaxation times were found to increase from 16.5 ± 3 ms to 130 ± 12 ms in the case of an ^{15}N, ^{13}C, ~70% ^{2}H labeled 37-kDa trpR/unlabeled DNA complex (correlation time of 14.5 ± 0.9 ns) (Yamazaki *et al.*, 1994a). The discrepancy between calculated and experimental values (160 ms versus 130 ms) is likely the result of the residual 30% protonation.

Substitution of deuterons for protons has a second major benefit as well. The significant decrease in carbon linewidths in deuterated proteins facilitates the use of constant-time carbon acquisition periods (Santoro and King, 1992; Vuister and Bax, 1992). For many experiments, such as the CT-HNCA and CT-HN(CO)CA, this period is set to $1/J_{CC}$, where J_{CC} is the one-bond ^{13}C–^{13}C homonuclear scalar coupling (Yamazaki *et al.*, 1994a,b). In addition to improving resolution by allowing significantly longer acquisition times than would otherwise be possible, the deleterious effects of passive one-bond carbon couplings on spectral resolution and sensitivity can be eliminated (Section 3.2).

A wide variety of triple-resonance pulse schemes have been modified to take advantage of reduced ^{13}C relaxation rates in highly deuterated proteins. Table 1 provides a listing of many of these experiments. In general, only minor changes are required to adapt a pulse sequence for use on deuterated proteins. Primarily these changes include the use of deuterium decoupling (Section 3.4) and, in some cases, the incorporation of various elements such as constant-time indirect detection periods, which likely would result in unacceptable sensitivity losses in experiments on protonated proteins. Specific modifications tailored to particular samples are also possible, such as pulse schemes developed for use on highly deuterated, site-protonated proteins including deuterated, methyl-protonated (Gardner *et al.*, 1996; Rosen *et al.*, 1996) and deuterated, Hα-protonated (Yamazaki *et al.*, 1997) molecules.

The fact that it is possible, at least in principle, to generate deuterated samples with high occupancy of protons at the labile sites in the molecule has resulted in the design of many experiments that are of the "out-and-back" variety where magnetization both originates and is detected on NH spins (Yamazaki *et al.*, 1994a,b). These experiments are performed on samples dissolved in H_2O and benefit by the relatively good dispersion of cross-peaks in ^{15}N, NH correlation spectra. Alternatively, a different strategy is required for samples in which the Hα site is protonated in proteins that are otherwise highly deuterated. In this case "out-and-back" experiments have been developed where the role of the NH spin discussed above is replaced by the Hα nucleus (Yamazaki *et al.*, 1997). Experiments of this type are best conducted in D_2O. It must be noted that a drawback of deuteration in all experiments is that the T_1 relaxation times of the remaining protons are significantly increased, requiring the use of longer recycle delays than typically employed in experiments performed with protonated molecules (Nietlispach *et al.*, 1996; Markus *et al.*, 1994; Yamazaki *et al.*, 1994a).

3.2. Backbone Chemical Shift Assignment

A large number of experiments for assignment of backbone chemical shifts have been developed for use on deuterated proteins, with many of these pulse schemes analogous to versions used for protonated samples. In addition to the 4D HN(COCA)NH described first by Bax and co-workers (Grzesiek *et al.*, 1993b) and the suite of experiments for correlating backbone $^{13}C\alpha$, ^{15}N, NH, and sidechain $^{13}C\beta$ chemical shifts designed by Yamazaki *et al.* (1994a,b), several other groups have published pulse schemes optimized for use on deuterated molecules. Experiments include the HN(CA)CO (Matsuo *et al.*, 1996a,b), HN(CA)NH (Ikegami *et al.*, 1997), alternative HN(COCA)NH sequences (Matsuo *et al.*, 1996a; Shirakawa *et al.*, 1995), as well as schemes providing backbone correlations with residue-selective editing (Dötsch *et al.*, 1996). To illustrate a number of the features that are particular to pulse schemes optimized for use on deuterated

Table 1
Triple-resonance NMR Methods with Versions Optimized for Use with Highly Deuterated Proteins

Method	References
Backbone	
Out-and-back	
3D CT-HNCA	Yamazaki *et al.* (1994b)
3D CT-HN(CO)CA	Yamazaki *et al.* (1994a)
3D HN(COCA)CB	Yamazaki *et al.* (1994a)
3D HN(CA)CB	Yamazaki *et al.* (1994a)
4D HNCACB	Yamazaki *et al.* (1994a)
3D CT-HN(COCA)CB	Shan *et al.* (1996)
3D CT-HN(CO)CA	Shan *et al.* (1996)
3D HN(CA)CO	Matsuo *et al.* (1996a,b)
3D HACAN	Yamazaki *et al.* (1997)
3D HACACO	Yamazaki *et al.* (1997)
3D HACACB	Yamazaki *et al.* (1997)
3D HACA(N)CO	Yamazaki *et al.* (1997)
Straight-through	
3D, 4D HN(COCA)NH	Grzesiek *et al.* (1993b), Shirakawa *et al.* (1995), Matsuo *et al.* (1996a)
4D HN(CA)NH	Ikegami *et al.* (1997)
4D HBCB/HACANNH	Nietlispach *et al.* (1996)
4D HBCB/HACA(CO)NNH	Nietlispach *et al.* (1996)
Sidechain	
Out-and-back	
2D H_2N-HSQC	Farmer and Venters (1996)
3D $H_2N(CO)C_{\gamma/\beta}$	Farmer and Venters (1996)
3D $H_2N(COC_{\gamma/\beta})C_{\beta/\alpha}$	Farmer and Venters (1996)
2D $H(N_{\varepsilon/\eta})C_{\delta/\varepsilon}$	Farmer and Venters (1996)
2D $H(N_{\varepsilon/\eta}C_{\delta/\varepsilon})C_{\gamma/\beta}$	Farmer and Venters (1996)
2D $H(N_{\varepsilon})C_{\zeta}$	Farmer and Venters (1996)
$H(N_{\varepsilon/\eta})C_{\zeta}$	Farmer and Venters (1996)
Straight-through	
3D C(CC)(CO)NH	Farmer and Venters (1995)
4D HCC(CO)NH	Nietlispach *et al.* (1996)
3D (H)C(CO)NH-TOCSY	Gardner *et al.* (1996)

samples, we provide a brief description of the CT-HNCA (Yamazaki *et al.*, 1994b). This experiment correlates the chemical shifts of intraresidue ^{15}N–NH spin pairs of residue *i* with the $^{13}C\alpha$ shifts of residues *i* and $(i-1)$. Figure 8 presents the pulse sequence of the CT-HNCA experiment (Yamazaki *et al.*, 1994b); it is clear that many of the magnetization transfer steps are identical to those employed in the

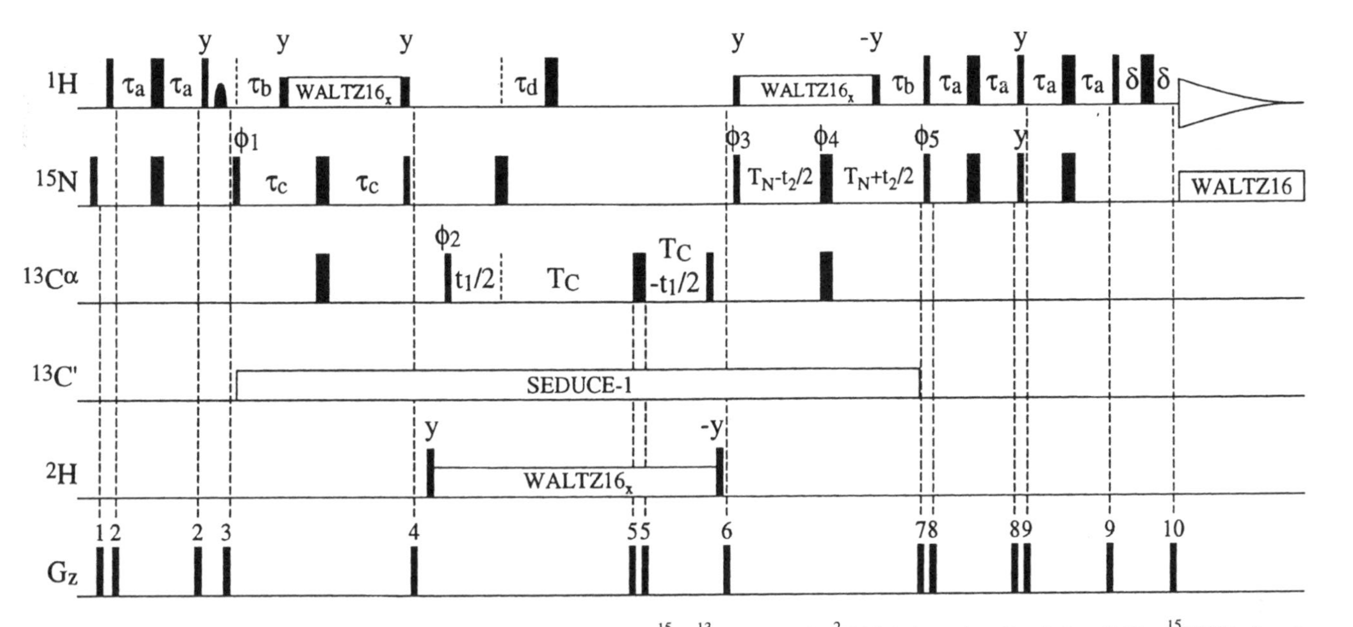

Figure 8. Pulse sequence for the CT-HNCA experiment for use on uniformly ^{15}N, ^{13}C, fractionally ^{2}H labeled proteins. Correlations linking ^{15}N,NH spin pairs with intra- (*i*) and inter-residue ($i-1$) $^{13}C_{\alpha}$ chemical shifts are provided by this experiment. Reprinted with permission from Yamazaki *et al.* (1994b) and modified by the addition of 90° pulses on both sides of the ^{2}H WALTZ-16 decoupling field.

HNCA experiment developed for ^{15}N, ^{13}C, ^{1}H samples (Grzesiek and Bax, 1992; Ikura *et al.*, 1990; Kay *et al.*, 1990). Schematically, these steps can be diagrammed as

$$NH(i) \xrightarrow{J_{NH}} {}^{15}N(i) \xrightarrow{J_{NC}} {}^{13}C\alpha(i,i-1)\,[t_1] \xrightarrow{J_{NC}} {}^{15}N(i)\,[t_2] \xrightarrow{J_{NH}} NH(i)\,[t_3]$$

where t_1, t_2, and t_3 are acquisition periods and the active couplings involved in each transfer step are indicated above the arrows. Fourier transformation of the resultant time domain data set gives a frequency domain map with both intra- (one bond) and inter- (two bond) correlations of the form (^{13}Cα, ^{15}N, NH). It is noteworthy that both ^{13}Cα and ^{15}N chemical shifts are recorded in a constant-time manner. During the carbon constant-time period, evolution of magnetization related to the passive one-bond ^{13}C–^{13}C scalar coupling proceeds according to the relation, $\cos^N(2\pi J_{CC}T_C)$, where $N = 0$ for Gly and 1 for all other amino acids. Because J_{CC} values connecting aliphatic carbons are essentially uniform, it is possible to refocus the effects of such couplings by choosing $2T_C$ to be a multiple of $1/J_{CC}$ (Santoro and King, 1992; Vuister and Bax, 1992). In practice, $2T_C$ is set to $1/J_{CC}$ (28 ms) so as to minimize signal attenuation caused by relaxation. In this regard, deuteration is critical, for in proteins of even modest size transverse relaxation of ^{13}Cα carbons can become limiting. For example, consider a protein with a correlation time of 14.5 ns, such as the 37-kDa trpR/DNA complex studied by Jardetzky and co-workers (Zhang *et al.*, 1994; Arrowsmith *et al.*, 1990a) and Yamazaki *et al.* (1994a,b). Assuming average ^{13}Cα T_2 values of 16 ms and 130 ms for protonated and deuterated samples, respectively, relaxation during the constant time period attenuates the final signal by factors of 5.8 and 1.2, respectively. In the case of the trpR/DNA complex described above, deuteration was critical for the complete assignment of backbone chemical shifts using triple-resonance methods.

A number of additional features relating to the sensitivity of the pulse scheme of Fig. 8 are worthy of comment. First, signal-to-noise can be improved by minimizing saturation and/or dephasing of the water signal (Kay *et al.*, 1994; Stonehouse *et al.*, 1994; Grzesiek and Bax, 1993). This is achieved by ensuring that water magnetization is placed along the $+z$ axis prior to the application of homospoil gradients and immediately prior to signal detection. Maintaining a reservoir of water magnetization is critical, especially in experiments performed on highly deuterated proteins where magnetization originates on NH spins. In this case, exchange between water and labile sites can increase the rate at which NH magnetization is replenished (Stonehouse *et al.*, 1994), circumventing to some extent the limitations imposed by long T_1 values on repetition delays. Second, as described in some detail in Section 3.4, it is essential that deuterium decoupling be employed during periods in which transverse carbon magnetization is present. The deuterium decoupling element is sandwiched between ^{2}H 90° pulses that place ^{2}H magnetization collinear with the decoupling field (Farmer and Venters, 1995; Muhandiram *et al.*, 1995). In this manner, the magnetization is restored to the $+z$ axis after decoupling, minimizing lock instability. Note the analogous use of ^{1}H 90° pulses flanking the WALTZ-16 decoupling elements in the CT-HNCA sequence, which minimizes scrambling of water caused by nonintegral multiples of the WALTZ scheme (Kay *et al.*, 1994). Third, an enhanced sensitivity pulsed field gradient approach is employed to select for ^{15}N magnetization during the nitrogen constant-time period. The use of gradients to select for coherence transfer pathways and a discussion of the classes of experiments and the molecular masses for which the enhanced sensitivity method is beneficial are deferred until Section 3.5.

As a final point of interest, it is noteworthy that an ^{1}H 180° pulse is applied at a time of $t_1/2 + \tau_d$ after the start of the carbon constant-time period. By choosing τ_d to be $1/(4J_{CH})$, where J_{CH} is the magnitude of the one-bond ^{13}C–^{1}H scalar coupling, carbon magnetization from ^{13}C–^{1}H spin pairs evolves for a net duration of $1/(2J_{CH})$ during this constant-time interval and is not refocused into observable signal by the application of the remaining pulses in the sequence. In the case where the experiment is performed on perdeuterated molecules, this pulse is not necessary. However, in applications to fractionally deuterated proteins it may be possible to observe signals from both ^{13}C–^{1}H and ^{13}C–^{2}H pairs, arising from residues with high levels of mobility and hence decreased carbon relaxation rates. The doubling of signals degrades resolution in the carbon dimension of the HNCA spectrum and application of this pulse is thus recommended in these cases.

As described above, many of the triple-resonance pulse schemes that are currently employed for the assignment of deuterated proteins record backbone and sidechain carbon chemical shifts and in these cases deuterium decoupling during carbon evolution is necessary. However, in special classes of experiments it is possible to eliminate deuterium decoupling through the use of magnetization transfer schemes that rely on carbon cross-polarization. For example, in a version of the HN(COCA)NH experiment proposed by Shirakawa *et al.* (1995), magnetization is transferred from ^{13}CO to ^{13}Cα spins and subsequently from ^{13}Cα to ^{15}N using cross-polarization sequences. This circumvents the need for deuterium decoupling, which would otherwise be necessary during the relay from carbon to nitrogen if INEPT-based magnetization transfer methods had been employed (Grzesiek *et al.*, 1993b).

Recently, Yamazaki's group has described a suite of pulse sequences for use with highly deuterated, Cα-protonated samples (Yamazaki *et al.*, 1997). The experiments are of particular value in cases where rapid exchange with water precludes the use of NH-based schemes. In a manner analogous to the CT-HNCA sequence described above, carbon chemical shift is recorded in a constant-time manner. To minimize the rapid decay of carbon magnetization during this period, double- and zero-quantum ^{13}C, ^{1}H coherences are established. It is straightforward to show that in the macromolecule limit the relaxation of these two-spin coherences proceeds in a manner that is independent of the large ^{13}C–^{1}H one-bond dipolar interaction (Griffey and Redfield, 1987). In the case of a protonated sample the advantages of using double- and zero-quantum coherences are offset to some extent by relaxation of the participating proton with proximal proton spins. Additionally, homonuclear proton couplings that evolve during the constant-time carbon evolution period further attenuate the signal (Grzesiek and Bax, 1995). Because the molecules prepared by Yamazaki *et al.* are deuterated at all non-Cα positions, pulse schemes that make use of this approach enjoy benefits that are seldom realized in applications involving protonated samples. For example, increases in relaxation times of approximately a factor of 4.5 (double/zero versus carbon single quantum) have been realized for the carboxy-terminal domain of the α subunit of *E. coli* RNA polymerase at 10 °C (correlation time of 17 ns).

3.3. Sidechain Chemical Shift Assignment

Two of the most often used experiments for the assignment of aliphatic sidechain chemical shifts in protonated, ^{15}N, ^{13}C-labeled proteins are the (H)C(CO)NH- and H(CCO)NH-TOCSY (Grzesiek *et al.*, 1993a, Logan *et al.*, 1993, 1992; Montelione *et al.*, 1992). In these pulse schemes, magnetization originating on sidechain protons is relayed via a carbon TOCSY step to the backbone Cα position and finally transferred to the ^{15}N, NH spins of the subsequent residue. The experiments provide correlations linking either aliphatic protons or carbons with backbone amide shifts or in the case of a 4D sequence developed by Fesik and co-workers (Logan *et al.*, 1992) connectivities are established between all four groups of spins (^{13}C, ^{1}H, ^{15}N, NH). The large number of transfer steps involved in the relay of magnetization from sidechain to backbone sites in these experiments limits their utility to proteins or protein complexes with molecular masses on the order of 20 kDa or less. Nietlispach *et al.* (1996) have developed a number of experiments for sidechain assignment in fractionally deuterated proteins and suggest that deuteration levels on the order of 50% provide a good balance between the need for aliphatic protons and the requirement of decreased carbon relaxation rates.

Farmer and Venters have developed a number of experiments for sidechain assignment in highly deuterated proteins. In one such scheme, a simple modification to the original (H)C(CO)NH-TOCSY is made allowing magnetization to originate on (deuterated) aliphatic carbon sites (Farmer and Venters, 1995). The utility of this sequence has been demonstrated in spectacular fashion on a perdeuterated sample of human carbonic anhydrase (HCA II, 29 kDa) allowing essentially complete assignment of sidechain carbon chemical shifts. A number of experiments for assignment of sidechain ^{15}N/NH resonances in highly deuterated samples have also been published by this group (Farmer and Venters, 1996). These triple-resonance sequences rely on correlating sidechain ^{15}N–NH pairs with carbon shifts that have been previously assigned using the experiment(s) discussed above. Because deuteration limits the number of available NOEs for structure determination, it is important that as many of the remaining protons in the molecule as possible be assigned. In this context, the assignment of Arg, Gln, and Asn labile sidechain protons is crucial.

3.4. Deuterium Decoupling

In the extreme narrowing limit, the ^{13}C spectrum of an isolated ^{13}C–^{2}H spin pair consists of a triplet, with each multiplet component separated from its nearest neighbor by J_{CD}, where J_{CD} is the magnitude of the one-bond ^{13}C–^{2}H scalar coupling. As the tumbling time of the molecule to which the ^{13}C–^{2}H pair is attached increases, the ^{2}H T_1 relaxation time decreases and the multiplet structure collapses in the vicinity of the T_1 minimum. In the slow correlation time limit, linewidths of the components narrow and the multiplet component structure becomes decidedly asymmetric as a result of cross-correlation between ^{13}C–^{2}H dipolar and ^{2}H quadrupolar relaxation interactions (Murali and Rao, 1996; Grzesiek and Bax, 1994; London *et al.*, 1994; Kushlan and LeMaster, 1993b). Finally, at very long correlation times the outer components disappear completely.

The deleterious effects of deuterium spin flips on ^{13}C spectra of deuterated proteins were recognized over two decades ago by Browne *et al.* (1973), who suggested the use of high-power deuterium decoupling as a means of reducing carbon linewidths. However, at that time the power levels required for efficient decoupling could not be achieved. The increase in ^{2}H T_1 relaxation times at the magnetic field strengths typically in use today coupled with the higher decoupling power levels that are available on commercial spectrometers have led to the successful use of deuterium decoupling in triple resonance NMR applications, first demonstrated by Bax and co-workers in 1993 (Grzesiek *et al.*, 1993b).

3.5. The Use of Enhanced Sensitivity Pulsed Field Gradient Coherence Transfer Methods for Experiments on Large Deuterated Proteins

It has long been recognized that pulsed field gradients could be used both to select for desired coherence transfer pathways and reject others (Bax *et al.*, 1980; Maudsley *et al.*, 1978) and to reduce the artifact contact in spectra (Keeler *et al.*, 1994). However, the use of this technology in high-resolution NMR spectroscopy had to await the development of probes with actively shielded gradient coils in the early 1990s; since then, the use of gradients has become widespread. Many of the initial applications to macromolecules employed gradients to select for coherence transfer pathways (Boyd *et al.*, 1992; Davis *et al.*, 1992; Tolman *et al.*, 1992), following on the pioneering work of Ernst and co-workers (Maudsley *et al.*, 1978) and Bax *et al.* (1980). A significant limitation associated with these early experiments is that only one of the two paths that normally contribute to the observed signal in non-gradient-based experiments was observed, reducing sensitivity (Kay, 1995a,b). More recently, building on the enhanced sensitivity nongradient methods of Rance and co-workers (Palmer *et al.*, 1991b; Cavanagh and Rance, 1988), gradient-based pulse schemes with pathway selection that do not suffer from the abovementioned sensitivity losses have been developed (Muhandiram and Kay, 1994; Schleucher *et al.*, 1994, 1993; Kay *et al.*, 1992b). In the absence of relaxation and pulse imperfections, the proposed methods are a full factor of 2 more sensitive than their counterparts that employ gradients for coherence transfer selection but do not make use of enhanced sensitivity and a factor of $\sqrt{2}$ more sensitive than nongradient experiments.

The sensitivity advantages associated with enhanced sensitivity pulsed field gradient coherence transfer selection in triple-resonance-based NH-detected spectroscopy have been described for applications to fully protonated, ^{15}N, ^{13}C labeled molecules ranging in size from approximately 10 to 20 kDa (Muhandiram and Kay, 1994). On average, gains of approximately 15–25% were noted in relation to nongradient methods. The decrease in sensitivity relative to the expected theoretical enhancement of $\sqrt{2}$ is largely the result of relaxation losses that occur during the increased number of delays in these experiments. As these losses are most severe for large molecules, it is important to establish the molecular mass limit above which the sensitivity advantages of this method are marginal. In this regard Shan *et al.* have compared signal-to-noise ratios of HNCO spectra recorded both with and without enhanced sensitivity methods (Shan *et al.*, 1996). An ^{15}N, ^{13}C, ~80% ^{2}H sample of the PLCC SH2 domain was prepared in either 0, 15, or 30% glycerol; the steep viscosity dependence of glycerol with temperature permits convenient manipulation of the overall correlation time of the SH2 domain, allowing spectra to be recorded as a function of "effective molecular mass." Statistically significant sensitivity gains were observed for correlation times as large as 21 ns. It is noteworthy that the enhancements in triple-resonance-based spectra are likely to be larger than in ^{15}N–NH HSQC spectra, as the ^{15}N chemical shift is not recorded in a constant-time manner in the latter experiment. Thus, additional delays must be included during ^{15}N evolution to allow for the application of coherence transfer selection gradients without the introduction of phase distortions resulting from chemical shift evolution. Finally, signal-to-noise advantages mentioned above in the context of highly deuterated proteins will be larger than for protonated molecules. As discussed in Section 1.1, deuteration increases transverse relaxation times of NH protons by approximately a factor of 2 (Venters *et al.*, 1996; Markus *et al.*, 1994). This has a particularly significant effect on the efficiency of sensitivity enhancement-based experiments, which, relative to other classes of experiments, rely on increased delay times during which NH magnetization evolves. It must be emphasized that even in the absence of notable sensitivity gains there are advantages in gradient-based coherence transfer selection methods, including artifact and solvent suppression and minimization of phase cycling.

3.6. Deuterium Isotope Effects on ^{13}C and ^{15}N Chemical Shifts

Along with the benefits that deuterium substitution provides in terms of the reduction of heteronuclear transverse relaxation rates comes the not so desired perturbation of the chemical shifts of ^{15}N and ^{13}C nuclei. These so-called deuterium isotope shifts are significant for ^{15}N and ^{13}C spins located within at least three bonds of the site of deuteration (Hansen, 1988). In an effort to quantitate the magnitude of one- [$^{1}\Delta$ ^{13}C(^{2}H)], two- [$^{2}\Delta$ ^{13}C(^{2}H)], and three-bond [$^{3}\Delta$ ^{13}C(^{2}H)] deuterium isotope effects on ^{13}C chemical shifts, Venters *et al.* (1996) and Gardner *et al.* (1997) have compared chemical shifts of ^{13}Cα and ^{13}Cβ carbons in protonated and highly deuterated versions of the same molecule. Based on results from HCA II (Venters *et al.*, 1996), a single ^{1}H $\rightarrow$ ^{2}H substitution produces chemical shift changes of -0.29 ± 0.05, -0.13 ± 0.02, and -0.07 ± 0.02 ppm for a ^{13}C nucleus one, two, or three bonds away, respectively. Gardner *et al.* have measured average values for $^{1}\Delta$ ^{13}C(^{2}H) and $^{2}\Delta$ ^{13}C(^{2}H) of -0.25 and -0.1 ppm, respectively. These effects are additive, resulting in total shifts of over 1 ppm for ^{13}C nuclei at sites with many proximal deuterons such as is the case for sidechains of long aliphatic amino acids. Slightly smaller changes are typical for backbone nuclei, on the order of -0.3 ppm (^{15}N) and -0.5 ppm (^{13}Cα) for highly deuterated proteins dissolved in H_2O (Gardner and Kay, 1997; Garrett *et al.*, 1997; Venters *et al.*, 1996; Grzesiek *et al.*, 1993b; Kushlan and LeMaster, 1993b). The ^{13}Cα deuterium isotope shifts are weakly dependent on secondary structure (LeMaster *et al.*, 1994). In a recent study of the EIN protein, ^{13}Cα resonances were shifted upfield by an average of 0.50 ± 0.08 ppm for residues in α-helical regions of the protein compared with 0.44 ± 0.08 ppm for amino acids in β-strands (Garrett *et al.*, 1997). In contrast to the ^{2}H-isotope shift observed for aliphatic carbon resonances, no significant changes were observed between NH or carbonyl chemical shifts in a comparison between protonated and highly deuterated PLCC SH2 domains.

Given the additivity and only weak structural dependence of deuterium isotope effects, one can reliably transfer chemical shift assignments between fully protonated and perdeuterated molecules (Venters *et al.*, 1996). This facilitates the use of "secondary shift"-based identification of secondary structure elements (Wishart and Sykes, 1994) from the ^{13}C chemical shifts of deuterated proteins (Constantine *et al.*, 1997; Gardner *et al.*, 1997; Venters *et al.*, 1996). Note that transferability of carbon chemical shifts (i.e., the ability to calculate ^{13}C shifts in a protonated molecule from assignments obtained in a perdeuterated system and vice versa) is a prerequisite for combining data from deuterated and protonated proteins where chemical shifts obtained from a highly deuterated sample are used to assign NOE cross-peaks recorded on a highly protonated molecule (Constantine *et al.*, 1997; Garrett *et al.*, 1997; Yu *et al.*, 1997; Venters *et al.*, 1996).

In the case of partially deuterated systems the presence of multiple deuterium-containing isotopomers at each site can significantly complicate spectra. Recall that in the case of the CT-HNCA experiment recorded on fractionally deuterated proteins, correlations involving $^{13}C\alpha$ sites that are protonated can be removed from spectra in a straightforward fashion by insertion of a single ^{1}H pulse. Signals from deuterated sites are unaffected by this pulse (Section 3.2). In the case of fractionally deuterated aliphatic sidechain sites the situation is somewhat more complex. Consider, for example, the high-resolution ^{13}C–^1H constant-time HSQC spectrum of a deuterated, methyl-protonated sample of the PLCC SH2 domain produced using pyruvate as a carbon source (Section 2.2.2) shown in Fig. 9a. Three peaks are observed for each Ala and Val methyl group, corresponding to the three possible proton-containing methyl isotopomers: CH_3, CH_2D, and CHD_2. Each methyl component is separated by deuterium isotope shifts of approximately 0.3 ppm for ^{13}C and 0.02 ppm for ^1H (Gardner *et al.*, 1997). A smaller number of isotopomers are observed for Met (two) and Thr (one), reflecting differences between these amino acids in the biosynthetic incorporation of protons from the protonated pyruvate carbon source.

Many of the multidimensional experiments that are used for assignment of chemical shifts and for measuring distance restraints are recorded with relatively short acquisition times in the indirect-detection dimensions (4–5 ms in carbon dimensions, for example), principally because of the lengthy acquisition times associated with collecting large (*n*D) data sets. In addition, in some applications acquisition times are constrained further by evolution involving passive homonuclear couplings. As such, these experiments are chiefly limited by poor digital resolution and cross-peaks from individual isotopomers are usually not resolved (Gardner *et al.*, 1997; Nietlispach *et al.*, 1996). However, in higher resolution spectra it is readily apparent that the deuterium isotopomers do significantly degrade spectral resolution, even in a well-dispersed spectrum recorded on a 12-kDa protein (Fig. 9a). As discussed in detail previously (Gardner *et al.*, 1997), the best solution to the problem is to eliminate all but CH_3 isotopomers during protein expression and in this regard Gardner and Kay have developed a strategy for producing highly deuterated, (Val, Leu, Ile $\delta 1$)-methyl-protonated proteins using appropriately labeled amino acids and amino acid precursors (Section 2.2.3). It is also possible to remove CH_2D and CHD_2 groups from ^{13}C–^1H correlation spectra using multipulse NMR methods although regrettably the spin alchemy employed does not restore the full complement of protons to the methyl groups that the biosynthetic approach described by Gardner and Kay (1997) provides. Nevertheless, the filtering schemes that have been used to suppress CH_2D and CHD_2 signals illustrate a number of important features regarding deuterium decoupling and purging of unwanted coherences that the reader may find of interest and with

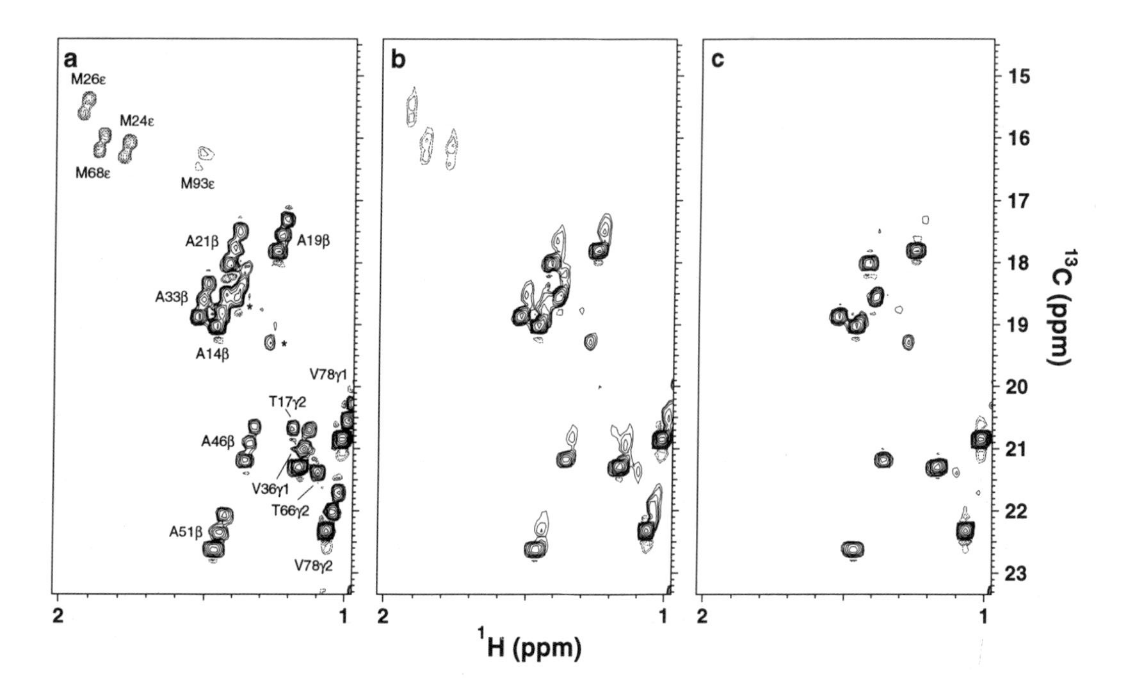

Figure 9. Removal of $^{13}CH_2D$ and $^{13}CHD_2$ cross-peaks in ^{13}C–^{1}H correlation spectra. All panels show a section of the methyl region from ^{13}C–^{1}H constant-time HSQC spectra recorded on an ^{15}N, ^{13}C, ^{2}H, ^{1}H-methyl (from ^{1}H-pyruvate)-labeled sample of the PLCC SH2 domain bound to a phosphotyrosine-containing peptide (Gardner *et al.*, 1997). (a) ^{2}H decoupling applied throughout the constant-time ^{13}C chemical shift evolution period (set to $1/J_{CC} \sim 28$ ms) for maximum resolution of each methyl isotopomer (CH_3:CH_2D:CHD_2 from downfield to upfield for Ala and Val; CH_2D:CHD_2 for Met; CHD_2 for Thr). (b) Minimizing cross-peak intensities from deuterium-containing isotopomers by eliminating ^{2}H decoupling during ^{13}C chemical shift detection. (c) Elimination of cross-peaks from deuterium-containing isotopomers by removing deuterium decoupling as in (b) and application of the purging scheme described in the text. Reprinted with permission from Gardner *et al.* (1997).

this in mind a brief description of the approach that we have developed for retaining only CH_3 isotopomers is in order.

Sections 3.2 and 3.4 considered the importance of deuterium decoupling during the evolution of carbon magnetization. In the absence of decoupling deuterium spin flips mix carbon multiplet components associated with different deuterium spin states, leading to significant broadening and hence attenuation of cross-peaks. This feature can be put to good use in the separation of CH_3 isotopomers from CH_2D and CHD_2 groups in the present application. By eliminating deuterium decoupling during the constant-time delay period where carbon chemical shift is recorded, cross-peaks originating from deuterium-containing methyl groups are significantly attenuated, as illustrated in Fig. 9b. It is clear that in most experiments recorded on deuterated molecules where the idea is to maximize signal intensity from CD_n groups, deuterium decoupling is essential. To further reduce the intensity of cross-peaks from CH_2D methyl types, it is possible to actively purge magnetization from these groups. This is achieved by allowing carbon magnetization to evolve for a period of $1/(2J_{CH})$ from the start of the carbon constant-time period. At this point magnetization from CH_3, CH_2D, and CHD_2 groups is given by terms of the form, $C_{TR}I_{Z^i}I_{Z^j}$ $(i \neq j)$, $C_{TR}I_{Z^i}$, C_{TR}, respectively, where C_{TR} is transverse carbon magnetization and I_{Zk} refers to the z component of magnetization associated with methyl proton k. Application of an ^{1}H 90_x90_ϕ pulse pair where the phase ϕ is cycled $(x,-x)$ with no change in the receiver phase eliminates magnetization from CH_2D methyl groups. Additional purging of signals arising from partially deuterated methyls occurs by applying an ^{2}H purge (90°) pulse at a time of $1/(4J_{CD})$ after the start of the constant-time carbon evolution period, where J_{CD} is the one-bond ^{13}C–^2H coupling constant. As described in some detail previously (Gardner *et al.*, 1997), the outer components of a CD triplet evolve at a rate of $\pm 2\pi J_{CD}$, whereas in the case of a CD_2 spin system the lines closest to and farthest from the center line evolve with frequencies of $\pm 2\pi J_{CD}$ and $\pm 4\pi J_{CD}$, respectively (neglecting ^{2}H spin flips). Note that the central lines do not evolve. Given the range of frequencies over which different multiplet components evolve, it is not possible to eliminate all lines completely with a single purge pulse. In Fig. 9c a compromise delay of $1/(4J_{CD})$ has been employed and it is clear that excellent purging of CH_2D and CHD_2 groups has been achieved.

4. IMPACT OF DEUTERATION ON STRUCTURE DETERMINATION

4.1. Structure Determination of Perdeuterated Proteins

Current NMR solution structure determination methods rely heavily both on NOE-based interproton distance restraints and on scalar-coupling-based dihedral angle restraints (Wüthrich, 1986). The quality of any given structure is, of course, heavily influenced by both the total number and the accuracy and precision of the input restraints (Hoogstraten and Markley, 1996; James, 1994; Zhao and Jardetzky, 1994; Clore *et al.*, 1993; Liu *et al.*, 1992). Because the number of restraints can vary significantly with deuteration levels and with the type of deuterium strategy employed, it is important to evaluate how each of the different deuteration approaches affects the quality of protein structures determined by NMR.

On a positive note the use of relatively high levels of deuteration (>75%) can improve the accuracy of NOE-derived interproton distance measurements, achieved largely through a reduction of spin diffusion. That is, by eliminating proton C that relays magnetization between protons A and B, the A–B separation can be measured more accurately. In addition, because the linewidths of the remaining protons in a deuterated molecule can be significantly narrowed (Section 1), overlap is reduced and, in the case of NH–NH cross-peaks, in particular, appreciable gains in sensitivity have been noted (Pachter *et al.*, 1992; LeMaster and Richards, 1988; Torchia *et al.*, 1988a,b). This leads to further improvements in the accuracy of distance measurements. Recently, 4D ^{15}N-, ^{15}N-edited NOESY experiments were developed (Grzesiek *et al.*, 1995; Venters *et al.*, 1995b) and data sets with high sensitivity and resolution can be recorded using samples of modest protein concentrations (1 mM). Deuteration also facilitates the use of longer NOE mixing times, allowing the measurement of larger distances than would be possible in protonated systems (Venters *et al.*, 1995b).

However, these benefits do come at a cost. Deuteration reduces the concentration of protons that would normally be available for providing NOE-based distance restraints, decreasing the total amount of structural information that can be used for analysis. In the most extreme case of a fully deuterated protein, the remaining protons derive exclusively from exchangeable sites such as amides and hydroxyls and only a subset of these will have sufficiently slow exchange rates and well-dispersed chemical shifts for analysis in NOESY experiments.

To investigate the impact of perdeuteration on the number of distance restraints that can be obtained from NOE experiments, we have used a data base of crystal structures solved to better than 2.5 Å that includes over 200 nonhomologous proteins (Heringa *et al.*, 1992) and tabulated the number of protons within 5 Å of each backbone amide proton. In the case of a fully protonated molecule, on average, 15.7 ± 2.0 protons are located within a 5-Å radius of each backbone NH. In contrast, only 2.5 ± 0.4 protons are within 5 Å of a backbone amide proton in the case of a fully deuterated molecule dissolved in H_2O. Two factors contribute to the sixfold reduction in the number of potential internuclear distance restraints involving backbone NHs that are available in a perdeuterated protein. Most obvious, the decrease in NOEs simply reflects the loss of protons that accompanies deuteration, as illustrated in Fig. 10 where a comparison of the distribution and numbers of protons within protonated and perdeuterated PLCC SH2 domains is provided. Perdeuteration leads to a fivefold reduction in the number of protons in this molecule, from over 750 to approximately 60 sidechain and 90 backbone NHs. In addition, the spatial location of the majority of the remaining protons further lowers the number of possible NOE-based distance restraints. Many backbone and sidechain amide protons are involved in hydrogen-bonding networks in the core of secondary structure elements and are thus separated by distances of greater than 5 Å from protons on other α-helices or β-sheets. This is particularly problematic for α-helices, given the clustering of backbone NH protons in the centers of these structures (Gardner *et al.*, 1997).

Studies using either experimentally derived distance restraints or distances obtained on the basis of previously determined structures have established that backbone NH–NH distances of 5 Å or less will not, in general, be sufficient for calculating accurate global folds (Gardner *et al.*, 1997; Smith *et al.*, 1996; Venters *et al.*, 1995b). Structures generated in this manner typically have backbone heavy atom root-mean-square deviations (RMSD) on the order of 8 Å relative to reference structures. In addition, the precision of these folds is also quite poor, especially for proteins containing a high percentage of α-helix. It is noteworthy that Venters, Farmer, and co-workers have calculated well-defined global folds of HCA II using data sets generated from the known structure of the molecule where all backbone and sidechain NH NOEs within 7 Å are included (Venters *et al.*, 1995b). Unfortunately, with the current generation of spectrometers and with limitations imposed on sample concentration related to issues of solubility, aggregation, viscosity, or expense of production, it seems unlikely that large numbers of NH–NH distances beyond 5 Å will be observed for most systems. For example, only a small fraction (4%) of the possible NOEs connecting NH backbone protons separated by more than 5 Å were measured in a study of a highly deuterated PLCC SH2 domain (1.9 mM) in complex with target peptide (Gardner *et al.*, 1997). In general, therefore, additional NOE restraints are required to obtain overall folds of proteins.

4.2. Improving the Quality of Structures from Perdeuterated Systems: Additional Distance Restraints

As discussed above, the generation of accurate global folds of perdeuterated proteins requires that backbone NH–NH NOEs be supplemented by additional restraints. One obvious source of distance information derives from sidechain amide protons of Arg, Asn, and Gln. Note that these sidechain positions can, in principle, be fully protonated in a perdeuterated protein dissolved in H_2O. In addition to developing experiments for the site-specific assignment of sidechain amides, Farmer and Venters have also demonstrated the utility of sidechain NOEs in improving both the precision and accuracy of global fold determination from a limited NOE restraint set (Farmer and Venters, 1996). Figure 11a illustrates the distribution of sidechain NH protons from Arg, Asn, Gln, and Trp in the PLCC SH2 domain. Many of these protons are located in the interior of the molecule where

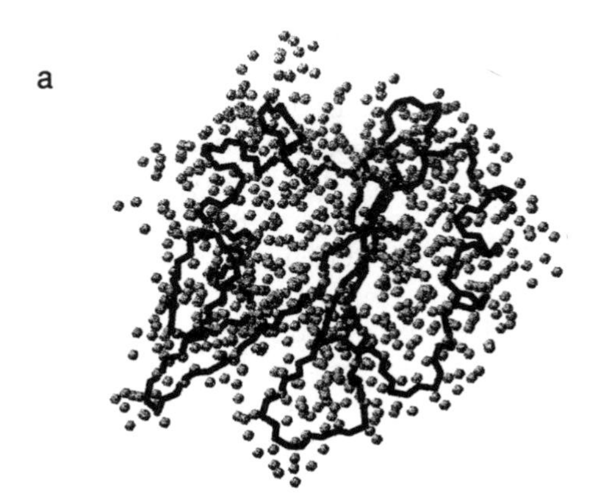

Figure 10. Effect of perdeuteration on the total number and distribution of protons within a protein. Protons from residues 10 to 100 in the PLCC SH2 domain are indicated by gray spheres while the protein N, Cα, C backbone is represented by a heavy black line (Pascal *et al.*, in preparation). (a) Fully protonated protein, containing a total of 756 protons. (b) Perdeuterated protein, containing 87 backbone amide and 62 sidechain amide protons on Arg, Asn, Gln, and Trp residues. The backbone and sidechain amide protons are represented by light and dark gray spheres, respectively. Figures were generated with the program MOLMOL (Koradi *et al.*, 1996).

they are sufficiently close to other NH protons to provide measurable distance restraints. However, a significant percentage are located on the exterior of the protein where they are likely prone to rapid exchange with solvent and poor chemical shift dispersion. Nevertheless, in the case of perdeuterated HCA II, Farmer and Venters were able to assign sidechain amide ^{1}H and ^{15}N chemical shifts for over 80% of the Asn and Gln residues and all of the Arg ε positions.

A second and more important source of additional NOE restraints derives from protonated methyl sites in otherwise highly deuterated proteins (Section 2.2). Several important factors enter into the choice of methyl groups as sites of protonation. First, the methyl region of ^{13}C–^{1}H correlation spectra is often reasonably well resolved and fast rotation about the methyl symmetry axis leads to narrow ^{13}C and ^{1}H linewidths, even in cases of large proteins (Kay *et al.*, 1992a). Analysis of the distribution of methyl-containing amino acids in proteins establishes that Ala, Val, Ile, and Leu are the most common residues in protein interiors and that Val and

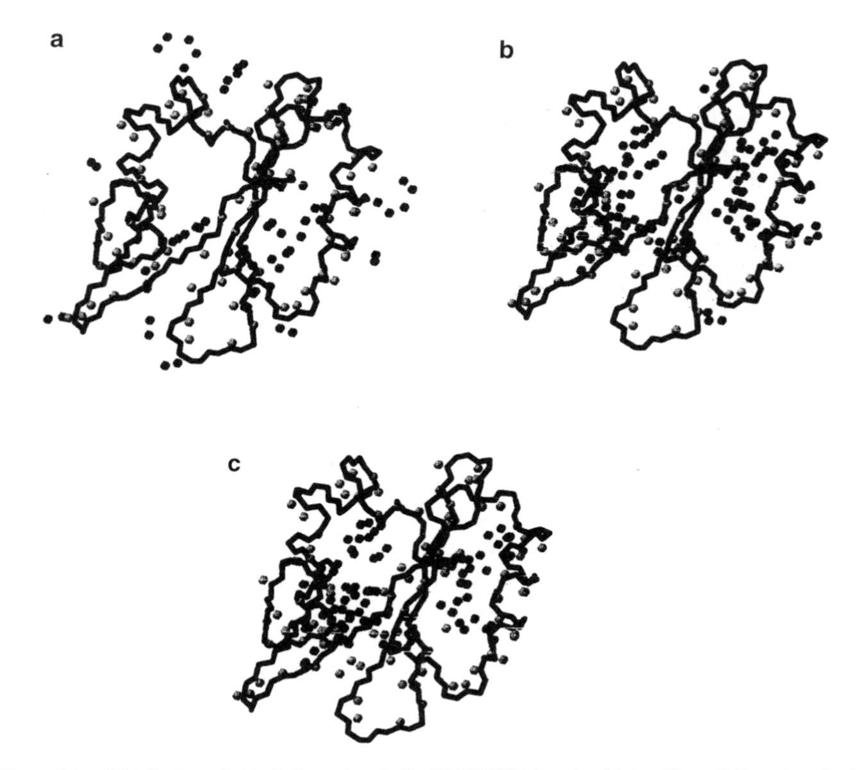

Figure 11. Distribution of sidechain protons in the PLCC SH2 domain obtained from different methods used to produce site-protonated, highly deuterated proteins. (a) Sidechain amide protons of Arg, Asn, Gln, and Trp available using any of the labeling strategies described in the text. (b) The 108 methyl protons of Ala, Val, Leu, and Ile ($\gamma 2$ methyl only) produced using ^{1}H-pyruvate-based labeling (Rosen *et al.*, 1996). (c) Labeling scheme obtained using the method of Gardner and Kay based on the addition of [2,3-$^2H_2$]-^{15}N, ^{13}C-Val, and [3,3-$^2H_2$]-^{13}C 2-ketobutyrate to D_2O medium with ^{2}H,^{13}C-glucose as the carbon source. Ninety methyl protons are produced with this approach. In all figures, backbone amide protons are represented by light gray spheres while additional protons are drawn as dark gray spheres. Molecular graphics were generated with the program MOLMOL (Koradi *et al.*, 1996).

Leu are two of the three most abundant amino acids at molecular interfaces (Janin *et al.*, 1988). Therefore, protonated methyl groups are often within 5 Å of methyl or NH protons of other residues on different secondary structure elements. Figure 11b highlights the distribution of the methyl groups of Ala, Val, Ile ($\gamma 2$ only), and Leu in the PLCC SH2 domain. Recall that this pattern of protonation is produced using D_2O-based media supplemented with ^{1}H-pyruvate (Section 2.2.2) (Rosen *et al.*, 1996). A similar distribution of methyl groups is observed for other patterns of site-specific protonation using deuterated, site-protonated or fully protonated forms of these amino acids (Gardner and Kay, 1997; Metzler *et al.*, 1996b; Smith *et al.*, 1996). For example, Fig. 11c shows the aliphatic proton distribution in a sample generated by the procedure of Gardner and Kay (1997) where methyls from Val, Leu, and Ile ($\delta 1$ only) are protonated.

The utility of methyl site-specific protonation in providing additional distance restraints is readily appreciated by analyzing distances in a large set of nonhomologous proteins solved to high resolution by X-ray diffraction techniques (Heringa *et al.*, 1992). For example, an average of 5.1 ± 1.9 backbone amide protons and a total of 2.8 ± 1.5 Val γ, Ile $\delta 1$ $(i, i \neq j)$ and Leu δ methyl groups are within 6 Å of a given Ile $\delta 1$ (j) methyl. NOEs between methyl protons in highly deuterated molecules where only the Val γ, Ile $\delta 1$, and Leu δ methyl groups are protonated (Section 2.2.3) involve residues with a median separation of 30 amino acids, as opposed to 2 and 3 for amide–amide and amide–methyl NOEs, respectively.

Given that intermethyl NOEs contain information constraining the location of residues that are so widely separated in the primary sequence, it is not surprising that they can significantly improve the quality of structures in relation to those calculated using distance restraints solely between backbone amide protons. Figure 12 illustrates this clearly by comparing several sets of structures of the PLCC SH2 domain generated using different subsets of NOE information. Figure 12a is a schematic representation of the secondary structure of this domain, highlighting the location of the methyl groups of Ala, Val, Ile ($\gamma 2$ only), and Leu at the interfaces between the central β-sheet and the two flanking α-helices. Figure 12b shows structures determined using only experimentally observed backbone NH–NH distances and loose (ϕ,ψ) dihedral angle restraints established from experiments performed on an ^{15}N, ^{13}C, ^{2}H (^{1}H-methyl Ala, Val, Ile($\gamma 2$), Leu) sample of this protein (Gardner *et al.*, 1997). Structures produced from such limited data are poorly defined as evidenced by the 3.5 Å RMSD of the 18 structures in Fig. 12b to the mean structure. More importantly, the experimentally determined structures have a different shape than the high-precision reference structure (bold line in Fig. 12b), because of the lack of long-range structural restraints and the use of a repulsive-only van der Waals potential. This distorts the overall fold, resulting in an RMSD of 7.7 Å between the experimental structures and the reference structure derived from data recorded on a fully protonated sample!

The accuracy and precision of the experimentally determined PLCC SH2 structures can be significantly improved by including distance restraints involving methyl groups obtained from 4D ^{13}C, ^{13}C-edited (Vuister *et al.*, 1993) and ^{13}C, ^{15}N-edited (Muhandiram *et al.*, 1993) NOESY experiments recorded on the ^{15}N, ^{13}C, ^{2}H, methyl-protonated sample described above. An additional 175 methyl–NH and methyl–methyl distance restraints are obtained from these two data sets and in combination with the NH–NH and dihedral angle restraints used to produce the structures of Fig. 12b generate the folds illustrated in Fig. 12c. The experimentally determined structures in Fig. 12c superimpose on the average and the reference structures with backbone RMSD values of 2.4 and 3.1 Å, respectively. Note that the precision of the structures of Fig. 12c is considerably lower than for the reference structures determined from a fully protonated sample of the PLCC SH2 domain complexed with a 12-residue phosphopeptide (Fig. 12d). Similar conclusions regarding the importance of methyl NOEs have been noted in the case of simulations where tables of suitable interproton distances derived from crystal

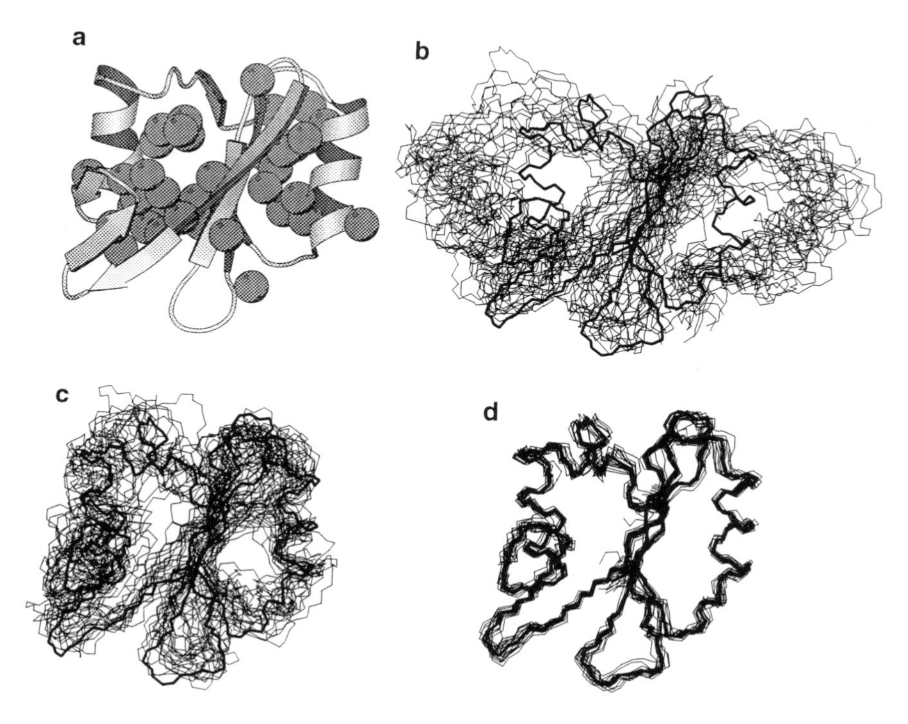

Figure 12. Structures of the PLCC SH2 domain obtained using different sets of distance restraints. (a) Schematic view showing ordered secondary structure elements identified by Pascal *et al.* (1994). Pyruvate-derived methyl groups are shown in a space-filling representation, highlighting their position between the central β-sheet and flanking α-helices. (b) Structures obtained on the basis of NH–NH NOEs and dihedral angle restraints established from $^{13}C\alpha$ and $^{13}C\beta$ chemical shifts (1.1 distance, 0.7 dihedral restraint/residue). (c) Structures generated by including methyl–methyl and NH–methyl restraints as well as the restraints in (b) (2.7 distance, 0.7 dihedral restraint/residue). (d) Reference structures, generated using an average of 16.0 distance and 1.7 dihedral angle restraints per residue, obtained from data recorded on a fully protonated sample (Pascal *et al.*, in preparation). Each bundle has 18 independently refined structures (thin lines) and is fit to the mean reference structure (thick line) using the backbone atoms of the secondary structure elements shown in (a). Panel (a) is drawn using MOLSCRIPT (Kraulis, 1991); all other panels were generated using MOLMOL (Koradi *et al.*, 1996). Reprinted with permission from Gardner *et al.* (1997).

structures are constructed and used to generate global folds of proteins (Gardner and Kay, 1997; Gardner *et al.*, 1997; Metzler *et al.*, 1996b; Smith *et al.*, 1996).

To date, most attention regarding potential sites of protonation in highly deuterated proteins has focused on methyl groups. From a structural perspective, aromatic residues are excellent candidates for protonation as well, as these amino acids are frequently important components of hydrophobic protein cores. However, the poor dispersion of aromatic ^{1}H and ^{13}C chemical shifts, the strong coupling between aromatic carbons, and the efficient transverse relaxation of aromatic ring spins complicate the use of these residues in a manner analogous to the use of methyl-containing amino acids described above. In addition, the large pseudoatom corrections required for aromatic-based distance restraints significantly lower the precision of the structural data available using these sites. In lieu of these problems, Smith *et al.* (1996) have suggested the use of ambiguously assigned NOEs (Nilges *et al.*, 1997; Nilges, 1995) involving protonated aromatic residues in determining the global folds of highly deuterated proteins.

4.3. Improving the Quality of Structures from Highly Deuterated Systems

4.3.1. NOE-Derived Distance Restraints

As of mid-1997, the largest monomeric protein structures that have been determined by NMR are of systems of approximately 30 kDa and include the 259-residue EIN protein (Garrett *et al.*, 1997) and the 245-residue ErmAm rRNA methyltransferase (Yu *et al.*, 1997). In both cases, highly (≥75%) deuterated samples were used for backbone and sidechain assignment as well as for the identification of proximal NH pairs from ^{15}N, ^{15}N-edited 4D NOESY experiments. The high sensitivity and resolution of this experiment, in particular, facilitate the rapid assignment of secondary structural elements. Despite the utility of deuteration in the assignment stages, the majority of interproton distances and all of the dihedral angle restraints used in structure determination were derived from experiments recorded on fully protonated samples.

Applications involving still larger proteins will likely be compromised by severe overlap and low signal intensity in spectra recorded on protonated molecules, necessitating the use of highly deuterated, amino-acid-protonated (Metzler *et al.*, 1996b; Smith *et al.*, 1996) (Section 2.2.1) or site-protonated (Gardner *et al.*, 1997, 1996; Rosen *et al.*, 1996) (Sections 2.2.2 and 2.2.3) proteins. However, as demonstrated so far (Gardner *et al.*, 1997; Metzler *et al.*, 1996b; Smith *et al.*, 1996), at best global folds of only moderate (2–3 Å) precision and accuracy can be obtained from a combination of methyl–methyl, methyl–NH, NH–NH NOEs and loose backbone dihedral angle restraints. Unfortunately, addition of sidechain NH–NH and sidechain NH–methyl restraints is unlikely to significantly improve the precision of structures beyond the limit of 2–3 Å. Clearly, more information is required to improve the quality of structures derived from data collected exclusively on samples of highly deuterated proteins.

4.3.2. Dihedral Angle Restraints

As described briefly above, backbone dihedral angle restraints (*see* chapter 7) can be obtained indirectly from the chemical shifts of nuclei that are assigned during the initial stages of the structure determination process. The difference ($\Delta\delta$) between the chemical shifts of backbone proton and carbon nuclei from their random coil values is secondary structure dependent (Spera and Bax, 1991; Wishart *et al.*, 1991). In a highly deuterated, ^{13}C-labeled protein, the $\Delta\delta$ values calculated from $^{13}C\alpha$, $^{13}C\beta$, and ^{13}CO chemical shifts can be used to quickly and qualitatively identify whether a residue is in an α-helix or β-strand conformation using the chemical shift index (CSI) method (Wishart and Sykes, 1994) after suitable correction for deuterium isotope effects on ^{13}C chemical shifts (Section 3.6). Recently, Metzler, Farmer, Venters, and co-workers have suggested using a function of the form $[\Delta\delta\ (^{13}C\alpha) - \Delta\delta\ (^{13}C\beta)]$ to identify secondary structural elements (Metzler *et al.*, 1996a, 1993; Venters *et al.*, 1996). A recent evaluation of the use of $^{13}C\alpha$ and $^{13}C\beta$ chemical shifts for secondary structure element identification based on a study of 14 proteins suggests that this approach can be used to correctly identify approximately 75% of α-helical and 50% of β-strand residues (Luginbühl *et al.*, 1995). However, in our experience the CSI method is a much more reliable indicator of secondary structure than this study suggests.

Chemical shift information can also be incorporated more quantitatively and directly into the structure determination process. For example, it is possible to perform direct structure refinement against $^{13}C\alpha$ and $^{13}C\beta$ chemical shifts (Kuszewski *et al.*, 1995b) or against (ϕ, ψ) dihedral angle restraints derived from $^{13}C\alpha$ shifts (Celda *et al.*, 1995). It may also be possible, in the case of partially protonated proteins, to refine against proton chemical shifts (Kuszewski *et al.*, 1995a). It is noteworthy that improvements in quantum mechanical calculations of chemical shifts as a function of (ϕ, ψ) will allow the use of tighter restraints than presently possible, significantly improving the quality of structures derived from sparse data sets (Le *et al.*, 1995; Oldfield, 1995).

The NOE-derived distance restraints and the loose (ϕ, ψ) restraints available from $^{13}C\alpha$ and $^{13}C\beta$ chemical shifts can be supplemented by additional dihedral angle restraints from measured coupling constants defining sidechain torsion angles. Recently, Konrat *et al.* (1997) and Bax and co-workers (Hu and Bax, 1997; Hu *et al.*, 1997) have developed triple-resonance pulse schemes for measuring χ_1 angles based on recording ^{15}N–$^{13}C\gamma$ and ^{13}CO–$^{13}C\gamma$ three-bond scalar couplings. Hennig *et al.* (1997) have described an experiment for measuring homonuclear three-bond couplings correlating $^{13}C\alpha$ and $^{13}C\delta$ spins in highly deuterated proteins, allowing determination of *trans*, *gauche*, or averaged $\chi 2$ rotameric states.

4.3.3. Bond Vector Restraints from Dipolar Couplings and Diffusion Anisotropy

A major shortcoming of NOE- and dihedral angle-based structural restraints is that they provide information only on local features such as short-range (≤5 Å) distances and angles between bonds that are proximal in sequence. Restraints between nuclei separated by distances in excess of 5 Å are difficult to obtain. In many cases, macromolecules consist of discrete domains or other structural elements separated by distances that are longer than those that can be measured using NOE methods. Recently a number of approaches have been described that address this limitation by providing restraints describing the relative orientation of domains in a manner that is independent of the distance between these modules (long range order).

One method for measuring long-range order in macromolecules is based on the well-known result that molecules with anisotropic magnetic susceptibilities will orient in an external magnetic field, B_o, with the degree of orientation a function of B_o^2 (Bastiaan *et al.*, 1987). Thus, the dipolar interaction between two spins, which in the high-field limit scales as $(3\cos^2\theta - 1)$ where θ is the angle between the vector joining these spins and B_o, no longer averages to zero (Tolman *et al.*, 1995). As a result, small residual splittings are observed. In the case of an axially symmetric susceptibility, these splittings depend on B_o^2, the size of the susceptibility anisotropy, and the angle between the dipole vector and the unique axis of the molecular susceptibility tensor. It is clear that the size of these splittings will be largest in molecules with large anisotropic magnetic susceptibility tensors including proteins with paramagnetic centers (Tolman *et al.*, 1995) or duplex DNA where smaller contributions to the susceptibility from each base pair add coherently (Kung *et al.*, 1995).

Residual dipolar contributions (*see* Volume 17, chapter 8) to splittings of ^{15}N resonances of ^{15}N–NH spin pairs and ^{13}C resonances of $^{13}C\alpha$–$^{1}H\alpha$ pairs in a number of proteins and protein–DNA complexes have been measured (Tjandra and Bax, 1997; Tjandra *et al.*, 1997b, 1996; Tolman *et al.*, 1995). Because the measured splittings are comprised of a field-invariant scalar coupling term, a dynamic frequency shift contribution that can be calculated, and a dipolar term that scales quadratically with field, dipolar contributions are readily obtained from a field dependent study. Differences in ^{15}N–NH splittings of between 0 and approximately –0.2 Hz have been measured for ubiquitin from spectra recorded at 600 and 360 MHz (Tjandra *et al.*, 1996). In contrast, dipolar couplings ranging from 1 to –5 Hz have been measured at 750 MHz in the case of the paramagnetic protein cyanometmyoglobin (Tolman *et al.*, 1995). The orientational dependence of the dipolar couplings with respect to the components of the susceptibility tensor can be exploited in structure refinement. In this regard a pseudoenergy potential has recently been developed for direct refinement against observed dipolar couplings during restrained molecular dynamics calculations and used in the structure determination of a GATA-1/DNA complex (Tjandra *et al.*, 1997b).

A second approach for obtaining long-range order in macromolecules is based on the measurement of ^{15}N relaxation times, T_1 and T_2. In the case of an isotropically tumbling molecule without internal motions, the ^{15}N T_1/T_2 ratio is uniform. In contrast, for a molecule with an axially symmetric diffusion tensor the T_1/T_2 ratio is a function of the angle between the N–NH bond vector and the unique axis of the diffusion tensor. In cases where the diffusion tensor is axially symmetric, therefore, it is possible to measure the orientation of each amide bond vector with respect to the diffusion frame and to use this information as an additional structural restraint, in an analogous manner to the use of dipolar couplings described above. Tjandra *et al.* (1997a) have shown that in the case of EIN ($D_{//} / D_{\perp} \sim 2$, where $D_{//}$ and $D_{\perp}$ are the parallel and perpendicular components of the diffusion tensor), ^{15}N T_1/T_2 data are important in defining the relative orientation of the two domains in the molecule.

4.3.4. Other Structural Information

In addition to the experiments described above, studies involving deuterium exchange (Wüthrich, 1986), paramagnetic perturbation of either chemical shifts (Guiles *et al.*, 1996; Gochin and Roder, 1995) and/or relaxation rates (Gillespie and Shortle, 1997; Kosen *et al.*, 1986), and chemical cross-linking (Das and Fox, 1979) can also provide useful distance restraints. Further information can be obtained from statistical analyses of data bases of previously determined protein structures. Residue-dependent preferences of various structural parameters, such as the values of backbone and sidechain dihedral angles or distances between pairwise combinations of amino acid residues, have been quantified and tables of these preferences converted into "knowledge-based" potentials of mean force to provide an energy scale to assign to values of these distances or dihedral angles (Jernigan and Bahar, 1996; Sippl, 1995). Such knowledge-based potentials have been used in a wide variety of applications, including error detection in experimentally determined structures, refinement of protein structures in combination with experimentally determined restraints (Kuszewski *et al.*, 1997, 1996; Skolnick *et al.*, 1997; Aszódi *et al.*, 1995), and *ab initio* structure prediction. One major advantage of knowledge-based potentials lies in their representation of structural information that is dependent on the entire range of molecular forces, including solvation effects. These potentials are thus better able to discriminate between favorable and unfavorable protein conformations than the commonly utilized potentials for van der Waals interactions. This, in turn, results in more efficient searches of conformational space and the generation of final structures with improved packing (Kuszewski *et al.*, 1996). A number of recent applications involving the use of knowledge-based dihedral angle potentials in structure calculations include studies of a specific

GAGA factor–DNA complex (Omichinski *et al.*, 1997) and the EIN protein (Garrett *et al.*, 1997).

In combination, the use of NOE and dihedral angle restraints, chemical shift deviations from random coil values, dipolar couplings, and protein structure data bases, will facilitate structure determination of higher molecular weight proteins using ^{2}H-based strategies with improved precision and accuracy.

5. USE OF DEUTERATION TO STUDY PROTEIN DYNAMICS

NMR spectroscopy can provide a wealth of information about molecular dynamics extending over a wide range of motional time scales. Studies to date have focused mostly on backbone dynamics, largely through measurement of backbone ^{15}N relaxation properties in uniformly ^{15}N labeled proteins (Nicholson *et al.*, 1996; Palmer, 1993). An attractive feature of backbone ^{15}N relaxation studies relates to the fact that the data can be easily analyzed. For example, the relaxation of a two-spin ^{15}N–NH spin pair can be described simply in terms of the ^{15}N–NH dipolar interaction and to a smaller extent by the ^{15}N chemical shift anisotropy (Kay *et al.*, 1989). Although interference between these two relaxation mechanisms does occur, methods have been developed that can effectively remove this complication (Kay *et al.*, 1992c; Palmer *et al.*, 1992).

The situation is, in general, more complex for the study of motional properties of sidechains. Although the use of carbon relaxation to probe sidechain dynamics appears to be an obvious approach, ^{13}C relaxation methods are not without some rather significant problems. First, most sidechain positions are of either the CH_2 or CH_3 variety and cross correlation between ^{13}C–^{1}H dipolar interactions can be significant in these cases (Kay *et al.*, 1992a; Palmer *et al.*, 1991a). Although the interpretation of such effects provides powerful insight into molecular dynamics (Werbelow and Grant, 1977), several experiments must often be performed where the relaxation behavior of the individual multiplet components is monitored. In addition, relaxation contributions from neighboring spins can complicate the extraction of accurate dynamics parameters. Second, differential relaxation of multiplet components occurring during the course of the pulse sequences used to measure ^{13}C relaxation times results in the transfer of magnetization from ^{13}C to ^{1}H in a manner that may not reflect the equilibrium intensity of each carbon multiplet component. Although pulse sequences which minimize this effect in the case of methyl groups have been described (Kay *et al.*, 1992a), they do not eliminate the problem completely. Third, in the case of applications involving uniformly ^{13}C labeled samples, both scalar and dipolar ^{13}C–^{13}C couplings must be taken into account (Engelke and Rüterjans, 1995; Yamazaki *et al.*, 1994c). Although this problem is eliminated through the use of molecules in which only a select number of sites are labeled, the information content available in such systems is, of course, much less.

A recent series of papers by LeMaster and Kushlan have addressed many of the abovementioned limitations in the use of ^{13}C spectroscopy to study sidechain dynamics (LeMaster and Kushlan, 1996; Kushlan and LeMaster, 1993a). Using a suitable strain of *E. coli* grown on medium containing either [2-^{13}C]-glycerol or [1,3-$^{13}C_2$]-glycerol, LeMaster and Kushlan have shown that it is possible to produce proteins in which isotopic enrichment is largely restricted to alternating carbon sites and in this manner eliminate the deleterious effects of carbon–carbon couplings. In addition, complications arising from the presence of more than one proton attached to a given ^{13}C spin are removed through the use of approximately 50% random fractional deuteration in concert with pulse schemes that select for ^{13}C–^{1}H two-spin spin systems. The method has been applied to investigate the dynamics of *E. coli* thioredoxin, a small protein of 108 residues (LeMaster and Kushlan, 1996).

To this point in the review all applications discussed have made use of deuteration simply as a method for removing proton spins, thereby improving the relaxation properties of the remaining NMR-active nuclei (Sections 2–4) or simplifying analysis of ^{13}C relaxation times in terms of molecular dynamics (Section 5). These deuterons can be thought of as passive participants in the experiments in that none of their spectroscopic properties are recorded. It is possible, however, to make use of fractionally deuterated protein samples in quite a different way than has been previously described to obtain information about protein dynamics. In this regard, it is noteworthy that ^{2}H NMR has enjoyed a rich history in the study of biomolecular motion through measurement of deuterium relaxation and line-shape parameters (Vold and Vold, 1991; Keniry *et al.*, 1983; Jelinski *et al.*, 1980; Seelig, 1977). Applications have largely focused on liquid-crystalline and solid-state samples (Keniry, 1989). The low sensitivity of ^{2}H direct detect spectra coupled with broad linewidths and poor chemical dispersion has, until recently, restricted the application of high-resolution solution-state ^{2}H NMR methods to a few relaxation studies of site-specifically deuterated proteins for estimate of molecular rotational correlation times (Schramm and Oldfield, 1983; Johnson *et al.*, 1989). However, the power of ^{2}H relaxation as a probe of molecular dynamics, largely the result of the fact that the relaxation is dominated by the well-understood quadrupolar interaction, has stimulated interest in developing solution-state-based methods for measuring ^{2}H relaxation rates that do not suffer from the problems mentioned above.

Recently, a triple-resonance method has been described for the measurement of ^{2}H spin relaxation times, T_1 and $T_{1\rho}$, in fractionally deuterated ^{15}N, ^{13}C labeled proteins (Muhandiram *et al.*, 1995). The idea is to select for methylene or methyl groups with only a single attached deuteron, allow relaxation of magnetization proportional to either D_z (T_1) or D_y($T_{1\rho}$) to proceed for a delay time, T, during the course of the experiment and encode this decay in cross-peak intensities measured in ^{13}C–^{1}H constant-time correlation spectra. Thus, the approach makes use of a series of magnetization transfer steps in which signal originating on a sidechain (^{13}CHD, Yang *et al.*, 1998; $^{13}CH_2D$, Muhandiram *et al.*, 1995) proton is transferred to the attached carbon and subsequently to the deuteron bound to the same carbon. At this point the relaxation proceeds for a defined time, *T*, and the signal is subsequently returned to the originating proton for detection. The flow of magnetization during the course of the experiment can be represented by

$$^{1}\mathrm{H} \xrightarrow{J_{\mathrm{CH}}} {}^{13}\mathrm{C} \xrightarrow{J_{\mathrm{CD}}} {}^{2}\mathrm{H}\,(T) \xrightarrow{J_{\mathrm{CD}}} {}^{13}\mathrm{C}\,(t_1) \xrightarrow{J_{\mathrm{CH}}} {}^{1}\mathrm{H}\,(t_2)$$

where the active couplings involved in each INEPT-based transfer are indicated above the arrows and t_1 and t_2 denote periods during which ^{13}C and ^{1}H chemical shifts are recorded. A set of 2D (^{13}C,^{1}H) correlation spectra are obtained where the time profile of the intensity of a cross-peak arising from a $^{13}CH_2D$ methyl, for example, is related to the relaxation time (T_1 or $T_{1\rho}$) of the attached deuteron. In practice, because of the magnetization transfer steps that are involved, the decay of operators of the form $I_zC_zD_z$ (T_1) or $I_zC_zD_y$ ($T_{1\rho}$) is measured during the relaxation time, *T*. T_1 and $T_{1\rho}$ values of pure deuterium magnetization can be obtained by recording an additional experiment in which the decay of the two-spin order, I_zC_z, is measured and subtracting this rate from the decay rates of $I_zC_zD_z$ and $I_zC_zD_y$ according to

$$1/T_1(D_z) = 1/T_1(I_zC_zD_z) - 1/T_1(I_zC_z)$$

$$1/T_{1\rho}(D_y) = 1/T_{1\rho}(I_zC_zD_y) - 1/T_1(I_zC_z).$$

Cross correlation between the many different relaxation mechanisms that could potentially complicate the decay of the triple spin terms described above has been examined in detail and shown not to contribute in a measurable way to the decay of the magnetization (Yang and Kay, 1996).

The deuterium relaxation methods described above have been applied to study the sidechain dynamics of the PLCC SH2 domain, in both the presence and absence of target peptide (Kay *et al.*, 1996). Remarkably, certain residues in the hydrophobic binding region of this SH2 domain which are important for the specificity of its interaction with peptide, are highly mobile in both peptide free and complexed states. A comparison of the dynamics of the PLCC SH2 domain with the amino-terminal SH2 domain from the Syp phosphatase has provided insight into the origin of the very different peptide-binding properties of these two highly homologous structures (Kay *et al.*, 1998).

Recently, Pervushin *et al.* (1997) have described 2D heteronuclear experiments for measuring transverse and longitudinal ^{2}H relaxation rates in ^{15}NHD groups (Asn and Gln) in uniformly ^{15}N labeled proteins. Samples are dissolved in a 1:1 mixture of H_2O/D_2O to give maximal concentrations of ^{15}NHD groups with deuteration at either E or Z positions. These pulse sequences are closely related to the ^{13}C–^2H relaxation experiments described above. Interestingly, because the direction of the bond connecting the amide nitrogen and the $^2H^Z$ deuteron is nearly parallel to the $C\beta$–$C\gamma$ or $C\gamma$–$C\delta$ bond in Asn and Gln, respectively, rotation about χ_2 (Asn) or χ_3 (Gln) affects the transverse relaxation of $^2H^Z$ much less than $^2H^E$. Thus, the relaxation rates for $^2H^Z$ are predicted to be larger than for $^2H^E$; this has been observed in relaxation studies of the 70-residue *ftz* (fushi-tarazu) homeodomain and its complex with a 14-base-pair DNA.

6. CONCLUDING REMARKS

The present review has highlighted many of the important current advances in the use of deuteration for both structural and dynamics studies of proteins and protein complexes. Outstanding areas of investigation include the further development of labeling schemes for site-specific incorporation of protons in highly deuterated proteins, new pulse sequence methodology that is customized for the labeling methods that will be introduced, and the design of new classes of experiments that provide structural parameters complementary to the existing NOE- and scalar-coupling-based classes of restraints. It is also clear that further increases in magnetic fields, resulting in gains in dispersion and sensitivity for multidimensional NMR applications, and future improvements in spectrometer hardware will also impact significantly on the scope of problems that can be studied. The future promises to be exciting indeed.

ACKNOWLEDGMENTS. The authors thank Julie Forman-Kay (Hospital for Sick Children, Toronto) for helpful discussions and a critical reading of the manuscript and Drs. Dan Garrett (NIH) and Steve Fesik (Abbott Laboratories) for providing preprints. This work was supported by a grant from the Medical Research Council of Canada. K.H.G. gratefully acknowledges a postdoctoral fellowship from the Helen Hay Whitney Foundation.

REFERENCES

Agback, P., Maltseva, T. V., Yamakage, S.-I., Nilsson, F. P. R., Földesi, A., and Chattopadhyaya, J., 1994, *Nucl. Acids Res.* **22:**1404–1412.

Allerhand, A., Doddrell, D., Glushko, V., Cochran, D. W., Wenkert, E., Lawson, P. J., and Gurd, F. R. N., 1971, *J. Am. Chem. Soc.* **93:**544–546.

Anglister, J., 1990, *Q. Rev. Biophys.* **23:**175–203.

Arrowsmith, C. H., Pachter, R., Altman, R. B., Iyer, S. B., and Jardetzky, O., 1990a, *Biochemistry* **29:**6332–6341.

Arrowsmith, C. H., Treat-Clemons, L., Szilágyi, L., Pachter, R., and Jardetzky, O., 1990b, *Makromol. Chem. Macromol. Symp.* **34:**33–46.

Aszódi, A., Gradwell, M. J., and Taylor, W. R., 1995, *J. Mol. Biol.* **251**:308–326.
Bastiaan, E. W., Maclean, C., Van Zijl, P. C. M., and Bothner-By, A. A., 1987, *Annu. Rep. NMR Spectrosc.* **19**:35–77.
Batey, R. T., Cloutier, N., Mao, H., and Williamson, J. R., 1996, *Nucl. Acids Res.* **24**:4836–4837.
Bax, A., 1994, *Curr. Opin. Struct. Biol.* **4**:738–744.
Bax, A., De Jong, D. E., Mehlkopf, A. F., and Smidt, J., 1980, *Chem. Phys. Lett.* **69**:567–570.
Boyd, J., Soffe, N., John, B., Plant, D., and Hurd, R., 1992, *J. Magn. Reson.* **98**:660–664.
Brodin, P., Drakenberg, T., Thulin, E., Forsén, S., and Grundström, T., 1989, *Protein Eng.* **2**:353–358.
Browne, D. T., Kenyon, G. L., Packer, E. L., Sternlicht, H., and Wilson, D. M., 1973, *J. Am. Chem. Soc.* **95**:1316–1323.
Cavanagh, J., and Rance, M., 1988, *J. Magn. Reson.* **88**:72–85.
Celda, B., Biamonti, C., Arnau, M. J., Tejero, R., and Montelione, G. T., 1995, *J. Biomol. NMR* **5**:161–172.
Clore, G. M., and Gronenborn, A. M., 1994, *Methods Enzymol.* **239**:349–363.
Clore, G. M., Robien, M. A., and Gronenborn, A. M., 1993, *J. Mol. Biol.* **231**:82–102.
Constantine, K. L., Mueller, L., Goldfarb, V., Wittekind, M., Metzler, W. J., Yanuchas, J. J., Robertson, J. G., Malley, M. F., Friedrichs, M. S., and Farmer, B. T., II, 1997, *J. Mol. Biol.* **267**:1223–1246.
Crespi, H. L., and Katz, J. J., 1969, *Nature* **224**:560–562.
Crespi, H. L., Rosenberg, R. M., and Katz, J. J., 1968, *Science* **161**:795–796.
Das, M., and Fox, C. F., 1979, *Annu. Rev. Biophys. Bioeng.* **8**:165–193.
Davis, A. L., Keeler, J., Laue, E. D., and Moskau, D., 1992, *J. Magn. Reson.* **98**:207–216.
Dötsch, V., Matsuo, H., and Wagner, G., 1996, *J. Magn. Reson. B* **112**:95–100.
Eisenstein, E., 1991, *J. Biol. Chem.* **266**:5801–5807.
Engelke, J. and Rüterjans, H., 1995, *J. Biomol. NMR* **5**:173–182.
Farmer, B. T., II, and Venters, R. A., 1995, *J. Am. Chem. Soc.* **117**:4187–4188.
Farmer, B. T., II, and Venters, R. A., 1996, *J. Biomol. NMR* **7**:59–71.
Földesi, A., Yamakage, S.-I., Nilsson, F. P. R., Maltseva, T. V., and Chattopadhyaya, J., 1996, *Nucl. Acids Res.* **24**:1187–1194.
Galimov, E. M., 1985, *The Biological Fractionation of Isotopes*, Academic Press, Orlando.
Gardner, K. H., and Kay, L. E., 1997, *J. Am. Chem. Soc.* **119**:7599–7600.
Gardner, K. H., Konrat, R., Rosen, M. K., and Kay, L. E., 1996, *J. Biomol. NMR* **8**:351–356.
Gardner, K. H., Rosen, M. K., and Kay, L. E., 1997, *Biochemistry* **36**:1389–1401.
Garrett, D. S., Seok, Y., Liao, D., Peterkofsky, A., Gronenborn, A. M., and Clore, G. M., 1997, *Biochemistry* **36**:2517–2530.
Gillespie, J. R., and Shortle, D., 1997, *J. Mol. Biol.* **268**:158–169.
Gochin, M., and Roder, H., 1995, *Protein Sci.* **4**:296–305.
Griffey, R. H., and Redfield, A. G., 1987, *Q. Rev. Biophys.* **19**:51–82.
Grzesiek, S., and Bax, A., 1992, *J. Magn. Reson.* **96**:432–440.
Grzesiek, S., and Bax, A., 1993, *J. Am. Chem. Soc.* **115**:12593–12594.
Grzesiek, S., and Bax, A., 1994, *J. Am. Chem. Soc.* **116**:10196–10201.
Grzesiek, S., and Bax, A., 1995, *J. Biomol. NMR* **6**:335–339.
Grzesiek, S., Anglister, J., and Bax, A., 1993a, *J. Magn. Reson. B* **101**:114–119.
Grzesiek, S., Anglister, J., Ren, H., and Bax, A., 1993b, *J. Am. Chem. Soc.* **115**:4369–4370.
Grzesiek, S., Wingfield, P., Stahl, S., Kaufman, J. D., and Bax, A., 1995, *J. Am. Chem. Soc.* **117**:9594–9595.
Guiles, R. D., Sarma, S., DiGate, R. J., Banville, D., Basus, V. J., Kuntz, I. D., and Waskell, L., 1996, *Nature Struct. Biol.* **3**:333–339.
Hansen, A. P., Petros, A. M., Mazar, A. P., Pederson, T. M., Rueter, A., and Fesik, S. W., 1992, *Biochemistry* **31**:12713–12718.
Hansen, P. E., 1988, *Prog. NMR Spectrosc.* **20**:207–255.

Hennig, M., Ott, D., Schulte, P., Löwe, R., Krebs, J., Vorherr, T., Bermel, W., Schwalbe, H. and Griesinger, C., 1997, *J. Am. Chem. Soc.* **119:**5055–5056.
Heringa, J., Sommerfeldt, H., Higgins, D., and Argos, P., 1992, *CABIOS* **8:**599–600.
Homer, R. J., Kim, M. S., and LeMaster, D. M., 1993, *Anal. Biochem.* **215:**211–215.
Hoogstraten, C. G., and Markley, J. L., 1996, *J. Mol. Biol.* **258:**334–348.
Hsu, V. L., and Armitage, I. M., 1992, *Biochemistry* **31:**12778–12784.
Hu, J.-S., and Bax, A., 1997, *J. Am. Chem. Soc.* **119:**6360–6368.
Hu, J.-S., Grzesiek, S., and Bax, A., 1997, *J. Am. Chem. Soc.* **119:**1803–1804.
Ikegami, T., Sato, S., Wälchli, M., Kyogoku, Y. and Shirakawa, M., 1997, *J. Magn. Reson.* **124:**214–217.
Ikura, M., Kay, L. E., and Bax, A., 1990, *Biochemistry* **29:**4659–4667.
James, T., 1994, *Methods Enzymol.* **239:**416–439.
Janin, J., Miller, S., and Chothia, C., 1988, *J. Mol. Biol.* **204:**155–164.
Jelinski, L. W., Sullivan, C. E., and Torchia, D. A., 1980, *Nature* **284:**531–534.
Jernigan, R. L., and Bahar, I., 1996, *Curr. Opin. Struct. Biol.* **6:**195–209.
Johnson, R. D., La Mar, G. N., Smith, K. M., Parish, D. W., and Langry, K. C., 1989, *J. Am. Chem. Soc.* **111:**481–485.
Kalbitzer, H. R., Leberman, R., and Wittinghofer, A., 1985, *FEBS Lett.* **180:**40–42.
Katz, J. J., and Crespi, H. L., 1966, *Science* **151:**1187–1194.
Kay, L. E., 1995a, *Curr. Opin. Struct. Biol.* **5:**674–681.
Kay, L. E., 1995b, *Prog. Biophys. Mol. Biol.* **63**:277–299.
Kay, L. E., Torchia, D. A., and Bax, A., 1989, *Biochemistry* **28:**8972–8979.
Kay, L. E., Ikura, M., Tschudin, R., and Bax, A., 1990, *J. Magn. Reson.* **89:**496–514.
Kay, L. E., Bull, T., Nicholson, L. K., Griesinger, C., Schwalbe, H., Bax, A., and Torchia, D., 1992a, *J. Magn. Reson.* **100:**538–558.
Kay, L. E., Keifer, P., and Saarinen, T., 1992b, *J. Am. Chem. Soc.* **114:**10663–10665.
Kay, L. E., Nicholson, L. K., Delaglio, F., Bax, A., and Torchia, D. A., 1992c, *J. Magn. Reson.* **97:**359–375.
Kay, L. E., Xu, G. Y., and Yamazaki, T., 1994, *J. Magn. Reson. A* **109:**129–133.
Kay, L. E., Muhandiram, D. R., Farrow, N. A., Aubin, Y., and Forman-Kay, J. D., 1996, *Biochemistry* **35**:362–368.
Kay, L. E., Muhandiram, D. R., Wolf G., Shoelson, S. E., and Forman-Kay, J. D., 1998, *Nat. Struct. Biol.* **5:**156–163.
Keeler, J., Clowes, R. T., Davis, A. L., and Laue, E. D., 1994, *Methods Enzymol.* **239**:145–207.
Keniry, M. A., 1989, *Methods Enzymol.* **176:**376–386.
Keniry, M. A., Rothgeb, T. M., Smith, R. L., Gutowsky, H. S., and Oldfield, E., 1983, *Biochemistry* **22:**1917–1926.
Konrat, R., Muhandiram, D. R., Farrow, N. A., and Kay, L. E., 1997, *J. Biomol. NMR* **9:**409–422.
Koradi, R., Billeter, M., and Wüthrich, K., 1996, *J. Mol. Graphics* **14:**51–55.
Kosen, P. A., Scheck, R. M., Nadevi, H., Basus, V. J., Manogaran, S., Schmidt, P. G., Oppenheimer, N. J., and Kuntz, I. D., 1986, *Biochemistry* **25:**2356–2364.
Kraulis, P. J., 1991, *J. Appl. Crystallogr.* **24:**946–950.
Kung, H. C., Wang, K. Y., Goljer, I., and Bolton, P. H., 1995, *J. Magn. Reson. B* **109:**323–325.
Kushlan, D. M., and LeMaster, D. M., 1993a, *J. Am. Chem. Soc.* **115:**11026–11027.
Kushlan, D. M., and LeMaster, D. M., 1993b, *J. Biomol. NMR* **3:**701–708.
Kuszewski, J., Gronenborn, A. M., and Clore, G. M., 1995a, *J. Magn. Reson. B* **107:**293–297.
Kuszewski, J., Qin, J., Gronenborn, A. M., and Clore, G. M., 1995b, *J. Magn. Reson. B* **106:**92–96.
Kuszewski, J., Gronenborn, A. M., and Clore, G. M., 1996, *Protein Sci.* **5:**1067–1080.
Kuszewski, J., Clore, G. M., and Gronenborn, A. M., 1997, *J. Magn. Reson.* **125:**171–177.
Le, H., Pearson, J. G., de Dios, A. C., and Oldfield, E., 1995, *J. Am. Chem. Soc.* **117:**3800–3807.
LeMaster, D. M., 1989, *Methods Enzymol.* **177:**23–43.

LeMaster, D. M., 1990, *Q. Rev. Biophys.* **23:**133–173.
LeMaster, D. M., 1994, *Prog. NMR Spectrosc.* **26:**371–419.
LeMaster, D. M., 1997, *J. Biomol. NMR* **9:**79–93.
LeMaster, D. M., and Kushlan, D. M., 1996, *J. Am. Chem. Soc.* **118:**9255–9264.
LeMaster, D. M., and Richards, F. M., 1988, *Biochemistry* **27:**142–150.
LeMaster, D. M., LaIuppa, J. C., and Kushlan, D. M., 1994, *J. Biomol. NMR* **4:**863–870.
Liu, Y., Zhao, D., Altman, R., and Jardetzky, O., 1992, *J. Biomol. NMR* **2:**373–388.
Logan, T. M., Olejniczak, E. T., Xu, R. X., and Fesik, S. W., 1992, *FEBS Lett.* **314:**413–418.
Logan, T. M., Olejniczak, E. T., Xu, R. X., and Fesik, S. W., 1993, *J. Biomol. NMR* **3:**225–231.
London, R. E., LeMaster, D. M., and Werbelow, L. G., 1994, *J. Am. Chem. Soc.* **116:**8400–8401.
Luginbühl, P., Szyperski, T., and Wüthrich, K., 1995, *J. Magn. Reson. B* **109:**229–233.
Markley, J. L., Putter, I., and Jardetzky, O., 1968, *Science* **161:**1249–1251.
Markus, M. A., Kayie, K. T., Matsudaira, P., and Wagner, G., 1994, *J. Magn. Reson. B* **105:**192–195.
Martin, M. L., and Martin, G. J., 1990, in *NMR Basic Principles and Progress*, Vol. 23, eds., Springer-Verlag, Berlin, pp. 1–61.
Matsuo, H., Kupce, E., Li, H., and Wagner, G., 1996a, *J. Magn. Reson. B* **111:**194–198.
Matsuo, H., Li, H., and Wagner, G., 1996b, *J. Magn. Reson. B.* **110:**112–115.
Maudsley, A. A., Wokaun, A., and Ernst, R. R., 1978, *Chem. Phys. Lett.* **55:**9–14.
McIntosh, L. P., and Dahlquist, F. W., 1990, *Q. Rev. Biophys.* **23:**1–38.
Metzler, W. J., Constantine, K. L., Friedrichs, M. S., Bell, A. J., Ernst, E. G., Lavoie, T. B., and Mueller, L., 1993, *Biochemistry* **32:**13818–13829.
Metzler, W. J., Leiting, B., Pryor, K., Mueller, L., and Farmer, B. T., II, 1996a, *Biochemistry* **35:**6201–6211.
Metzler, W. J., Wittekind, M., Goldfarb, V., Mueller, L., and Farmer, B. T., II, 1996b, *J. Am. Chem. Soc.* **118:**6800–6801.
Montelione, G. T., Lyons, B. A., Emerson, D. S., and Tashiro, M., 1992, *J. Am. Chem. Soc.* **114:**10974–10975.
Moore, P. B., 1979, *Methods Enzymol.* **59:**639–655.
Muhandiram, D. R., and Kay, L. E., 1994, *J. Magn. Reson. B* **103:**203–216.
Muhandiram, D. R., Xu, G. Y., and Kay, L. E., 1993, *J. Biomol. NMR* **3:**463–470.
Muhandiram, D. R., Yamazaki, T., Sykes, B. D., and Kay, L. E., 1995, *J. Am. Chem. Soc.* **117:**11536–11544.
Murali, N., and Rao, B. D. N., 1996, *J. Magn. Reson. A* **118:**202–213.
Nicholson, L. K., Kay, L. E. and Torchia, D. A., 1996, in *NMR Spectroscopy and Its Application to Biomedical Research* (S. K. Sarkar, ed.), Elsevier, Amsterdam, pp. 241–280.
Nietlispach, D., Clowes, R. T., Broadhurst, R. W., Ito, Y., Keeler, J., Kelly, M., Ashurst, J., Oschkinat, H., Domaille, P. J., and Laue, E. D., 1996, *J. Am. Chem. Soc.* **118:**407–415.
Nilges, M., 1995, *J. Mol. Biol.* **245:**645–660.
Nilges, M., Macias, M. J., O'Donoghue, S. I., and Oschkinat, H., 1997, *J. Mol. Biol.* **269:**408–422.
Oda, Y., Nakamura, H., Yamazaki, T., Nagayama, K., Yoshida, M., Kanaya, S., and Ikehara, M., 1992, *J. Biomol. NMR* **2:**137–147.
Oldfield, E., 1995, *J. Biomol. NMR* **5:**217–225.
Omichinski, J. G., Pedone, P. V., Felsenfeld, G., Gronenborn, A. M., and Clore, G. M., 1997, *Nature Struct. Biol.* **4:**122–132.
Ono, A., Makita, T., Tate, S., Kawashima, E., Ishido, Y., and Kainosho, M., 1996, *Magn. Reson. Chem.* **34:**S40–S46.
Pachter, R., Arrowsmith, C. H., and Jardetzky, O., 1992, *J. Biomol. NMR* **2:**183–194.
Palmer, A. G., III, 1993, *Curr. Opin. Biotechnol.* **4:**385–391.
Palmer, A. G., III, Wright, P. E., and Rance, M., 1991a, *Chem. Phys. Lett.* **185:**41–46.
Palmer, A. G., III, Cavanagh, J., Wright, P. E., and Rance, M., 1991b, *J. Magn. Reson.* **93:**151–170.

Palmer, A. G., III, Skelton, N. J., Chazin, W. J., Wright, P. E., and Rance, M., 1992, *Mol. Phys.* **75:**699–711.
Pascal, S. M., Singer, A. U., Gish, G., Yamazaki, T., Shoelson, S. E., Pawson, T., Kay, L. E., and Forman-Kay, J. D., 1994, *Cell* **77:**461–472.
Pascal, S. M., Singer, A. U., Kay, L. E., and Forman-Kay, J. D., 1997, in preparation.
Pervushin, K., Wider, G., and Wüthrich, K., 1997, *J. Am. Chem. Soc.* **119:**3842–3843.
Reisman, J., Jariel-Encontre, I., Hsu, V. L., Parello, J., Guiduschek, E. P., and Kearns, D. R., 1991, *J. Am. Chem. Soc.* **113:**2787–2789.
Reisman, J. M., Hsu, V. L., Jariel-Encontre, I., Lecou, C., Sayre, M. H., Kearns, D. R., and Parello, J., 1993, *Eur. J. Biochem.* **213:**865–873.
Rosen, M. K., Gardner, K. H., Willis, R. C., Parris, W. E., Pawson, T., and Kay, L. E., 1996, *J. Mol. Biol.* **263:**627–636.
Santoro, J., and King, G. C., 1992, *J. Magn. Reson.* **97:**202–207.
Sattler, M., and Fesik, S. W., 1996, *Structure* **4:**1245–1249.
Schleucher, J., Sattler, M., and Griesinger, C., 1993, *Angew. Chem. Int. Ed. Engl.* **32:**1489–1491.
Schleucher, J., Schwendinger, M., Sattler, M., Schmidt, P., Schedletzky, O., Glaser, S. J., Sorensen, O. W., and Griesinger, C., 1994, *J. Biomol. NMR* **4:**301–306.
Schramm, S., and Oldfield, E., 1983, *Biochemistry* **22:**2908–2913.
Seeholzer, S. H., Cohn, M., Putkey, J. A., Means, A. R., and Crespi, H. L., 1986, *Proc. Natl. Acad. Sci. USA* **83:**3634–3638.
Seelig, J., 1977, *Quart. Rev. Biophys.* **10:**363–418.
Shan, X., Gardner, K. H., Muhandiram, D. R., Rao, N. S., Arrowsmith, C. H., and Kay, L. E., 1996, *J. Am. Chem. Soc.* **118:**6570–6579.
Shirakawa, M., Wälchli, M., Shimizu, M. and Kyogoku, Y., 1995, *J. Biomol. NMR* **5:**323–326.
Sippl, M., 1995, *Curr. Opin. Struct. Biol.* **5**:229–235.
Skolnick, J., Kolinski, A., and Ortiz, A. R., 1997, *J. Mol. Biol.* **265:**217–241.
Smith, B. O., Ito, Y., Raine, A., Teichmann, S., Ben-Tovim, L., Nietlispach, D., Broadhurst, R. W., Terada, T., Kelly, M., Oschkinat, H., Shibata, T., Yokoyama, S., and Laue, E. D., 1996, *J. Biomol. NMR* **8:**360–368.
Spera, S., and Bax, A., 1991, *J. Am. Chem. Soc.* **113:**5490–5492.
Stonehouse, J., Shaw, G. L., Keeler, J., and Laue, E. D., 1994, *J. Magn. Reson. A* **107:**178–184.
Tjandra, N., and Bax, A., 1997, *J. Magn. Reson.* **124:**512–515.
Tjandra, N., Grzesiek, S., and Bax, A., 1996, *J. Am. Chem. Soc.* **118:**6264–6272.
Tjandra, N., Garrett, D. S., Gronenborn, A. M., Bax, A., and Clore, G. M., 1997a, *Nature Struct. Biol.* **4:**443–449.
Tjandra, N., Omichinski, J. G., Gronenborn, A. M., Clore, G. M., and Bax, A., 1997b, *Nature Struct. Biol.* **4:**732.
Tolbert, T. J., and Williamson, J. R., 1996, *J. Am. Chem. Soc.* **118:**7929–7940.
Tolman, J. R., Chung, J., and Prestegard, J. H., 1992, *J. Magn. Reson.* **98:**462–467.
Tolman, J. R., Flanagan, J. M., Kennedy, M. A., and Prestegard, J. H., 1995, *Proc. Natl. Acad. Sci. USA* **92:**9279–9283.
Torchia, D. A., Sparks, S. W., and Bax, A., 1988a, *J. Am. Chem. Soc.* **110:**2320–2321.
Torchia, D. A., Sparks, S. W., and Bax, A., 1988b, *Biochemistry* **27:**5135–5141.
Tsang, P., Wright, P. E., and Rance, M., 1990, *J. Am. Chem. Soc.* **112:**8183–8185.
Venters, R. A., Huang, C.-C., Farmer, B. T., II, Trolard, R., Spicer, L. D., and Fierke, C. A., 1995a, *J. Biomol. NMR* **5:**339–344.
Venters, R. A., Metzler, W. J., Spicer, L. D., Mueller, L., and Farmer, B. T., II, 1995b, *J. Am. Chem. Soc.* **117:**9592–9593.
Venters, R. A., Farmer , B. T., II, Fierke, C. A., and Spicer, L. D., 1996, *J. Mol. Biol.* **264:**1101–1116.
Vold, R. R., and Vold, R. L., 1991, *Adv. Magn. Opt. Reson.* **16:**85–171.

Vuister, G. W., and Bax, A., 1992, *J. Magn. Reson.* **98:**428–435.
Vuister, G. W., Clore, G. M., Gronenborn, A. M., Powers, R., Garrett, D. S., Tschudin, R., and Bax, A., 1993, *J. Magn. Reson. B* **101:**210–213.
Waugh, D. S., 1996, *J. Biomol. NMR* **8:**184–192.
Werbelow, L. G., and Grant, D. M., 1977, *Adv. Magn. Reson.* **9:**189–299.
Wishart, D. S., and Sykes, B. D., 1994, *J. Biomol. NMR* **4:**171–180.
Wishart, D. S., Sykes, B. D., and Richards, F. M., 1991, *J. Mol. Biol.* **222:**311–333.
Wüthrich, K., 1986, *NMR of Proteins and Nucleic Acids*, Wiley, New York.
Yamazaki, T., Lee, W., Arrowsmith, C. H., Muhandiram, D. R., and Kay, L. E., 1994a, *J. Am. Chem. Soc.* **116:**11655–11666.
Yamazaki, T., Lee, W., Revington, M., Mattiello, D. L., Dahlquist, F. W., Arrowsmith, C. H., and Kay, L. E., 1994b, *J. Am. Chem. Soc.* **116:**6464–6465.
Yamazaki, T., Muhandiram, D. R., and Kay, L. E., 1994c, *J. Am. Chem. Soc.* **116:**8266–8278.
Yamazaki, T., Tochio, H., Furui, J., Aimoto, S., and Kyogoku, Y., 1997, *J. Am. Chem. Soc.* **119:**872–880.
Yang, D., and Kay, L. E., 1996, *J. Magn. Reson. B* **110:**213–218.
Yang, D., Mittermaier, T., Mok, Y. K., and Kay, L. E., 1998, *J. Mol. Biol.* **276:**939–954.
Yu, L., Petros, A. M., Schnuchel, A., Zhong, P., Severin, J. M., Walter, K., Holzman, T. F., and Fesik, S. W., 1997, *Nature Struct. Biol.* **4:**483–489.
Zhang, H., Zhao, D., Revington, M., Lee, W., Jia, X., Arrowsmith, C., and Jardetzky, O., 1994, *J. Mol. Biol.* **238:**592–614.
Zhao, D., and Jardetzky, O., 1994, *J. Mol. Biol.* **239:**601–607.

3

NMR of Perdeuterated Large Proteins

Bennett T. Farmer II and Ronald A. Venters

1. INTRODUCTION AND HISTORICAL PERSPECTIVE

1.1. Assignment and Structural Studies of Larger Proteins

1.1.1. ^{1}H Only

Ever since the first sequential ^{1}H NMR assignment of a protein was reported (Wagner and Wüthrich, 1982), considerable effort has been expended to increase the size of proteins that can be studied by high-resolution NMR spectroscopy. To date, ^{1}H homonuclear 2D NMR has provided both chemical-shift assignments and three-dimensional solution-state structures at atomic resolution for numerous peptides and small proteins (Barlow *et al.*, 1993; Davis *et al.*, 1993; Senn and Klaus, 1993; Kallen *et al.*, 1991). However, ^{1}H homonuclear 2D methodologies are fundamentally constrained to a maximum protein size of ~10 kDa. Although protein size, in kilodaltons, is typically used to categorize the ability of NMR to provide three-dimensional structures, for ^{1}H homonuclear 2D NMR, a more relevant parameter is the total number of protons in the protein.

Bennett T. Farmer II • Macromolecular NMR, Pharmaceutical Research Institute, Bristol-Myers Squibb, Princeton, New Jersey 08543-4000. **Ronald A. Venters** • Duke University Medical Center, Durham, North Carolina 27710.

Biological Magnetic Resonance, Volume 16: Modern Techniques in Protein NMR, edited by Krishna and Berliner. Kluwer Academic / Plenum Publishers, 1999.

As the number of protons in the protein increases, ^{1}H homonuclear 2D NMR begins to fail because of insufficient spectral resolution. In this case, spectral resolution is a function of both the proton linewidths and the intrinsic chemical-shift dispersion. The number of proton resonances increases linearly with molecular mass; therefore, more resonances are added to a fixed chemical-shift range, eventually leading to resonance crowding. Confidence in resonance assignments degrades as the density of resonances (peaks per ppm) begins to approach the real spectral resolution.

Similarly, proton linewidths for a globular protein increase approximately linearly with molecular mass. However, spectral resolution is less affected by increased linewidths than by increased crowding. For this reason, it is the extent of resonance crowding that mostly determines whether increasingly larger, unlabeled proteins are amenable to complete study by ^{1}H homonuclear 2D NMR. One can reduce the extent of resonance crowding by an increase in the effective proton chemical-shift dispersion. Because the effective 2D proton chemical-shift dispersion is related to the square of the magnetic field strength, ^{1}H homonuclear 2D NMR experiments, when applied to unlabeled proteins, should be run at the highest possible field strength.

1.1.2. ^{1}H/^{13}C/^{15}N

Within a decade, this molecular-mass barrier (~10 kDa) was increased to ~25 kDa both by ^{13}C/^{15}N isotopic labeling techniques (Venters *et al.*, 1991; Ikura *et al.*, 1990; Westler *et al.*, 1988) and by concomitant advances in NMR spectrometer design and multidimensional/multinuclear NMR pulse sequences (Muhandiram and Kay, 1994; Bax and Grzesiek, 1993; Logan *et al.*, 1993; Clubb *et al.*, 1992; Montelione *et al.*, 1992). Pioneering advances in heteronuclear NMR pulse-sequence design have led to the chemical-shift assignments of ^{13}C, ^{15}N, and ^{1}H nuclei in proteins by exploiting one- and two-bond heteronuclear and homonuclear scalar-coupling constants. In these experiments, the combined chemical-shift dispersion of three separate (3D) or four separate (4D) nuclei provides a dramatic increase in overall spectral resolution. In addition, isotopic labeling with ^{13}C and ^{15}N allows one to make use of more sophisticated spectral editing techniques (Dötsch *et al.*, 1996; Farmer *et al.*, 1995; Grzesiek and Bax, 1993a; Yamazaki *et al.*, 1993) that lead to a practical increase in spectral resolution. Both ^{13}C and ^{15}N isotopic labeling and the current arsenal of heteronuclear *n*D experiments benefit mostly the NMR structure determination of proteins > 10 kDa and < 25 kDa: for such proteins, NMR can now provide a complete structure determination. In addition, one should not overlook the more modest yet equally important benefits of these combined methodologies to the NMR structure determination of smaller proteins: For such proteins, the benefits are both increased throughput and increased structural precision.

Although $^{13}C/^{15}N$ isotopic labeling yields increased spectral resolution largely through multinuclear/multidimensional NMR experiments, this increased spectral resolution comes at the price of a marked decrease in sensitivity per unit acquisition time. This point is most clearly illustrated by a study using ^{12}C reverse labeling of aromatic residues (Vuister *et al.*, 1994). As the protein size increases from 10 kDa to 25 kDa, key backbone correlation experiments, e.g., the HNCACB (Wittekind and Mueller, 1993) and CBCA(CO)NH (Grzesiek and Bax, 1992), experience an exponentially progressive decrease in sensitivity. In fact, it is the exponential nature of this decrease in sensitivity that teases out the nuclei whose decreased relaxation times most dramatically affect overall sensitivity. These critical nuclei have relaxation times that, for a particular experiment, begin to approach the duration of pertinent delay intervals. By pertinent delay intervals, we mean those delays during which the nucleus in question exists as a transverse spin operator in some *n*-spin, generally single-quantum coherence.

Consequently, sensitivity in the key backbone correlation experiments should be most affected by decreases in carbon and H_N T_2 relaxation times. For noncarbonyl carbon nuclei, the most significant contribution to T_2 relaxation comes from the strong dipolar coupling to any directly attached proton(s) (Yamazaki *et al.*, 1994a,b; Brown *et al.*, 1973; Grzesiek *et al.*, 1993). For amide protons, a significant contribution (~40%) to T_2 relaxation comes from dipolar coupling to surrounding aliphatic protons (Markus *et al.*, 1994). Implicit in the previous two statements is the observation that significant increases in both carbon and H_N relaxation times in a protein can be achieved by eliminating most, if not all, carbon-bound protons. Practically, this is accomplished by perdeuterative isotopic labeling (Venters *et al.*, 1995a; LeMaster, 1994). In this technique, one perdeuterates all nonexchangeable sites in a protein.

1.2. Deuteration

1.2.1. History of Deuteration

Since the mid-1960s, efforts to simplify protein 1D NMR spectra have involved deuteration in the following ways (Crespi *et al.*, 1968; Markley *et al.*, 1968; Katz and Crespi, 1966): (1) uniform deuteration of a protein, (2) selective deuteration in an otherwise protonated protein, (3) selective protonation in an otherwise deuterated protein. In the mid-1980s, these same deuteration methods were being similarly exploited in homonuclear 2D ^{1}H COSY and NOESY studies on otherwise non-isotopically enriched proteins. Deuteration was shown to effectively extend the size of proteins that could be studied by high-resolution ^{1}H homonuclear 2D NMR. Deuteration achieved this goal by decreasing both the number of observable resonances and, in most cases, the linewidths of these resonances (Pachter *et al.*,

1992; Reisman *et al.*, 1991; Arrowsmith *et al.*, 1990; Feeney *et al.*, 1989; LeMaster and Richards, 1988; Torchia *et al.*, 1988; Searle *et al.*, 1986).

High levels of deuteration offered an additional benefit to the ^{1}H homonuclear NOESY experiment: markedly reduced spin-diffusion (Tsang *et al.*, 1990). Predictably, in a deuterated protein, sensitivity in the NOESY experiment was shown to benefit significantly from the combined effect of both narrower linewidths and longer, permissible mixing times (Pachter *et al.*, 1992; Reisman *et al.*, 1991; LeMaster and Richards, 1988; Torchia *et al.*, 1988). Proteins as large as 25 kDa have been partially assigned based on NOESY data collected on selectively deuterated protein analogues (Arrowsmith *et al.*, 1990).

Individually, both deuteration and ^{13}C/^{15}N isotopic labeling have been shown to increase the size of protein amenable to NMR structural studies. Initially, deuteration found its niche in the study of proteins by ^{1}H homonuclear 1D and 2D NMR. In contrast, ^{13}C/^{15}N isotopic labeling has achieved a more broad-based appeal, mainly because of the almost simultaneous advent of 3D and 4D heteronuclear NMR experiments, especially the NOESY variants. Recently, the combined use of perdeuteration and ^{13}C/^{15}N isotopic labeling, in conjunction with heteronuclear 3D and 4D experiments, has yielded another quantum increase in the size of proteins for which NMR can provide both chemical-shift assignments and tertiary structural information (Constantine *et al.*, 1997; Garrett *et al.*, 1997; Farmer *et al.*, 1996; Shan *et al.*, 1996; Yamazaki *et al.*, 1994a).

To date, our groups have taken advantage of perdeuteration to obtain NMR resonance assignments and/or structural information on the following proteins: UDP-*N*-acetylenolpyruvylglucosamine reductase (MurB, 38.5 kDa) (Constantine *et al.*, 1997; Farmer *et al.*, 1996), human carbonic anhydrase II (HCA II, 29.1 kDa) (Venters *et al.*, 1996), the methionine repressor protein MetJ (22 kDa), and the L78K mutant of *E. coli* thioredoxin (11.7 kDa) (deLorimier *et al.*, 1996). Other groups have utilized either perdeuteration or random fractional deuteration to study a variety of proteins including: Shc phosphotyrosine binding domain (Zhou *et al.*, 1995); Bcl-x_L (Muchmore *et al.*, 1996), a ternary complex of two tandem dimers of *trp* repressor (Shan *et al.*, 1996); a ternary complex of *trp* repressor with the corepressor 5-methyltryptophan and a 20-bp trp-operator DNA fragment (Yamazaki *et al.*, 1994b); PLCC SH2 and the murine Crk protein (Rosen *et al.*, 1996); HIV-1 Nef protein (Grzesiek *et al.*, 1996); and the N-terminal domain of enzyme I from the *E. coli* phosphoenolpyruvate:sugar phosphotransferase system (Garrett *et al.*, 1997).

1.2.2. Advantages Using Modern Assignment Strategies

Because the deuteron gyromagnetic ratio (γ_D) is 6.5-fold smaller than the proton γ_H, a deuteron is 42-fold (γ_H^2/γ_D^2) less effective than a proton at causing dipolar relaxation of both the directly attached X nucleus and the surrounding proton nuclei. Relaxation of noncarbonyl carbon nuclei is dominated by dipolar interaction with any attached proton(s) (Browne *et al.*, 1973); relaxation of amide protons has a significant component (40%) arising from dipolar interaction with the surrounding aliphatic protons (Markus *et al.*, 1994). Therefore, the incorporation of ^{2}H into $^{13}C/^{15}N$-labeled proteins dramatically increases both the carbon and H_N T_2 relaxation times. The increase in these T_2 relaxation times leads to much greater sensitivity in key amide-detected $^{1}H/^{13}C/^{15}N$ triple-resonance correlation experiments, e.g., HN(CO)CA/CB, HNCA/CB, HN(CACO)NH, HNCO, and HN(CA)CO (Constantine *et al.*, 1997; Shan *et al.*, 1996; Venters *et al.*, 1995a; Yamazaki *et al.*, 1994a; Grzesiek *et al.*, 1993). With these multidimensional heteronuclear NMR experiments, backbone ^{1}H, ^{15}N, ^{13}CO, $^{13}C_{\alpha}$, and $^{13}C_{\beta}$ chemical shifts have been obtained for $^{13}C/^{15}N$-labeled and perdeuterated proteins exceeding 35 kDa (Constantine *et al.*, 1997; Farmer *et al.*, 1996; Shan *et al.*, 1996; Yamazaki *et al.*, 1994a).

1.2.3. Perdeuteration versus Random Fractional Deuteration

Complete aliphatic deuterium enrichment provides optimum sensitivity in key amide-detected, out-and-back heteronuclear NMR experiments used to obtain both backbone and sidechain $^{13}C/^{15}N$ chemical-shift assignments. Complete aliphatic deuterium enrichment achieves this optimum state both by maximizing carbon and H_N T_2 relaxation times and by establishing a uniform ^{2}H isotopic environment for ^{13}C and ^{15}N nuclei. A uniform ^{2}H isotopic environment minimizes losses in both sensitivity and resolution arising from the multibond ^{2}H isotope effect (Venters *et al.*, 1996; Hansen, 1988). In contrast, random fractional deuteration of uniformly ^{13}C-labeled proteins gives rise to all possible CH_nD_m isotopomers for each aliphatic carbon group. For a particular aliphatic carbon, its ^{13}C chemical shift depends not only on its isotopomeric state (n and m values) but also on the isotopomeric state of aliphatic carbon groups both one and two bonds removed. Because of the multibond ^{2}H isotope effect, random fractional deuteration produces a broad distribution of ^{13}C chemical shifts for aliphatic carbons, especially methylene and methyl groups, thereby decreasing both sensitivity and resolution in the aforementioned heteronuclear NMR experiments.

Random 50% fractional deuteration has been proposed as optimum for obtaining sidechain ^{13}C and ^{1}H resonance assignments in large proteins (Nietlispach *et al.*, 1996). Clearly, complete deuteration precludes the assignment of sidechain ^{1}H resonances. However, no experimental data have been published as to whether sidechain ^{13}C resonances can be assigned more effectively using either a completely deuterated protein sample (Farmer and Venters, 1995) or a random 50% fractionally deuterated sample (Nietlispach *et al.*, 1996). For assigning sidechain ^{13}C resonances, the completely deuterated approach has two main disadvantages: (1) the $^{13}C(D)$ T_1 [~2.6 s for $C_{\alpha}(D)$ in ^{2}H-HCA II; Farmer and Venters, 1995] is about fourfold longer than the $^{13}C(H)$ T_1 (Kushlan and LeMaster, 1993); (2) magnetization transfer starts on carbon, which is intrinsically fourfold less sensitive than proton. However, the random fractional deuteration approach has the following disadvantages: (1) broad ^{13}C resonance linewidths arising from the distributed ^{2}H isotope effect, (2) decreased magnetization transfer efficiency during isotropic mixing, $H_C \rightarrow C$ INEPT and $C\alpha \rightarrow CO$ INEPT magnetization transfers, (3) decreased sensitivity in the terminal $^{15}N-^{1}H$ reverse INEPT subsequence because of decreased $^{1}H_N$ T_2 values.

Although random fractional deuteration does indeed allow for the assignment of sidechain ^{1}H resonances, the utility of such assignments is questionable unless NOE data on a similarly labeled protein sample yield a sufficient number of uniquely assignable NOEs. Random fractional deuteration might be quite effective when applied to a small protein, regardless of the overall rotational correlation time, because small proteins, unlike larger ones, do not demand a high level of spectral resolution. Because larger proteins demand much higher spectral resolution, random fractional deuteration may be ill suited to their study. For larger proteins, selective protonation (Metzler *et al.*, 1996b; Rosen *et al.*, 1996; Smith *et al.*, 1996; Oda *et al.*, 1992) (see Section 4.3.1) offers a more powerful and less limited approach both to making and then to exploiting sidechain ^{1}H and ^{13}C resonance assignments.

In 2D homonuclear NOE experiments (LeMaster and Richards, 1988; Torchia *et al.*, 1988), as previously mentioned, random fractional deuteration has been shown to increase both resolution and sensitivity. Longer mixing times can be used in NOESY experiments because spin-diffusion pathways are reduced both in number and in efficiency. For larger proteins, 2D proton–proton NOESY experiments provide insufficient spectral dispersion; therefore, $^{13}C/^{15}N$ labeling is required in conjunction with multidimensional, heteronuclear-separated NOESY experiments. Both longer mixing times and increased spectral dispersion are realized in the 3D ^{15}N-separated and the 4D $^{15}N/^{15}N$-separated NOESY experiments when utilizing a combination of $^{13}C/^{15}N$ labeling (or just ^{15}N) and perdeuteration of all nonexchangeable proton sites. Perdeuteration precludes the use of any ^{13}C-separated NOESY experiments.

For any ^{13}C-separated NOESY experiment on larger proteins, random fractional deuteration is also problematic. With this labeling scheme, the distributed ^{2}H isotope effect leads to severe ^{13}C and $^{1}H_C$ resonance broadening. This dual resonance broadening will undoubtedly render the 4D $^{13}C/^{15}N$-separated and 4D $^{13}C/^{13}C$-separated NOESY experiments of limited utility in the study of larger, random fractionally deuterated proteins. NOE data involving $^{1}H_C$ spins are extremely important because of the limited number and type of NOE interactions measurable in a perdeuterated protein (*vide infra*). Therefore, an isotopic labeling scheme has been devised that achieves the collective advantages of both random fractional deuteration and perdeuteration: type-specific $^{1}H/^{13}C$ labeling within an otherwise deuterated and uniformly ^{15}N-labeled protein (Metzler *et al.*, 1996b; Smith *et al.*, 1996; Oda *et al.*, 1992). Type-specific protonation retains many benefits of perdeuteration while providing a greater number and type of NOE interactions.

2. ISOTOPIC LABELING

2.1. Perdeuteration Using Acetate

2.1.1. Growth Conditions

High-level expression of many proteins has been achieved in *E. coli* by constructing vectors that contain the protein gene subcloned behind a phage T7 RNA polymerase promoter vector (Rosenberg *et al.*, 1987). Protein production is then initiated by the addition of isopropyl-D-thiogalactopyranoside (IPTG), inducing a chromosomal copy of T7 RNA polymerase. The polymerase, in turn, begins the actual transcription of the protein gene (Studier and Moffatt, 1986).

Uniformly $^{13}C/^{15}N$-labeled proteins can be obtained with this expression system using defined medium containing 3 g/liter sodium [1,2-$^{13}C_2$, 99%] acetate as the sole carbon source and 1 g/liter [^{15}N, 99%] ammonium chloride or ammonium sulfate as the sole nitrogen source (Venters *et al.*, 1991). Other carbon sources can be used, most notably glucose and algal hydrolysates. In addition to these isotopically labeled compounds, the defined medium contains M9 salts (Sambrook *et al.*, 1989), 2 mM $MgSO_4$, 1 μM $FeCl_3$, 10 ml/liter vitamin mixture (contains 0.1 mg/ml each of biotin, choline chloride, folic acid, niacinamide, D-pantothenate, and pyridoxal; 0.01 mg/ml riboflavin), 5 mg/liter thiamine, 100 μM $CaCl_2$, and an appropriate antibiotic. Some proteins possess additional requirements for expression. For instance, the expression of HCA II requires the addition of 50 μM $ZnSO_4$, as zinc is bound at the active site of this protein. Likewise, the expression of methionine repressor protein, MetJ, requires the addition of methionine.

We can extend this procedure to include deuterium labeling by using the medium described above, with the exception that H_2O is replaced by D_2O (Venters *et al.*, 1995a). Figure 1 shows the general procedure for deuterium incorporation using an *E. coli* expression system. Employing this procedure, we have purified the 29-kDa enzyme HCA II from BL21(DE3)pACA cells grown in minimal [$^1H_3$]-acetate media containing either 50, 75, or 98.8% D_2O. Subsequent to purification, we analyzed the molecular mass of these proteins using a Fisons-VG Quattro BQ triple quadrupole mass spectrometer equipped with a pneumatically assisted electrostatic ion source (Venters *et al.*, 1995a). The mass of the protein grown in 98.8% D_2O indicates that simply growing cells in D_2O leads to high levels of deuteration, in this case 96%. In addition, for the HCA II protein isolated from cells grown in minimal acetate

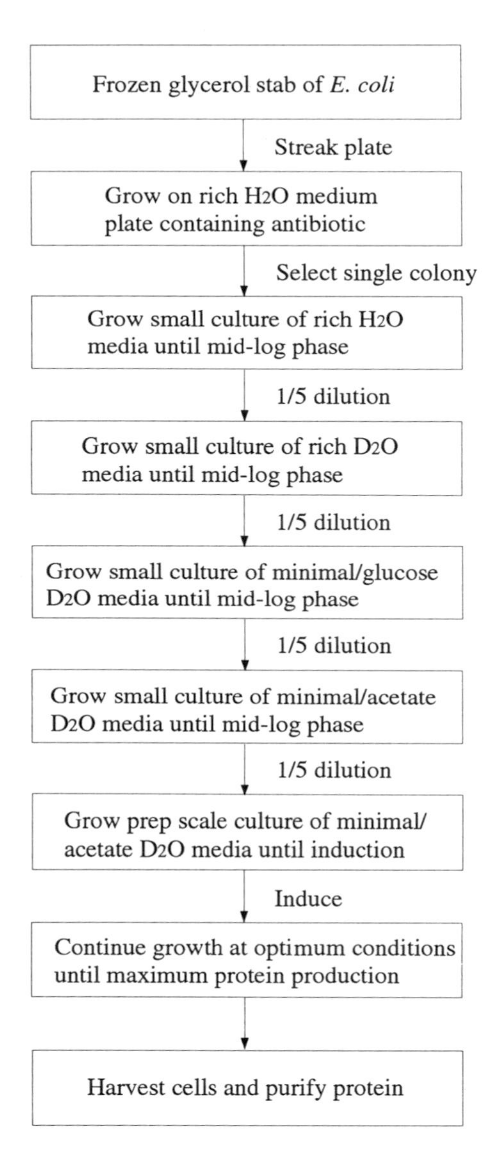

Figure 1. Generalized growth flow chart of *E. coli* in D_2O for the purpose of biosynthetic labeling. Unless otherwise indicated, all steps are performed at 37 °C and pH 7.0. When glucose is used as the primary carbon source, replace all steps containing acetate with glucose.

media containing either 50 or 75% D_2O, the protein mass indicates that deuterium was incorporated at nonexchangeable sites to levels of, respectively, 49 and 73%. Therefore, with acetate as the sole carbon source, the level of aliphatic deuteration in a protein depends linearly on the percentage of D_2O in the growth medium. This observation opens up the strong possibility that proteins can be easily prepared with any desired level of deuteration.

In our experience, it is important to optimize growth conditions in defined acetate medium for maximum protein yields. Conditions that should be optimized include the following: A_{600} at time of induction; induction time; growth temperature; antibiotic levels; and pH. Doubling times and induction times were found to increase substantially for cells grown in essentially 100% D_2O. In addition to the increased doubling time, an appreciable lag was observed on dilution of cells from minimal/glucose/D_2O medium into minimal/acetate/D_2O medium. Therefore, it is not unusual for cell growths to require 48–72 h from inoculation to cell harvest. Several other factors must also be considered when expressing proteins using deuterated media. All deuterated media must be sterile filtered, they must never be autoclaved, and they should be used immediately after preparation so as to minimize $^1H/^2H$ exchange. Finally, we have consistently observed that cells grown in D_2O medium do not completely lyse using the standard conditions of 0.25 mg/ml lysozyme and 0.1% Triton X-100. Therefore, the initial lysis step has been supplemented with a lysis step employing a French press.

2.1.2. Reintroducing Exchangeable Protons

For proteins initially expressed in completely deuterated solvent, exchange-protected amide groups may remain largely deuterated after purification. The complete reprotonation of all amide groups is necessary to obtain maximum sensitivity for all residues in amide-detected heteronuclear 3D and 4D NMR experiments. Insufficient sensitivity may lead to "missing spin systems," the bane of NMR studies on large proteins (Wagner, 1993). Ideally, this reprotonation can be accomplished by first unfolding and then subsequently refolding the protein in the presence of H_2O; however, not all proteins, e.g., MurB (Constantine *et al.*, 1997), are amenable to such harsh treatment. In many such cases where unfolding/refolding cannot be tolerated for either functional and/or spectroscopic reasons, a sufficient level of amide reprotonation may be achieved by mildly destabilizing the protein in one or more of the following ways: high pH; low concentration of denaturants; elevated temperatures. Finally, in the case of MurB, it was observed that even those amide groups classified as highly exchange protected by traditional D_2O exchange experiments were sufficiently reprotonated by the simple act of protein purification in H_2O-based buffers (Constantine *et al.*, 1997).

2.1.3. Residual Aliphatic Protonation

In *E. coli*, acetate is converted into acetyl-CoA. Acetyl-CoA is then directly used by the cells in the tricarboxylic acid cycle to obtain energy for the biosynthesis of amino acids (Gottschalk, 1986). A thorough examination of the glutamic acid biosynthetic pathway in *E. coli* reveals that the Glu C_γ carbon is derived directly from acetyl-CoA (Lehninger *et al.*, 1993). Because the acetyl moiety in the acetyl-CoA complex is derived from the externally supplied, fully protonated, ^{13}C-labeled acetate, the Glu C_γ carbon should remain fully protonated in the expressed protein. Furthermore, glutamic acid is the biosynthetic precursor of both glutamine and arginine: Glu is converted to Gln by a single-step process of amination whereas a complex, multistep process is required to convert Glu to Arg. The Gln C_γ carbon has been observed to remain fully protonated in HCA II protein expressed from media containing protonated acetate and D_2O (Farmer and Venters, 1996). By inference, the Glu C_γ also remains fully protonated in this HCA II protein. For the Arg C_γ carbon, the CH_2, CHD, and CD_2 isotopomers occur in a ratio of 5 : 2 : 3. These data indicate that, during the biosynthetic conversion of Glu to Arg, the eventual Arg C_γ carbon undergoes some exchange with solvent ^2H atoms. These residual, high-level protonation sites should be readily eliminated by using fully deuterated ^{13}C-labeled acetate as the sole carbon source instead of using protonated ^{13}C-labeled acetate.

2.1.4. Perdeuteration Using Glucose

^{13}C-labeled glucose can be used instead of acetate as the sole carbon source in the bacterial expression of $^2H/^{13}C/^{15}N$-labeled proteins. In fact, compared with acetate media, cell growth can be as much as 50% faster in glucose media; protein expression yields may also be slightly higher in glucose media. However, because glucose is utilized to synthesize amino acids, especially aromatic ones, through complex metabolic pathways, proteins expressed using $^1H/^{13}C$ glucose are less likely to contain completely random ^2H labeling (LeMaster, 1994). This problem can be eliminated by the use of $^2H/^{13}C$ glucose.

For most laboratories, the aspect of cost must also be considered when choosing a carbon source. Given present pricing strategies, $^2H/^{13}C$-labeled glucose is approximately five times more expensive than $^1H/^{13}C$-labeled acetate and two times more expensive than $^2H/^{13}C$-labeled acetate. In most instances, selective protonation at the glutamic acid, glutamine, and arginine C_γ positions is not detrimental either to making resonance assignments or to obtaining other types of structural information (Constantine *et al.*, 1997; Farmer *et al.*, 1996; Venters *et al.*, 1996). Therefore, $^2H/^{13}C$-labeled glucose should be used only in cases where the difference in yields makes up for the difference in cost or in cases involving selective protonation.

2.2. Selective Protonation

Because protein structure determination by NMR is based primarily on "loose," pairwise ^1H–^1H NOEs, the quality of calculated structures improves as both the number and accuracy of these interactions increase (Clore *et al.*, 1993). This is especially true for long-range interactions. We have previously stated that perdeuteration dramatically increases the sensitivity in 4D amide-detected NOESY experiments. This increase in sensitivity allows the detection of extremely long-range backbone/backbone, backbone/sidechain, and sidechain/sidechain H_N–H_N NOEs. Based on such NOE data alone, global folds can be calculated for globular proteins at least as large as 30 kDa. Both the accuracy and precision of structures determined solely from H_N–H_N NOE data can be improved with NOEs involving selectively protonated groups and/or residues within an otherwise perdeuterated protein (Metzler *et al.*, 1996b; Rosen *et al.*, 1996; Smith *et al.*, 1996).

Kay and co-workers have obtained selective methyl-group protonation in otherwise perdeuterated proteins in the following way: Proteins are overexpressed in *E. coli* grown on defined D_2O medium containing protonated pyruvate as the sole carbon source (Rosen *et al.*, 1996). Pyruvate, like acetate, enters the TCA cycle after being converted into acetyl-CoA (Gottschalk, 1986); therefore, the deuteration profile should closely resemble that found in defined acetate medium. In contrast to acetate, however, the methyl group of pyruvate is used directly in the synthesis of Ala, Val, Leu, and Ile($\gamma 2$) methyl groups. These researchers have determined both that the resulting proteins maintain high levels of uniform deuteration at most carbon sites and that the methyl groups of Ala, Val, Leu, and the $\gamma 2$ methyl group of Ile are highly protonated (Rosen *et al.*, 1996). As was similarly observed with acetate metabolism (Farmer and Venters, 1996), the C_γ position of Glu, Gln, Pro, and Arg also remain highly protonated with pyruvate metabolism.

Another approach to selective protonation has been to provide nonauxotrophic *E. coli* with several $^{13}C/^{15}N$- or ^{15}N-labeled protonated amino acids in defined D_2O medium containing a labeled nitrogen source, such as $^{15}NHSO_4$ or $^{15}NH_4Cl$, and either [^2H,^{12}C]- or [^2H,^{13}C]-glucose (Metzler *et al.*, 1996b; Smith *et al.*, 1996). Glucose is used in place of acetate to minimize breakdown of the supplied amino acids. Furthermore, [^2H]-glucose is required to prevent significant aromatic protonation. Farmer and co-workers have demonstrated this methodology both using $^1H/^{13}C/^{15}N$-labeled Ile/Leu/Val amino acids in the ^{15}N-labeled N-terminal SH3 domain (Metzler *et al.*, 1996b) and using ^{15}N-labeled Tyr in $^2H/^{13}C/^{15}N$-labeled MurB (Constantine *et al.*, 1997; B.T. Farmer II *et al.*, unpublished results). Laue and co-workers examined ^{15}N-labeled Phe/Tyr and $^1H/^{13}C/^{15}N$-labeled Ile/Leu/Val selective protonation. Both of these groups found essentially no scrambling of protons to other residue types. A high-level deuteration on C_α was also observed as a result of exchange with D_2O solvent during biosynthesis (Crespi *et al.*, 1968).

3. PROTEIN ASSIGNMENT

3.1. Backbone Assignment Using Perdeuterated Proteins

3.1.1. T_2 Relaxation: Effect on Resonance Assignment Approaches

Fully protonated $^{13}C/^{15}N$-labeled proteins are routinely assigned based on heteronuclear 3D experiments that correlate various intraresidue (i) and sequential ($i-1$) resonances (Metzler *et al.*, 1996a; Muhandiram and Kay, 1994; Grzesiek and Bax, 1993a; Wittekind and Mueller, 1993; Olejniczak *et al.*, 1992; Seip *et al.*, 1992; Bax and Ikura, 1991; Kay *et al.*, 1990a). The resonance types used in one approach are the H_α, C_α, and backbone CO. The HNCA and HN(CO)CA experiments are collectively used both to identify $C_\alpha(i)$ and $C_\alpha(i-1)$ resonances and to correlate each $C_\alpha(i)$ and $C_\alpha(i-1)$ resonance pair with its amide ($H_N(i)$, $N(i)$) resonance group. Similarly, the HN(CA)HA and HN(COCA)HA experiments are collectively used to assign each $H_\alpha(i)$ and $H_\alpha(i-1)$ resonance pair to its amide ($H_N(i)$, $N(i)$) resonance group. The HNCO is used to correlate each CO($i-1$) resonance with its ($H_N(i)$, $N(i)$) resonance group; the HCACO experiment is then used to correlate each CO(i) resonance with its ($H_\alpha(i)$, $C_\alpha(i)$) resonance group. Finally, backbone resonances are assigned to residues in the protein sequence by comparing these sequential and intraresidue H_α, C_α, and CO chemical shifts.

This particular approach has several problems. First, neither the H_α nor the CO chemical shift is at all indicative of residue type; therefore, stretches of backbone spin systems linked by these chemical shifts are more difficult to align with the protein sequence. Second, CO resonances exhibit a limited chemical-shift dispersion (~12 ppm). Third, the HCACO, because of spectral overlap between the H_α resonances and the water resonance, has historically been acquired on protein samples in 100% D_2O buffer. However, resonances from data sets collected in 90% H_2O/10% D_2O buffer may not register properly in chemical shift with resonances from data sets collected in 100% D_2O buffer. Finally, the (H_α, C_α) resonance groups are less disperse in chemical shift than the (H_N, N) resonance groups. Although the latter two problems can be addressed by replacing the HCACO with the HN(CA)CO experiment (Löhr and Rüterjans, 1995), the HN(CA)CO is markedly less sensitive than the HCACO.

In contrast, C_β resonances exhibit far greater intrinsic chemical-shift dispersion than do CO resonances. Furthermore, (C_β, C_α) chemical shifts are indicative of residue type; therefore, an amide-based spin system can be assigned at least to a particular group of residue types, if not to just a single residue type (Friedrichs *et al.*, 1994; Grzesiek and Bax, 1993a). Consequently, an alternate assignment approach uses the CBCA(CO)NH and HNCACB experiments to assign each {$C_\alpha(i)$, $C_\beta(i)$, $C_\alpha(i-1)$, $C_\beta(i-1)$} resonance quartet to its amide ($H_N(i)$, $N(i)$) resonance group. The HNCA and HN(CO)CA experiments are now largely superfluous. In the HNCACB experiment, non-Gly C_α correlations are opposite in sign both to Gly C_α correlations and to all C_β correlations. In the CBCA(CO)NH experiment, all correlations, both C_α and C_β, have the same sign; therefore, it can be problematic to distinguish between C_α and C_β resonances for Ser and some Thr residues. Although the HN(CO)CACB preserves the $C_{\alpha/\beta}$ sign discrimination as it exists in the HNCACB, the CBCA(CO)NH has remained, for protonated $^{13}C/^{15}N$-labeled proteins, the experiment of choice. This is largely because of the following attributes in the CBCA(CO)NH experiment: (1) constant-time $C_{\alpha/\beta}$ evolution, (2) sufficient sensitivity, (3) less involved experimental setup.

Using this approach the HBHA(CO)NH and HBHANH experiments are collectively used to assign each $\{H_\alpha(i), H_{\beta/\beta'}(i), H_\alpha(i-1), H_{\beta/\beta'}(i-1)\}$ resonance quartet (pentet or sextet, depending on both the multiplicity and the degeneracy of the H_β protons) to its amide ($H_N(i)$, $N(i)$) resonance group. The HN(CA)HA and HN(COCA)HA experiments are now largely superfluous. In the HBHANH experiment, non-Gly H_α correlations are opposite in sign both to Gly H_α correlations and to all $H_{\beta/\beta'}$ correlations. In the HBHA(CO)NH experiment, all correlations, both H_α and $H_{\beta/\beta'}$, have the same sign; therefore, it can be problematic to distinguish between H_α and $H_{\beta/\beta'}$ resonances for both Ser and Thr residues. For smaller, protonated $^{13}C/^{15}N$-labeled proteins, e.g., the SH3 domain of Grb2 (Wittekind *et al.*, 1994), the 3D ^{15}N-separated ^1H–^1H TOCSY (Fesik and Zuiderweg, 1988) has been used to establish intraresidue $H_{aliphatic}$-to-H_N correlations. However, in this regard, the HBHANH is less susceptible to ϕ and χ_1 conformationally induced variations in intensity of both H_α and $H_{\beta/\beta'}$ correlations.

Chemical-shift dispersion in the H_α protons is largely based on secondary structure (Wishart and Sykes, 1994): Helices contain a predominance of downfield-shifted H_α protons; extended structures contain a predominance of upfield-shifted H_α protons. H_α chemical-shift dispersion is, to a large extent, independent of residue type. For these reasons, H_α resonances complement C_α and C_β resonances in establishing unique, chemical-shift-based links between independent amino-acid spin systems. Although the $H_{\beta/\beta'}$ protons are both less influenced by structure and more influenced by residue type, both their chemical shift and their multiplicity on a given residue can help narrow down the possible choices for residue type.

No pair of resonances is more diagnostic of residue type than C_α and C_β. Consider that a spin system i and its sequential spin system $i-1$ are part of a linked stretch of spin systems. In this case, $C_\beta(i-1)$ chemical shifts, obtained from the more sensitive CBCA(CO)NH spectrum, can be used to restrict significantly the possible residue types identifiable with the sequential spin system. Such information, if extracted for each residue i in the linked stretch of spin systems, would facilitate a unique alignment of this stretch with the protein sequence. In cases where the $i-1$ and $i+1$ spin systems are not yet linked unambiguously, intraresidue C_β chemical-shift data may be required to establish these links (Venters *et al.*, 1996). Unfortunately, the HNCACB experiment may not yield sufficient C_β sensitivity on fully protonated proteins larger than ~30 kDa because of rapid C_α and H_N T_2 relaxation rates.

Relative to the HNCACB experiment, increased C_α T_2 relaxation rates have an even greater detrimental effect on sensitivity in the HBHANH experiment: Virtually no H_α or H_β correlations will be observed for large protonated proteins. Although the HBHA(CO)NH fares somewhat better in this regard than the HBHANH, the HBHA(CO)NH data cannot be used to link spin systems in the absence of the complementary HBHANH data. Therefore, in large, fully protonated $^{13}C/^{15}N$-labeled proteins, spin systems would, in general, have to be linked through the following restricted set of resonance types: C_α, H_α, and perhaps CO.

The process of sequential resonance assignment relies on a battery of heteronuclear 3D experiments, all of which must operate above their respective level of minimum sensitivity. This level is breached in larger protonated proteins because of their more rapid C_α and H_N T_2 relaxation rates. As a result, NMR studies on large, $^{13}C/^{15}N$-labeled proteins are faced with the following scenario: The number of unique spin systems is increasing, spectral resolution is rapidly decreasing, and fewer resonance types can be detected for each spin system. There are more spin systems to link, yet we are far less able to link these spin systems reliably. There are more and shorter fragments of linked spin systems to align, yet, regardless of fragment size, we are now less able to establish a unique alignment.

To the above problems, deuteration is a natural solution. Deuteration has been shown to decrease both C_α and H_N T_2 relaxation rates. In this way, both sensitivity and resolution are increased. Deuteration, in conjunction with $^{13}C/^{15}N$ isotopic labeling, also increases the number of resonance types with which to link spin systems: The set of such resonance types is now defined to include $\{C_\alpha, C_\beta, C_\gamma,$ CO, H_N, N$\}$ (Constantine *et al.*, 1997). Spin systems are also linked into larger fragments more easily and more reliably; in turn, these larger fragments are uniquely aligned more readily.

3.1.2. T_2 Relaxation: Experimental Measurement

Figure 2 shows the calculated T_2 relaxation times as a function of τ_c for all backbone nuclei in both a fully protonated protein and its perdeuterated equivalent. The calculations include both CSA and dipolar contributions (Yamazaki *et al.*, 1994c; Constantine *et al.*, 1993). The theoretical data in Fig. 2 would indicate that perdeuteration dramatically increases the T_2 relaxation times both for C_α nuclei (Yamazaki *et al.*, 1994a) and for H_N nuclei (Markus *et al.*, 1994). To confirm the validity of these theoretical data, we have calculated and measured the C_α T_2 values specifically for both protonated (C_α:H) and deuterated (C_α:D) HCA II. In these calculations, we use an isotropic rotational correlation time (τ_c) of 11.4 ns; we also neglect both internal motion and all types of exchange.

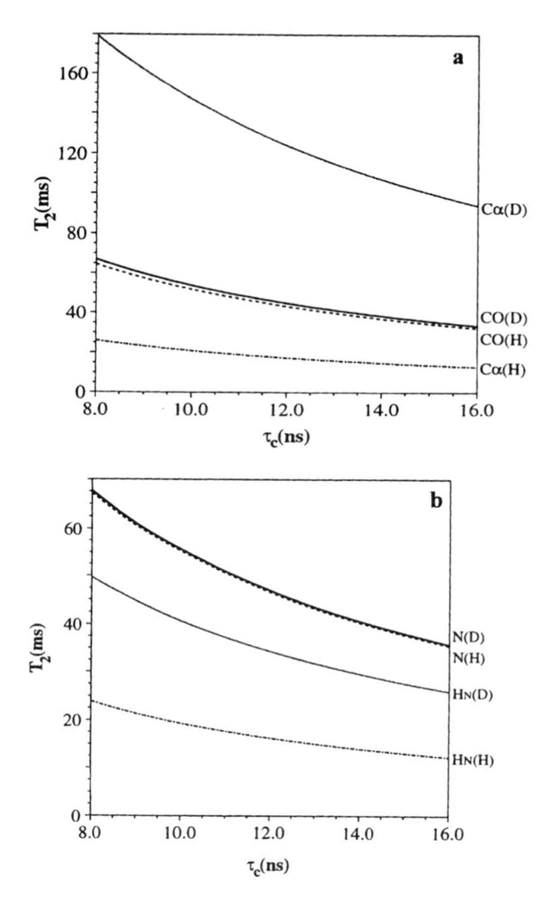

Figure 2. Calculated T_2 relaxation times are plotted against the isotropic rotational correlation time (τ_c) for all backbone nuclei in both a fully protonated protein (:H) and its perdeuterated equivalent (:D). (**a**) C_α(D), ...; C_α(H), ·–·–; CO(D),——; CO(H),– – –. (**b**) H_N(D), ...; H_N(H), ·–·–; N(D), ——; N(H), – – –. T_2(N) relaxation times were calculated with the following parameters: $\Delta\sigma_N = -160$ ppm, $r_{N\text{-}HN} = 1.02$ Å, $r_{N\text{-}H\alpha} = 2.12$ Å, $r_{N\text{-}CO} = 1.49$ Å, and $r_{N\text{-}C\alpha} = 1.49$ Å. T_2(CO) and T_{1calc}(CO) relaxation times were calculated with the following parameters: $\Delta\sigma_{CO} = 102$ ppm (Haberkorn *et al.*, 1981; Naito *et al.*, 1981), $r_{CO\text{-}C\alpha} = 1.54$Å (Jönsson and Kvich, 1972), $r_{CO\text{-}N} = 1.49$ Å (Jönsson and Kvich, 1972), $r_{CO\text{-}H\alpha} = 2.16$ Å, and $r_{CO\text{-}HN}$ = (2.24 Å, 3.30 Å). $T_2(C_\alpha)$ relaxation times were calculated with the following parameters: $\Delta\sigma_{C\alpha} = 29.5$ ppm for Gly (Haberkorn *et al.*, 1981) and 21.5 ppm for Ala (Naito *et al.*, 1981), $r_{C\alpha\text{-}D\alpha} = 1.05$ Å, $r_{C\alpha\text{-}H\alpha} = 1.09$ Å, $r_{C\alpha\text{-}H\beta} = 2.16$ Å, $r_{C\alpha\text{-}CO} = 1.54$ Å, $r_{C\alpha\text{-}N} = 1.49$ Å, $r_{C\alpha\text{-}C\beta} = 1.54$ Å (Ala only), and $r_{C\alpha\text{-}HN}$ = (2.24 Å, 3.00 Å). $T_2(H_N)$ relaxation times represent an average of values from α-helices and parallel β-sheets. For an α-helix, each amide proton was assumed to interact with four other amide protons at the following distances: 2.8, 2.8, 4.2, and 4.2 Å. For a parallel β-sheet, each amide proton was assumed to interact with three other amide protons at the following distances: 4.2, 4.2, and 4.0 Å. In both cases, CSA contributions to the amide proton linewidth for $\tau_c = 11.4$ ns were estimated at 3.8 Hz (Grzesiek and Bax, 1994).

Table 1
Calculated and Average Experimental Nuclear T_2 Relaxation Times in ^{2}H/^{13}C/^{15}N-Labeled (:D) and Calculated T_2 Times in ^{1}H/^{13}C/^{15}N-Labeled (:H) HCA II[a]

		Calculated	
Nucleus	Experimental (:D)	(:D)	(:H)
$C\alpha$ (Ala)	124 ± 12 (9)	131	18
$C\alpha$ (Gly)	104 ± 11 (13)	93	
CO	47 ± 3 (161)	48	45
N	52 ± 7 (184)	49	49
N(Hz)	43 ± 5 (187)		
H_N	29 ± 7 (186)	31	19
H_N (α-helix)	24 ± 3 (20)	25	21
H_N (β-sheet)	29 ± 5 (36)	36	17

[a]Experimental values are reported as mean ± SD (n), with all individual measurements lying within 2.5 SD of the mean. C_α and CO experimental T_2 relaxation times were calculated from $T_{1\rho}$ measurements. T_2 values were calculated for an isotropic rotational correlation time (τ_c) of 11.4 ns and a proton frequency of 600 MHz.

For HCA II, the Ala $T_2(C_\alpha{:}H)$ and $T_2(C_\alpha{:}D)$ are calculated to be, respectively, 18 and 131 ms. The average $T_2(H_N{:}H)$ and $T_2(H_N{:}D)$ of α-helical and β-sheet regions of the protein are calculated to be, respectively, 19 and 31 ms. T_2 relaxation times for ^{13}CO and ^{15}N nuclei have been similarly calculated for both ^{1}H-HCA II and ^{2}H-HCA II. The calculated T_2 relaxation data for HCA II are summarized in Table 1. The data in Fig. 2 and Table 1 would indicate that perdeuteration does not affect to any significant extent the T_2 relaxation of inphase ^{15}N and ^{13}CO magnetization. This stands in stark contrast to the significant effect that perdeuteration has on the T_2 relaxation of C_α and H_N magnetization. Table 1 also includes experimental T_2 relaxation data obtained on ^{2}H-HCA II. The experimental data for each type of nuclear magnetization agree well with the respective calculated data.

Direct measurements of carbon T_2 relaxation are hindered in uniformly ^{13}C-labeled proteins both by large carbon–carbon scalar couplings (Yamazaki *et al.*, 1994c) and by limited chemical-shift dispersion between coupled carbon spins. To minimize these problems, $T_{1\rho}$ relaxation times have been measured using a weak ^{13}C RF spinlocking field for both C_α and CO magnetization; the $T_{1\rho}$ times are then used to calculate the respective T_2 values (Venters *et al.*, 1996; Yamazaki *et al.*, 1994c). To further minimize these problems, $T_2(C_\alpha{:}D)$ values in ^{2}H-HCA II have been reported for only Ala and Gly residues. In these two residues, the C_α nucleus is both easily and effectively decoupled from all homonuclear scalar interactions. $T_{1\rho}$ measurements are also less susceptible to exchange effects. However, if any exchange process differentially affects either $T_2(C_\alpha{:}D)$ or $T_2(C_\alpha{:}H)$, then $T_{1\rho}$-extrapolated T_2 values cannot be used to accurately predict relative gains in experimental sensitivity related to perdeuteration. If $T_2(C_\alpha)$ is significantly affected in a given protein by one or more exchange processes, then $T_{1\rho}$-extrapolated T_2 values also cannot be used to guide the optimum pulse-sequence design.

Other groups have reported both $T_2(C_\alpha)$ and $T_2(H_N)$ values consistent with the aforementioned HCA II values. In the 37-kDa *trp* repressor–DNA complex, $T_2(C_\alpha)$ for nonglycine residues has been observed following high-level deuteration (~70%) to increase on average from 16.5 ms to 130 ms (Yamazaki *et al.*, 1994a). Additionally, the $T_2(H_N)$ for villin 14T has been shown to increase on average from 19 ms to 33 ms on perdeuteration (Markus *et al.*, 1994). These large increases in both the C_α and H_N T_2 relaxation times should lead to a dramatic increase in sensitivity for current multidimensional, heteronuclear scalar-correlation experiments.

3.1.3. Sensitivity

Perdeuteration, based on the foregoing analysis of relaxation, is expected to increase sensitivity in all amide-detected NOE and scalar-correlation experiments. This expectation holds true even for the most basic scalar-correlation experiment, the ^1H–^{15}N HSQC. For the gradient-enhanced, sensitivity-enhanced ^1H–^{15}N HSQC (Kay *et al.*, 1992a), sensitivity has been experimentally measured on both perdeuterated (^2H-HCA II) and fully protonated (^1H-HCA II) HCA II (Venters *et al.*, 1996). On average, perdeuterating HCA II increases sensitivity in the ^1H–^{15}N HSQC 2.5-fold. This average gain in sensitivity can be further partitioned on the basis of secondary structure: Following perdeuteration, residues in a β-sheet show a 2.8-fold gain in sensitivity; residues in an α-helix show only a 1.7-fold gain.

The expected increase in sensitivity related to perdeuteration, S_r, can be theoretically calculated from transfer functions that include both scalar coupling and T_2 relaxation. For the gradient-enhanced, sensitivity-enhanced ^1H–^{15}N HSQC,

$$\begin{aligned}
S_r(D/H) = {} & [\exp(-(2\tau^D + 2\delta_2)/T_2(H_N{:}D)) \,/\, \exp(-(2\tau^H + 2\delta_2)/T_2(H_N{:}H))] \,* \\
& [\{\exp(-2\tau^D/T_2(HN_{MQ}{:}D)) * \exp(-2\tau^D/T_2(H_N{:}D)) + \\
& \exp(-2\tau^D/T_2(H_N{:}D)) * \exp(-2\tau^D/T_{1s}(H_N{:}D))\} \,/ \\
& \{\exp(-2\tau^H/T_2(HN_{MQ}{:}H)) * \exp(-2\tau^H/T_2(H_N{:}H)) + \\
& \exp(-2\tau^H/T_2(H_N{:}H)) * \exp(-2\tau^H/T_{1s}(H_N{:}H))\}] \,* \\
& [\{\sin^2(2\pi J_{HN}\tau^D) \,/\, \sin^2(2\pi J_{HN}\tau^H)\} \,/\, \cos^2(2\pi J_{HNHA}\tau^H)] \qquad (1)
\end{aligned}$$

$$S_{rN}(D/H) \sim S_r(D/H) * [\exp(-2\delta_1 * \langle 1/(T_2(N{:}D))\rangle) / \exp(-2\delta_1 * \langle 1/(T_2(N{:}D))\rangle)] * [F(0, \langle 1/(T_2(N{:}D))\rangle, t_{max}(^{15}N), 1) / F(0, \langle 1/T_2(N{:}H))\rangle, t_{max}(^{15}N), 1)] \quad (2)$$

where

$$\tau^D = (\pi J_{NH})^{-1} \tan^{-1}(\pi J_{NH} T_2(H_N{:}D)) \quad (3a)$$

$$\tau^H = (\pi J_{NH})^{-1} \tan^{-1}(\pi J_{NH} T_2(H_N{:}H)) \quad (3b)$$

$$\langle 1/(T_2(N)\rangle = (T_2(N) * T_2(NH_z))^{-1/2} \quad (3c)$$

$$F(A, B, t, 1) = [(B/t)/(A^2 + B^2)] * \{1 - (\cos(At) + (A/B)\sin(At))e^{-tB}\} \quad (3d)$$

τ, δ_1, and δ_2 are as defined in Kay *et al.* (1992a). τ^H refers to the value used for ^{1}H-HCA II; τ^D, to the value used for ^{2}H-HCA II. The constant delay δ_2 is the combined duration of the final coherence refocusing gradient and its associated recovery period; the constant delay δ_1 is the combined duration of the initial coherence defocusing gradient and its associated recovery period. Note that S_r(D/H) is approximate for the following reasons: Differences in $T_{1ns}(H_N{:}H)$ and $T_{1ns}(H_N{:}D)$ relaxation times have been neglected for the relaxation delay; Eq. (1) deals only empirically with the T_2 relaxation of N_y and N_xH_z terms, which interconvert during the ^{15}N t_1 evolution period.

For the ^{1}H–^{15}N HSQC experiments on ^{13}C/^{15}N-labeled ^{1}H-HCA II and ^{2}H-HCA II, relevant acquisition parameters are as follows: $t_{max}(^{15}N) = 51.0$ ms; $\delta_1 = 5.5$ ms; $\delta_2 = 1.2$ ms; and $\tau^D = \tau^H = 2.5$ ms. Additionally, J_{HN} is set to 92 Hz; J_{HNHA} is set to 4.5 Hz for an α-helix and 9.0 Hz for a β-sheet. Inserting these values into Eq. (1), we obtain that the perdeuteration of HCA II, a 29-kDa protein with τ_c = 11.4 ns at 30°C, leads to a theoretical signal-to-noise gain in the gradient-enhanced, sensitivity-enhanced ^{1}H–^{15}N HSQC experiment of 1.4 for residues in α-helices and 3.0 for residues in β-sheets. These theoretical results are in excellent agreement with the aforementioned experimental results.

As described above, the sequential assignment of proteins relies on two key heteronuclei: the α carbon and the amide nitrogen. As a result, the two most important coupling constants in triple-resonance backbone scalar-correlation experiments are $^1J_{NC\alpha}$ (~11 Hz) and $^2J_{NC\alpha}$ (~5–8 Hz). The HNCA is the most basic triple-resonance scalar-correlation experiment that makes use of both $^1J_{NC\alpha}$ and $^2J_{NC\alpha}$ to establish sequential links between residues. The sensitivity of the HNCA is determined mainly by $T_2(H_N)$ and $T_2(N)$. However, both $T_2(C_\alpha)$ and carbon homonuclear scalar couplings have a significant effect on the achievable resolution in the ^{13}C dimension and at least a measureable effect on the overall experimental sensitivity. Resolution in the C_α dimension can be augmented by using a constant-time C_α evolution period (CT-HNCA; Yamazaki *et al.*, 1994b). Decreased sensitivity caused by a short $T_2(C_\alpha)$ can be overcome by perdeuteration.

Figure 2 illustrates that $T_2(C_\alpha)$ becomes quite short for nondeuterated, large proteins: on the order of 18 ms for a 30-kDa globular protein. Therefore, in the CT-HNCA experiment, increased resolution comes at the cost of significantly decreased sensitivity. Because the constant-time C_α evolution period must be set to $1/J_{CC}$ (~27.8 ms) to allow all carbon homonuclear couplings to refocus, a nondeuterated, 30-kDa globular protein would incur an approximately sixfold loss in sensitivity related to $T_2(C_\alpha)$ alone. For such a protein, perdeuteration has been shown, in Fig. 2 and Table 1, to increase $T_2(C_\alpha)$ to ~131 ms. In the CT-HNCA experiment, this increase in $T_2(C_\alpha)$ dramatically reduces the overall sensitivity loss. In addition, short $T_2(C_\alpha)$ values in fully protonated proteins lead to significant losses in sensitivity in experiments that transfer magnetization through the C_α spin, e.g., HNCACB, HN(CO)CACB, and HN(CACO)NH.

The HNCACB experiment is diagrammed in Fig. 3 (N_{CC} = 1). For this experiment, perdeuteration leads to a gain in sensitivity that can be estimated theoretically using Eqs. (4)–(6):

$$S_{r\alpha}(D/H) = S_r(D/H) * [\{\sin({}^1J_{NC\alpha}\pi\Delta_1{}^D)\cos({}^2J_{NC\alpha}\pi\Delta_1{}^D)\} / \{\sin({}^1J_{NC\alpha}\pi\Delta_1{}^H)\cos({}^2J_{NC\alpha}\pi\Delta_1{}^H)\}]^2 * [\{\cos(\pi J_{CC}\Delta_2{}^D) / \cos(\pi J_{CC}\Delta_2{}^H)\}^2 * \{\exp(-2\Delta_2{}^D/T_2(C_\alpha{:}D)) / \exp(-2\Delta_2{}^H/T_2(C_\alpha{:}H))\}] [F(\pi J_{CC}, 1/T_2(C_\alpha{:}D), t_{max}({}^{13}C), 1) / F(\pi J_{CC}, 1/T_2(C_\alpha{:}H), t_{max}({}^{13}C), 1)] \quad (4)$$

$$S_{r\beta}(D/H) = S_r(D/H) * [\{\sin({}^1J_{NC\alpha}\pi\Delta_1{}^D)\cos({}^2J_{NC\alpha}\pi\Delta_1{}^D)\} / \{\sin({}^1J_{NC\alpha}\pi\Delta_1{}^H)\cos({}^2J_{NC\alpha}\pi\Delta_1{}^H)\}]^2 * [\{\sin(\pi J_{CC}\Delta_2{}^D) / \sin(\pi J_{CC}\Delta_2{}^H)\}^2 * \{\exp(-2\Delta_2{}^D/T_2(C_\alpha{:}D)) / \exp(-2\Delta_2 H/T_2(C_\alpha{:}H))\}] [F(\pi J_{CC}, 1/T_2(C_\beta{:}D), t_{max}({}^{13}C), 2) / F(\pi J_{CC}, 1/T_2(C_\beta{:}H), t_{max}({}^{13}C), 2)] \quad (5)$$

where

$$\Delta_2 D = (\pi J_{CC})^{-1} \tan^{-1}(\pi J_{CC} T_2(C_\alpha{:}D)) \quad (6a)$$

$$\Delta_2 H = (\pi J_{CC})^{-1} \tan^{-1}(\pi J_{CC} T_2(C_\alpha{:}H)) \quad (6b)$$

$$\Delta_1 D = \{\pi({}^1J_{NC\alpha} + {}^2J_{NC\alpha})\}^{-1} \tan^{-1}(\pi({}^1J_{NC\alpha} + {}^2J_{NC\alpha})T_2(N{:}D)) \quad (6c)$$

$$\Delta_1 H = \{\pi({}^1J_{NC\alpha} + {}^2J_{NC\alpha})\}^{-1} \tan^{-1}(\pi({}^1J_{NC\alpha} + {}^2JN_{C\alpha})T_2(N{:}H)) \quad (6d)$$

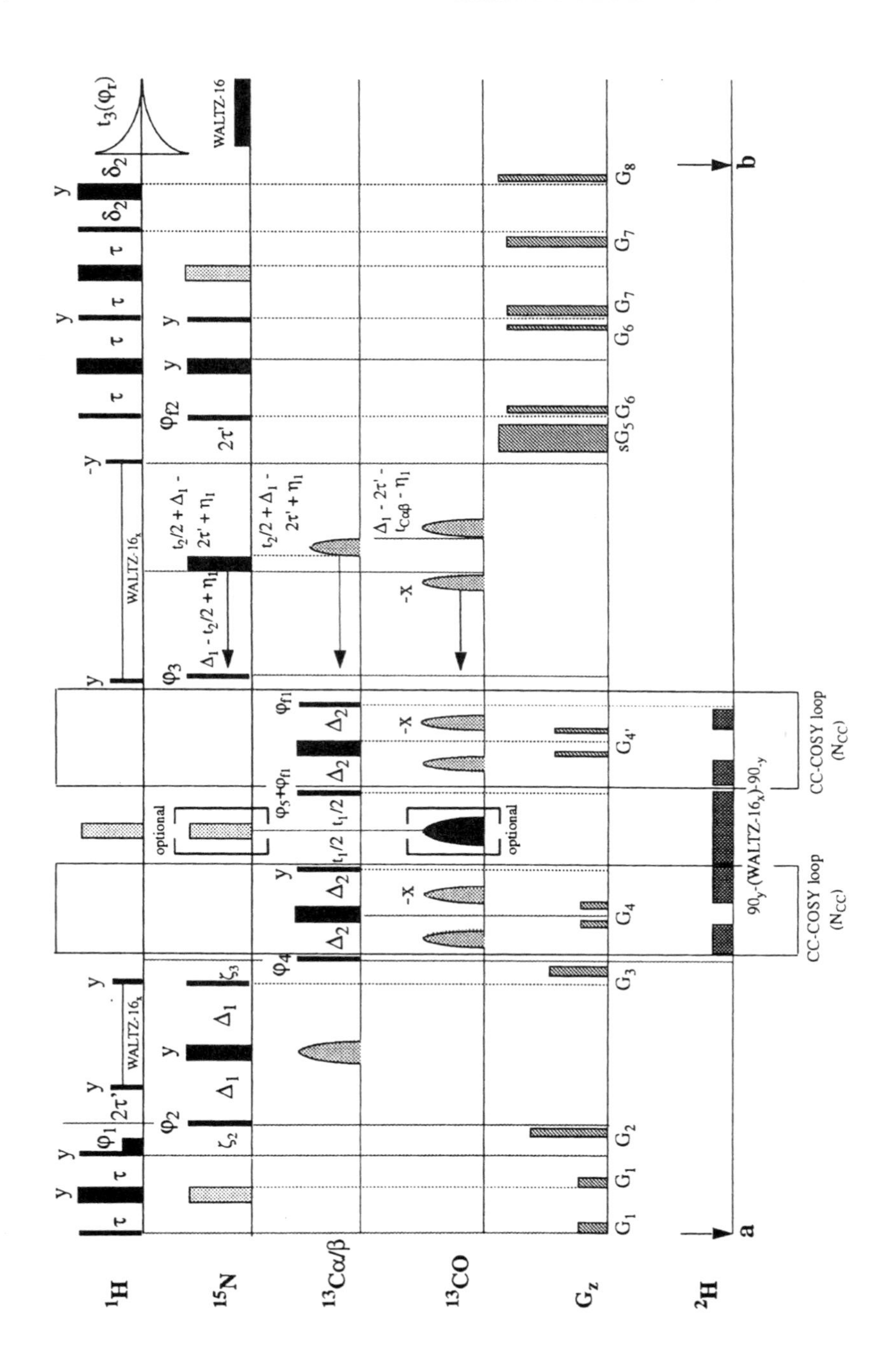

^1H
^{15}N
$^{13}C\alpha/\beta$
^{13}CO
G_z
^2H
WALTZ-16
WALTZ-16$_x$
optional
CC-COSY loop
(N_{CC})
90_y-(WALTZ-16$_x$)-90_{-y}
a
b

Figure 3. The 3D gradient-enhanced HNCACB and HN(CX)$_n$CY pulse sequences with ^{2}H decoupling, minimized H_2O saturation, and sensitivity optimization (Constantine *et al.*, 1997; Venters *et al.*, 1996). N_{CC} = 1 for the HNCACB experiment, N_{CC} = 2 for the HN(CACB)CG experiment, N_{CC} = 3 for the HN(CACBCG)CD experiment, and so on. Ninety-degree pulses are represented by wide lines; simple 180° pulses, by black rectangles; and $90_x240_y90_x$ composite inversion pulses, by diagonally striped rectangles. Unless otherwise indicated, all pulses have phase *x*. Broadband ^{2}H decoupling is achieved by WALTZ-16 modulation of the ^{2}H RF [$\gamma B_4(^2H)$ = 758 Hz]. Each diagonally striped box on the ^{2}H line represents the sequence element 90_y-(WALTZ-16_x)-90_{-y} applied to ^{2}H. It is not necessary in this application for WALTZ-16 to complete an integral number of cycles. All ^{13}C pulses are generated from a single RF channel because the fourth channel is set to ^{2}H. For this reason, all CO pulses must be generated on the third RF channel as frequency-shifted pulses (Patt, 1992). Complex data are collected in t_1 (States *et al.*, 1982) and in t_2 (Palmer *et al.*, 1991), with FIDs for $\varphi_{f1} = (x, (-1)^{Ncc}*y)$ and [$\varphi_{f2} = (x, -x)$; $s = (+1, -1)$ for G_{10}] being sorted separately. The phase cycle is as follows: $\varphi_1 = -x + \Delta\varphi_1$; $\varphi_2 = 8(x), 8(-x)$; $\varphi_3 = x, -x$; $\varphi_4 = 4(x), 4(-x)$; $\varphi_5 = 2(-y), 2(y)$; and $\varphi_r = 2(x, -x), 4(-x, x), 2(x, -x)$. $\Delta\varphi_1$ = 3.0°. Delay values used with ^{2}H-HCA II were as follows (in ms; Venters *et al.*, 1996): $\tau = 2.63$, $\tau' = 2.8$, $\Delta_1 = 11$, $\Delta_2 = 3$, $\eta_1 = 0$, $\zeta_2 = 2$, $\zeta_3 = 1.6$, and $\delta_2 = 0.7$. All solid-black ^{13}C pulses represent full-power, rectangular pulses. The cross-hatched, cone-shaped pulses are G3 inversion pulses of duration 320 μs for $^{13}C_{\alpha/\beta}$ and 806.4 μs for ^{13}CO. The solid-black, cone-shaped pulse on ^{13}CO, which occurs in the middle of the ^{13}C t_1 evolution period, is a cosine-modulated, Gaussian (64-step, 5σ) inversion pulse of duration 192 μs. The ^{13}C carrier is set at 44 ppm for all $^{13}C_{\alpha/\beta}$ pulses. Additional experimental details are given in Venters *et al.* (1996) and Constantine *et al.* (1997).

$$F(A, B, t, 2) = (2tB)^{-1}\{1 - e^{-tB}\} + F(A, B, t, 1)/2 \tag{6e}$$

Note that Eqs. (4) and (5) are valid only for $t_{max}(^{13}C) < 1/(2J_{CC})$ because of limitations in Eqs. (3d) and (6e). $S_{r\alpha}(D/H)$ in Eq. (4) represents, for a non-Gly C_α correlation in an HNCACB experiment, the optimum signal intensity obtained on a perdeuterated protein relative to that obtained on the equivalent, protonated protein. $S_{r\beta}(D/H)$ in Eq. (5) represents the same quantity for a methylene C_β correlation. Both $S_{r\alpha}(D/H)$ and $S_{r\beta}(D/H)$ depend strongly on the T_2 relaxation times of various coherences. As was demonstrated with $S_r(D/H)$ [Eq. (1)], if the overall correlation time (τ_c) of the protein is known, these relaxation times can be estimated reasonably accurately for both perdeuterated and protonated proteins.

Figure 4 plots, as a function of τ_c, the calculated $S_r(D/H)$, $S_{r\alpha}(D/H)$, and $S_{r\beta}(D/H)$ values for both α-helical and β-sheet regions of a protein. For HCA II, with τ_c ~11.4 ns at 30°C, the optimum C_α signal intensity in the HNCACB experiment is increased, on perdeuteration, 1.8-fold for α-helical regions and 3.8-fold for β-sheet regions. Similarly, on perdeuteration, the optimum C_β signal intensity is increased 4.3-fold (α-helix) and 9.3-fold (β-sheet). Intraresidue correlation peaks are not distinguished from interresidue ones because the Δ_1 delay [Eqs. (6c) and (6d)] has been optimized to achieve the maximum sum of their intensities. The HN(CO)CACB experiment yields $S_{r\alpha}(D/H)$ and $S_{r\beta}(D/H)$ values that are almost quantitatively identical to those obtained in the HNCACB experiment as neither $T_2(N)$ nor $T_2(CO)$ is significantly affected by perdeuteration.

Although it is important to know the expected relative gain in sensitivity on perdeuteration of a particular protein, it is equally important to know the expected absolute sensitivity for that perdeuterated protein. In this regard, ^{2}H-HCA II will serve as our reference perdeuterated protein at τ_c ~11.4 ns. Table 2 presents average experimental signal-to-noise ratios measured for both intraresidue and sequential C_α and C_β correlations from HNCACB data collected on ^{2}H-HCA II [1.6 mM protein sample, 37-h acquisition time, 600-MHz spectrometer, 30 °C, τ_f = 6 ms (nomenclature of Yamazaki *et al.*, 1994a). Equations (1), (4), and (5) can be modified (by replacing protonated relaxation values with deuterated values at τ_c = 11.4 ns) to yield sensitivity scaling factors for perdeuterated proteins larger than HCA II. If one assumes a similar protein concentration and sample temperature as for HCA II, a maximum acquisition time of 5 days for an HNCACB experiment, and a minimum required signal-to-noise of 5:1 for the intraresidue C_β correlation, the data in Table 2 suggest that perdeuteration, from the perspective of sensitivity but not necessarily resolution/dispersion, should permit the backbone resonance assignment of proteins having a maximum τ_c of ~32 ns. For a globular protein at a sample temperature of 30°C, a τ_c of 32 ns corresponds to a protein molecular mass of ~82 kDa. In line with this prediction, resonance assignments have just recently been obtained for a ^{2}H/^{13}C/^{15}N-labeled 64-kDa *trp* repressor–operator complex

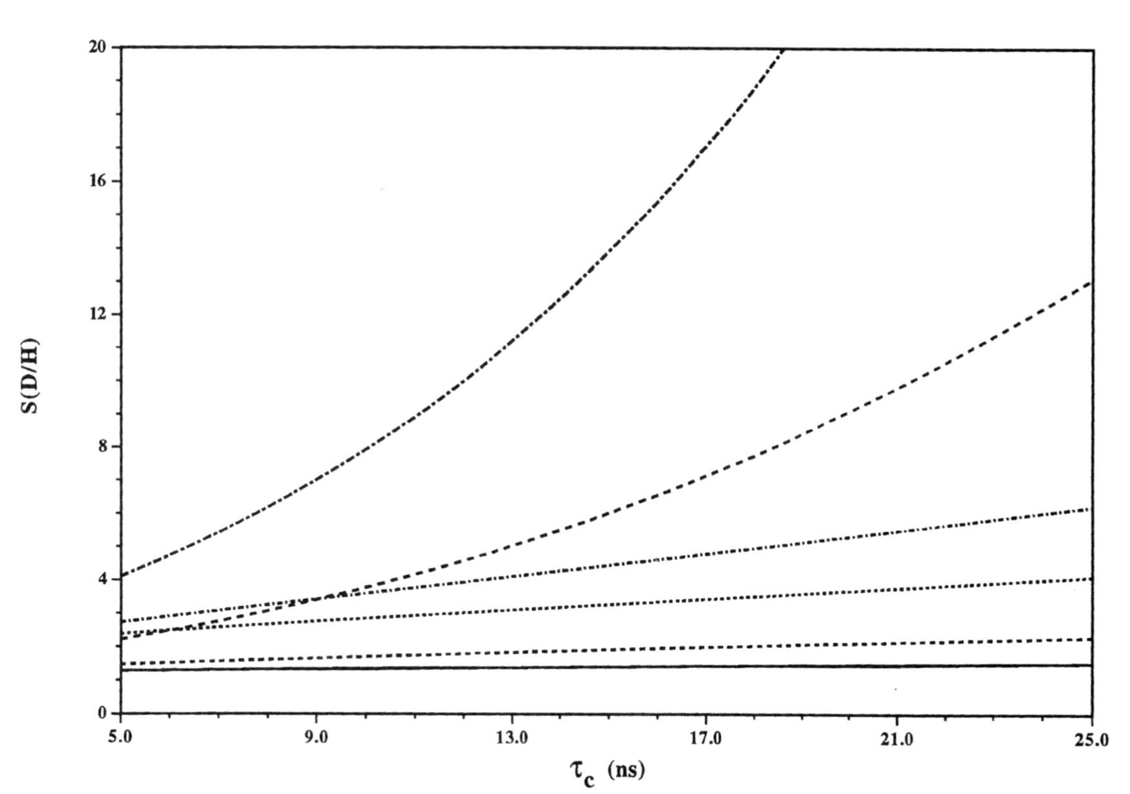

Figure 4. S_r(D/H), $S_{r\alpha}$(D/H), and $S_{r\beta}$(D/H) are plotted as a function of τ_c for amide protons in both α-helices and β-sheets. (——), S_r(D/H) for α-helix; - - - -, $S_{r\alpha}$(D/H) for α-helix; ------, S_r(D/H) for β-sheet; -·-·-·-, $S_{r\alpha}$(D/H) for β-sheet, - - - -; $S_{r\beta}$(D/H) for α-helix, -·-·-·-; $S_{r\beta}$(D/H) for β-sheet. These theoretical data were calculated using Eqs. (1)–(6) in the text. Parameters used to calculate nuclear relaxation times are as described in the legend to Fig. 2.

Table 2
Average Signal-to-Noise Ratios in the HNCACB Spectrum of $^{2}H/^{13}C/^{15}N$-Labeled HCA II

HNCACB	Signal-to-noise
C_{α}	134:1
$C_{\alpha}(-1)$	64:1
C_{β}	59:1
$C_{\beta}(-1)$	30:1

consisting of two tandem dimers of the *trp* repressor protein (108 residues per monomeric subunit) (Shan *et al.*, 1996).

3.1.4. Special Experiments for Larger Proteins

Our cumulative experience indicates that, in general, the backbone H_N, N, C_{α}, C_{β}, and CO resonances can be sequentially assigned for perdeuterated $^{13}C/^{15}N$-labeled proteins using a strategy based on the HNCACB, HN(CO)CACB, HN(CACO)NH, HNCO, and HN(CA)CO experiments. These experiments, when modified to work optimally with perdeuterated proteins, provide the sensitivity required to assign proteins much larger than 30 kDa. However, particular experience with the 38-kDa protein MurB (Constantine *et al.*, 1997; Metzler *et al.*, 1996b) suggests that resolution, not sensitivity, becomes the limiting factor with proteins of this size. This section describes several "special" experiments that can be adopted to improve resolution in perdeuterated proteins.

Sequential links between amino-acid spin systems (RIDs in the terminology of Friedrichs *et al.*, 1994) must be established in deuterated proteins based on a comparison of intraresidue (for spin system *i*) and sequential (for spin system *j*) ^{13}C chemical shifts (C_{α}, C_{β}, CO, and perhaps C_{γ}). In the absence of any Pro residues, of any ^{13}C chemical-shift overlap, and of any missing resonances, one could in theory sequentially assign any protein by this approach, regardless of complexity. However, in practice, fragments of only up to 10–12 linked RIDs are achieved because of the following factors: Pro residues; missing peaks; intrinsic $C_{\alpha/\beta}$ dispersion; achievable ^{13}C resolution. The latter three factors become increasingly problematic with larger proteins.

One is then faced with the task of aligning these *m*-RID fragments with the primary sequence of the protein. This alignment is done by comparing the $C_{\alpha/\beta}$ chemical shifts for every RID in the *m*-RID fragment with reference $C_{\alpha/\beta}$ chemical shifts for the corresponding residue in every possible *m*-residue sequential stretch within the protein. If only a single preferred alignment is found, the *m*-RID fragment can then be assigned to that particular m-residue sequential stretch. As m increases, so does the probability that the m-RID fragment will yield a single preferred alignment within the protein (Friedrichs *et al.*, 1994). However, if m is small, e.g., $m = 2$–5, and if there are several stretches in the primary sequence comprised of different amino acids with sufficiently similar $C_{\alpha/\beta}$ chemical shifts, one can be faced with a number of statistically indistinguishable alignments for the m-RID fragment. Moreover, the frequency and severity of this occurrence are expected to increase with the size of the protein.

In this sequential assignment process, deuteration can be viewed as the "great equalizer." What deuteration taketh away in terms of nuclei (namely, $H_{\alpha/\beta}$) with which to establish unique RID–RID links, deuteration giveth back in terms of greater opportunity for ^{13}C editing. More stringent ^{13}C editing can be used to limit possible residue types for each RID. As the possible residue types for each RID become more limited, the probability increases that an m-RID fragment will yield a single preferred alignment within the protein. Indeed, the editing capability intrinsic to the HN(CACB)CG experiment has already been documented as key to the sequential backbone assignment of MurB (Constantine *et al.*, 1997). In addition to the HN(CACB)CG experiment, we have also combined the useful features of the CT-HN(COCA)CB experiment (Yamazaki *et al.*, 1994b) with those of the 2D C_{aro}-edited ^1H–^{13}C CT-HSQC (Grzesiek and Bax, 1993a). The result is a 3D gradient-enhanced, sensitivity-enhanced C_{aro}-edited (or CO-edited) constant-time HN(COCA)CB experiment with broadband ^2H decoupling and water-flipback pulses, denoted CT-HN(COCA)CB$^{CO/Caro}$. The editing feature of these two experiments allows us to separately identify both Asn/Asp residue groups (CO editing) and Phe/Tyr/Trp/His residue groups (C_{aro} editing). The latter group can be further partitioned into Phe/Tyr and Trp/His residues based on the absolute C_β chemical shift.

Figure 5 depicts the CT-HN(COCA)CB$^{CO/Caro}$experiment. This experiment has been applied to perdeuterated $^{13}C/^{15}N$-labeled HCA II (^2H-HCA II). The results are presented in Table 3. Except for His residues, the CT-HN(COCA)CB$^{CO/Caro}$ experiment has worked quite well on ^2H-HCA II. Why His residues preferentially yield insufficient signal is not known. However, it is known that the CT C_β t_1 evolution period is expected to yield proportionally lower sensitivity for residues experiencing proportionally more chemical-shift exchange at the C_β nucleus. His, with its ability to undergo rapid tautomerization on its imidazole ring, may simply be more susceptible to this chemical-shift exchange.

By adding more CC COSY steps to the HNCACB (see Fig. 3) with $\Delta_2 = 6.25$ ms, one can readily extend this sequence to the C_δ ($N_{CC} = 3$) and even C_ε ($N_{CC} = 4$) carbons. These generic HN(CX)$_n$CY experiments edit amino acids based on the branch point of the sidechain. For example, Ile and Leu residues, whose C_β nuclei exhibit similar chemical shifts, have different sidechain branch points: Ile branches at C_β, Leu branches at C_γ. Therefore, Ile will give, at best, a weak signal in the

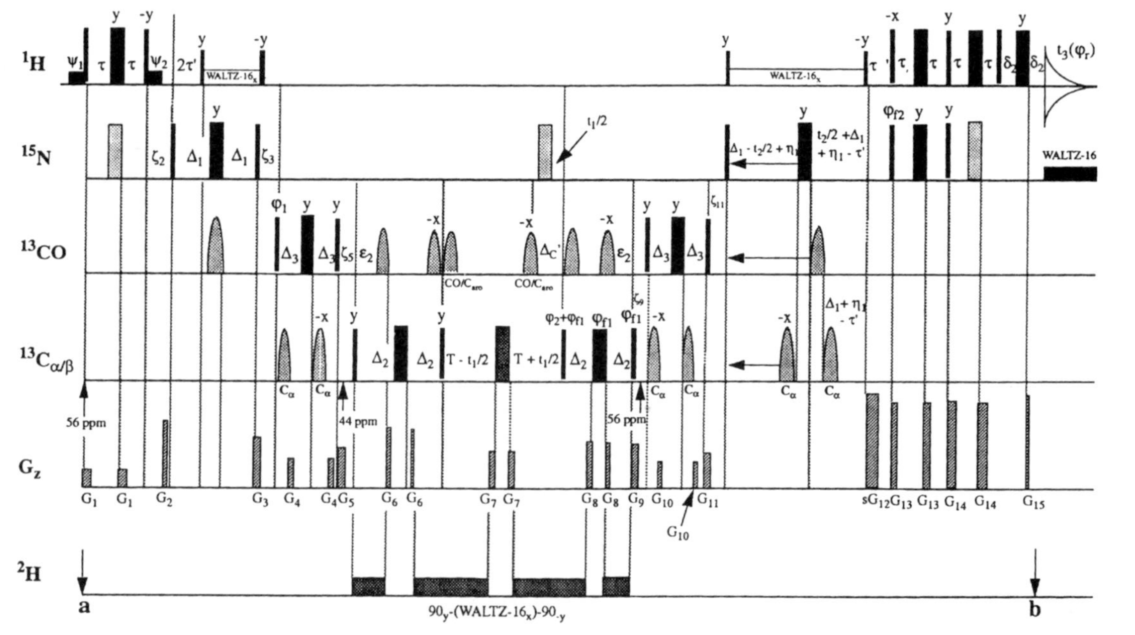

Figure 5. The CT-HN(COCA)CB$^{CO/Caro}$ pulse sequence. See the legend to Fig. 3 for details on diagram nomenclature, ^{13}C RF generation, ^{2}H broadband decoupling, and t_1/t_2 complex data collection. All cross-hatched, cone-shaped ^{13}C pulses are G3 inversion pulses. The two G3-shaped selective ^{13}C pulses (on the ^{13}CO line) during the CT t_1 evolution period have an inversion profile centered either at 125 ppm for C_{aro} editing or at 175 ppm for CO editing. All solid-black $^{13}C_{\alpha/\beta}$ pulses are nonselective, rectangular RF pulses (14 μs); the vertically striped $^{13}C_{\alpha/\beta}$ 180° pulse during the CT t_1 evolution period has a null either at the aromatic carbons for C_{aro} editing (69.2-μs pulse length) or at the carbonyl carbons for CO editing (45.2-μs pulse length). The phase cycle is as follows: $\psi_1 = 170°$; $\psi_2 = 71.5°$; $\varphi_1 = x, -x$; $\varphi_2 = 2(-y), 2(y)$; and $\varphi_r = 2(x, -x)$. Delay values used with ^{2}H-HCA II were as follows (in ms): $\tau = 2.6$, $\tau' = 2.8$, $\Delta_1 = 12.5$, $\Delta_2 = 6.25$, $\Delta_3 = 9.1$, $\zeta_2 = 2$, $\zeta_3 = 1.6$, $\zeta_5 = 1.5$, $\varepsilon_2 = 4.5$, $T = 13.5$, $\zeta_{11} = 1.5$, $\eta_1 = 0$, and $\delta_2 = 0.9$. C_{aro}/CO editing is accomplished by separately storing two FIDs collected with the following Δ_C' delay values: $\Delta_C' = (0.0, 11.3)$ ms for C_{aro} editing and $\Delta_C' = (0.0, 9.1)$ ms for CO editing. Δ_C' is cycled ahead of φ_{f1} and φ_{f2}.

Table 3
Efficacy of the CT-HN(COCA)CB$^{CO/Caro}$ Experiment in Resolving Both Asx and Aromatic Spin Systems in $^{2}H/^{13}C/^{15}N$-Labeled HCA II

Asn	8 out of 10 residues were observed; both N62 and N67 were not observed
Asp	17 out of 19 residues were observed; both D41 and D179 are N-terminal to a Pro
Phe	11 out of 12 residues were observed; only F20 was not observed
Tyr	6 out of 8 residues were observed; Y193 is N-terminal to a Pro; Y51 was not observed
Trp	5 out of 7 residues were observed; W5 and W244 were not observed
His	7 out of 12 residues were observed; H3, H4, H64, H94, and H119 were not observed

HN(CACB)CG experiment; however, Leu can give a strong signal in this experiment. Furthermore, because the weak signal from Ile originates from three-spin, single-quantum coherences, the resulting pattern of correlations (three positive and one negative) indicates that the overall coherence transfer pathway has traversed a sidechain branch point.

3.2. Sidechain Assignments

3.2.1. Aliphatic ^{13}C

Either the HCCH-TOCSY experiment (Bax *et al.*, 1990) or the HC(CC)(CO)NH experiment (Montelione *et al.*, 1992) can be used to assign sidechain carbons in moderately sized, fully protonated $^{13}C/^{15}N$-labeled proteins. In larger proteins, the HCCH-TOCSY may have a more limited utility because of increased spectral overlap brought on by type-specific "resonance clumping." We define type-specific "resonance clumping" as the tendency for $^{1}H/^{13}C$ correlations to segment less based on secondary and tertiary structural factors and more based on the particular amino acid type. Resonance clumping is a major problem in proteins that contain few aromatic residues.

Extending the HCCH-TOCSY from three dimensions to four is not likely to mitigate, at least to any large extent, the problem of spectral overlap endemic to the sidechain resonances of larger proteins. This assertion clearly does not hold for NOESY experiments, especially the ^{13}C-separated ones. NOESY experiments differ from the HCCH-TOCSY in two fundamental ways: (1) NOESY experiments are almost never used to assign the majority of sidechain resonances but are, rather, analyzed once the sidechain resonances have been assigned by some other method (e.g., HCCH-TOCSY); (2) NOESY experiments are not limited to intraresidue correlations. The latter difference is the major reason why a 4D $^{13}C/^{13}C$-separated NOESY is more powerful than its 3D analogue. However, the HCCH-TOCSY yields only intraresidue correlations; and must be analyzed in the absence of existing sidechain resonance assignments.

In contrast to the HCCH-TOCSY, the HC(CC)(CO)NH experiment does not suffer nearly as much from resonance clumping: The chemical shift of the amide proton–nitrogen pair is far less dependent on the amino acid type. However, rapid carbon and H_N relaxation can render the HC(CC)(CO)NH experiment too insensitive to be useful on proteins with long rotational correlation times. Whereas we have argued that a 4D HCCH-TOCSY would have minimal benefits over a 3D HCCH-TOCSY, we would conversely argue that a 4D HC(CC)(CO)NH (Clowes *et al.*, 1993), with sufficient sensitivity, is more suitable than its 3D analogue(s).

Two 3D experiments, the (H)C(CC)(CO)NH and H(C)(CC)(CO)NH, must be collected to identify all carbon and proton resonances belonging to a particular amino-acid sidechain. With C_α and C_β carbons already assigned, it is straightforward to assign most of the remaining sidechain carbon resonances: Only Leu C_γ and $C_{\delta 1}$ resonances are typically ambiguous in their assignment. However, assigning the proton resonances is far more difficult in the absence of the attached carbon chemical shift, which is provided only with the 4D experiment. Additionally, the acquisition parameters for the ^1H_C and ^{13}C dimensions in the 4D HC(CC)(CO)NH can be set up to match exactly those for the donor dimensions of the 4D $^{13}C/^{15}N$-separated and $^{13}C/^{13}C$-separated NOESY experiments. In this way, correlations identified in the 4D NOESY experiments are more readily and accurately assigned based on correlations in the 4D HC(CC)(CO)NH experiment.

Although perdeuteration has been seen to provide clear benefits in assigning backbone resonances, it would appear to obviate the use of either the HCCH-TOCSY or HC(CC)(CO)NH experiment to assign sidechain resonances because both experiments start with aliphatic proton magnetization. However, only the HCCH-TOCSY detects aliphatic proton magnetization; the HC(CC)(CO)NH detects amide proton magnetization. Therefore, we have applied a variant of the HC(CC)(CO)NH experiment to deuterated proteins so as to obtain sidechain ^{13}C resonance assignments. This experiment, C(CC)(CO)NH (Farmer and Venters, 1995), is identical to the (H)C(CC)(CO)NH except that magnetization transfer starts not on proton but rather on carbon.

Theoretical calculations suggest that the C(CC)(CO)NH experiment applied to a 29-kDa perdeuterated, $^{13}C/^{15}N$-labeled protein provides significant gains in sensitivity over the (H)C(CC)(CO)NH applied to the fully protonated, $^{13}C/^{15}N$-labeled protein: an ~3.5-fold increase for methine groups and an ~7-fold increase for methylene groups. Experimental data on HCA II have confirmed this overall increase in sensitivity afforded by the C(CC)(CO)NH experiment. However, the exact magnitude of the sensitivity increase could not be ascertained because the (H)C(CC)(CO)NH experiment on ^1H-HCA II yielded essentially no observable correlations.

These theoretical calculations on relative sensitivity include the effect of a shorter carbon T_1 relaxation time for protonated versus deuterated carbon atoms. The theoretical, dipolar-induced T_1 relaxation time for deuterated ^{13}C nuclei is shorter than might be expected because of the I=1 nuclear spin number for ^2H

(Kushlan and LeMaster, 1993): The calculated T_1 value is ~4.4 s for deuterated methine groups and ~2.4 s for fully deuterated methylene groups. We have qualitatively measured the ^{2}H-HCA II C_α T_1 relaxation time using 1D saturation recovery: The approximate relaxation time is 2.6 s. Therefore, as dipolar mechanisms become increasingly ineffective at relaxing the ^{13}C magnetization, other mechanisms, most likely random-field fluctuations (e.g., those related to dissolved O_2), will begin to make initially a more significant contribution and finally a dominant contribution to the carbon T_1 relaxation.

Sidechain ^{13}C correlation data from the C(CC)(CO)NH experiment are both essential to the assignment of Arg sidechain amine and Glu sidechain amide protons (Farmer and Venters, 1996) and quite useful in further delineating possible residue types that are $i - 1$ to the correlated amide proton–nitrogen resonance pair. These sidechain ^{13}C chemical shifts, corrected for the ^{2}H isotope effect (Venters *et al.*, 1996; Hansen, 1988), can also be used to facilitate the resonance assignment in an HCCH-TOCSY applied to the fully protonated protein. Finally, such ^{13}C chemical-shift data could be used in a more restricted fashion to facilitate the assignment of sidechain methyl groups in a 4D ^{13}C/^{15}N-separated NOESY applied to the ILV type-specifically protonated protein (Metzler *et al.*, 1996b).

3.2.2. Aliphatic $^1H_N/^{15}N$

Traditionally, sidechain NH_x groups are assigned by identifying NOE interactions to sidechain aliphatic protons; however, in perdeuterated proteins there are no such protons. Alternatively, these resonances can be assigned by making use of through-bond correlations between the aliphatic $^1H_N/^{15}N$ resonances and sidechain ^{13}C resonances. Because the sidechain ^{13}C resonances of a perdeuterated protein can be assigned, as described above, with the C(CC)(CO)NH and $HN(CX)_nCY$ experiments, the assignment of the sidechain $^1H_N/^{15}N$ resonances becomes straightforward using a family of gradient-enhanced, sensitivity-enhanced pulse sequences. These sequences include: NH_2-filtered 2D ^{1}H–^{15}N HSQC (H_2N-HSQC), 3D $H_2N(CO)C_{\gamma/\beta}$, and 3D $H_2N(COC_{\gamma/\beta})C_{\beta/\alpha}$ for Gln and Asn sidechain amide groups; 2D $H(N_{\varepsilon/\zeta})C_{\delta/\varepsilon}$ and $H(N_{\varepsilon/\zeta}C_{\delta/\varepsilon})C_{\gamma/\delta}$ for Arg and Lys sidechain amine groups; and 2D refocused $H(N_\varepsilon)C_\zeta$ and non-refocused $H(N_\eta)C_\zeta$ for Arg sidechain guanidino groups (Farmer and Venters, 1996; Yamazaki *et al.*, 1995; Vis *et al.*, 1994). All of these sequences make use of water-flipback pulses.

The sidechain $^1H_N/^{15}N$ assignments obtained from these studies can be used in conjunction with the mainchain amide group assignments to analyze 4D ^{15}N/^{15}N-separated NOESY data, as will be described herein. Recent studies have demonstrated that the global fold of a perdeuterated protein can be calculated using only H_N–H_N NOE restraints (Grzesiek *et al.*, 1995; Venters *et al.*, 1995b). It has also been demonstrated that the quality of the resultant global fold can be improved by the inclusion of restraints both to sidechain H_N groups (Venters *et al.*, 1995b, and unpublished observations) and to selectively protonated methyl and aromatic groups in otherwise perdeuterated proteins (Gardner and Kay, 1997; Metzler *et al.*, 1996b; Rosen *et al.*, 1996; Smith *et al.*, 1996). NOEs involving sidechain protons appear invaluable in tying down elements of secondary structure in the correct tertiary topology.

3.3. ^{2}H Isotope Shifts

The substitution of a deuteron for a proton induces isotope effects in nuclei that are separated from the site of substitution by as many as four bonds (Hansen, 1988; Majerski *et al.*, 1985). Therefore, ^{13}C chemical shifts measured on a perdeuterated protein [^{13}C(D) shifts] will be substantially different than those measured on the fully protonated equivalent [^{13}C(H) shifts] (Garrett *et al.*, 1997; Venters *et al.*, 1996). To convert ^{13}C(D) shifts to approximate ^{13}C(H) shifts, the total ^{2}H isotope effect, ΔC(D), must be calculated for each carbon atom in a perdeuterated protein. To this end, we have measured ΔC_{α}(D) and ΔC_{β}(D) for all residue types by comparing the $C_{\alpha/\beta}$ chemical shifts observed in perdeuterated HCA II with those observed in fully protonated HCA II. Table 4 presents the results from our analysis of these data. A statistical analysis indicates that the total ^{2}H isotope effect can be accurately predicted for most C_{α} and C_{β} nuclei in a perdeuterated protein (Venters *et al.*, 1996). The one-bond ^{2}H isotope effect on Gly C_{α} carbons has been separately measured: $^{1}\Delta C_{gly}(D) = -0.39 \pm 0.04$ ppm. In cases where sensitivity limitations may preclude the direct measurement of $C_{\alpha/\beta}$(H) chemical shifts, $C_{\alpha/\beta}$(D) chemical shifts can be corrected based on the data presented in Table 4. $C_{\alpha/\beta}$(H) chemical

Table 4
Total Deuterium Isotope Shifts for C_{α} and C_{β} Nuclei

$^{1}\Delta C(D)$ (one bond) = -0.29 ± 0.05
$^{2}\Delta C(D)$ (two bond) = -0.13 ± 0.02
$^{3}\Delta C(D)$ (three bond) = -0.07 ± 0.02
$^{1}\Delta C_{gly}(D) = -0.39 \pm 0.04$

	ΔC_{α}	ΔC_{β}
Asn, Asp, Ser, His, Phe, Trp, Tyr, Cys	−0.55	−0.71
Lys, Arg, Pro	−0.69	−1.11
Gln, Glu, Met	−0.69	−0.97
Ala	−0.68	−1.00
Ile	−0.77	−1.28
Leu	−0.62	−1.26
Thr	−0.63	−0.81
Val	−0.84	−1.20

shifts estimated in this way can be subsequently used to characterize both helical and extended elements of secondary structure. Additionally, sidechain $^{13}C(D)$ chemical shifts (see Section 3.2.1) can be used to estimate corresponding $^{13}C(H)$ shifts. These estimated sidechain $^{13}C(H)$ shifts can then be used to help assign $^{13}C/^{15}N$- and $^{13}C/^{13}C$-separated NOESY spectra on either a type-specifically protonated (Metzler *et al.*, 1996b) or a fully protonated protein.

3.4. Secondary Structure

3.4.1. C_α and C_β Chemical Shifts

Backbone chemical shifts are predictably affected by the secondary structure of the protein (Wishart and Sykes, 1994; Metzler *et al.*, 1993; Spera and Bax, 1991). The C_α and carbonyl carbons experience an upfield shift in extended structures and a downfield shift in helical structures. Both the C_β carbon and the H_α proton chemical shifts exhibit the opposite correlation, namely, a downfield shift in extended structures and an upfield shift in helices. These shifts have been advantageously utilized in several related approaches to predict the secondary structure of a number of proteins (Constantine *et al.*, 1997; Venters *et al.*, 1996; Wishart *et al.*, 1994; Metzler *et al.*, 1993).

The first protocol uses only C_α and C_β chemical shifts (Metzler *et al.*, 1993) measured either directly from a protonated protein or from a perdeuterated protein followed by a correction for the total ^2H isotope effect (Constantine *et al.*, 1997; Venters *et al.*, 1996). Observed C_α chemical shifts are subtracted from C_α random-coil values; observed C_β chemical shifts are subtracted from C_β random-coil values. The difference between these two chemical-shift deviations (ΔC_α minus ΔC_β) is calculated and subsequently smoothed with a three-point binomial function, yielding $(\Delta C_\alpha - \Delta C_\beta)_{\text{smoothed}}$ (Metzler *et al.*, 1996a). Values for $(\Delta C_\alpha - \Delta C_\beta)_{\text{smoothed}}$ are subsequently plotted against the protein sequence. A cluster of positive $(\Delta C_\alpha - \Delta C_\beta)_{\text{smoothed}}$ values is indicative of a helical region; a cluster of negative values is indicative of a region of extended structure.

The second commonly utilized method for secondary structure determination from chemical-shift analysis is the CSI protocol. The CSI method relies not only on the C_α and C_β shifts but also on the carbonyl carbon and H_α shifts as well (Wishart and Sykes, 1994). This protocol uses a two-step digital filtering process. In the first step, a chemical-shift index (CSI) of $-1, 0$, or 1 is assigned to each residue based on the chemical shift of a particular nucleus relative to an appropriate random-coil value. In the second step, secondary structural elements are identified by a clustering of like chemical-shift indices. With this CSI protocol, helices are indicated by a cluster of chemical-shift indices having value 1 for C_α and carbonyl carbons and -1 for H_α. In contrast, β-strands are indicated by a cluster of chemical-shift indices having value -1 for C_α and carbonyl carbons and 1 for H_α and C_β.

A consensus secondary structure can be determined by a "majority rules" algorithm (Wishart and Sykes, 1994). In general, the CSI protocol tends to be more conservative than the first protocol (Metzler *et al.*, 1993) in defining regions of secondary structure.

3.4.2. Use as Restraints in Global Fold Determination

In perdeuterated proteins, information on secondary structure is important to a global-fold determination for at least two reasons. First, this information can be used to anticipate spin-diffusion effects in 4D $^{15}N/^{15}N$-separated NOESY data collected at long mixing times. For instance, H_N–H_N NOEs observed within an α-helix are far more subject to spin diffusion than those observed within a β-sheet. Second, this information can be transformed into loose dihedral-angle restraints; programs such as X-PLOR (Brunger, 1992) and DYANA (Guentert *et al.*, 1997) readily utilize such restraints in structure calculations.

4. GLOBAL FOLD DETERMINATION

4.1. 4D ^1H_N–^1H_N NOESY

4.1.1. 4D NN-NOESY Sequence

Protein structures are determined by NMR largely on the basis of proton–proton NOEs. With the advent of ^{13}C and ^{15}N isotopic labeling has come the introduction of 4D heteronuclear NOESY experiments (Metzler *et al.*, 1996a; Vuister *et al.*, 1993; Zuiderweg *et al.*, 1991; Kay *et al.*, 1990b): These experiments greatly facilitate the assignment of NOEs in larger proteins by taking advantage of the increase in available chemical-shift dispersion (Clore and Gronenborn, 1991). The assignment of NOEs is further simplified by perdeuteration: Perdeuteration both limits the available types of proton–proton NOE interactions and provides a dramatic increase in spectral sensitivity. In a perdeuterated protein, only solvent exchangeable protons remain: All other protons are largely replaced with deuterons.

Perdeuteration decreases both the amide proton T_2 (by ~40%) and the amide proton T_{1s} relaxation rates; furthermore, perdeuteration dramatically reduces the occurrence of spin diffusion, thereby permitting the use of rather long NOE mixing times (Constantine *et al.*, 1997; Garrett *et al.*, 1997; Venters *et al.*, 1995b; Tsang *et al.*, 1990). If these benefits of perdeuteration are exploited optimally, the 4D $^{15}N/^{15}N$-separated NOESY experiment (Grzesiek *et al.*, 1995; Venters *et al.*, 1995b), depicted in Fig. 6, is expected on theoretical grounds (Venters *et al.*, 1996; Tsang *et al.*, 1990) to achieve a maximum seven-fold increase in signal-to-noise on a perdeuterated protein relative to its protonated equivalent. The following section

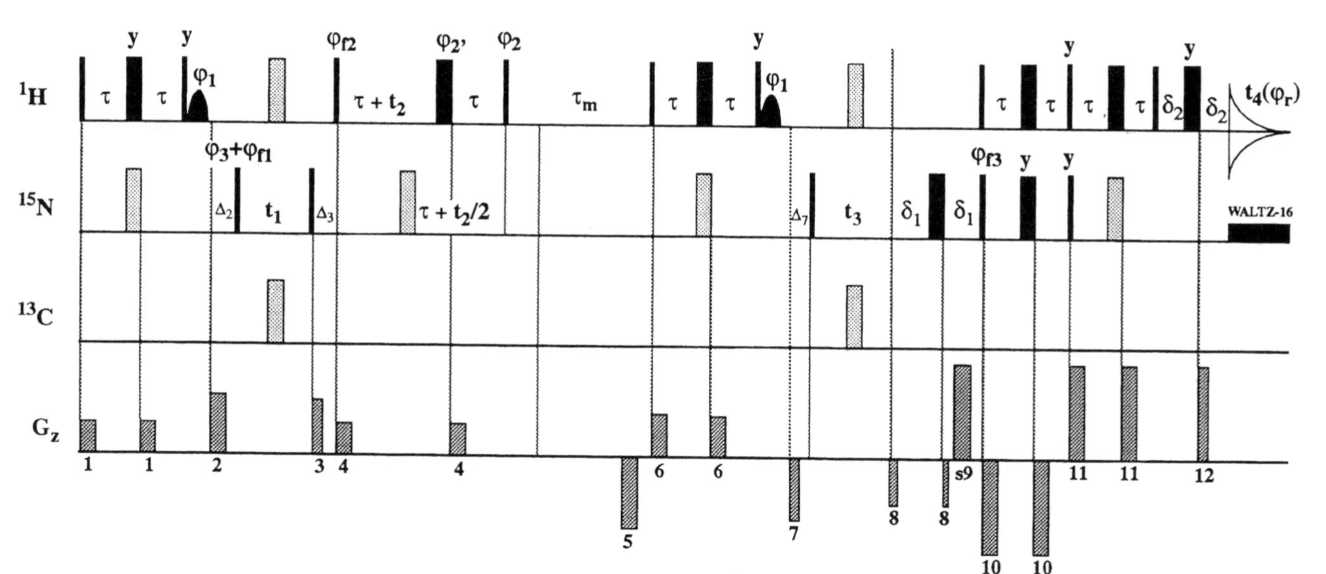

Figure 6. The 4D $^{15}N/^{15}N$-separated NOESY pulse sequence. Ninety-degree pulses are represented by wide lines; simple 180° pulses, by black rectangles; and $90_x240_y90_x$ composite inversion pulses, by diagonally striped rectangles. Unless otherwise indicated, all pulses have phase x. The general phase cycle is as follows: $\varphi_1 = 4°$; $\varphi_3 = x, -x$; and $\varphi_r = x, -x$. Active water-flipback through sine-modulation in t_2 is achieved with these additional phase settings: $\varphi_{f2} = y$; $\varphi_2 = -y$; and $\varphi_2' = x$. Conversely, passive water-flipback with RADAR-θ is achieved with these additional phase settings: $\varphi_{f2} = [x,y]$; $\varphi_2 = y + \theta$; and $\varphi_2' = y$. Amide proton–proton NOE data on $^2H/^{15}N$-labeled HCA II were collected both with sine-modulation in t_2 and with the following additional delay values (in ms): $\tau = 2.6$, $\Delta_2 = \Delta_3 = \Delta_7 = 1.0$, $\delta_1 = 2.5$, $\delta_2 = 0.9$, and $\tau_m = 200$.

details several key features of the 4D $^{15}N/^{15}N$-separated NOESY experiment as it is applied to a perdeuterated protein.

4.1.2. Experimental Optimization

Several groups have documented the importance of maximizing the steady-state level of water magnetization throughout the course of multidimensional, amide-proton-based correlation and NOESY experiments (Mueller *et al.*, 1995; Kay *et al.*, 1994; Li and Montelione, 1994; Grzesiek and Bax, 1993a,b). This detail is important to the 4D $^{15}N/^{15}N$-separated NOESY experiment, especially when it is applied to perdeuterated proteins. In a perdeuterated protein, the paucity of protons dramatically increases the importance of each potential amide proton–proton NOE.

The experimental approaches to maximize the steady-state level of water magnetization can be delineated into three broad groups: (1) active water-flipback (Kay *et al.*, 1994; Grzesiek and Bax, 1993b), (2) passive RADAR water-flipback (Mueller *et al.*, 1995; Stonehouse *et al.*, 1995) (RADAR = *Ra*diation *D*amping *A*ssisted *R*ecovery), and (3) T_1 water recovery (Li and Montelione, 1994). Although the latter approach is the most general, it is also the most inefficient. Active water-flipback, which to some extent uses selective proton pulses to restore water magnetization to the $\pm z$ axis, is most commonly found in out-and-back, amide-proton-based correlation experiments, e.g., the HNCACB. Passive water-flipback, which uses RADAR, is best suited to NOESY with $\tau_m \geq \tau_d$ (radiation-damping time constant): This approach has recently been incorporated into a 4D $^{15}N/^{15}N$-separated NOESY (Grzesiek *et al.*, 1995; Venters *et al.*, 1995b). If the water proton T_2 is much shorter than τ_d, the RADAR technique becomes increasingly less effective in preserving water magnetization throughout the course of the NOESY.

Active water-flipback, as currently implemented in triple-resonance correlation experiments (Kay *et al.*, 1994; Grzesiek and Bax, 1993b), is unsuitable for the 4D $^{15}N/^{15}N$-separated NOESY: $^1H_{donor}$ quadrature artifacts are produced by the shaped, water-selective pulse that must be placed at the start of the mixing period (so as to rotate transverse water magnetization back to the $+z$ axis) (B.T. Farmer II and R.A. Venters, unpublished observations). However, active water-flipback in the 4D $^{15}N/^{15}N$-separated NOESY can be achieved by an alternate approach: sine-modulation and nonquadrature detection in the $^1H_{donor}$ dimension. This alternate approach to active water-flipback leads both to no observable $^1H_{donor}$ quadrature artifacts and to less saturation of water magnetization compared with the RADAR approach.

Figure 6 presents the sine-modulated 4D $^{15}N/^{15}N$-separated gradient-enhanced, sensitivity-enhanced (GESE) NOESY pulse sequence with active water-flipback. The proton carrier must be on resonance with the water signal except during the following periods: the relaxation delay, any ^{15}N evolution time, τ_m, and the $^{1}H_N$ explicit acquisition time. As a matter of practice, however, we leave the proton carrier on resonance with the water signal at all times. In this pulse sequence, $C_{f2} - 2C_{2'} + C_2 = 0$ is a required condition so that the water magnetization is always returned to the $+z$ axis at the start of the mixing period. This phase condition leads to sine-modulation of the $^{1}H_{donor}$ t_2 interferogram. Concomitantly, this phase condition disallows any non-concerted phasecycling of proton pulses both bounding and within the $^{1}H_{donor}$ t_2 evolution time, thereby preventing the cancellation and/or TPPI shifting (Marion *et al.*, 1989a) of F_2 axial peaks.

Because almost all H_N protons resonate downfield of the water resonance, observing H_N–H_N NOEs is minimally impacted by either the $^{1}H_{donor}$ non-quadrature detection or the $C_{f2} - 2C_{2'} + C_2 = 0$ phasecycle constraint. F_2 axial peaks are most conveniently removed by time-domain, low-frequency digital filtering (Marion *et al.*, 1989b) of the t_2 interferogram. The absence of F_2 quadrature detection in the sine-modulated 4D NOESY experiment is a non-issue provided that $sw_{F2} = 2*[\max(\delta_{HN}) - \delta_{H2O} + 0.2]$ ppm. The 0.2 ppm allows for the finite bandwidth in t_2 of the time-domain, low-frequency digital filter. The sine-modulated 4D $^{15}N/^{15}N$-separated NOESY sequence (see Fig. 6) and the published 4D $^{15}N/^{15}N$-separated NOESY sequence (Grzesiek *et al.*, 1995; Venters *et al.*, 1995b) are identical in the following characteristics: both afford the same t_2 max per unit acquisition time; both lead to the same size of time-domain data per t_2 increment; both lead to the same of size of spectral matrix per unit of F_2 resolution. In addition, both sequences give rise to approximately the same magnitude, albeit a different distribution, of F_2 axial peaks.

How the two types of NOESY experiments differ lies mainly in the degree to which water magnetization is preserved at different points within the respective pulse sequences. Figure 7 illustrates this difference in retained water magnetization. At all mixing times, the 4D $^{15}N/^{15}N$-separated NOESY with active water-flipback (sine-modulation in t_2) leads to less saturation of the water magnetization than does the NOESY with either RADAR-45 or RADAR-0 passive water-flipback. For RADAR-0 data in Fig. 7, $C_{f2} = x$ and $C_{2'} = C_2 = y$: all water magnetization is essentially transverse at the start of τ_m. For RADAR-45 data, $C_{f2} = x$, $C_{2'} = y$ and $C_2 = y + 45^o$:$(2)^{-1/2}$ of the total water magnetization is aligned along $+z$ at the start of τ; a similar amount remains transverse at the same point. Therefore, radiation damping requires more time with RADAR-0 than with RADAR-45 to return all water magnetization back to the $+z$ axis. The available time may be severely constrained by a short water proton T_2 (~22 ms for the data in Fig. 7).

In the NOESY with RADAR passive water-flipback, the level of axial peak intensity could be reduced by applying a two-step, axial-peak-suppression subcycle to the phase of the $90^o(^{1}H)$ pulse immediately preceding the mixing period. However, this additional level of phase cycling has two practical problems. First, it extends the minimum number of transients per FID from 2 to 4; therefore, less

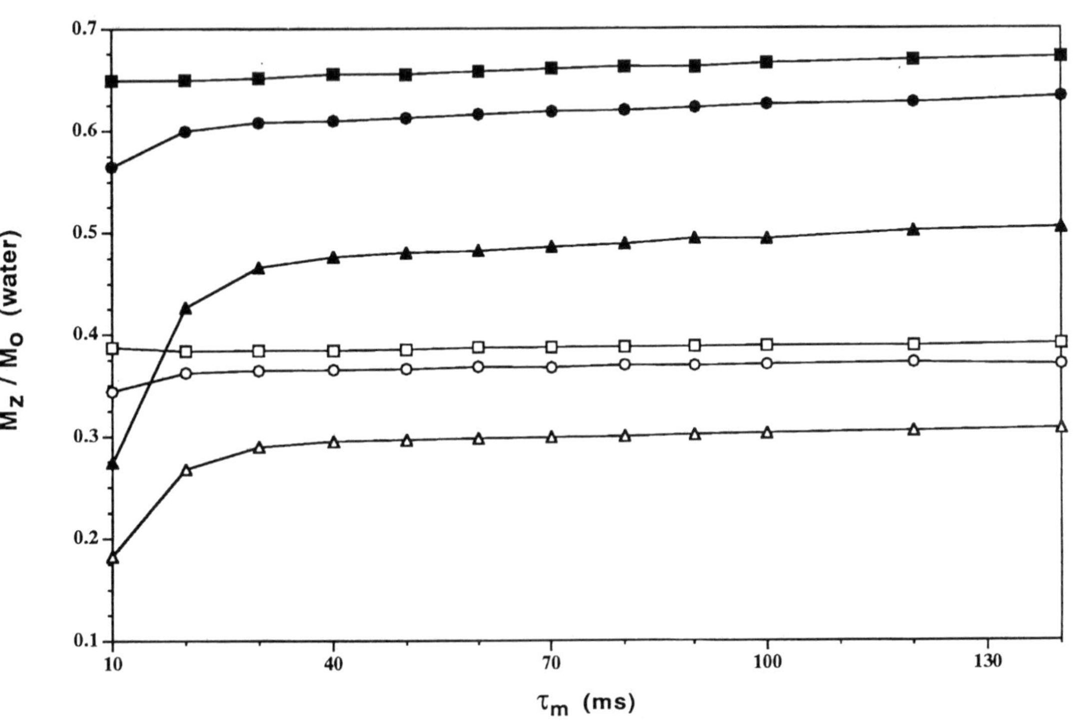

Figure 7. The fraction of remaining water magnetization $(M_z/M_o)_{water}$, as a function of the mixing time τ_m, is measured, using the basic pulse sequence in Fig. 6, either immediately after τ_m (filled-in symbols) or immediately prior to the t_4 acquisition time (open symbols) for the following pulse-sequence variants: (1) t_2 sine-modulated (filled-in and open squares), (2) RADAR-45 (filled-in and open circles), and (3) RADAR-0 (filled-in and open triangles). The data were measured at 600 MHz (Varian UnityPlus) on a 1 mM sample of $^{2}H/^{13}C/^{15}N$-labeled MurB at 30 °C in the presence of 0.5 M d_2-glycine (Constantine *et al.*, 1997; Farmer *et al.*, 1996). The water proton T_2 for this sample has been measured to be ~22 ms (B.T. Farmer II, unpublished data).

real spectral resolution is obtained per unit acquisition time. Second, for every even (or odd) t_2 time point, $(2)^{-1/2}$ of the total water magnetization is aligned along the $+z$ axis; however, for every odd (or even) t_2 time point, $(2)^{-1/2}$ of the total water magnetization is now aligned along the $-z$ axis! For these two cases, RADAR passive water-flipback leads to a different level of overall water suppression in the explicit acquisition dimension: a higher level when the partial water magnetization is aligned along the $+z$ axis, a lower level when the partial water magnetization is aligned along the $-z$ axis. Such differential water suppression has been observed to produce F_1 quadrature artifacts in DQF-COSY data (Mattiello *et al.*, 1996).

4.2. Structure Calculations

4.2.1. Generating NOE and Dihedral Restraint Files

The first step in any NMR-based structure calculation, regardless of its goal, is to provide the chosen program with the following details: primary molecular structure, NOE-derived distance restraints, and dihedral-angle restraints. Primary molecular structures includes both the amino acid sequence and any known disulfide bonds. For a perdeuterated protein, the NOE-derived distance restraints are generated from the assignment and quantitation of the 4D $^{15}N/^{15}N$-separated NOESY data. Similarly, for a perdeuterated protein, dihedral-angle restraints are generated from the secondary structure analysis using backbone $^{13}C_\alpha$ and $^{13}C_\beta$ chemical shifts or by explicitly measuring coupling constants (*see* chapter 7). The absence of H_α protons in a perdeuterated protein precludes measuring both ϕ based on $^3J_{HNH\alpha}$ (Billeter *et al.*, 1992) and ψ based on $\{H_\alpha\text{–}C_\alpha\}_{i-1}$ and $\{HN\text{–}N\}_i$ dipolar cross-correlation effects (Reif *et al.*, 1997).

Interproton NOE peak intensitites can be converted into distance restraints by several different approaches. In all approaches, the proton van der Waals radius defines the lower bound for any distance restraint to be ~1.8 Å. In the most basic approach, the upper bound for a distance restraint is determined from the maximum interproton distance for which any NOE peak intensity is expected to be observed, usually 5 Å in the case of fully protonated proteins. In a second approach, NOE peak intensities that correspond to known interproton distances, e.g., the $H_N(i)$–$H_N(i\pm1)$ distance in an α-helix (2.5 Å) or a β-sheet (4.4 Å), can be used to classify other NOEs, based on relative peak intensity, usually into three distance-restraint bins (Braun *et al.*, 1983): strong (2–3 Å), medium (2–4 Å) and weak (2–5 Å). Each bin has a different upper bound for the distance restraint. In a third approach, a cross-relaxation rate is calculated using an isolated spin-pair approximation (Macura and Ernst, 1980) for each interproton NOE interaction. This cross-relaxation rate is then converted, by reference to cross-relaxation rates measured for known interproton distances, into a target distance. That target distance is then scaled by some "fudge" factor to yield the distance-restraint upper bound. This third approach has been used to analyze the 4D $^{15}N/^{15}N$-separated NOESY data collected on $^{2}H/^{15}N$-labeled HCA II and a core packing mutant of *E. coli* thioredoxin, L78K-TRX (deLorimier *et al.*, 1996).

This analysis begins by measuring the peak volume of each assigned NOE correlation and symmetry-related partner. The volume is also measured for both, when possible, associated autocorrelation (i.e., diagonal) peaks. In some cases, only one autocorrelation was sufficiently resolved to yield a valid peak volume. Therefore, each interproton NOE interaction is characterized by at least three and sometimes four peak volumes. The cross-relaxation rate for each such interaction is then calculated first by dividing the sum of the correlation peak volumes by the sum of the autocorrelation peak volumes and then by dividing that quantity by the NOE mixing time (Macura *et al.*, 1986). Each cross-relaxation rate can then be converted into a distance-restraint upper bound using the following equation:

$$UB_i = S * (CR_r / CR_i)^{1/6} * d_r \tag{7}$$

where UB_i is the upper bound for the interproton interaction of interest, S is a scaling factor to account for inaccuracies in NOE peak volumes, CR_r is the cross-relaxation rate for reference interproton NOE interactions of known distance, CR_i is the cross-relaxation of the interproton NOE interaction of interest, and d_r represents the distance between the two reference protons.

As described earlier, proton and carbon chemical-shift data, obtained during the NMR assignment process, can be used to discern regions of well-defined secondary structure. The type of secondary structure, either α-helix or β-sheet, can then be used to define loose backbone dihedral-angle restraints: $\phi = -60 \pm 30°$ and $\psi = -60 \pm 30°$ for α-helices; $\phi = -129 \pm 30°$ and $\psi = 124 \pm 30°$ for β-sheets. A similar approach has previously been used in DNA structure calculations to predefine the helical handedness (Gronenborn and Clore, 1989).

4.2.2. Metric Matrix Distance Geometry Subembedding and Simulated Annealing

Global folds have been calculated for perdeuterated HCA II and perdeuterated L78K-thioredoxin (TRX) using both NOE-derived distance restraints obtained *solely* from the 4D $^{15}N/^{15}N$-separated NOESY experiment and chemical-shift-derived backbone dihedral-angle restraints. The HCA II NOESY data were collected over 10 days on a 2.8 mM perdeuterated, ^{15}N-labeled protein sample using a 200-ms mixing time; the L78K-TRX NOESY data were collected over 10 days on a 4.4 mM perdeuterated, ^{15}N-labeled protein sample using a 400-ms mixing time. Figure 8 presents the $^{1}H_N/^{15}N$ donor plane for residue V25 in L78K-TRX. Various NOE correlation peaks are annotated with both resonance assignments and crystal-structure-derived (Katti *et al.*, 1990) distances. V25 is located in a core β-sheet of L78K-TRX. The data presented in Fig. 8 clearly indicate that V25 makes long-range spatial contact with

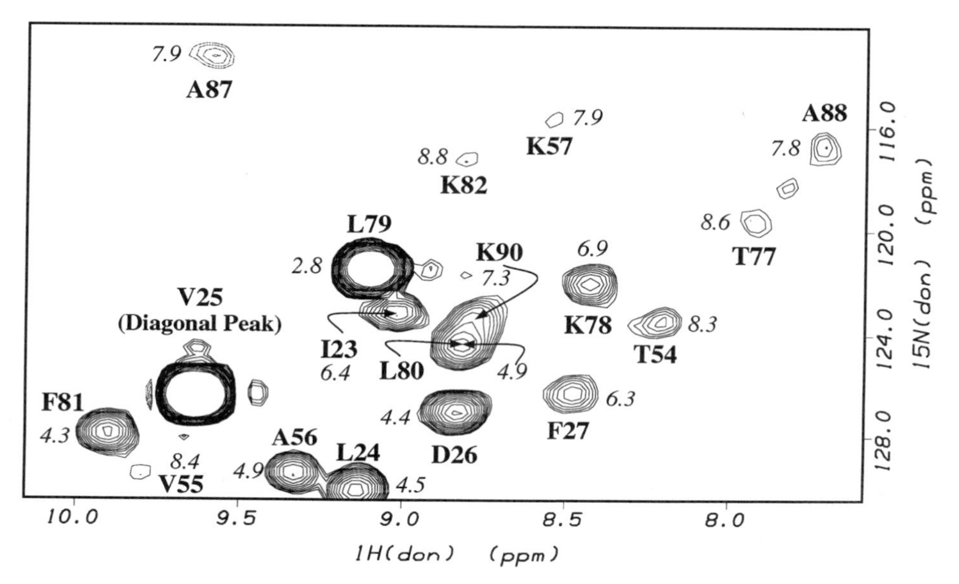

Figure 8. ^{1}H (donor)/^{15}N (donor) plane from the 4D ^{15}N/^{15}N-separated NOESY data set (w sine-modulation) collected on a 4 mM sample of the L78K mutant of *E. coli* thioredoxin.

three other regions of the protein: residues 54–57, 77–82, and 87–90. The same data also reveal that the 4D NOESY on L78K-TRX is measuring extremely large H_N–H_N distances, some of which are approaching 9 Å. NOE buildup curves on L78K-TRX indicate that NOE peak intensities are not appreciably affected by spin diffusion even at mixing times > 400 ms.

Based on the aforedescribed distance and dihedral-angle restraints, a family of embedded substructures (all H_N, N, CO, C_α, C_β, C_γ atoms) can be calculated in X-PLOR (Brunger, 1992) using a metric-matrix distance-geometry protocol (Clore *et al.*, 1993). Table 5 presents statistics on the H_N–H_N NOE restraints used to calculate structures of L78K-TRX and HCA II. A linear template coordinate set with ideal local geometry was used as the starting point for the substructure embeds. After the embedding process was completed, sidechains were built onto the substructures using a random-coil template. The resulting structures were then regularized using a simulated annealing protocol (Clore *et al.*, 1993). Finally, the NMR structures were compared against the crystal structure of the respective wild-type protein (Hakansson *et al.*, 1992; Katti *et al.*, 1990). Crystal structures appropriate for comparison with the NMR structures were obtained with X-PLOR by first building on protons and then energy minimizing.

Table 5
Total Number of NOEs Assigned from 4D ^1H_N–^1H_N NOESY Data Collected on a 4 mM Sample of Perdeuterated ^{15}N-labeled L78K-thioredoxin and a 2.8 mM Sample of Perdeuterated ^{15}N-Labeled HCA II

Protein	Backbone/backbone	Backbone/sidechain	Sidechain/sidechain
L78K-TRX	502	100	5
+ILVY[a]		58	134
HCA II	693	211	30
+ILVY		104	224

[a]+ILVY indicates additional NOEs that could theoretically be expected from data collected on samples containing selectively protonated Ile, Leu, Val, and Tyr residues at a 4-Å cutoff distance.

4.2.3. Examination of Structure Families and Comparison with X-Ray

H_N–H_N NOE restraints alone were used as described above to calculate 50 structures for L78K-TRX and 20 structures for HCA II. In each case, an average backbone RMSD (H_N, N, CO, C_α) was calculated for the ensemble by comparing the mean structure with each individual structure. For L78K-TRX, the average backbone RMSD was 2.0 Å; for HCA II, the average backbone RMSD was 2.9 Å. Subsequently, the mean structure was compared with the respective reference crystal structure. For L78K-TRX, the mean-to-reference backbone RMSD was 1.9 Å; for HCA II, the mean-to-reference backbone RMSD was 4.2 Å.

Visually comparing each mean structure with its respective reference crystal structure (data not shown) clearly demonstrates that H_N–H_N NOEs, obtained on a perdeuterated protein, are sufficient in and of themselves to determine *at least* the protein global fold, if not more. For example, the NMR-determined backbone of L78K-TRX overlays almost exactly with the reference crystal structure: All helices and β-sheets are both present and placed in their proper tertiary organization. For HCA II, the first 25 residues show very few NOEs to the remainder of the protein; therefore, these residues are not positioned correctly in the mean structure. However, the rest of the NMR-determined HCA II protein displays all of the correct secondary structural elements placed in their proper tertiary organization.

4.3. Selective Protonation

4.3.1. Rationale

Although perdeuteration permits the measurement of quite large H_N–H_N distances, perdeuteration incurs rather severe limitations in terms of both the number and type of interproton interaction that can be observed. Recently, selective protonation has been proposed by several groups (Constantine *et al.*, 1997; Metzler *et al.*, 1996b; Rosen *et al.*, 1996; Smith *et al.*, 1996; Oda *et al.*, 1992) as a method both to maintain the increased sensitivity and decreased extent of spin diffusion, clear benefits wrought by perdeuteration, and to increase both the number and type of NOE interaction available in the otherwise perdeuterated protein. In all cases, selective protonation has involved methyl and/or aromatic protons: Both methyl groups and aromatic rings tend to be located in the hydrophobic core of a protein. In addition, NOEs involving these sidechain groups should be quite useful in tying down elements of secondary structure. Table 5 presents the additional number and type of NOE restraint theoretically available from selective protonation of Ile, Leu, Val, and Tyr residues in both L78K-TRX and HCA II.

Methyl groups, albeit insufficiently disperse in terms of proton chemical shifts, retain favorable T_2 relaxation characteristics even in the presence of ^{13}C labeling (Kay *et al.*, 1992b); therefore, ^{13}C labeling has accompanied selective protonation of methyl groups in otherwise perdeuterated proteins so as to increase the available chemical-shift dispersion. In contrast to methyl groups, aromatic groups are not ^{13}C labeled for the following reasons: In aromatic groups of large proteins, ^{13}C isotopic labeling dramatically reduces both sensitivity and proton resolution (Vuister *et al.*, 1994); compared with methyl groups, ^{13}C chemical shifts in aromatic rings are less beneficial to establishing a unique, residue-level NOE assignment.

4.3.2. Theoretical Improvements in Structure Calculation

Structure calculations using both experimental and theoretical data (Gardner and Kay, 1997; Metzler *et al.*, 1996b; Rosen *et al.*, 1996; Smith *et al.*, 1996) indicate that a significant improvement in both the precision and accuracy of the resultant structures can be achieved if H_N–H_N NOE restraints are supplemented with NOE restraints involving the selectively protonated methyl (ILV) and aromatic (Y) groups. Although we certainly do not dispute that the latter, additional restraints improve the quality of the calculated protein global folds, our experimental data indicate that, for the most part, these improvements have been overstated. This overstatement arises mainly because the comparison structures, i.e., those calculated solely from H_N–H_N NOE restraints, are significantly poorer in quality than we have obtained, both for HCA II and for L78K-TRX, using *experimental data exclusively*. In several studies (Rosen *et al.*, 1996; Smith *et al.*, 1996), these comparison structures were generated using only backbone H_N–H_N NOE restraints with a maximum observable cutoff distance of 5 Å. Yet experimental data, obtained in our laboratories on both the 29-kDa protein HCA II and the 12-kDa L78K mutant of thioredoxin, clearly indicate that NOEs can routinely be observed among both backbone *and* sidechain H_N nuclei whose interproton distance is > 7 Å. Furthermore, the exclusion of H_N–H_N sidechain/sidechain and sidechain/mainchain interactions is especially damaging: Such interactions not only are readily available (Farmer and Venters, 1996) but also are *absolutely* essential for organizing elements of secondary structure into their correct tertiary topology (Venters *et al.*, 1995b).

5. CONCLUDING REMARKS

NMR has come a long way in its pursuit of structural information on proteins. Along the way, deuteration has progressed from its initial role of "spectral simplifier" to its current role of "sensitivity enhancer." In this regard, deuteration has clearly increased the size of protein for which sequential backbone resonance assignments can be obtained. These assignments, even for large proteins, can be used quite effectively in a drug-discovery environment (Farmer *et al.*, 1996; Shuker *et al.*, 1996). But what about tertiary structural information on large proteins? Can deuteration help in this effort?

We would answer this question with a resounding "yes". Experimental amide proton–proton NOE data clearly can determine the global fold of at least two compact, globular proteins (L78K-TRX and HCA II). These data involve both backbone and sidechain H_N protons; furthermore, these data discern proton–proton interactions out to > 7 Å. Once a global fold has been determined, the experimental path diverges. One may want greater structural detail on the whole protein or one may want greater structural detail only on a region of the protein, e.g., the active site of an enzyme. If the active site is enriched in particular amino acids, type-specific protonation can be used to probe semiselectively either the intramolecular structural details of this site or the intermolecular structural details in the presence of a bound ligand.

Regardless of one's further structural needs, some form of proton reintroduction is necessary in larger proteins to obtain more than just a backbone global fold. We favor ILV and FY type-specific protonation. But even such type-specifically protonated large proteins do not appear to yield sufficient additional NOEs for a high-resolution structure. So, where to now, St. Peter?

ACKNOWLEDGMENTS. B.T.F. would like to acknowledge Drs. Keith Constantine, William Metzler, and Luciano Mueller and R.A.V. would like to acknowledge Drs. Leonard Spicer, Carol Fierke, Donald G. Davis, and Robert deLorimier both for their valued collaborative effort on and for many stimulating discussions about NMR applied to perdeuterated, large proteins. The Duke University NMR Center was established with grants from the NIH, NSF, and the North Carolina Biotechnology Center, which are gratefully acknowledged.

REFERENCES

Arrowsmith, C. H., Pachter, R., Altman, R. B., Iyer, S. B., and Jardetzky, O., 1990, *Biochemistry* **29**:6332.
Barlow, P. N., Steinkasserer, A., Norman, D. G., Kieffer, B., Wiles, A. P., Sim, R. B., and Campbell, I. D., 1993, *J. Mol. Biol.* **232**:268.
Bax, A., and Grzesiek, S., 1993, *Acc. Chem. Res.* **26**:131.
Bax, A., and Ikura, M., 1991, *J. Biomol. NMR* **1**:99.
Bax, A., Clore, G. M., and Gronenborn, A. M., 1990, *J. Magn. Reson.* **88**:425.
Billeter, M., Neri, D., Otting, G., Qian, Y., and Wüthrich, K., 1992, *J. Biomol. NMR* **2**:257.
Braun, W., Wider, G., Lee, K.H., and Wüthrich, K., 1983, *J. Mol. Biol.* **169**:921.
Browne, D. T., Kenyon, G. L., Packer, E. L., Sternlicht, H., and Wilson, D.M., 1973, *J. Am. Chem. Soc.* **95**:1316.
Brunger, A. T., 1992, *X-PLOR Version 3.1: A system for X-ray crystallography and NMR*, Yale University Press, New Haven, CT.
Clore, G. M., and Gronenborn, A., 1991, *Science* **252**:1390.
Clore, G. M., Robien, M. A., and Gronenborn, A. M., 1993, *J. Mol. Biol.* **231**:82.
Clowes, R. T., Boucher, W., Hardman, C. H., Domaille, P. J., and Laue, E. D., 1993, *J. Biomol. NMR* **3**:349.
Clubb, R. T., Thanabal, V., and Wagner, G., 1992, *J. Magn. Reson.* **97**:213.
Constantine, K. L., Friedrichs, M. S., Goldfarb, V., Jeffrey, P. D., Sheriff, S., and Mueller, L., 1993, *Proteins Struct. Funct. Genet.* **15**:290.
Constantine, K. L., Mueller, L., Goldfarb, V., Wittekind, M., Metzler, W. J., Yanchunas, J., Jr., Robertson, J. G., Malley, M. F., Friedrichs, M. S., and Farmer, B. T., II, 1997, *J. Mol. Biol.* **267**:1223.
Crespi, H. L., Rosenberg, R. M., and Katz, J. J., 1968, *Science* **161**:795.
Davis, J. H., Bradley, E. K., Miljanich, G. P., Nadasdi, L., Ramachandran, J., and Basus, V. J., 1993, *Biochemistry* **32**:7396.
deLorimier, R. M., Hellinga, H., and Spicer, L. D., 1996, *Protein Sci.* **5**:2552.
Dötsch, V., Oswald, R. E., and Wagner, G., 1996, *J. Magn. Reson. Ser. B* **110**:107.
Farmer, B. T., II, and Venters, R. A., 1995, *J. Am. Chem. Soc.* **117**:4187.
Farmer B. T., II, and Venters, R. A., 1996, *J. Biomol. NMR* **7**:59.
Farmer, B. T., II, Lavoie, T. B., Mueller, L., and Metzler, W. J., 1995, *J. Magn. Reson. Ser. B* **107**:197.
Farmer, B. T., II, Constantine, K. L., Goldfarb, V., Friedrichs, M. S., Wittekind, M., Yanchunas, J., Jr., Robertson, J. G., and Mueller, L., 1996, *Nature Struct. Biol.* **3**:995.
Feeney, J., Birdsall, B., Akiboye, J., Tendler, S. J. B., Jimenez Barbero, J., Ostler, G., Arnold, J. R. P., Roberts, G. C. K., Kuhn, A., and Roth, K., 1989, *FEBS Lett.* **248**:57.
Fesik, S. W., and Zuiderweg, E. R. P., 1988, *J. Magn. Reson.* **78**:588.
Friedrichs, M. S., Mueller, L., and Wittekind, M., 1994, *J. Biomol. NMR* **4**:703.
Gardner, K. H., and Kay, L. E., 1997, *J. Am. Chem. Soc.* **119**:7599.
Garrett, D. S., Seok, Y.-J., Liao, D.-I., Peterkofsky, A., Gronenborn, A. M., and Clore, G. M., 1997, *Biochemistry* **36**:2517.
Gottschalk, G., 1986, *Bacterial Metabolism*, Springer-Verlag, Berlin, p. 99.
Gronenborn, A. M., and Clore, G. M., 1989, *Biochemistry* **28**:5978.
Grzesiek, S., and Bax, A., 1992, *J. Am. Chem. Soc.* **114**:6291.
Grzesiek, S., and Bax, A., 1993a, *J. Biomol. NMR* **3**:185.
Grzesiek, S., and Bax, A., 1993b, *J. Am. Chem. Soc.* **115**:12593.
Grzesiek, S., and Bax, A., 1994, Deuterium, relaxation, and protein NMR, presented at the *XVIth International Conference on Magnetic Resonance in Biological Systems*, Veldhoven, The Netherlands.
Grzesiek, S., Anglister, J., Ren, H., and Bax, A., 1993, *J. Am. Chem. Soc.* **115**:4369.
Grzesiek, S., Wingfield, P., Stahl, S., Kaufman, J. D., and Bax, A., 1995, *J. Am. Chem. Soc.* **117**:9594.

Grzesiek, S., Stahl, S. J., Wingfield, P. T., and Bax, A., 1996, *Biochemistry* **35:**10256.
Guentert, P., Mumenthaler, C., and Wüthrich, K., 1997, *J. Mol. Biol.* **273:**283.
Haberkorn, R. A., Stark, R. E., van Willigen, H., and Griffin, R. G., 1981, *J. Am. Chem. Soc.* **103:**2534.
Hakansson, K., Carlsson, M., Svensson, L.A., and Liljas, A., 1992, *J. Mol. Biol.* **227:**1192.
Hansen, P. E., 1988, *Prog. NMR Spectrosc.* **20:**207.
Ikura, M., Kay, L. E., and Bax, A., 1990, *Biochemistry* **29:**4659.
Jönsson, P. G., and Kvich, A., 1972, *Acta Crystallogr. Sect. B* **28:**1827.
Kallen, J., Spitzfaden, C., Zurini, M. G., Wider, G., Widmer, H., Wüthrich, K., and Walkinshaw, M. D., 1991, *Nature* **353:**276.
Katti, S. K., LeMaster, D. M., and Eklund, H., 1990, *J. Mol. Biol.* **212:**167.
Katz, J. J., and Crespi, H. L., 1966, *Science* **151:**1187.
Kay, L. E., Ikura, M., Tschudin, R., and Bax, A., 1990a, *J. Magn. Reson.* **89:**496.
Kay, L. E., Bax, A., Clore, G. M., and Gronenborn, A. M., 1990b, *Science* **249:**411.
Kay, L. E., Keifer, P., and Saarinen,T., 1992a, *J. Am. Chem. Soc.* **114:**10663.
Kay, L. E., Bull, T. E., Nicholson, L. K., Griesinger, C., Schwarbe, H., Bax, A., and Torchia, D. A., (1992b), *J. Magn. Reson.* **100:**538.
Kay, L. E., Xu, G. Y., and Yamazaki, T., 1994, *J. Magn. Reson. Ser. A* **109:**129.
Kushlan, D. M., and LeMaster, D. M., 1993, *J. Am. Chem. Soc.* **115:**11026.
Lehninger, A. L., Nelson, D. L., and Cox, M. M., 1993, *Principles of Biochemistry*, 2nd ed., Worth Publishers, New York.
LeMaster, D. M., 1994, *Prog. NMR Spectrosc.* **26:**371.
LeMaster, D. M., and Richards, F. M., 1988, *Biochemistry* **27:**142.
Li, Y.-C., and Montelione, G., 1994, *J. Magn. Reson. Ser. B* **105:**45.
Logan, T. M., Olejniczak, E. T., Xu, R. X., and Fesik, S. W., 1993, *J. Biomol. NMR* **3:**225.
Löhr, F., and Rüterjans, H., 1995, *J. Biomol. NMR* **6:**189.
Macura, S., and Ernst, R. R., 1980, *Mol. Phys.* **41:**95.
Macura, S., Farmer, B. T., II, and Brown, L. R., 1986, *J. Magn. Reson.* **70:**493.
Majerski, Z., Zuanic, M., and Metelko, B., 1985, *J. Am. Chem. Soc.* **107:**1721.
Marion, D., Ikura, M., Tschudin, R., and Bax, A., 1989a, *J. Magn. Reson.* **85:**393.
Marion, D., Kay, L. E., Sparks, S. W., Torchia, D. A., and Bax, A., 1989b, *J. Am. Chem. Soc.* **111:**1515.
Markley, J. L., Putter, I., and Jardetzky, O., 1968, *Science* **161:**1249.
Markus, M. A., Dayie, K. T., Matsudairat, P., and Wagner, G., 1994, *J. Magn. Reson. Ser. B* **105:**192.
Mattiello, D. L., Warren, W. S., Mueller, L., and Farmer, B. T., II, 1996, *J. Am. Chem. Soc.* **118:**3253.
Metzler, W. J., Constantine, K. L., Friedrichs, M. S., Bell, A. J., Ernst, E. G., Lavoie, T. B., and Mueller, L., 1993, *Biochemistry* **32:**13818.
Metzler, W. J., Leiting, B., Pryor, K., Mueller, L., and Farmer, B. T., II, 1996a, *Biochemistry* **35:**6201.
Metzler, W. J., Wittekind, M., Goldfarb, V., Mueller, L., and Farmer, B. T., II, 1996b, *J. Am. Chem. Soc.* **118:**6800.
Montelione, G. T., Lyons, B. A., Emerson, S. D., and Tashiro, M. J., 1992, *J. Am. Chem. Soc.* **114:**10974.
Muchmore, S. W., Sattler, M., Liang, H., Meadows, R. P., Harlan, J. E., Yoon, H. S., Nettesheim, D., Chang, B. S., Thompson, C. B., Wong, S.-L., Ng, S.-C., and Fesik, S. W., 1996, *Nature* **381:**335.
Mueller, L., Legault, P., and Pardi, A., 1995, *J. Am. Chem. Soc.* **117:**11043.
Muhandiram, D. R., and Kay, L. E., 1994, *J. Magn. Reson. Ser. B* **103:**203.
Naito, A., Ganapathy, S., and McDowell, C. A., 1981, *J. Chem. Phys.* **74:**5393.
Nietlispach, D., Clowes, R. T., Broadhurst, R. W., Ito, Y., Keeler, J., Kelly, M., Ashurst, J., Oschkinat, H., Domaille, P. J., and Laue, E. D., 1996, *J. Am. Chem. Soc.* **118:**407.
Oda, Y., Nakamura, H., Yamazaki, T., Nagayama, K., Yoshida, M., Kanaya, S., and Ikehara, M., 1992, *J. Biomol. NMR* **2:**137.
Olejniczak, E. T., Xu, R. T., Petros, A. M., and Fesik, S.W., 1992, *J. Magn. Reson.* **100:**444.
Pachter, R., Arrowsmith, C. H., and Jardetzky, O., 1992, *J. Biomol. NMR* **2:**183.

Palmer, A. G., III, Cavanagh, J., Wright, P. E., and Rance, M., 1991, *J. Magn. Reson.* **93:**151.
Patt, S. L., 1992, *J. Magn. Reson.* **96:**94.
Reif, B., Hennig, M., and Griesinger, C., 1997, *Science* **276:**1230.
Reisman, J., Jariel-Encontre, I., Hsu, V. L., Parello, J., Geiduscek, P., and Kearns, D. R., 1991, *J. Am. Chem. Soc.* **113:**2787.
Rosen, M. K., Gardner, K. H., Willis, R. C., Parris, W. E., Pawson, T., and Kay, L. E., 1996, *J. Mol. Biol.* **263:**627.
Rosenberg, A. H., Lade, B. N., Chui, D. S., Lin, S. W., Dunn, J. J., and Studier, F. W., 1987, *Gene* **56:**125.
Sambrook, S., Fritsch, E. F., and Maniatis, T., 1989, *Molecular Cloning: A Laboratory Manual*, Cold Spring Harbor Laboratory Press, Cold Spring Harbor, NY.
Searle, M. S., Hammond, S. J., Birdsall, B., Roberts, G. C. K., Feeney, J., King, R. W., and Griffiths, D. V., 1986, *FEBS Lett.* **194:**165.
Seip, S., Balbach, J., and Kessler, H., 1992, *J. Magn. Reson.* **100:**406.
Senn, H., and Klaus, W., 1993, *J. Mol. Biol.* **232:**907.
Shan, X., Gardner, K. H., Muhandiram, D. R., Rao, N. S., Arrowsmith, C. H., and Kay, L. E., 1996, *J. Am. Chem. Soc.* **118:**6570.
Shuker, S. B., Hajduk, P. J., Meadows, R. P., and Fesik, S. W., 1996, *Science* **274:**1531.
Smith, B. O., Ito, Y., Raine, A., Teichmann, S., Ben-Tovim, L., Nietlispach, D., Broadhurst, R. W., Terada, T., Kelly, M., Oschkinat, H., Shibata, T., Yokoyama, S., and Laue, E. D., 1996, *J. Biomol. NMR* **8:**360.
Spera, S., and Bax, A., 1991, *J. Am. Chem Soc.* **113:**5490.
States, D. J., Haberkorn, R. A., and Ruben, D. J., 1982, *J. Magn. Reson.* **48:**286.
Stonehouse, J., Clowes, R. T., Shaw, G. L., Keeler, J., and Laue, E. D., 1995, *J. Biomol. NMR* **5:**226.
Studier, F. W., and Moffatt, B. A., 1986, *J. Mol. Biol.* **189:**113.
Torchia, D. A., Sparks, S. W., and Bax, A., 1988, *J. Am. Chem. Soc.* **110:**2320.
Tsang, P., Wright, P. E., and Rance, M., 1990, *J. Am. Chem. Soc.* **112:**8183.
Venters, R. A., Calderone, T. L., Spicer, L. D., and Fierke, C. A., 1991, *Biochemistry* **30:**4491.
Venters, R. A., Huang, C.-C., Farmer, B. T., II, Trolard, R., Spicer, L. D., and Fierke, C. A., 1995a, *J. Biomol. NMR* **5:**339.
Venters, R. A., Metzler, W. J., Spicer, L. D., Mueller, L., and Farmer, B. T., II, 1995b, *J. Am. Chem. Soc.* **117:**9592.
Venters, R. A., Farmer, B. T., II, Fierke, C. A., and Spicer, L. D., 1996, *J. Mol. Biol.* **264:**1101.
Vis, H., Boelens, R., Mariani, M., Stroop, R., Vorgias, C. E., Wilson, K. S., and Kaptein, R., 1994, *Biochemistry* **33:**14858.
Vuister, G. W., Clore, G. M., Gronenborn, A. M., Powers, R., Garrett, D. S., Tschudin, R., and Bax, A., 1993, *J. Magn. Reson. Ser. B* **101:**210.
Vuister, G. W., Kim, S.-J., Wu, C., and Bax, A., 1994, *J. Am. Chem. Soc.* **116:**9206.
Wagner, G., 1993, *J. Biomol. NMR* **3:**375.
Wagner, G., and Wüthrich, K., 1982, *J. Mol. Biol.* **155:**347.
Westler, W. M., Stockman, B. J., and Markley, J. L., 1988, *J. Am. Chem. Soc.* **110:**6256.
Wishart, D. S., and Sykes, B. D., 1994, *J. Biomol. NMR* **4:**171.
Wittekind, M., and Mueller, L., 1993, *J. Magn. Reson. Ser. B* **101:**201.
Wittekind, M., Mapelli, C., Farmer, B. T., II, Suen, K. L., Goldfarb, V., Tsao, J., Lavoie, T., Barbacid, M., Meyers, C. A., and Mueller, L., 1994, *Biochemistry* **33:**13531.
Yamazaki, T., Forman-Kay, J. D., and Kay, L. E., 1993, *J. Am. Chem. Soc.* **115:**11054.
Yamazaki, T., Lee, W., Arrowsmith, C. H., Muhandiram, D. R., and Kay, L. E., 1994a, *J. Am. Chem. Soc.* **116:**11655.
Yamazaki, T., Lee, W., Revington, M., Mattiello, D. L., Dahlquist, F. W., Arrowsmith, C. H., and Kay, L. E., 1994b, *J. Am. Chem. Soc.* **116:**6464.
Yamazaki, T., Muhandiram, R., and Kay, L. E., 1994c, *J. Am. Chem. Soc.* **116:**8266.

Yamazaki, T., Pascal, S. M., Singer, A. U., Forman-Kay, J. D., and Kay, L. E., 1995, *J. Am. Chem. Soc.* **117:**3556.

Zhou, M. M., Ravichandran, K. S., Olejniczak, E. T., Petros, A. M., Meadows, R. P., Sattler, M., Harlan, J. E., Wades, W. S., Burakoff, S. J., and Fesik, S. W., 1995, *Nature* **378:**584.

Zuiderweg, E. R. P., Petros, A. M., Fesik, S. W., and Olejniczak, E. T., 1991, *J. Am. Chem. Soc.* **113:**370.

4

Recent Developments in Multidimensional NMR Methods for Structural Studies of Membrane Proteins

Francesca M. Marassi, Jennifer J. Gesell, and Stanley J. Opella

1. INTRODUCTION TO THE STRUCTURAL BIOLOGY OF MEMBRANE PROTEINS

Membrane proteins perform a variety of important biological functions, including intercellular and transmembrane signaling, ion and small molecule transport, electron transport, and catalysis. It is thus not surprising that over 30% of protein sequences encoded in the genomes of organisms, varying in complexity from bacteria to humans, correspond to hydrophobic membrane proteins (Fraser *et al.*, 1995; Goffeau, 1995). However, compared with the wealth of primary sequence and genetic information available for membrane proteins, very little is known about their three-dimensional structures. The majority of structural studies performed to date on membrane proteins have relied on FTIR, Raman, and CD spectroscopy, as well as conventional biochemical and mutagenesis experiments, which have limited

---

Francesca M. Marassi, Jennifer J. Gesell, and Stanley J. Opella • Department of Chemistry, University of Pennsylvania, Philadelphia, Pennsylvania 19104.

Biological Magnetic Resonance, Volume 16: Modern Techniques in Protein NMR, edited by Krishna and Berliner. Kluwer Academic / Plenum Publishers, 1999.

or no spatial resolution. As a result, of the approximately 9000 proteins with structures deposited in the Brookhaven National Laboratories Protein Data Bank, less than 1% are membrane proteins, and most of those are in a few families of similar proteins. However, the situation is likely to improve in the near future through the implementation of NMR spectroscopy in structural studies of membrane proteins (Opella, 1997).

The three-dimensional structures of membrane proteins that have been determined by X-ray crystallography, electron microscopy, or NMR spectroscopy typically have limited resolution compared with that obtained for globular proteins using these same methods (Almeida and Opella, 1997; Gesell *et al.*, 1997; MacKenzie *et al.*, 1997; Pebay-Peyroula *et al.*, 1997; Song *et al.*, 1996; Tsukihara *et al.*, 1996; Iwata *et al.*, 1995; McDermott *et al.*, 1995; Schirmer *et al.*, 1995; Unwin, 1995, 1993; Kuhlbrandt *et al.*, 1994; Picot *et al.*, 1994; van de Ven *et al.*, 1993; Cowan *et al.*, 1992; Henry and Sykes, 1992; Jap *et al.*, 1991; Weiss *et al.*, 1991; Deisenhofer *et al.*, 1985). There are two main types of folds in membrane proteins: the α-helix bundle and the hollow β-barrel. Bacteriorhodopsin is the prototype of the helical class (Pebay-Peyroula *et al.*, 1997; Henderson and Unwin, 1975), which includes the light-harvesting centers, cytochrome *c* oxidase, and the photosynthetic reaction centers. These proteins are characterized by multiple transmembrane α-helices, interconnecting loops with varying degrees of structure, and large cytoplasmic and extracellular globular domains. The transmembrane helices are relatively long with 20–27 amino acids, and, for the most part, the residues buried in the hydrophobic membrane are uncharged. Aromatic residues, especially tryptophan, appear to cluster at the interfacial membrane regions. OmpF porin is typical of the β-barrel class that includes the remaining porins and α-hemolysin, with 14 to 18 β-strands in their barrels. Each strand typically has 7–9 amino acids in the membrane-buried region, with alternating hydrophobic and polar residues. To date, definitely one and possibly two membrane proteins do not conform to the criteria of either of these families. Prostaglandin H_2 is unique in that it does not contain any transmembrane regions. Instead, four amphipathic helices lie in the plane of the bilayer and serve to anchor the protein to the membrane surface (Garavito *et al.*, 1994). In addition, it has been suggested that the acetylcholine receptor is composed of a pentameric helical bundle surrounded by a β-barrel (Unwin, 1993). A major goal for structural biology is to expand the data base of atomic resolution structures of membrane proteins, as a first step toward understanding the molecular basis of their functions.

Membrane proteins continue to pose a difficult challenge for the conventional methods of structural biology, X-ray crystallography and NMR spectroscopy, for a number of reasons associated with their hydrophobic character. Because membrane proteins are often toxic to the bacterial cells commonly used for protein overexpression (Grisshammer and Tate, 1995), it is essential to utilize fusion protein expression systems to obtain the large amounts of isotopically labeled samples required for high-resolution structural studies. Once purified, membrane proteins tend to aggregate, often irreversibly and nonspecifically, because a significant portion of their exposed area is hydrophobic. Detergents, chaotropic salts, and organic solvents can be used to avoid aggregation, but their presence complicates conventional chromatographic techniques, making isolation and purification more difficult than for globular proteins. Finally, because these proteins are native to, functional in, and stabilized by the anisotropic environment of the biological membrane, the topology of membrane association is an essential component of their structure. Mimicking the hydrophobic membrane environment complicates the preparation of both X-ray-quality crystals and rapidly reorienting samples for solution NMR studies.

The goal of this review is to present recent progress in NMR spectroscopy for the determination of membrane protein structure. The recent development of mutant cell strains (Miroux and Walker, 1996), new expression systems (Prive and Kaback, 1996), and the availability of new lipids for membrane protein reconstitution have opened the field of structural biology to a multitude of membrane proteins, previously unavailable for examination. These advances, together with the parallel development of new spectroscopic methods, which take advantage of uniform isotopic labeling schemes for both solution NMR and solid-state NMR studies, provide a powerful arsenal with which to tackle the problem of membrane protein structure determination.

High-resolution NMR spectroscopy of liquids and solids provides two independent but complementary methods for structure determination of membrane proteins. To obtain high-resolution spectra, solution NMR methods require samples that reorient rapidly enough to average the operative nuclear spin interactions to their isotropic values and give narrow resonances. To obtain such samples, micelle-forming lipids are used to solubilize the proteins. The bound lipid molecules add to the effective mass of the polypeptides, and this size increase results in slower overall reorientation rates and, consequently, rapid relaxation rates, which are deleterious to solution NMR experiments. Unlike solution NMR spectroscopy, solid-state NMR methods do not have any size limitations imposed by molecular reorientation rates and nuclear spin relaxation. Instead the proteins are immobilized in lipid bilayers, and high-resolution spectra are obtained by using radiofrequency irradiations in combination with magic angle sample spinning or uniaxial sample orientation of the protein with respect to the magnetic field.

2. BIOLOGICAL EXPRESSION AND CHEMICAL SYNTHESIS OF PROTEINS AND PEPTIDES

NMR structural studies typically require samples containing 0.5–2.0 mg of isotopically labeled proteins. Smaller peptides can be prepared by automated solid-phase peptide synthesis, although this becomes impractical for larger proteins where efficient expression systems are essential, and precludes the preparation of uniformly labeled samples. Because membrane proteins tend to target and congest the membranes of the bacterial cells in which they are expressed, they usually act as toxic, antibacterial agents. A variety of expression systems, all involving the use of fusion proteins, have been designed to address this problem. The fusion partner serves to keep the hydrophobic polypeptide away from the bacterial membranes, either by keeping it in solution or by sequestering it in inclusion bodies. The fusion partner also helps with the isolation and purification of the target protein, an important feature because all NMR experiments require high levels of protein purity. Using fusion protein expression systems, we have successfully expressed and purified tens of milligrams of membrane proteins with between 25 and 122 residues, and all indications are that these same fusion systems can be extended to substantially larger proteins.

The expression of membrane proteins in bacteria offers a number of advantages. It is easier to manipulate expressed rather than synthetic hydrophobic polypeptides, perhaps because procedures that prevent aggregation can be implemented more readily. However, the primary motivation for expressing membrane proteins is the variety of isotopic labeling schemes that can be incorporated in the experimental strategy. Bacterial expression allows both selective and uniform labeling, whereas automated synthesis only permits specific labeling at a few sites in the protein. For selective labeling by amino acid type, the bacteria harboring the protein gene are grown on defined media, where only the amino acid of interest is labeled and all others are not. Uniform labeling, where all of the nuclei of one or several types (^{15}N, ^{13}C, ^{2}H) are incorporated in the protein, is accomplished by growing the bacteria on defined medium containing ^{15}N-labeled ammonium sulfate, ^{13}C-labeled glucose, D_2O, or any combination of these. The availability of uniformly labeled samples shifts the burden from sample preparation to spectroscopy where complete spectral resolution is the starting point for structure determination. Uniform ^{15}N labeling was first implemented for solid-state NMR spectroscopy (Cross *et al.*, 1982) and later applied to solution NMR (Bogusky *et al.*, 1985) where the structure determination of expressed proteins has been facilitated by the resolution available in multidimensional spectra (Marion *et al.*, 1989; Fesik and Zeiderweg, 1988).

Heteronuclei, such as ^{13}C and ^{15}N, are important sources of chemical-shift dispersion in the backbone sites of helices, and open the door for systematic assignment schemes that rely on uniformly ^{13}C and ^{15}N labeled proteins for both solution (Kay *et al.*, 1990; Wuthrich, 1986) and solid-state NMR (Gu *et al.*, 1998). In micelle samples, where each polypeptide has the equivalent of about 20 kDa added to it by the associated lipid molecules, the overall reorientation rate in solution becomes a limiting factor. Even in cases where the ^{1}H/^{15}N two-dimensional correlation spectra are well resolved, the relaxation properties of polypeptides with more than about 50 residues are unfavorable for many experiments. In these cases the ability to uniformly label with ^{2}H, in addition to ^{15}N or ^{13}C, is essential and significantly extends the range of proteins that can be investigated (Veglia *et al.*, 1998; Shon and Opella, 1989).

3. SOLUTION NMR SPECTROSCOPY

3.1. Sample Preparation

Solution NMR methods rely on rapid molecular reorientation for line narrowing. Lipid micelles provide a very effective model membrane environment for the study of membrane proteins (Brown, 1979), and small micelles are isotropic and afford relatively rapid reorientation of the protein without the deleterious effects of organic solvents, which can denature proteins or promote helix formation distorting the secondary structure (Nelson and Kallenbach, 1989; Liebes *et al.*, 1975; Coni *et al.*, 1970; Doty *et al.*, 1954). For the proteins examined so far, similar structural features have been found in micelle and bilayer samples. The first step in solution NMR structural studies of membrane proteins is the preparation of homogeneous, well-behaved micelle samples. Careful handling of the protein throughout the purification or synthesis is essential, and subtle changes in the sample preparation protocol can have a significant impact on the quality of the final sample. The primary goal in the preparation of micelle samples is to reduce the effective rotational correlation time of the protein as much as possible, so that resonances will have the narrowest achievable linewidths. We have carefully optimized the conditions needed for several types of micelles including those formed from sodium dodecyl sulfate (SDS), dodecylphosphocholine (DPC), and dihexanoylphosphocholine (DHPC) (Gesell *et al.*, 1997; Veglia *et al.*, 1998; Opella *et al.*, 1994). These three lipids offer flexibility in the choice of headgroup and length of the acyl chains. We have found that lipid concentrations much higher than the critical micelle concentration are essential for preparing samples that yield NMR spectra with narrow linewidths. It appears that peptides associated with the membrane surface do not have as strict requirements for sample preparation as do those with hydrophobic membrane-spanning helices. Generally, salt concentration, pH, and temperature also affect the quality of the NMR spectra obtained from peptides and proteins in micelles.

Even with highly optimized micelle samples, solution NMR experiments of unlabeled peptides with as few as 25 residues are problematic because of the broad linewidths and limited chemical-shift dispersion. In contrast, uniformly ^{15}N labeled peptides generally yield well-resolved two-dimensional heteronuclear correlation spectra. This is the starting point for the three-dimensional double and triple-resonance experiments used for measuring short-range internuclear distances and making resonance assignments. These data also provide qualitative indications of secondary structure, and enable full three-dimensional structures to bedetermined.Relaxation measurements enable protein dynamics to be described.

3.2. Multidimensional Experiments for Spectral Resolution and Resonance Assignment

The two-dimensional $^{1}H-^{15}N$ correlation spectrum is the base for most subsequent solution NMR experiments. These spectra reflect sample quality, and the presence of more than one peak for each amide site in the polypeptide is indicative of inhomogeneity and incomplete sample optimization (McDonnell and Opella, 1993). Figure 1 shows the $^{1}H-^{15}N$ correlation spectra of five different uniformly ^{15}N labeled membrane proteins ranging in size from 23 to 122 amino acids. The samples were optimized with respect to salt, pH, protein and lipid concentration, and lipid composition (Veglia *et al.*, 1998). In each spectrum, there is a single correlation resonance for each amide site, as well as for each indole nitrogen of the tryptophan sidechains.

The spectrum in Fig. 1B is from a sample of acetylcholine receptor (AChR) M2 peptide in DPC micelles. The sequences of the second transmembrane segments, M2, of ligand-gated ionotropic receptors are highly conserved. In the case of the AChR, mutagenesis and electrophysiological experiments indicate that M2 segments line the pore of the ion channel formed by the associated receptor subunits (Montal *et al.*, 1990; Kersh *et al.*, 1989; Leonardo *et al.*, 1988). The sequence of the AChR M2 peptide is N-EKMST AISVL LAQAV FLLLT SQR-C. Determining its structure in membrane environments is essential for understanding the molecular basis of ion channel activity in other M2 peptides. The sample was obtained by overexpressing the peptide fused to the C-terminus of glutathione *S*-transferase protein. The fusion protein was isolated by conventional chromatography techniques and the M2 peptide was then enzymatically cleaved from the fusion partner and purified using a combination of gel filtration and high-pressure liquid chromatography.

After resolving the resonances in the polypeptide backbone, the next step is to assign them to specific residues. Because even small peptides in micelles suffer from short relaxation times, the usual combination of ^{15}N-resolved ^{1}H TOCSY and ^{1}H NOESY spectra is inadequate for even the most basic steps of spin system identification and assignment, and triple-resonance assignment methods are required. Figure 2A shows HNCA and HNCOCA $^{13}C/^{1}H$ spectral strips for the ^{15}N resonances of the indicated amino acids of the uniformly ^{13}C and ^{15}N labeled AChR M2 peptide in DPC micelles.

This pair of experiments was used to sequentially assign the amide, as well as the backbone $^{13}C_{\alpha}$ resonances, in the peptide. A combination of ^{13}C- and ^{15}N-edited TOCSY spectra was then used to assign the sidechain ^{1}H resonances. The secondary

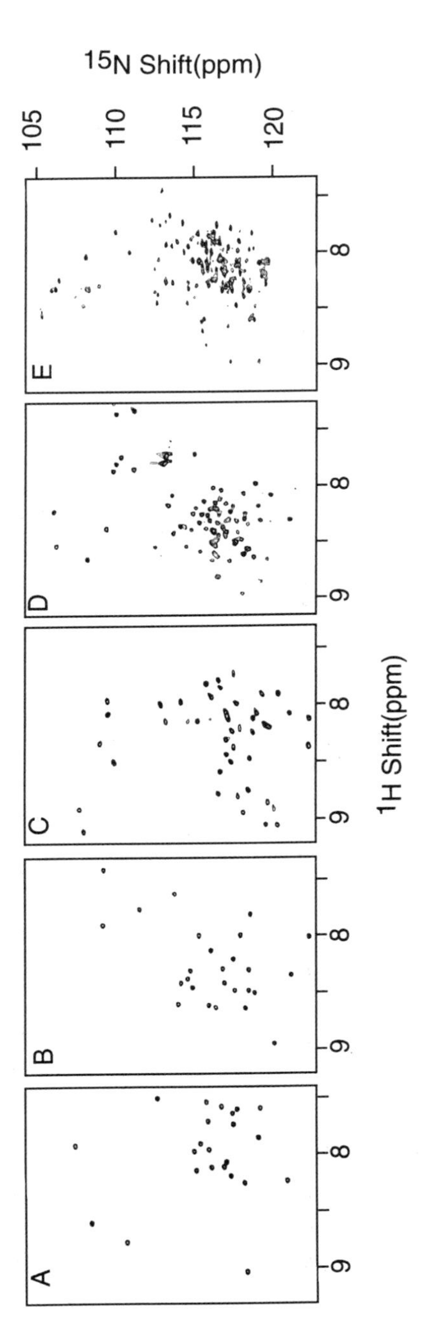

Figure 1. Two-dimensional $^1H/^{15}N$ heteronuclear correlation NMR spectra from uniformly ^{15}N labeled proteins in micelles. (A) Magainin 2 (23 residues). (B) Acetylcholine M2 (25 residues). (C) fd coat protein (50 residues). (D) HIV-1 Vpu (80 residues). (E) MerT protein (122 residues) (Veglia *et al.*, 1998).

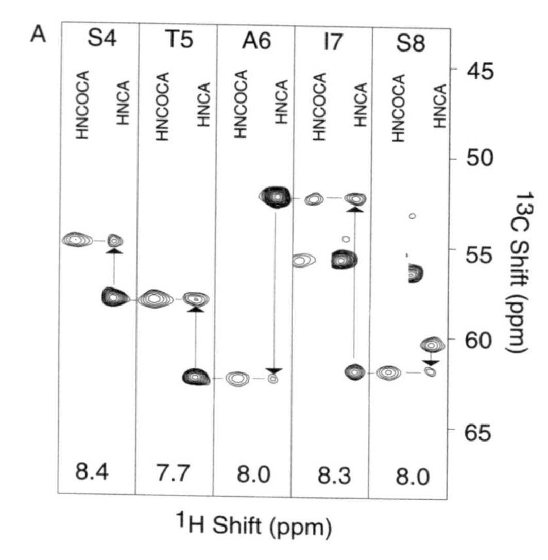

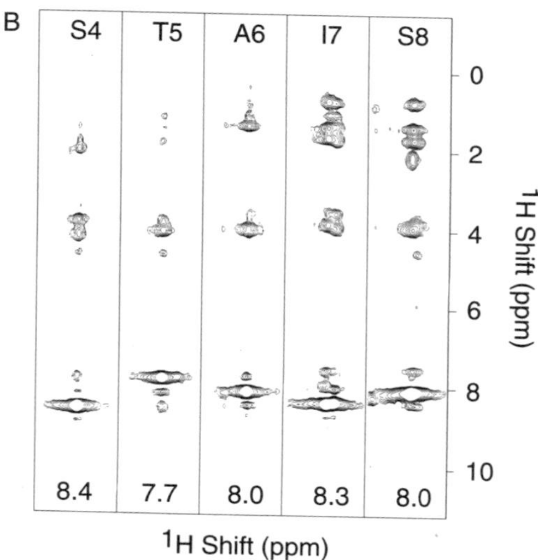

Figure 2. Two-dimensional spectral strips taken from three-dimensional spectra of uniformly $^{15}N/^{13}C$ labeled acetylcholine M2 peptide in DPC micelles. (A) Heteronuclear $^{15}N/^{13}C$ strips taken from HNCOCA and HNCA spectra. (B) Homonuclear $^{1}H/^{1}H$ nuclear Overhauser effect (NOE) strips.

structure of the protein can be inferred from the $^{13}C_{\alpha}$ and C_{α} ^{1}H resonance frequencies (Wishart and Sykes, 1994) as well as the NOEs from short-range interactions (Wuthrich, 1986). Figure 3 presents a summary of the short-range spectral parameters that describe the secondary structure of the AChR M2 peptide. The peptide is helical with the exception of a few residues at both termini.

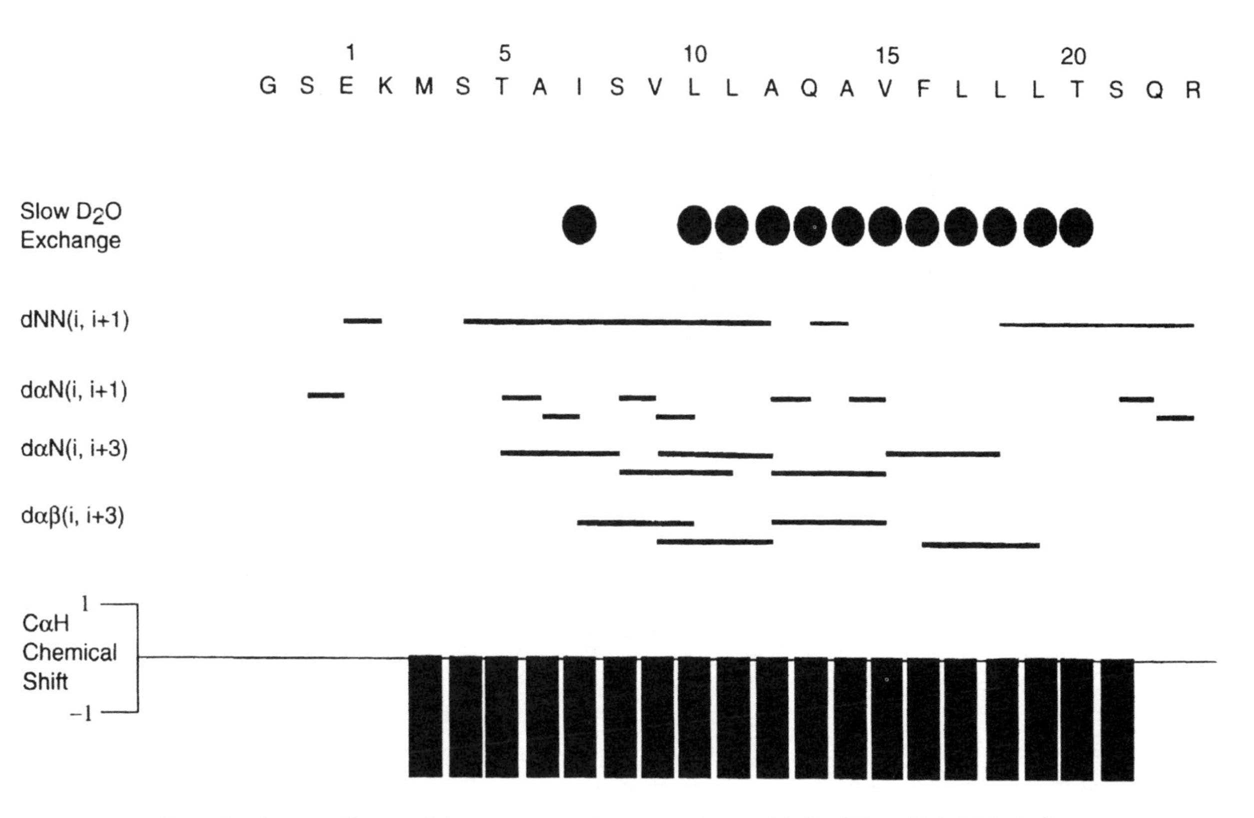

Figure 3. Summary diagram of short-range spectral parameters for acetylcholine M2 peptide in DPC micelles.

3.3. Structure Determination from Distance and Torsion Angle Constraints

Homonuclear ^{1}H NOEs are the primary source of restraints for structure determination by solution NMR spectroscopy. Figure 2B shows $^{1}H/^{1}H$ spectral strips taken from a three-dimensional ^{15}N-edited, NOESY spectrum. Each strip corresponds to the ^{15}N resonance frequency of the indicated amino acid in the AChR M2 sequence. The cross-peaks are between pairs of ^{1}H nuclei that are separated by less than 5 Å. To determine the three-dimensional structure of the peptide in DPC micelles, the cross-peaks are grouped into three classes of strong, medium, and weak intensity, corresponding to interhydrogen distances of 1.9–2.5, 1.9–3.5, and 3.0–5.0 Å, respectively. In addition to NOEs, torsion-angle restraints derived from HNHA experiments, which measure J couplings, were used in the calculations. Dihedral restraints for the torsion angle ϕ were specified to obtain a target value of $-60 \pm 25°$. Finally, the amide resonances detected in a two-dimensional $^{1}H-^{15}N$ correlation spectrum 1 h after dissolving the sample in D_2O at pH 4, were assigned hydrogen bond constraints by fixing the distance between the amide nitrogen of residue i and the carbonyl oxygen of residue $i-4$ to be 2.8 ± 0.5 Å, and fixing the distance between the amide hydrogen of residue i and the carbonyl oxygen of residue $i-4$ to be 1.8 ± 0.5 Å. Using the program XPLOR (Brunger, 1992), a family of structures was calculated based on these restraints. Starting from random-coil conformations, a protocol of distance geometry, simulated annealing, and energy minimization was used to calculate a set of 30 structures, of which the 10 with the

Figure 4. Superposition of the 10 lowest energy structures of the backbone heavy atoms of acetylcholine M2 peptide in DPC micelles. The structures were selected from a set of 30 calculated.

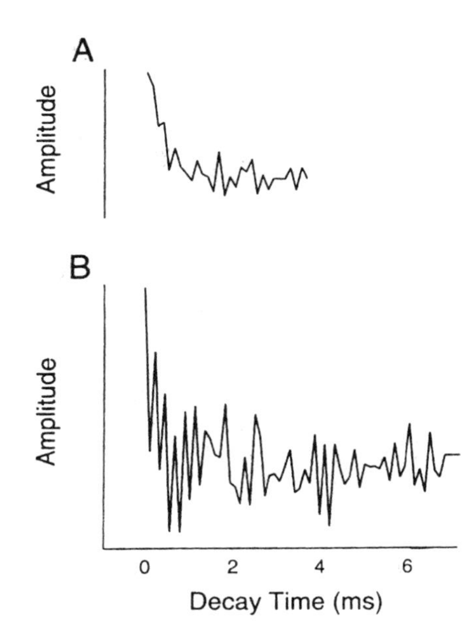

Figure 5. Free induction decays from a selected C_α resonance in the three-dimensional HNCA spectrum of a uniformly $^{15}N/^{13}C$ labeled sample of the 122-residue, integral membrane protein merT in SDS micelles (Veglia *et al.*, 1997). (A) Uniformly $^{15}N/^{13}C$ labeled sample. (B) Uniformly $^{15}N/^{13}C$ and ^{2}H labeled sample, with ^{2}H decoupting.

lowest energies are superimposed in Fig. 4; there are no distance restraint violations greater than 0.5 Å, or torsion-angle restraint violations greater than 5°.

In principle, the structure of the 25-residue AChR M2 peptide could be determined from triple-resonance, three-dimensional solution NMR experiments, on a single uniformly $^{15}N/^{13}C$ labeled sample. However, even a modest increase in size makes this approach intractable, and polypeptides with 50 or more residues require uniform deuteration. merT is a 122-residue integral membrane protein from the bacterial mercury detoxification system (Summers, 1986). Figure 5 compares two free induction decays, from a selected C_α resonance in the three-dimensional HNCA spectrum of a uniformly $^{15}N/^{13}C$ labeled sample of merT in SDS micelles, with (Fig. 5B) and without (Fig. 5A) uniform deuteration (Veglia *et al.*, 1998). The results are clear in indicating that, as for the case of globular proteins, deuteration increases the size limit of proteins that can be studied by solution NMR methods.

4. SOLID-STATE NMR SPECTROSCOPY

4.1. Solid-State NMR of Oriented Samples

As shown above, in favorable cases solution NMR methods can be successfully applied to membrane proteins in micelles, however it is highly desirable to determine their structures within the context of lipid bilayers where solution NMR methods fail completely. Solid-state NMR spectroscopy is well suited for peptides and proteins immobilized in phospholipid bilayers, and two complementary approaches are under development. One utilizes unoriented samples and relies on magic angle sample spinning (MAS) methods to obtain high-resolution spectra, and to make distance and torsion angle measurements via homo- or heteronuclear dipolar interactions (Feng *et al.*, 1996; Smith, 1996; Levitt *et al.*, 1990: Gullion and Schaefer, 1989). This approach utilizes samples labeled with stable isotopes only in one or a few selected sites, which not only makes the determination of complete structures laborious, but limits applications to peptides prepared by solid-phase synthesis, or small proteins with favorable distributions of amino acids, or selected regions of large proteins. The second solid-state NMR approach, which we are developing for determining membrane protein structure, takes advantage of the favorable spectroscopic properties exhibited by uniaxially oriented samples (Ramamoorthy *et al.*, 1996b; Cross and Opella, 1994; Opella *et al.*, 1987). When the direction of molecular orientation is parallel to that of the applied magnetic field, the resulting spectra of spin $S=\frac{1}{2}$ nuclei are characterized by single line resonances in all of the frequency dimensions (Opella and Waugh, 1977; Pausak *et al.*, 1973). Because the resonance frequencies depend on the orientations of the molecular sites relative to the axis of the magnetic field, they provide the basis for a method of structure determination (Opella *et al.*, 1987).

Solid-state NMR of oriented samples is independent of multidimensional solution NMR spectroscopy, which relies instead on the properties of isotropic resonances observed from rapidly reorienting molecules (Wuthrich, 1986). It also differs from the MAS experiments where the orientation dependence is lost by spinning the sample at the magic angle, and the spectra are characterized by isotropic average resonance frequencies. The spin interactions present at ^{15}N-labeled backbone sites in proteins include the ^{1}H chemical shift of the amide hydrogen, the ^{15}N chemical shift of the amide nitrogen, and the ^{1}H–^{15}N heteronuclear dipolar coupling of these two directly bonded nuclei. Generally, the strong ^{1}H–^{1}H dipolar couplings are decoupled, and the weak ^{15}N–^{15}N dipolar couplings are ignored in these experiments, following the principles of dilute spin NMR spectroscopy (Pines *et al.*, 1973) of uniformly ^{15}N labeled proteins (Cross *et al.*, 1982). In separate experiments these homonuclear couplings can provide distance measurements and an approach to making sequential resonance assignments (Ramamoorthy *et al.*, 1996a, 1995b; Cross *et al.*, 1983).

We have recently developed a family of two-, three-, and four-dimensional solid-state NMR experiments (Ramamoorthy *et al.*, 1996a, 1995a,b; Wu *et al.*, 1994) for high-resolution spectroscopy and structure determination of proteins (Jelinek *et al.*, 1995; Ramamoorthy *et al.*, 1995c). These enable fully resolved three-dimensional solid-state NMR correlation spectra to be obtained from uniformly ^{15}N labeled membrane proteins in phospholipid bilayers (Marassi *et al.*,

1997). In these spectra each resonance associated with an amide site is characterized by three frequencies, the ^{1}H chemical shift, ^{15}N chemical shift, and ^{1}H–^{15}N dipolar coupling, that are responsible for resolution among resonances and provide sufficient angular restrictions for the determination of the orientation of each peptide plane. Complete three-dimensional structures are then assembled from the orientations of contiguous peptide planes (Opella *et al.*, 1987).

Both glass-supported oriented bilayers (Marassi *et al.*, 1997; Ketchem *et al.*, 1993; Bechinger *et al.*, 1991) and magnetically oriented bicelles (Howard and Opella, 1996; Prosser *et al.*, 1996; Sanders *et al.*, 1994) containing membrane proteins are ideal for this approach, as both accomplish the principal requirements of immobilizing and orienting the protein. The conventional approach is to prepare planar lipid bilayers on glass slide supports, which are then oriented in the probe so that the bilayer normal is parallel to the field of the magnet. By using a stack of thin glass plates and wrapping the RF coil directly around the flat or square sample (Bechinger and Opella, 1991), it is possible to perform multidimensional solid-state NMR experiments on samples containing less than 1 mg of uniformly ^{15}N labeled protein. Because the proteins are effectively immobile on the relevant NMR time scales of the chemical shift and the dipolar coupling interactions (10 kHz), they behave spectroscopically like solids, and the NMR spectra are generally not complicated by the effects of motional averaging. The structures of a variety of membrane peptides and proteins have been investigated using this approach (Kovacs and Cross, 1997; Marassi *et al.*, 1997; Bechinger *et al.*, 1996; North *et al.*, 1995; Ulrich *et al.*, 1994; Ketchem *et al.*, 1993). In a variation of this approach, bilayer disks (bicelles) composed of mixtures of long- and short-chain phospholipids are magnetically oriented with their normal parallel to the field, by the addition of small amounts of lanthanide ions. These samples offer the potential of having solutions containing immobilized oriented peptides and proteins, and preliminary results are extremely promising.

4.2. Multidimensional Experiments for Spectral Resolution and Resonance Assignment Strategies

All of the features in the one-dimensional solid-state NMR spectra in Fig. 6 result from the ^{15}N chemical-shift interaction. The spectrum in Fig. 6D was obtained from an unoriented sample of uniformly ^{15}N labeled fd coat protein in phospholipid bilayers. The major coat protein of fd bacteriophage has 50 residues with the primary sequence N-AEGDD PAKAA FDSLQ ASATE YIGYA WAMVV VIVGA TIGIK LFKKF TSKAS-C. Newly synthesized copies of the protein are stored in the membrane of infected *E. coli* bacteria before incorporation into virus particles during extrusion through the membrane. The structure of the membrane-bound form of this protein has been determined in micelles by solution NMR spectroscopy (Almeida and Opella, 1997; McDonnell *et al.*, 1993; van de Ven *et*

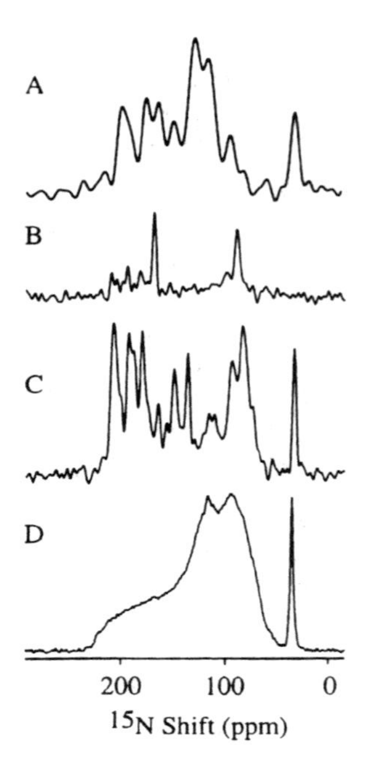

Figure 6. One-dimensional ^{15}N chemical shift solid-state NMR spectra of ^{15}N-labeled fd coat protein (8, 35). (A) Uniformly ^{15}N labeled protein in oriented bicelles (Howard and Opella, 1996). (B) Selectively ^{15}N Leu labeled protein in oriented bilayers (Marassi, *et al.*, 1997). (C) Uniformly ^{15}N labeled protein in oriented bilayers (Marassi, *et al.*, 1997). (D) Uniformly ^{15}N labeled protein in unoriented bilayer vesicles (Marassi, *et al.*, 1997).

al., 1993; Henry and Sykes, 1992). It has an amphipathic helix that lies in the plane of the membrane (residues 7 to 96), and a longer hydrophobic membrane-spanning helix (residues 27 to 44). A loop, showing some evidence of mobility, connects the two helices. and residues near the N- and C-termini are mobile. The narrow peak near 30 ppm results from the amino groups of the lysine sidechains and the N-terminus.

Most of the backbone sites are structured and immobile on the time scale of the ^{15}N chemical-shift interaction (10 kHz) and contribute to the characteristic amide powder pattern between about 220 and 60 ppm. Several backbone sites, including those near the N- and C-termini and some in the loop region between the two helices, are mobile and unstructured, resulting in the narrower resonance band centered at the isotropic resonance frequency near 120 ppm. No resolution among the resonances is feasible in the spectrum of an unoriented sample with multiple labeled sites without additional sample manipulations. Magic angle sample spinning narrows the resonances significantly and, in favorable cases, it is possible to resolve individual resonances in proteins, albeit at the expense of the orientational information. In contrast, sample orientation gives well-resolved spectra that retain the orientational information used for structure determination. In both magic angle sample spinning and oriented sample approaches, it is possible to utilize dipole couplings to make distance measurements.

The solid-state NMR spectrum of an oriented sample is strikingly different from that of an unoriented sample. The spectrum in Fig. 6C, obtained from an oriented sample of uniformly ^{15}N labeled fd coat protein in bilayers, displays significant resolution with identifiable peaks at frequencies throughout the range of the ^{15}N amide chemical shift anisotropy powder pattern. There are two resonances in the spectrum of selectively ^{15}N Leu labeled fd coat protein in oriented phospholipid bilayers (Fig. 6B). We have observed ^{15}N resonance linewidths narrower than 3 ppm from these samples, consistent with mosaic spreads of sample orientation of less than about 2°. The large 80 ppm frequency difference between the resonances from Leu-41, near the downfield end of the spectrum, and Leu-14, near the upfield end, demonstrates the effect of protein structure on these spectra. Leu-41, in the transmembrane helix, has its N–H bond approximately parallel to the field and to the σ_{33} component of the chemical shift tensor, while Leu-14, in the amphipathic in-plane helix, has its N–H bond perpendicular (McDonnell *et al.*, 1993).

The spectrum in Fig. 6A, obtained from a magnetically oriented bicelle sample of uniformly ^{15}N labeled fd coat protein, displays somewhat lower resolution compared with that of an oriented bilayer sample (Howard and Opella, 1996). Wobbling of the magnetically aligned bicelles about their axis of average orientation leads to motional averaging and, hence, scaling of the anisotropic interactions. In this example, the scaling is characterized by an order parameter of 0.70 ± 0.05, as determined by comparison with a rigid-lattice ^{15}N amide chemical shift powder pattern (Wu *et al.*, 1995).

Multidimensional NMR experiments, which are common practice for structure determination of proteins in solution, are only beginning to be used in solid-state NMR spectroscopy. In the solid-state NMR spectra of oriented proteins, multiple dimensions provide the means for resolving resonances in the spectra of uniformly labeled proteins, and enable the simultaneous measurement of multiple, orientational constraints, which are used for structure determination. The two-dimensional PISEMA (polarization inversion with spin exchange at the magic angle) (Wu *et al.*, 1994), and three-dimensional solid-state NMR correlation (Ramamoorthy *et al.*, 1995a) experiments are based on the principle of separated local field spectroscopy (Waugh, 1976), and utilize flip-flop, phase- and frequency-switched Lee–Goldburg homonuclear decoupling (Bielecki *et al.*, 1990; Mehring and Waugh, 1972; Lee and Goldburg, 1965) to provide line narrowing in the ^{1}H–^{15}N dipolar coupling and ^{1}H chemical shift dimensions. The sensitivity enhancement available from cross-polarization (Pines *et al.*, 1973) is so great that multidimensional solid-state and solution NMR spectra can be obtained on similar amounts of uniformly ^{15}N labeled protein in comparable time periods. Typically, high-quality two-dimensional spectra are obtained in overnight runs, whereas the three-dimensional data shown in Fig. 8 were obtained in 3 days, the same length of time as the corresponding solution NOESY-HMQC spectrum from a sample of the same uniformly ^{15}N labeled protein in micelles (Almeida and Opella, 1997).

The two-dimensional PISEMA spectra of three uniformly ^{15}N labeled proteins in oriented lipid bilayers, shown in Fig. 7, have many resolved resonances, each characterized by a single ^{1}H–^{15}N dipolar coupling frequency and an ^{15}N chemical shift frequency. There are some resonances with zero dipolar coupling frequencies. Those backbone sites that are mobile on the time scales of the ^{1}H–^{15}N dipolar and ^{15}N chemical shift interactions have resonances with zero dipolar and isotropic chemical shift frequencies. Because of the alignment of the ^{15}N chemical shift and the ^{1}H–^{15}N dipolar coupling tensors in the molecule, those orientations of amide groups in helices perpendicular to the field have very small or zero dipolar coupling frequencies associated with chemical shift frequencies near σ_{11} and σ_{22} (Ramamoorthy *et al.*, 1995c; Wu *et al.*, 1995; Bechinger *et al.*, 1993). Also, any magnetization transferred from the ^{1}H spin reservoir to the ^{15}N spin reservoir through cross-polarization before polarization inversion, which does not participate in spin exchange at the magic angle, appears at zero dipolar frequency.

The spectrum in Fig. 7A is from an oriented sample of uniformly ^{15}N labeled magainin peptide in bilayers. Magainin antibiotic peptides were originally identified in frog skin where they act against wound infections (Zasloff, 1987). Magainin-2, N-GIGKF LHSAK KFGKA FVGEI MNS-C, a typical member of this family, displays a broad range of protective activity, and is similar to other highly charged antibiotic peptides, found in various organisms, such as cecropin, bombolitin, mastoparan, and melittin. Magainin has been shown to be an amphipathic α-helix in membrane environments (Gesell *et al.*, 1997; Marion *et al.*, 1988), and our solid-state NMR experiments (Ramamoorthy *et al.*, 1995c; Bechinger *et al.*, 1991)

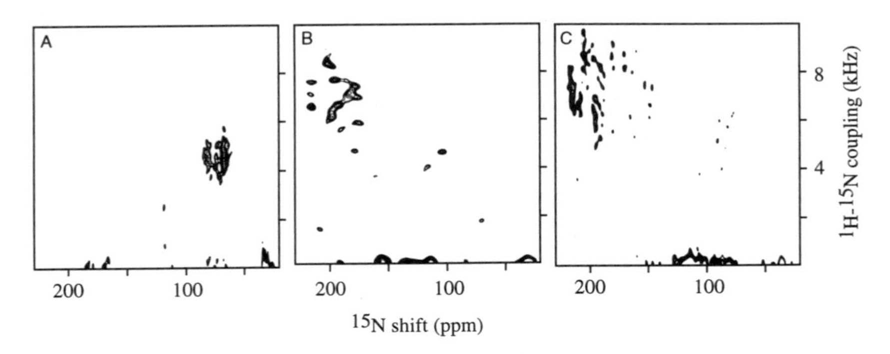

Figure 7. Two-dimensional solid-state ^{1}H–^{15}N dipolar coupling/^{15}N chemical shift correlation PISEMA (polarization inversion with spin exchange at the magic angle) spectra of oriented samples of uniformly ^{15}N labeled proteins in phospholipid bilayers. For each spectrum 64 t_1 values were incremented by 40.8 ms. (A) Magainin 2. (B) Acetylcholine M2. (C) fd coat protein.

show that the helix axis lies in the plane of the phospholipid bilayers suggesting that its mechanism of action may be different from other amphipathic peptides generally recognized as channel forming. In the PISEMA spectrum of magainin, all of the resonances have ^{15}N chemical shifts near the upfield end of the tensor, and ^{1}H–^{15}N dipolar couplings near $\nu_{\perp}$, as expected for an α-helix parallel to the bilayer surface.

Despite favorable linewidths of about 3 ppm and 250 Hz for the ^{15}N chemical shift and ^{1}H–^{15}N dipolar coupling, respectively, the resolution is limited by the narrow chemical shift and dipolar coupling frequency dispersions exhibited by uniformly labeled helices lying in the plane of the bilayer. Comparison with the PISEMA spectrum from a polycrystalline model peptide sample (Wu *et al.*, 1995) shows that the chemical shift and dipolar coupling dispersions available for transmembrane helices are nearly twice those available for in-plane helices such as magainin. The two-dimensional PISEMA spectrum of the AChR M2 peptide (Fig. 7B) is nearly fully resolved, in contrast to that of magainin, although both peptides are 25-residue amphipathic helices. The solution NMR structure of the AChR M2 peptide in DPC micelles was described above. In the PISEMA spectrum, all of the resonances have ^{15}N shifts near the downfield end of the spectrum, and ^{1}H–^{15}N dipolar couplings near $\nu_{\parallel}$, reflecting the transmembrane orientation of this peptide. The two-dimensional PISEMA spectrum of the 50-residue, uniformly ^{15}N labeled fd coat protein in oriented bilayers is shown in Fig. 7C. This spectrum exhibits many resolved resonances from both transmembrane and in-plane helices, spanning the entire frequency range available to the ^{15}N chemical shift and the ^{1}H–^{15}N dipolar coupling frequencies.

The PISEMA spectrum of magainin illustrates the need for additional frequency dimensions. The ^{1}H chemical shift provides advantages of both resolution and a third orientational constraint for structure calculation. The three-dimensional correlation ^{15}N shift/^{1}H–^{15}N dipolar coupling/^{1}H chemical shift spectrum of magainin in oriented bilayers has been completely resolved, and contains sufficient information for backbone structure determination. Figure 8 displays ^{1}H–^{15}N dipolar coupling/^{15}N chemical shift planes corresponding to the transmembrane and in-plane helical regions of fd coat protein.

The spectra in Figs. 8A and D are expansions of the two-dimensional PISEMA spectrum of uniformly ^{15}N labeled fd coat protein in bilayers shown in Fig. 7C. The two-dimensional planes in Figs. 8B, C, E, and F were extracted from the three-dimensional ^{1}H chemical shift/^{1}H–^{15}N dipolar coupling/^{15}N chemical shift correlation spectrum of uniformly ^{15}N labeled fd coat protein. These planes contain only those resonances with the selected ^{1}H chemical shift frequencies. In particular, the planes in Figs. 8C and F were selected at ^{1}H chemical shifts of 11.0 and 11.6 ppm so that they include the resonances of Leu-41 and Leu-14, respectively, as marked. Examination of two-dimensional planes throughout the ^{1}H chemical-shift range demonstrates that all of the resonances of this

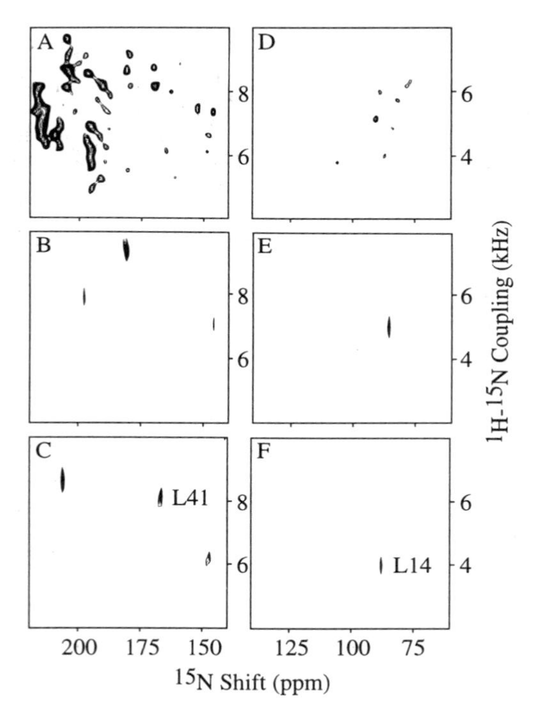

Figure 8. Two-dimensional ^{1}H–^{15}N heteronuclear dipolar coupling/^{15}N chemical shift spectral planes from spectra of an oriented sample of uniformly ^{15}N labeled fd coat protein in phospholipid bilayers. (A) and (D) are expansions of the two-dimensional PISEMA spectrum shown in Fig. 7C; (A) is the transmembrane helix region and (D) is the in-plane helix region. (B) Plane extracted from a three-dimensional correlation spectrum at ^{1}H chemical shift 7.4 ppm. (C) Plane extracted from a three-dimensional correlation spectrum at ^{1}H chemical shift 11.0 ppm; the resonance assigned to Leu-41 in the hydrophobic transmembrane helix is marked. (E) Plane extracted from a three-dimensional correlation spectrum at ^{1}H chemical shift 12.5 ppm. (F) Plane extracted from a three-dimensional correlation spectrum at ^{1}H chemical shift 11.6 ppm; the resonance assigned to Leu-14 in the amphipathic in-plane helix is marked. The three-dimensional correlation spectrum was obtained with 16 t_1 and 16 t_2 experiments with respective dwell times of 32.7 and 40.8 ms. The two leucine resonances in the three-dimensional spectrum were assigned by comparison with the two-dimensional PISEMA spectrum of an oriented sample of selectively ^{15}N Leu labeled fd coat protein in phospholipid bilayers.

uniformly ^{15}N labeled protein are fully resolved. This enables the measurement of the ^{1}H chemical shift, ^{1}H–^{15}N dipolar coupling, and ^{15}N chemical shift frequencies for all amide sites in the protein.

The very high spectral resolution available in three-dimensional correlation spectra of uniformly ^{15}N labeled proteins comes from several sources. The linewidths observed in the one-dimensional spectral slices taken from two-dimensional spectra are 1.2 ppm for the ^{1}H chemical shift, 250 Hz for the ^{1}H–^{15}N dipolar coupling, and 3 ppm for the ^{15}N chemical shift dimensions. These compare favorably with the values (0.8 ppm, 180 Hz, and 2–8 ppm) observed in spectra of single-crystal samples of ^{15}N-labeled dipeptides obtained in solenoidal coil probes. We anticipate that even better resolution will be observed with further spectroscopic development, and especially the use of very high field spectrometers. A useful index of spectral resolution is the ratio of the total spectral range available to the resonance linewidth. The observed ratios are about 10, 30, and 50 for the ^{1}H chemical shift, ^{1}H–^{15}N dipolar coupling, and ^{15}N chemical shift dimensions, respectively. These ratios are comparable to those observed along the ^{1}H and ^{15}N frequency axes of multidimensional solution NMR spectra of the same fd coat protein in micelles (Almeida and Opella, 1997); although the linewidths are considerably narrower in solution NMR spectra, the range of isotropic chemical shift frequencies is also correspondingly smaller.

Once the amide resonances are resolved and their frequencies measured, the final necessary step is to assign them to individual residues in the protein. Spin-exchange experiments, using either ^{1}H or ^{15}N nuclei, provide a general assignment strategy for solid-state NMR spectra of uniformly ^{15}N labeled proteins. Abundant spin exchange occurring among nearby ^{1}H nuclei in model peptides (Ramamoorthy *et al.*, 1996a), and dilute spin exchange among ^{15}N sites in both model peptides and proteins (Ramamoorthy *et al.*, 1995b; Cross *et al.*, 1983) have been demonstrated. Alternative assignment strategies that utilize uniformly ^{13}C and ^{15}N labeled proteins (Gu *et al.*, 1998), analogous to those used in solution NMR spectroscopy (Kay *et al.*, 1990), are also under development.

4.3. Structure Determination from Angular Constraints

In principle, a single three-dimensional correlation spectrum of an oriented sample of a uniformly ^{15}N labeled protein provides sufficient information for complete structure determination. The three frequencies, ^{1}H and ^{15}N chemical shift and ^{1}H/^{15}N dipolar coupling, measured for each resonance depend on the magnitudes and orientations of the principal elements of the spin-interaction tensors in the molecule, and on the orientation of the molecular site with respect to the direction of the applied magnetic field. Because the orientation of the bilayer membrane is fixed by the method for sample preparation, each frequency reflects the orientation of a specific site in the protein with respect to the bilayer membrane. Structures are calculated using the angular constraints extracted from the measured spectral frequencies (Opella and Stewart, 1989). The ^{13}C chemical shift and the ^{2}H quadrupolar coupling frequencies, measured from selectively or specifically labeled samples in separate experiments, also provide valuable structural information, and have been used together with the ^{15}N chemical shift and the ^{1}H/^{15}N dipolar coupling to determine the structure of the gramicidin channel at high resolution (Ketchem *et al.*, 1993).

The backbone structure of a protein can be described equivalently by vectors representing bonds between nonhydrogen atoms, as by the peptide planes formed by the individual rigid peptide bonds and their directly bonded atoms. The orientations of individual peptide planes of the AChR M2 peptide, including residues Val-9 to Leu-18, were derived from the PISEMA spectrum in Fig. 7B, and are presented in Fig. 9A. The resonances for each of these sites could be assigned by comparison with the PISEMA spectra from individual, specifically labeled synthetic, and selectively labeled recombinant M2 peptides. The orientations are described in terms of the polar angles α and β, defined in Fig. 9B (Tycko *et al.*, 1986). All sites have similar values for α and β near 0° and 90°, respectively. This is as expected for a membrane-spanning α-helix having its N–H bonds aligned nearly parallel to the magnetic field direction.

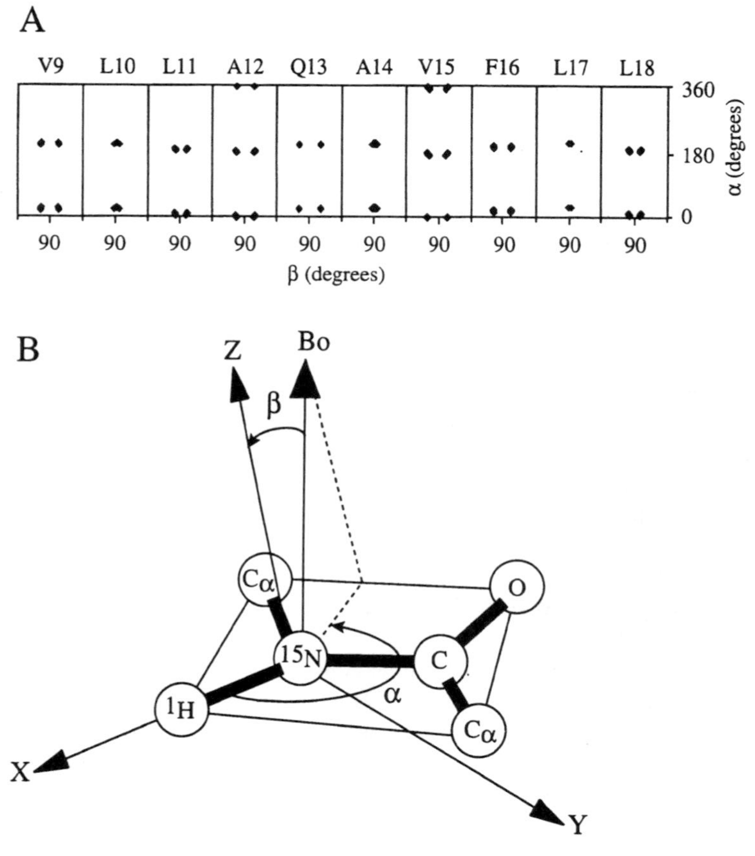

Figure 9. (A) α–β correlation plots summarizing the peptide plane orientations obtained for residues 9 to 18 of the acetylcholine M2 peptide, from the solid-state NMR data. (B) The polar angles α and β used to describe peptide plane orientation.

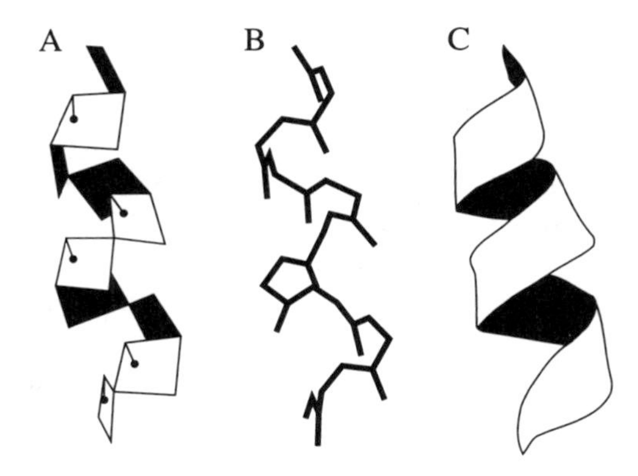

Figure 10. Solid-state NMR three-dimensional backbone structure of the acetylcholine M2 peptide in oriented bilayers (residues 9 to 18). The transmembrane helix structure was calculated using the ^{15}N chemical shift and ^{1}H–^{15}N dipolar coupling data from the PISEMA spectrum shown in Fig. 7B. Three equivalent representations are shown. (A) Peptide plane representation with the N–H bonds highlighted. (B) Vector backbone representation. (C) Ribbon backbone representation.

Once the orientations of the individual peptide planes are determined, they can be assembled through their common C_α site, with the only constraint of a fixed tetrahedral value around 110° for the $NC_\alpha C'$ angle. This procedure applied to residues 9 to 18 of the AChR M2 peptide leads to the α-helical, transmembrane structure shown in Fig. 10A. The equivalent representations of the structure as vectors through the backbone atoms (Fig. 10B) and as a ribbon (Fig. 10C) are also shown.

5. SUMMARY AND FUTURE PROSPECTS

Understanding the mechanism of action of membrane proteins, within the natural environment of the biological membrane, requires the same kind of high-resolution three-dimensional structures, currently achieved for their globular counterparts. Because the structure and function of membrane proteins are intimately related to their hydrophobic membrane environment, serious investigations of these species must, at some stage, focus on the entire supramolecular assembly. NMR spectroscopy is a powerful and versatile technique, well suited to this purpose, with the micelle and bilayer providing excellent model systems for multidimensional solution and solid-state NMR studies, respectively. Considering the recent progress in the fields of NMR spectroscopy as well as molecular biology, there is every reason to be optimistic that the data base of atomic resolution structures can be expanded to represent an increasing number of membrane proteins.

Uniform isotopic labeling, with the recent extension to uniform deuteration, all made possible by the development of efficient bacterial expression systems, greatly increases the range and size of membrane proteins whose structures can be determined by NMR spectroscopy. Both solution and solid-state NMR benefit from the multitude of isotopic labeling schemes available, with new multidimensional experiments taking full advantage of this capability. However, even highly optimized, uniformly $^{15}N/^{13}C/^{2}H$ labeled samples, and the highest magnetic fields available, cannot defeat the size limitation, imposed on solution NMR, by the fundamental correlation time problem. Solid-state NMR is different in this respect and has greater potential because it is not limited by slow correlation times. The solid-state NMR approach we have described relies on uniaxial sample orientation to achieve line narrowing, and the sample must be immobilized within the lipid bilayer. This ensures that the resonance linewidths will not degrade in the solid-state spectra of larger proteins. In solid-state NMR spectra, increased size is not accompanied by changes in relaxation parameters and linewidths, and sensitivity will only diminish because a smaller number of molecules will fit in a fixed sample volume. This is in contrast to solution NMR where shorter relaxation times and broader lines render it difficult to make correlations and to transfer magnetization. As the proteins get larger, there will be additional resonances in both types of spectra, shifting the onus of securing complete resolution to the spectroscopy. The recent success in obtaining completely resolved solid-state NMR spectra of immobile proteins extends the range of molecules that can have their structures determined by NMR spectroscopy to include membrane proteins in lipid bilayers. Although this approach is still at an early stage of development, the resolution and sensitivity observed in bilayer samples are already comparable to those seen in solution NMR spectra of the same ^{15}N-labeled proteins in micelles. The ^{1}H chemical shift resolution in particular stands to benefit from higher field spectrometers with the added bonus of greater sensitivity.

The determination of tertiary structure from orientationally dependent spectral frequencies is particularly direct and accurate. The orientation of each peptide plane, or a selected segment of a protein, is measured independently from the others. Because structures are assembled from a set of completely independent measurements, the method is free from errors generated by relative measurements (i.e., distance measurements), which propagate along the protein backbone. In other words, errors are not cumulative. This should prove particularly useful for resolving structure in those areas of proteins, such as some loops or hinges, where the distances to other reference features of the protein are too large to yield constraints. This approach is straightforward for amide sites that are completely immobilized and ordered on the relevant NMR time scales, as shown by each having a single orientation and a maximal order parameter of 1. No structural information is available on these time scales, for sites with isotropic resonances such as those at the N- and C-termini. However, it is possible to obtain crucial details about structure and dynamics of sites with limited motional averaging, whether related to local segmental or residue motions or because the entire sample undergoes some motion, as seen in magnetically oriented bicelles.

Taken together, these considerations demonstrate that NMR spectroscopy, especially solid-state NMR spectroscopy of oriented samples, is capable of giving detailed and reliable descriptions of the structure and dynamics of membrane proteins.

ACKNOWLEDGMENTS. We thank M. Montal for his many contributions to the studies of the channel-forming peptides. This research was supported by Grants RO1 GM29754, RO1 AI20770, and R37 GM24266 from the National Institutes of Health, Grant R823576 from the Environmental Protection Agency, and Grant DE-FG07-97ER62314 from the Department of Energy. The research utilized the Resource for Solid-State NMR of Proteins: An NIH Sponsored Resource supported by Grant P41 RR09793 from the Biomedical Resource Technology Program National Center for Research Resource. F.M.M. was supported by postdoctoral fellowship 9304FEN-1004-43344 from the Medical Research Council of Canada.

REFERENCES

Almeida, F., and Opella, S. J., 1997, *J. Mol. Biol.* **270:**481.
Bechinger, B., and Opella, S. J., 1991, *J. Magn. Reson.* **95:**585.
Bechinger, B., Kim, Y., Chirlian, L. E., Gesell, J., Neumann, J.-M., Montal, M., Tomich, J., Zasloff, M., and Opella, S. J., 1991, *J. Biomol. NMR* **1:**167.
Bechinger, B., Zasloff, M., and Opella, S. J., 1993, *Protein Sci.* **2:**2007.
Bechinger, B., Gierasch, L. M., Montal, M., Zasloff, M., and Opella, S. J., 1996, *Solid State NMR* **7:**185.
Bielecki, A., De Groot, H. J. M., Griffin, R. G., and Levitt, M. H., 1990, *Adv. Magn. Reson.* **14:**111.
Bogusky, M. J., Tsang, P., and Opella, S. J., 1985, *Biochem. Biophys. Res. Commun.* **127:**540.
Brown, L. R., 1979, *Biochim. Biophys. Acta* **557:**135.
Brunger, A. T., 1992, *A System for X-ray Crystallography and NMR*, Version 3.1, Yale University Press, New Haven, CT.
Coni, G., Patrone, E., and Brighetti, S. J., 1970, *J. Biol. Chem.* **245:**3335.
Cowan, S., Schirmer, T., Rummel, G., Steiert, M., Ghosh, R., Pauptit, R., Jansonius, J., and Rosenbusch, J., 1992, *Nature* **358:**727.
Cross, T. A., and Opella, S. J., 1994, *Curr. Opin. Struct. Biol.* **4:**574.
Cross, T. A., Di Verdi, J. A., and Opella, S. J., 1982, *J. Am. Chem. Soc.* **104:**1759.
Cross, T. A., Frey, M. H., and Opella, S. J., 1983, *J. Am. Chem. Soc.* **105:**7471.
Deisenhofer, J., Epp, O., Miki, K., Huber, R., and Michel, H., 1985, *Nature* **318:**618.
Doty, P., Holzer, A., Bradbury, J., and Blout, E. J., 1954, *J. Am. Chem. Soc.* **76:**4493.
Feng, X., Lee, Y. K., Sandstroem, D., Edn, M., Maisel, H., Sebald, A., and Levitt, M. H., 1996, *Chem. Phys. Lett.* **257:**314.
Fesik, S. W., and Zeiderweg, E. R. P., 1988, *J. Magn. Reson.* **78:**588.
Fraser, C. M., Gocayne, J., White, O., Adams, M., Clayton, R., Fleischmann, R., Bult, C., Kerlavage, A., Sutton, G., Kelley, J., Fritchman, J., Weidman, J., Small, K., Sandusky, M., Furhmann, J., Nguyen, D., Utterback, T., Saudek, D., Phillips, C., Merrick, J., Tomb, J., Dougherty, B., Bott, K., Hu, P., Lucier, T., Peterson, S., Smith, H., Hutchinson, C., and Venter, J. C., 1995, *Science* **270:**397.
Garavito, R. M., Picot, D., and Loll, P. J., 1994, *Curr. Opin. Struct. Biol.* **4:**529.
Gesell, J. G., Zasloff, M., and Opella, S. J., 1997, *J. Biomol. NMR* **9:**127.
Goffeau, A., 1995, *Science* **270:**445.

Grisshammer, R., and Tate, C. G., 1995, *Q. Rev. Biophys.* **28:**315.
Gu, Z. T., and Opella, S. J., 1998, *J. Magn. Reson.*, in press.
Gullion, T., and Schaefer, J., 1989, *J. Magn. Reson.* **81:**196.
Henderson, R., and Unwin, P. N. T., 1975, *Nature* **257:**28.
Henry, G. D., and Sykes, B. D., 1992, *Biochemistry* **315:**5284.
Howard, K. P., and Opella, S. J., 1996, *J. Magn. Reson. Ser. B* **112:**91.
Iwata, S., Ostermeier, C., Ludwig, B., and Michel, H., 1995, *Nature* **376:**660.
Jap, B., Walian, P., and Gehring, K., 1991, *Nature* **350:**167.
Jelinek, R., Ramamoorthy, A., and Opella. S. J., 1995, *J. Am. Chem. Soc.* **117:**12348.
Kay, L. E., Ikura, M., Tschudin, R., and Bax, A., 1990, *J. Magn. Reson.* **89:**496.
Kersh, G. J., Tomich, J. M., and Montal, M., 1989, *Biochem. Biophys. Res. Commun.* **162:**352.
Ketchem, R. R., Hu, W., and Cross, T. A., 1993, *Science* **261:**1457.
Kovacs, F. A., and Cross. T. A., 1997, *Biophys. J.* **73:**2511.
Kuhlbrandt, W., Wang, D. N., and Fujiyoshi, Y., 1994, *Nature* **367:**614.
Lee, M., and Goldburg, W. I., 1965, *Phys. Rev.* **A140:**1261.
Leonardo, R. J., Labarca, C. G., Chatnet, P., Davidson, N., and Lester, H., 1988, *Science* **242:**1578.
Levitt, M. H., Raleigh, D. P., Creuzet, F., and Griffin, R. G., 1990, *J. Chem. Phys.* **92:**6347.
Liebes, L. F., Zand, R., and Phillips, W., 1975, *Biochim. Biophys. Acta* **405:**27.
MacKenzie, K. R., Prestegard, J. H., and Engelman, D. H., 1997, *Science* **276:**131.
Marassi, F. M., Ramamoorthy, A., and Opella, S. J., 1997, *Proc. Natl. Acad. Sci. USA* **94:**8551.
Marion, D., Zasloff, M., and Bax, A., 1988, *FEBS Lett.* **227:**21.
Marion, D., Driscoll, P. C., Kay, L. E., Wingfield, P. E., Bax, A., Gronenborn, A. M., and Clore, G. M., 1989, *Biochemistry* **29:**6150.
McDermott, G., Prince, S., Freer, A., Hawthornthwaitelawless, A. M., Papis, M. Z., Cogdell, R. J., and Isaacs, N. W., 1995, *Nature* **374:**517.
McDonnell, P. A., and Opella, S. J., 1993, *J. Magn. Reson., Ser. B* **102:**120.
McDonnell, P. A., Shon, K., Kim, Y., and Opella, S. J., 1993, *J. Mol. Biol.* **233:**447.
Mehring, M., and Waugh, J. S., 1972, *Phys. Rev.* **B5:**3459.
Miroux, B., and Walker, J. E., 1996, *J. Mol. Biol.* **260:**289.
Montal, M., Montal, M. S., and Tomich, J. M., 1990, *Proc. Natl. Acad. Sci. USA* **87:**6929.
Nelson, J., and Kallenbach, N., 1989, *Biochemistry* **28:**5256.
North, C. L., Barranger-Matys, M., and Cafiso, D. S., 1995, *Biophys. J.* **69:**2392.
Opella, S. J., 1997, *Nature Struct. Biol.* NMR Supplement:845.
Opella, S. J., and Stewart, P. L., 1989, *Methods Enzymol.* **176:**242.
Opella, S. J., and Waugh, J. S., 1977, *J. Chem. Phys.* **66:**4919.
Opella, S. J., Stewart, P. L., and Valentine, K. G., 1987, *Q. Rev. Biophys.* **19:**7.
Opella, S. J., Kim, Y., and McDonnell, P., 1994, *Methods Enzymol.* **239:**536.
Pausak, S., Pines, A., Gibby, M. G., and Waugh, J. S., 1973, *J. Chem. Phys.* **59:**591.
Pebay-Peyroula, E., Rummel, G., Rosenbusch, J. P., and Landau, E. M., 1997, *Science* **277:**1676.
Picot, D., Loll, P., and Garavito, M., 1994, *Nature* **367:**243.
Pines, A., Gibby, M. G., and Waugh, J. S., 1973, *J. Chem. Phys.* **59:**569.
Prive, G. G., and Kaback, H. R., 1996, *J. Bioenerg. Biomembr.* **28:**29.
Prosser, R. S., Hunt, S. A., DiNatale, J. A., and Vold, R. R., 1996, *J. Am. Chem. Soc.* **118:**269.
Ramamoorthy, A., Wu, C. H., and Opella, S. J., 1995a, *J. Magn. Reson. Ser. B* **107:**88.
Ramamoorthy, A., Gierasch, L. M., and Opella, S. J., 1995b, *J. Magn. Reson. Ser. B* **109:**112.
Ramamoorthy. A., Marassi, F. M., Zasloff, M., and Opella, S. J., 1995c, *J. Biomol. NMR* **6:**329.
Ramamoorthy, A., Gierasch, L. M., and Opella, S. J., 1996a, *J. Magn. Reson. Ser. B* **111:**81.
Ramamoorthy, A., Marassi, F. M., and Opella, S. J., 1996b, in *Dynamics and the Problem of Recognition in Biological Macromolecules* (O. Jardetzky and J. Lefevre, eds.), Plenum, New York, pp. 237–255.

Sanders, C. R., Hare, B., Howard, K. P., and Prestegard, J. H., 1994, *Prog. NMR Spectrosc.* **26:**421.
Schirmer, T., Keller, T. A., Wang, Y. F., and Rosenbusch, J. P., 1995, *Science* **267:**512.
Shon, K., and Opella, S. J., 1989, *J. Magn. Reson.* **82:**193.
Smith, S. O., 1996, *Magn. Reson. Rev.* **17:**1.
Song, L., Hobaugh, M., Shustak, C., Cheley, S., Bayley, H., and Gouaux, J. E., 1996, *Science* **274:**1859.
Summers, A.O., 1986, *Annu. Rev. Microbiol.* **40:**607.
Tsukihara, T., Aoyama, H., Yamashita, E., Tomizaki, T., Yamaguchi, H., Shinzawa-Itoh, K., Nakashima, R., Yaono, R., and Yoshikawa, S., 1996, *Science* **272:**1136.
Tycko, R., Stewart, P. L., and Opella, S. J., 1986, *J. Am. Chem. Soc.* **108:**5419.
Ulrich, A. S., Watts, A., Wallat, I., and Heyn, M. P., 1994, *Biochemistry* **33:**5370.
Unwin, N., 1993, *J. Mol. Biol.* **229:**1101.
Unwin, N., 1995, *Nature* **373:**37.
van de Ven, F. J. M., Os, J. W. M., Aelen, J. M. A., Wymenga, S. S., Remerowski, M. L., Konings, R. N. H., and Hilbers, C. W., 1993, *Biochemistry* **32:**8322.
Veglia, G. L., Brabazon, D. M., Gesell, J. J., Ma, C., and Opella, S. J., 1998, manuscript submitted.
Waugh, J. S., 1976, *Proc. Natl. Acad. Sci. USA* **73:**1394.
Weiss, M., Abele, U., Weckesser, J., Welte, W., Schiltz, E., and Shulz, G. E., 1991, *Science* **254:**1627.
Wishart, D. S., and Sykes, B. D., 1994, *Methods Enzy. Mol.* **239:**363.
Wu, C. H., Ramamoorthy, A., and Opella, S. J., 1994, *J. Magn. Reson. Ser. A* **109:**270.
Wu, C. H., Ramamoorthy, A., Gierasch, L. M., and Opella, S. J., 1995, *J. Am. Chem. Soc.* **117:**6148.
Wuthrich, K., 1986, *NMR of Proteins and Nucleic Acids*, Wiley, New York.
Zasloff, M., 1987, *Proc. Natl. Acad. Sci. USA* **84:**5449.

II

Pulse Methods

5

Homonuclear Decoupling in Proteins

Ēriks Kupče, Hiroshi Matsuo, and Gerhard Wagner

1. INTRODUCTION

Soon after the fine structure of NMR resonances was detected and theoretically explained (Hahn and Maxwell, 1951; Gutowsky *et al.*, 1951), spin decoupling in NMR was suggested by Bloch (1954). It was subsequently demonstrated experimentally (Royden, 1954; Bloom and Shoolery, 1955) that applying a second RF field can be used to collapse multiplets into single lines. Decoupling became an important tool for assignment of NMR spectra. The theory of continuous wave (CW) decoupling was subsequently further developed by Anderson and Freeman (1962).

Introduction of noise decoupling (Ernst, 1966) made heteronuclear decoupling a routine and one of the most efficient methods of spectral simplification and sensitivity enhancement in NMR. Further development of various techniques of hetero decoupling was boosted by formulating a theoretical basis for wide-band decoupling (Waugh, 1982a,b). Composite pulse decoupling (CPD) sequences—MLEV (Levitt and Freeman, 1982; Levitt *et al.*, 1982), WALTZ (Shaka *et al.*, 1983a,b), GARP (Shaka *et al.*, 1985)—have been intensively used for 15 years.

Ēriks Kupče • Varian NMR Instruments, Walton-on-Thames, Surrey KT12 2QF, England. **Hiroshi Matsuo and Gerhard Wagner** • Department of Biological Chemistry and Molecular Pharmacology, Harvard Medical School, Boston, Massachusetts 02115.

Biological Magnetic Resonance, Volume 16: Modern Techniques in Protein NMR, edited by Krishna and Berliner. Kluwer Academic / Plenum Publishers, 1999.

Phase cycles and supercycles introduced by Levitt *et al.*, (1983) became an extremely useful tool for improvement of decoupling performance.

Before the advent of 2D NMR, homonuclear decoupling served mainly as an assignment tool. Only recently was it demonstrated that band-selective homonuclear decoupling can be employed as a spectral simplification and sensitivity enhancement technique. Furthermore, the lack of techniques for homonuclear band-selective decoupling was essentially slowing down development of several very important techniques for structural studies of ^{13}C-labeled peptides, such as the HNCA (Kay *et al.*, 1990). When it became clear that essentially any inversion pulse can be employed as a decoupling waveform, several groups started to explore the idea of using selective pulses for band-selective decoupling, namely, homonuclear C–C (McCoy and Mueller, 1992a–c) or H–H decoupling (Kupče and Freeman, 1993a) as well as heteronuclear C–H decoupling (Zuiderweg and Fesik, 1991; Eggenberger *et al.*, 1992). These techniques immediately found important applications in 2D and 3D NMR, mainly for band-selective ^{13}C (CO) decoupling (Yamazaki *et al.*, 1994a,b). Recently, further progress was made by introducing adiabatic homo-decoupling techniques (Kupče and Wagner, 1995). Because of low power requirements and extremely clean off-resonance performance of this method, further sensitivity enhancement and simplification of spectra was obtained by decoupling the spectral regions that are very close to the resonances to be observed, for instance the β-carbons.

The enhancement of resolution and sensitivity offered by homo-decoupling is also considered as a tool for decreasing dimensionality of NMR spectra. As pointed out by Bax and Grzesiek (1993), increasing the dimensionality is beneficial only if the resolution is limited by natural linewidth (T_2) and not by use of short acquisition times in the indirectly detected dimension.

The present chapter summarizes the recent applications of homonuclear band-selective decoupling in NMR of proteins.

2. BASIC ASPECTS OF SPIN DECOUPLING

The theory of spin decoupling has been extensively treated in the literature (Waugh, 1982b; Levitt *et al.*, 1983; Shaka and Keeler, 1987) and is only briefly repeated below. The commonly used basic theory of decoupling makes several relatively severe assumptions:

1. Spins I and S are separated to an extent that irradiation near the I spins has no direct effect on the S spins.
2. The I–S spin system is isolated, i.e., any I–I and S–S couplings are absent.
3. The irradiating field B_2 is strong compared with the J_{IS} coupling ($\gamma B_2 >> 2\pi J$).

4. Relaxation effects are ignored.

With these assumptions the total Hamiltonian in a doubly rotating reference frame is

$$H(t) = -\Delta\omega I_z + J_{IS} I_z S_z + H_{\mathrm{RF}}(t) \tag{1}$$

where $\Delta\omega$ is the separation between the RF field and the Larmor frequency of the I spin and $H_{\mathrm{RF}}(t)$ represents the interaction between the I spins and the RF field.

2.1. Coherent Decoupling

In the case of CW irradiation (coherent decoupling) the expression for the RF field component of the total Hamiltonian becomes very simple allowing expression of the effect of the decoupling field on S-spin spectra in simple terms (Anderson and Freeman, 1962). In the presence of the RF (decoupling) field the two forbidden transitions in the S spectrum become partially allowed and four lines can be observed. The two inner (parent) lines appear at frequencies $\Delta\omega_i$ from the center of the doublet and have intensities a_i given by

$$\Delta\omega_i = \pm \tfrac{1}{2} (\omega_{e+} - \omega_{e-}) \tag{2}$$

$$\omega_i = \tfrac{1}{2} [1 - \cos(\Theta_+ - \Theta_-)] \tag{3}$$

whereas the frequencies ($\Delta\omega_o$) and intensities (a_o) of the two outer (satellite) lines are given by

$$\Delta\omega_o = \pm \tfrac{1}{2} (\omega_{e+} + \omega_{e-}) \tag{4}$$

$$a_o = \tfrac{1}{2} [1 + \cos(\Theta_+ - \Theta_-)] \tag{5}$$

where ω_e is the effective field as seen by individual components of the I spin doublet centered at a (Larmor) frequency ω_L:

$$\omega_{e+-} = [(\omega_L \pm \tfrac{J}{2})^2 + \omega_2^2]^{1/2} \tag{6}$$

and Θ is the angle between the vectors of the effective field ω_e and the RF field ω_2:

$$\Theta = \arctan \left(\frac{\omega_2}{\Delta\omega}\right) \tag{7}$$

The separation between the two parent lines is usually regarded as a residual splitting (J_r) that can be related to the original J coupling via the scaling factor (λ):

$$J_r = \lambda \cdot J \tag{8}$$

where

$$\lambda \approx \frac{\Delta\omega}{\omega_e} = \cos(\Theta) \tag{9}$$

As is apparent from Eqs. (2) and (7–9), to obtain essentially complete decoupling ($\lambda < 0.01$) the RF amplitude ω_2 must be much larger than the mismatch $\Delta\omega$ between the irradiation frequency and the I-spin resonance frequency.

Figure 1 shows the effect of the decoupling field strength applied on resonance of I spins on the appearance of S-spin spectra. As the field strength increases, the intensities of the two satellite (outer) lines in S-spin spectrum decrease and their separation from the parent (inner) lines increases. For instance, if the decoupling field strength is $5J$, the sideband intensity is less than 1%. In many practical situations, this is sufficient and can be accepted as a reasonable setting for decoupling field strength.

The effect of the mismatch between the frequency of the decoupling field ω_2 and the Larmor frequency of the I spin ω_I is shown in Fig. 2. As can be appreciated from Fig. 2, the minimum for essentially complete decoupling is very sharp. In

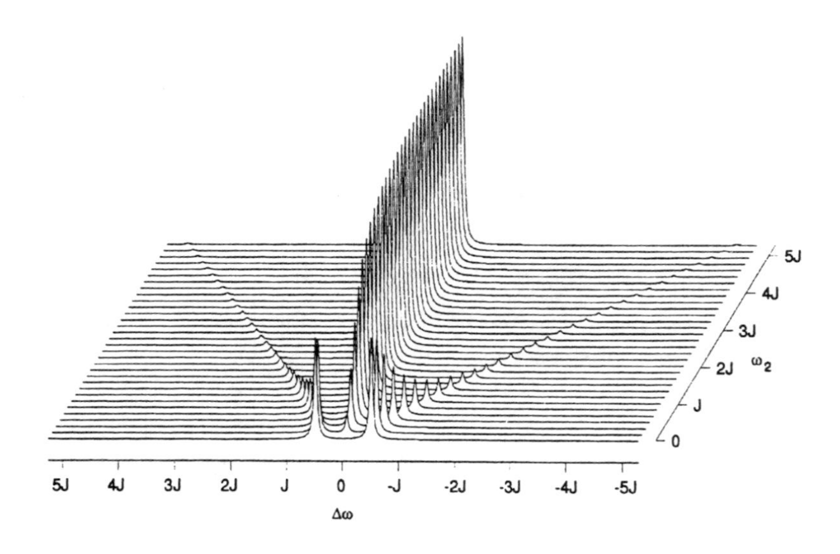

Figure 1. Simulation of the effect of decoupling field strength in a two-spin system. The spectrum of I spins is observed while a monochromatic (CW) decoupling field is applied at the exact frequency of the S spins. Both the frequency and the decoupler field strength are expressed in the units of J_{IS} coupling constants.

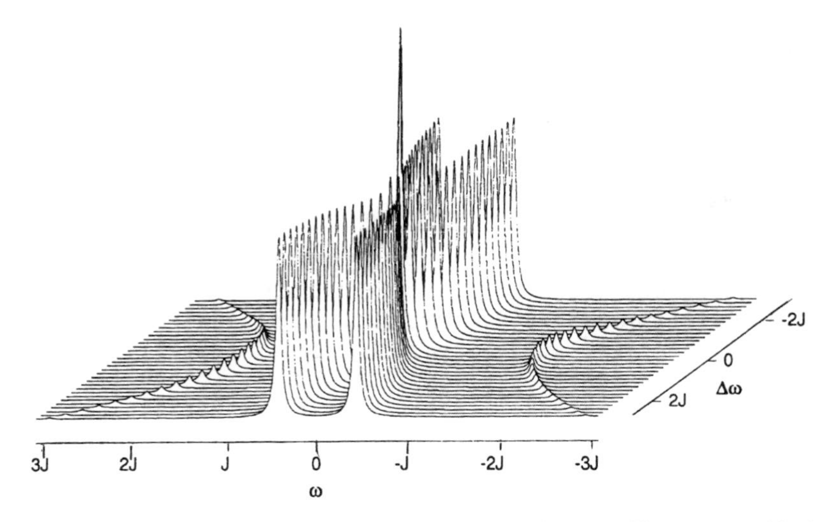

Figure 2. Simulation of the effect of decoupler offset in a two-spin system. The spectrum of I spins is observed while a monochromatic (CW) decoupling field of strength equal to $5J_{IS}$ is swept through the resonance of S spins. The frequency scales of spins I and S are given relative to the J_{IS} coupling.

practice, this allows decoupling of spins only within a region comparable to the magnitude of the J_{IS} coupling.

2.2. Band-Selective and Wide-Band Decoupling

From Fig. 2, the region corresponding to optimum conditions in the case of CW decoupling is very narrow. Obviously, CW decoupling becomes problematic if a band of frequencies needs to be covered as, for example, in the case of complex (homonuclear) spin systems, in addition to heteronuclear decoupling, where uniform decoupling over an extremely wide range of frequencies is usually required.

The theory of broadband decoupling based on the average Hamiltonian approach was introduced in the early 1980s (Waugh, 1982b) and has been treated in detail by Levitt *et al.* (1983) and Shaka and Keeler (1987). The quality of decoupling for complex (composite) pulse sequences and amplitude-modulated (AM) or frequency-modulated (FM) waveforms is traditionally quantified by calculating the offset dependence of the scaling factor λ given by

$$\lambda(\Delta\omega) = \frac{1}{t_c}\left[\frac{d\beta}{d(\Delta\omega)}\right]^{-1} \tag{10}$$

where t_c is the cycle time (Waugh, 1982a,b) and β is the offset-dependent overall rotation angle. Obviously, the scaling factor does not reflect the complications (decoupling sidebands) coming from the finite power of the RF field, effects initiated by homonuclear couplings, and RF field inhomogeneity. However, it can be used as a first approximation to assess the quality of a particular decoupling waveform.

2.3. Supercycles

The basic idea of spin decoupling is based on the principle that the spins to be decoupled are repeatedly inverted. Essentially any inversion pulse can be used as a basic building block of a decoupling waveform. As an example, the scaling factor for I-BURP-2 pulse (Geen and Freeman, 1991) is shown in Fig. 3a.

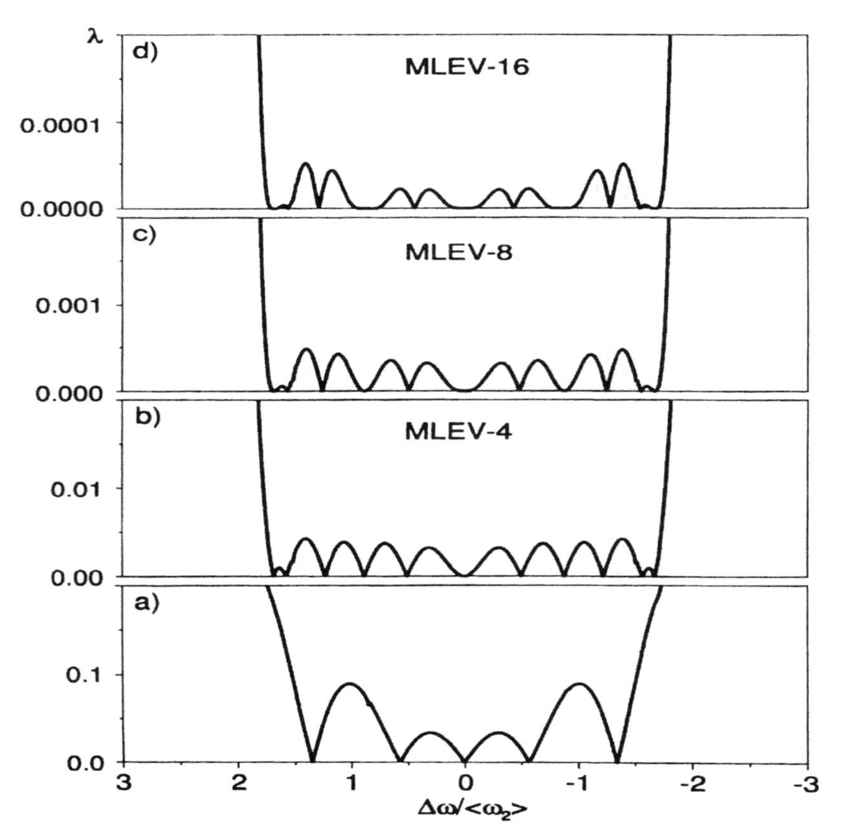

Figure 3. Effect of different phase-cycling schemes on the scaling factor (λ) in an I-BURP-2 decoupling experiment: (a) no phase cycling, (b) MLEV-4 cycle, (c) MLEV-8 cycle, and (d) MLEV-16 cycle.

Table 1
Most Frequently Used Decoupling Cycles and Supercycles

	Steps	Phases
MLEV-4 (a)	4	0,0,180,180
(b)	4	0,180,180,0
TPG-5	5	0,120,60,120,0
MLEV-8	8	0,0,180,180, 0,180,180,0
MLEV-16	16	0,0,180,180, 0,180,180,0, 180,180,0,0, 180,0,0,180
WALTZ-16	24	0,0,0,180,180,180,180,0,0,180,180,180, 0,180,180,0,0,0,0,180,180,0,0,0
(TPG-5) MLEV-4	20	0,120,60,120,0, 0,120,60,120,0, 180,300,240,300,180, 180,300, 240,300,180
(TPG-5) TPG-5	25	0,120,60,120,0, 120,240,180,240,120, 60,180,120,180,60, 120,240,180,240,120, 0,120,60,120,0
(WALTZ-16) MLEV-4	96	0,0,0,180,180,180,180,0,0,180,180,180, 0,180,180,0,0,0,0,180,180,0,0,0, 180,180,180,0,0,0,0,180,180,0,0,0, 180,0,0,180,180,180,180,0,0,180,180,180, 180,180,180,0,0,0,0,180,180,0,0,0, 180,0,0,180,180,180,180,0,0,180,180,180, 0,0,0,180,180,180,180,0,0,180,180,180, 0,180,180,0,0,0,0,180,180,0,0,0

Because it is very difficult to design a waveform that would invert to a very high degree of accuracy all spins in the desired range of frequencies, a common practice is to use expansion procedures (Basus *et al.*, 1979), known as decoupling cycles and supercycles (Levitt and Freeman, 1982; Levitt *et al.*, 1983) which can improve the quality of decoupling dramatically (see Fig. 3b–d). Extremely efficient phase cycles have also been suggested by Tycko *et al.* (1985). Even simple composite pulse sequences, e.g., WALTZ-16 (Shaka *et al.*, 1983a), can be used as expansion procedures (McCoy and Mueller, 1992a; Zuiderweg *et al.*, 1996). The decoupling *cycles* can be easily expanded into *supercycles* by nesting the simplest phase cycles (Tycko *et al.*, 1985; Fujiwara and Nagayama, 1988) or by using other expansion procedures (Levitt *et al.*, 1983). The most frequently used phase cycles are shown in Table 1.

2.4. Decoupling Sidebands

As discussed in Section 2.1, in the presence of the RF field some of the forbidden transitions become partially allowed. This increases the number of observable lines in the NMR introducing the decoupling sidebands. The intensities and appearance of the decoupling sidebands depend on the complexity of a particular decoupling waveform. In the case of adiabatic frequency swept decou-

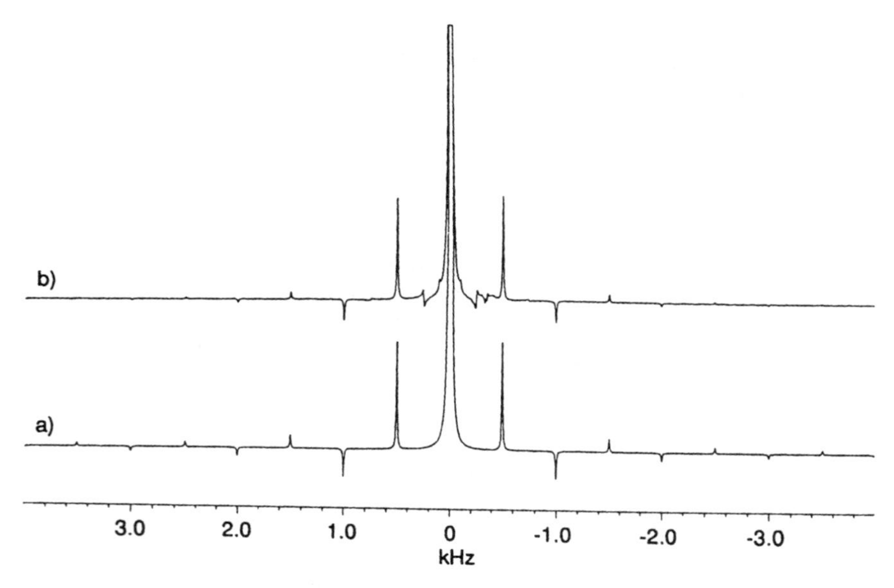

Figure 4. Comparison between the outer decoupling sidebands (a) predicted by simulation assuming an instantaneous flip and (b) observed experimentally for ^{13}C-labeled MeJ at a 500-MHz proton frequency on a Unity Plus spectrometer using WURST-40 decoupling waveform. In both cases, $J = 151$ Hz and a decoupling pulse length of 2 ms was used. Note that in the experiment, a faster decay of the intensities of the higher-order sidebands is caused by the finite length of the inversion pulses.

pling waveforms, the treatment of the decoupling sidebands is particularly simple (Slichter, 1996). Assuming an instantaneous flip, the modulation of the free induction decay can be approximated by a Fourier series:

$$M_y(t) = 0.5A_0 + \sum_{n=1}^{\infty} A_n \cos(2\pi nt/T_p) \tag{11}$$

where T_p is the length of the inversion (decoupling) pulse. The decoupled peak has the intensity

$$A_0 = 2\sin(\pi JT_p/2)/\pi JT_p \tag{12}$$

whereas the sideband intensities can be found from the higher-order Fourier coefficients:

$$A_n = C \cdot \sin(\pi JT_p/2) \cdot \cos(n\pi) \tag{13}$$

where

$$C = \frac{-4JT_p}{\pi[4n^2 - (JT_p)^2]} \tag{14}$$

In practice, the time necessary to invert spins at a given frequency is finite. This has the advantageous effect of reducing the intensities of higher-order sidebands (see Fig. 4).

2.5. Relaxation Effects

Although relaxation effects are usually excluded from the treatment, they become important if the J coupling is relatively small, i.e., if the J coupling becomes comparable to the transverse relaxation rate $1/T_2$. Obviously, if $1/T_2 >> J$, there is no benefit from the decoupling and sensitivity may even be reduced. Figure 5 shows

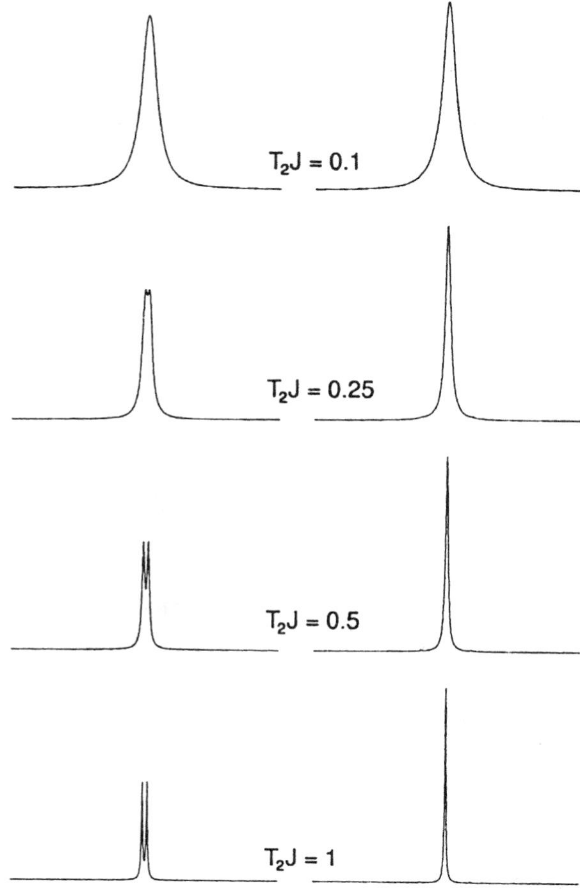

Figure 5. Simulation for a two-spin system of the effect of relaxation on the sensitivity gain that can be obtained by decoupling. As the transverse relaxation rate increases and the product of T_2J decreases, the sensitivity improvement from the decoupling becomes less obvious. In a real experiment, further losses occur as a result of time-shared decoupling mode and the presence of sidebands.

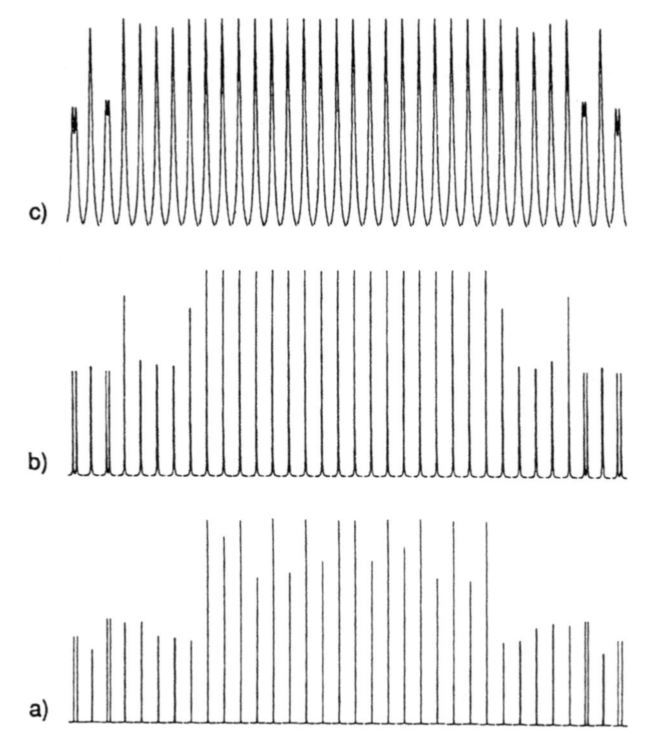

Figure 6. Simulation of the effect of the transverse relaxation on the decoupling profile of I-BURP-2 waveform assuming $J = 20$ Hz, an RF peak amplitude of 0.194 kHz, and relaxation times of (a) 3 s, (b) 0.3 s, and (c) 0.03 s. The decoupler frequency was swept from −132 Hz to 132 Hz in 8-Hz steps.

the dependence on the relaxation rate of the sensitivity enhancement related to decoupling. Coupled and decoupled spectra are compared for four values of $T_2 J$.

Relaxation has another advantageous effect, namely, it tends to alleviate the criteria for good decoupling by smoothing out the decoupling profile. This is because the residual splittings become more difficult to resolve at short T_2 relaxation times (see Fig. 6).

The relaxation time T_2 sets the practical (upper) limit for the length of the acquisition time and therefore (indirectly) for the length of the decoupling cycle. In principle, the duration of an elementary decoupling cycle should be much smaller than T_2, otherwise relaxation disturbs the process of compensation of the short-term imperfections and the performance of decoupling will degrade (Levitt *et al.*, 1983). This condition usually is most important if decoupling is used during the evolution period in the indirectly detected dimension, because this period tends to be relatively short, especially in 3D and 4D experiments. On the other hand, the elementary decoupling cycles in some schemes are relatively long. For example, SEDUCE decoupling (McCoy and Mueller, 1992a–c) is based on a 24-step elementary decoupling cycle (WALTZ-16) that is nested in a 4-step MLEV supercycle. Therefore, the detection period (and T_2) should be at least 192 times the length of the elementary inversion pulse used in such a decoupling scheme, to ensure that the decoupling performance meets the theoretical predictions.

Therefore, as the transverse relaxation rate becomes comparable with the I–S coupling constant, long decoupling cycles become unnecessary and inefficient. Short decoupling cycles, e.g., MLEV-4 or TPG-5, are sufficient in most situations when homonuclear H–H or C–C decoupling is required.

2.6. Decoupling at Several Frequencies

Some applications may require simultaneous band-selective decoupling at multiple frequencies (Eggenberger *et al.*, 1992; Kupče and Freeman, 1993a; Kupče and Wagner, 1996; Zhang *et al.*, 1996). Several decoupling waveforms can be combined into a single waveform according to the principle of superposition known from optics (e.g., Allen and Eberly, 1975):

$$\phi = c_1\phi_1 + c_2\phi_2 \tag{15}$$

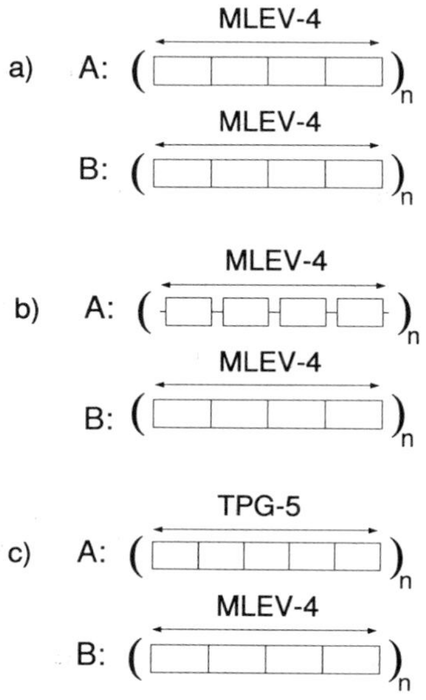

Figure 7. Examples of implementation of decoupling at two frequencies A and B that require different decoupling bandwidths. (a) Adiabatic waveforms allow use of equal pulse lengths for both frequencies. (b) With the planarly polarized waveforms, the pulse lengths will be different in accordance with the decoupling bandwidth and a fixed pulse-width-to-bandwidth product (BPP). As a result the decoupling waveform needs to be "fixed" by inserting appropriate delays or (c) by using matched supercycles.

In practice, this can be implemented either by interleaving the waveforms (Geen *et al.*, 1989) or, more conveniently, by combining the waveforms using a vector addition scheme (Tomlinson and Hill, 1973; Patt, 1992; Kupče and Freeman, 1993b).

In a general case, different decoupling bands may have different widths. Because planarly polarized waveforms have a fixed bandwidth-to-pulse-width product (BPP), this implies also a different pulse length for individual inversion pulses, unless the shapes of the decoupling waveforms have been chosen so that their BPP match the corresponding decoupling bandwidths (see Fig. 7a). So as to retain the symmetry of the *J*-refocusing, the shorter inversion pulses should be centered with respect to the longest pulse in the decoupling sequence, as shown in Fig. 7b. Alternatively, the same can be achieved by matching the length of the supercycle, as suggested by Eggenberger *et al.* (1992; see Fig. 7c).

Adiabatic pulses have variable BPP and the pulse length is essentially independent of the decoupling bandwidth. Therefore, the length of the decoupling pulse can be adjusted according to the magnitude of the *J* coupling. Usually it is convenient to keep T_p the same for all waveforms.

3. HOMONUCLEAR DECOUPLING

In the absence of the RF field, spins precess at their Larmor frequencies ω_L. However, if an RF field of constant amplitude ω_2 is applied off resonance ($\Delta\omega$), the precession occurs about the effective field, ω_e:

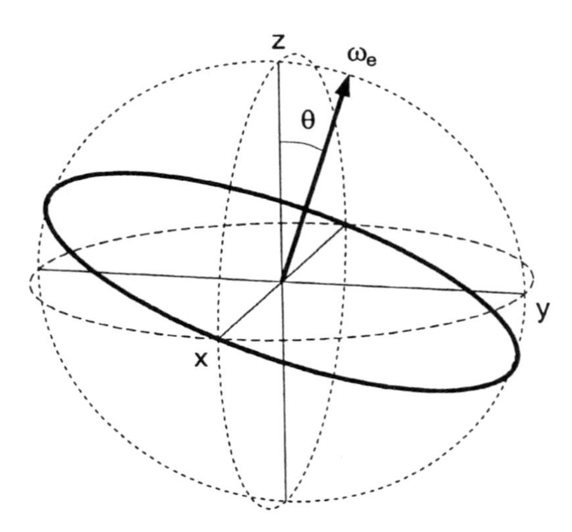

Figure 8. Evolution of magnetization in the presence of 0.1-kHz strong monochromatic (CW) RF field experienced by spins 250 Hz away from the RF field. The trajectory of spin magnetization is evolving in a reference frame that is tilted with respect to the usual rotating frame by an angle Θ = 21.8° and is aligned with the effective field ω_e. As a result, the magnetization vector spends more time along the *z* axis reducing the amount of the detectable signal.

$$\omega_e = \sqrt{\omega_2^2 + \Delta\omega^2} \tag{16}$$

which is tilted with respect to the H_0 by an angle Θ (see Fig. 8):

$$\Theta = \arctan(\frac{\omega_2}{\Delta\omega}) \tag{17}$$

This introduces a number of side effects, which in the case of heteronuclear decoupling are usually weak and in practice can be safely neglected. However, in the case of homonuclear decoupling these effects become much more disturbing and may seriously compromise the quality of the spectra. The most important of these side effects include (1) Bloch–Siegert shifts, (2) modulation sidebands, (3) numerous spurious peaks ("decoupling noise"), and (4) sensitivity loss related to time-shared decoupling.

3.1. Bloch–Siegert Effect

The presence of the decoupling field during the detection period in a homodecoupling experiment affects the orientation and strength of the effective field seen by spins and therefore disturbs the usual evolution of magnetization. If the RF field is applied during the detection period, this gives rise to a frequency shift (BS shift, ω_{BS}) that always points away from the RF field:

$$\omega_{BS} = \omega_e - \Delta\omega \approx \omega_2^2/2\Delta\omega \tag{18}$$

Band-selective decoupling usually requires amplitude-modulated (shaped) waveforms. In the case of adiabatic decoupling the calculation of ω_{BS} is further complicated by the fact that the frequency offset $\Delta\omega$ is continuously changing. Exact calculation of ω_{BS} can be done by numerical integration of the Bloch equations. In practice, this task can be simplified by introducing several approximations. First, it has been shown (Emsley and Bodenhausen, 1990) that to a good approximation a constant-amplitude RF field in Eq. (16) can be replaced by the RMS value of an amplitude-modulated RF field. Similarly, the instantaneous frequency of frequency-modulated waveforms can be replaced by the average (carrier) frequency of the waveform. And finally, if several RF fields are applied to the spin system, their effect can be regarded as independent and additive. This can be exploited to compensate for Bloch–Siegert effects (McCoy and Mueller, 1993). Indeed, if a compensating RF field is applied simultaneously and symmetrically to the decoupling field, the Bloch–Siegert shift cancels in the midpoint between the two fields (see Fig. 9b). Note, however, that the compensating field may introduce even larger ω_{BS} in certain areas. Such compensating fields also require a twofold higher RF peak amplitude, which is usually undesirable.

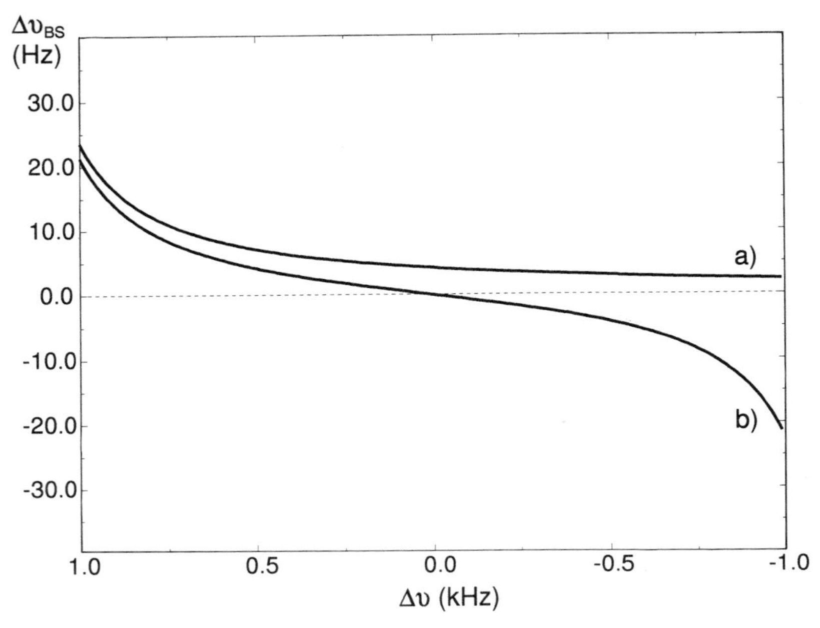

Figure 9. The Bloch–Siegert (BS) shift as a function of the distance from (a) a coherent decoupling field of 0.1 kHz applied at 1.2 kHz and (b) two simultaneous RF fields of 0.1 kHz applied at ±1.2 kHz. The BS shift cancels out only when both distances (both $|\omega_e|$) are equal. It is partially compensated on one side of the spectral region, but can be enhanced on the other side.

3.2. Modulation Sidebands

There are two basic kinds of sidebands present in homonuclear decoupled spectra: *decoupling sidebands*, which are known from heteronuclear decoupling theory, and *modulation sidebands*, which are introduced by the presence of the modulated RF (decoupling) field. Obviously, only the peaks that are being decoupled from resonances within the decoupling bandwidth will have the decoupling sidebands. On the other hand, the modulation sidebands affect all resonances present in a homo-decoupled spectrum.

As shown in Fig. 10, the direction and strength of the effective field change continuously under the influence of the modulated RF field. As a result, the pattern of the modulation sidebands depends on the RF modulation scheme and therefore can be quite complex. The intensity of the modulation sidebands decreases rapidly as the distance from the decoupling field increases and the strength of the RF field decreases.

A relatively simple pattern of modulation sidebands is observed in the case of adiabatic decoupling waveforms (see Fig. 11). As the inversion pulse of length T_p

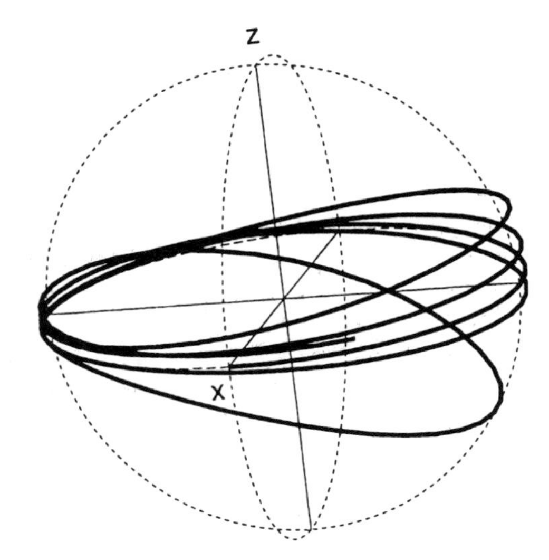

Figure 10. Modulation of the strength and direction of the effective field (ω_e) caused by a 5-ms-long I-BURP-2 pulse of 0.5-kHz peak amplitude as seen by spins 1 kHz away from the RF field.

is repeated many times during the decoupling, the modulation sidebands appear at frequencies of n/T_p ($n = 1, 2, \ldots$). Unlike for the Bloch–Siegert shift, the intensity of the modulation sidebands directly depends on the *peak* amplitude of the RF field. Therefore, it is advantageous to use waveforms that do not require high RF peak amplitudes. Interestingly, the higher order modulation sidebands essentially cancel out in the case of WURST-3 decoupling scheme (Fig. 11).

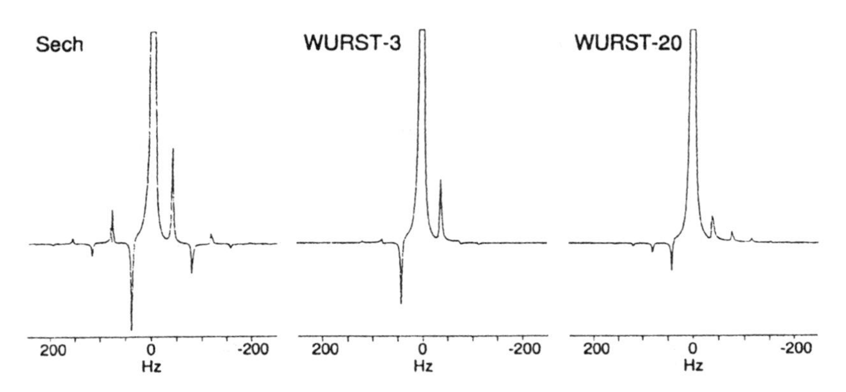

Figure 11. Modulation sidebands for hyperbolic secant, WURST-3, and WURST-20 decoupling waveforms as simulated by numerical integration of the Bloch equations. The intensities of the strongest sidebands depend on the peak amplitude of the decoupling waveform, which is the highest in the case of the hyperbolic secant pulse.

Because the J evolution is modulated by the same frequency (n/T_p), the modulation sidebands and decoupling sidebands coincide and may partially cancel out making the sideband intensities asymmetric (see Fig. 12).

Not only may the sidebands obstruct observation of minor signals in the vicinity of strong peaks, but they are also another source of sensitivity loss. Therefore, it is important to adjust the parameters of the decoupling sequence, so that the intensity of the sidebands is minimal. A low peak amplitude of the decoupling sequence appears to be particularly important in the case of homonuclear decoupling.

The appearance of the spectra can be improved by using different sideband suppression schemes. Unfortunately, such schemes do not restore the intensity of the central signal, leaving the problem of sensitivity loss unresolved.

As demonstrated by Weigelt *et al.* (1996), the modulation sidebands can be suppressed by shifting the origin of the decoupling waveform with respect to the beginning of the data acquisition. This is essentially equivalent to using the decoupling asynchronously with respect to the receiver. Unfortunately, this technique enhances the decoupling sidebands, which appear as residual sidebands in the final spectra. The decoupling sidebands are more difficult to suppress and require more sophisticated techniques (Kupče *et al.*, 1996). Postacquisition data processing, such as reference deconvolution techniques (Morris, 1988; Gibbs and

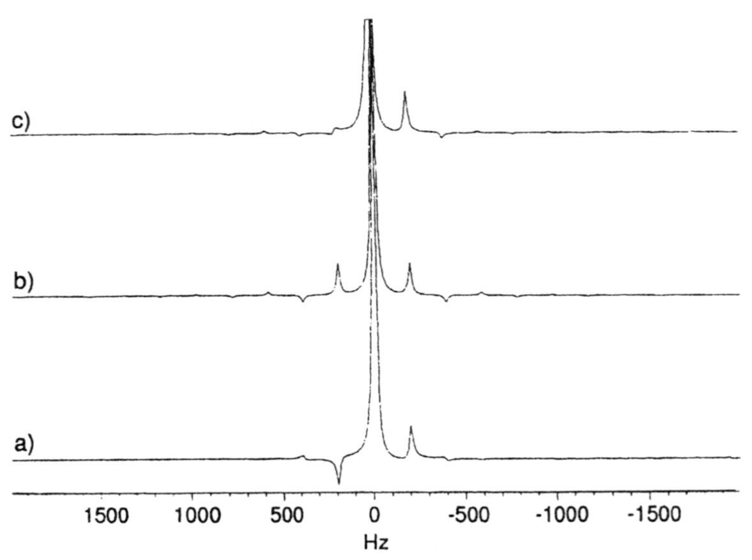

Figure 12. Partial cancellation (c) of the modulation sidebands (a) and decoupling sidebands (b) for WURST-2 decoupling simulated assuming a 5.12-ms-long decoupling pulse and a 100-Hz J coupling.

Morris, 1991), seems to be efficient only when used in combination with other sideband suppression techniques (Weigelt *et al.*, 1996).

3.3. Decoupling Noise

Apart from the modulation sidebands, the presence of the RF field may introduce numerous spurious peaks ("decoupling noise") in homo-decoupled spectra. The position and intensity of these peaks depend on the strength and complexity of both the decoupling field and the NMR signal. Although the decoupling noise can partially be averaged out by phase cycling the decoupling waveform with respect to the receiver, various spectrometer instabilities and the presence of strong resonances (water) usually make complete suppression very difficult.

The waveforms based on square wave modulation (composite pulse decoupling) have strong Fourier components outside the decoupling bandwidth associated with sudden switching of phase and/or amplitude within the decoupling waveform. Therefore, these waveforms inevitably create strong noise outside the decoupling bandwidth and for this reason cannot be used for band-selective homonuclear decoupling.

Band-selective decoupling methods based on use of selective pulses usually provide relatively clean off-resonance performance. However, even these waveforms create harmonics at frequencies

$$f = \frac{1}{t_s} \tag{19}$$

where t_s is the length of a single step in the decoupling waveform. Therefore, $1/t_s$ should be at least of the order of the spectral window or more.

Another source of noise is the observable magnetization within or in the vicinity of the decoupling bandwidth. This magnetization is captured (locked) by the decoupling (RF) field and appears at frequencies corresponding to the Fourier components present in the decoupling waveform and their higher-order harmonics. Therefore, even the magnetization that we are interested in can create spikes essentially anywhere within the spectral window. For this reason it is very important to minimize the amount of observable magnetization, retaining only the signals of interest. In most cases this can be conveniently achieved using magnetic field gradients, purging pulses exploiting RF inhomogeneity, or both. The quality of suppression of the unwanted magnetization becomes extremely important in the case of homo-nuclear H–H decoupling in water solutions.

And finally, the amount of locked magnetization depends on the strength of the RF field. Therefore, the lowest possible RF field levels should be used for homonuclear decoupling. For conventional (planarly polarized) inversion pulses, the relationship between the decoupling bandwidth and the RF power is fixed and the RF power directly depends on the decoupling bandwidth. These two parameters are related by the so-called "figure of merit" (Ξ) that has been introduced for heteronuclear wide-band decoupling techniques:

$$\Xi = \frac{\Delta\omega_d}{<\omega_2>} \tag{20}$$

where $\Delta\omega_d$ is the decoupling bandwidth and $<\omega_2>$ is the RMS amplitude of the RF (decoupling) field.

For adiabatic inversion pulses, the inversion bandwidth is proportional to the square root of $<\omega_2>$ (Kupče *et al.*, 1996):

$$\Delta\omega_d = C\sqrt{<\omega_2>} \tag{21}$$

As a result, the RF power can be kept at a minimum. This has a dramatic impact on the off-resonance performance of homo-nuclear adiabatic decoupling. A comparison of noise created by adiabatic (WURST-2) and conventional (G3) decoupling waveforms is shown in Fig. 13.

3.4. Time-Shared Decoupling

Homonuclear decoupling during acquisition requires a time-shared decoupling mode (Jesson *et al.*, 1973; Tomlinson and Hill, 1973). To prevent leaking of the strong RF signal from the transmitter into the receiver, homonuclear decoupling is usually applied in the form of short RF pulses of duration τ_p between sampling intervals of length τ_s as depicted in Fig. 14. In this mode of operation, the decoupling waveform is essentially split into a DANTE-like irradiation sequence (Morris and Freeman, 1978), which according to the theory of Fourier analysis (Walker, 1988) creates the central decoupling band (ω_0) as well as sidebands at frequencies ω_n that are multiples of the spectral width:

$$\omega_n = \omega_0 + n \cdot \Delta\omega_s \tag{22}$$

where $n = 0, \pm 1, \pm 2, \ldots$, and $\Delta\omega_s$ is the spectral width determined by the sampling rate $1/\tau$; τ is the dwell time given by the sum of τ_p and τ_n:

$$\tau = \tau_p + \tau_s \tag{23}$$

Obviously, unless the decoupler frequency is set at the edge of the spectrum, these sidebands fall outside the spectral window and in most cases can be neglected.

The reduction in sensitivity r introduced by time-shared decoupling is given (Jesson *et al.*, 1973) by

$$r = \sqrt{1-d} \tag{24}$$

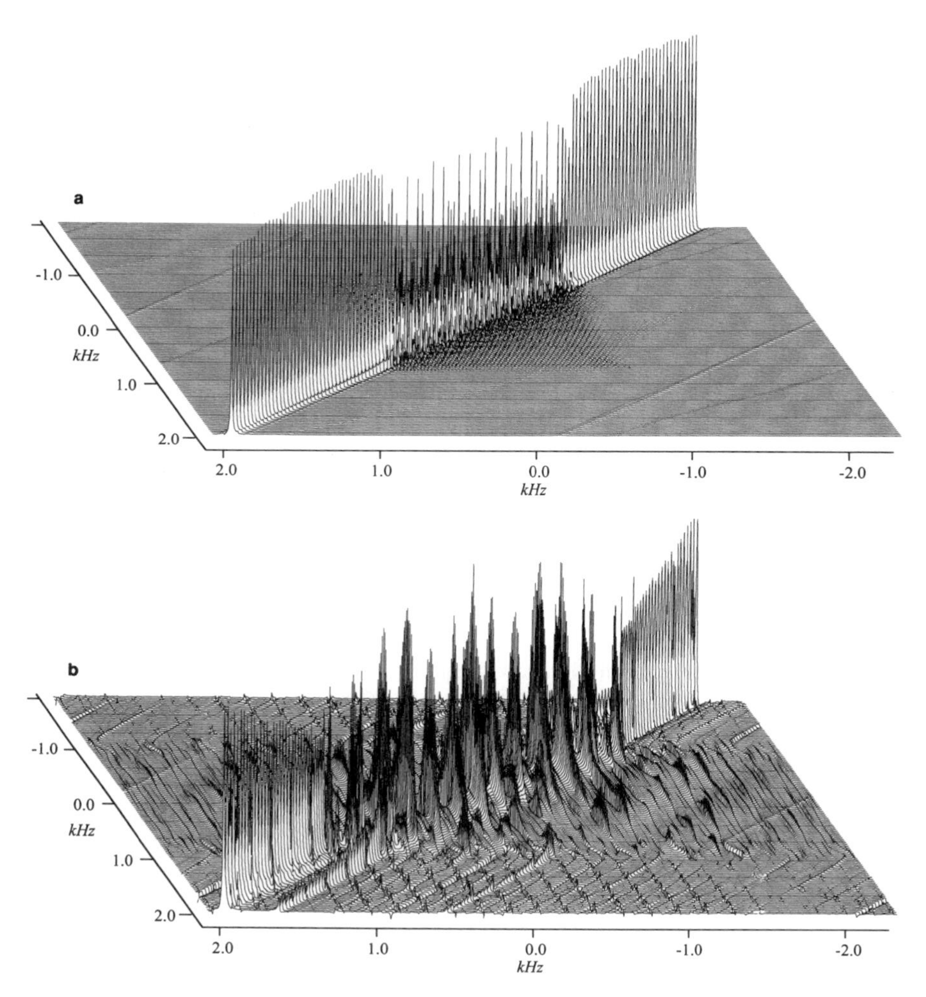

Figure 13. Comparison of spectral artifacts introduced in proton spectra of water by homonuclear time-shared decoupling at 20% duty cycle employing (a) WURST-2 pulses and (b) G3 pulses. The decoupling bandwidth was 1.2 kHz and the position of the water resonance was changed with respect to the decoupler frequency over a total bandwidth of 4 kHz. Reproduced from Kupče and Wagner (1995).

where d is the decoupler duty cycle:

$$d = \frac{\tau_p}{\tau} \tag{25}$$

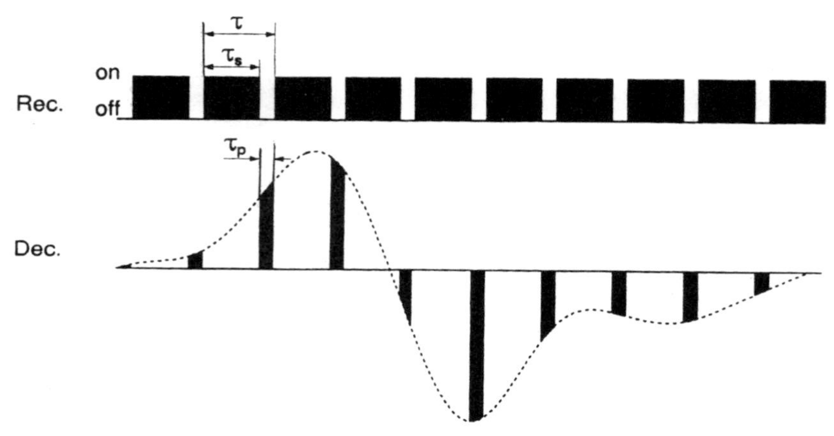

Figure 14. A schematic representation of time-shared homonuclear decoupling discussed in the text. The homo-decoupling is applied in the intervals between the sampling periods.

Therefore, the intrinsic sensitivity loss in the time-shared decoupling is caused by restricted sampling time. To minimize these sensitivity losses, the decoupler duty cycle should be set as low as possible. This in turn is limited by spectrometer hardware: (1) the time-resolution capabilities, (2) the available RF peak amplitude, and (3) the pulse ringdown times. For example, if the decoupler duty cycle is 5%, the corresponding sensitivity loss is essentially negligible (ca. 2.2%). Such a low decoupler duty cycle can only be afforded if the particular decoupling scheme used does not require high RF peak amplitude for the given decoupling bandwidth.

4. ADIABATIC DECOUPLING WAVEFORMS

As demonstrated above, adiabatic decoupling has many advantages over the methods based on the traditional planarly polarized waveforms. In this section we briefly introduce adiabatic pulses that are useful for homonuclear adiabatic decoupling in protein NMR.

4.1. The Adiabatic Condition

To achieve complete inversion of magnetization in an adiabatic sweep experiment, the frequency sweep rate must be slow compared with the variation of the tilt angle so that at any given time the following condition, known as the adiabatic condition (Abragam, 1961), is satisfied:

$$\omega_e >> | d\Theta/dt | \tag{26}$$

This condition can be quantified by introducing the adiabaticity factor Q (Baum *et al.*, 1985):

$$Q = \frac{\omega_e}{|\, d\Theta/dt \,|} \tag{27}$$

The adiabaticity factor can be expressed in terms of experimental parameters as follows (Kupče and Freeman, 1996a):

$$Q = \frac{(\omega_2^2 + \Delta\omega^2)^{3/2}}{|\, \omega_2(d\Delta\omega/dt) - \Delta\omega(d\omega_2/dt) \,|} \tag{28}$$

The second condition that must be satisfied requires the magnetization be aligned along the positive or negative Z-axis at the beginning and end of the sweep:

$$\Theta_{0,T} = \pm\pi/2 \tag{29}$$

This can easily be achieved by turning the RF amplitude on and off smoothly, so that at the beginning and end of the sweep ω_2 is close to zero (Chen *et al.*, 1987, Bohlen and Bodenhausen, 1993; Kupče and Freeman, 1995).

There are many waveforms that satisfy condition (29). One of the simplest and most efficient inversion pulses is the so-called WURST pulse (Kupče and Freeman, 1995), which employs a linear frequency sweep ("chirp"):

$$\Delta\omega(t) = kt \tag{30}$$

and an amplitude modulation function given by

$$\omega_2 = \omega_2(\max)[1 - |\sin(\beta t)|^n] \tag{31}$$

where k is the sweep rate, $\omega_2(\max)$ is the RF peak amplitude, and βt runs from $-\pi/2$ to $+\pi/2$. The index n determines the steepness of the rounding of the amplitude function at the beginning and end of the pulse. Although the value of n is not very critical, the optimum profile is obtained by setting n close to half of the product of the pulse width and sweep width.

4.2. Constant-Adiabaticity Pulses

The linear frequency sweep function becomes inefficient at low magnitudes of n [Eq. (31)] and with simple amplitude-modulated pulses, such as Gaussian (Bauer *et al.*, 1984) or Lorentzian pulses. Fortunately, the performance of such pulses can be improved by optimizing the profile of the frequency sweep (Kupče and Wagner, 1995, Kupče and Freeman, 1996a,b).

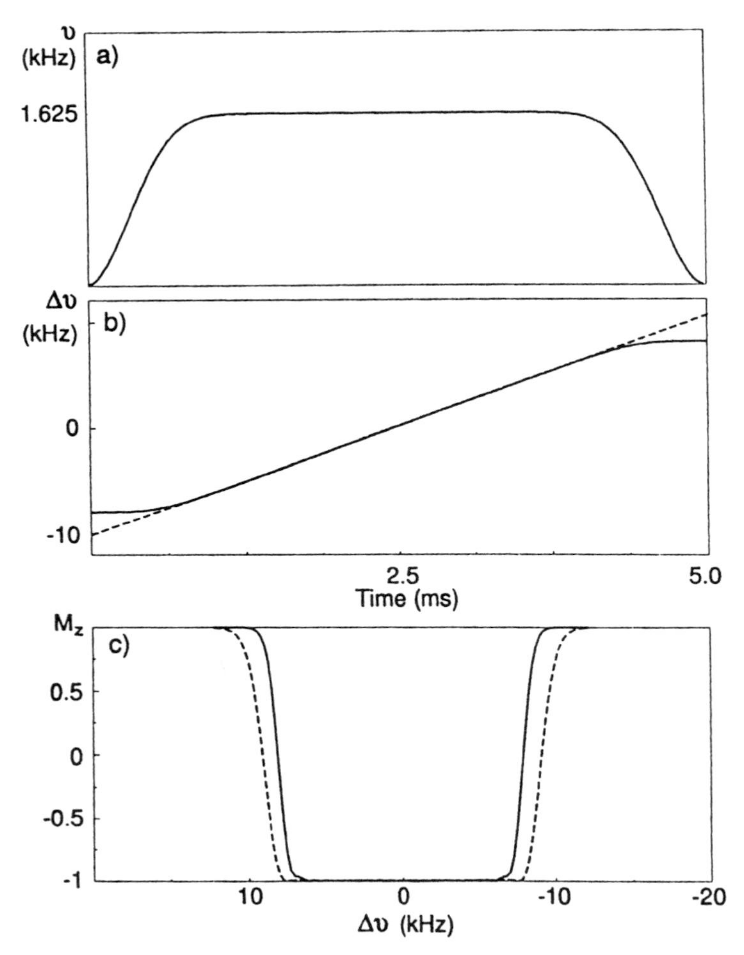

Figure 15. The amplitude (a), frequency (b), and inversion (c) profiles of the conventional (linear sweep) WURST-40 pulse (dashed line) and constant-adiabaticity (CA) WURST-40 pulse (solid line). Both pulses are 5.0 ms long and have equivalent adiabaticities on resonance (4.0) and peak RF amplitudes (1.625 kHz). Note that the CA WURST-40 inversion profile has a slightly steeper transition region and slightly narrower effective bandwidth relative to the original WURST-40 pulse.

By transforming Eq. (28) into a frequency-modulated frame rotating in synchronism with the RF frequency, we arrive at

$$Q_{\mathrm{FM}} = \frac{\omega_2^2}{| d\Delta\omega/dt |} \tag{32}$$

One of the most efficient ways of adiabatic sweep is to keep the minimum adiabaticity constant over the whole frequency sweep range. Therefore, the optimum frequency sweep rate can be obtained by setting

$$Q_{\mathrm{FM}} = q \tag{33}$$

which relates the optimum frequency sweep rate with the amplitude modulation function:

$$k(t) = \frac{[\omega_2(t)]^2}{q} \tag{34}$$

The most common pairs of amplitude and frequency modulation functions are given in Table 2. These pulses have constant adiabaticity in the frequency-modulated reference frame and may be called constant-adiabaticity (CA_{FM}) pulses. They should be distinguished from the constant-adiabaticity pulse (CAP) derived in the usual rotating frame by Baum *et al.* (1985). Because the CAP is a unique pulse, we may drop the FM index and use the abbreviation CA to denote the family of optimized sweep pulses that have constant adiabaticity in the frequency-modulated reference frame (Kupče and Freeman, 1996a,b; Tannus and Garwood, 1996). A typical example of CA pulse is the well-known hyperbolic secant pulse (McCall and Hahn, 1969; Allen and Eberly, 1975; Silver *et al.*, 1985).

Although CA sweep dramatically improves the inversion profiles of simple amplitude-modulated pulses, the effect is not as impressive in the case of WURST pulses with $n > 10$ (see Fig. 15). On the other hand, CA WURST pulses are more convenient to use, because the inversion bandwidth approximately corresponds to the frequency sweep range, whereas if a linear sweep is employed, definition of the inversion bandwidth is less straightforward.

Table 2
Pairs of Amplitude and Frequency Modulation Functions for Constant-Adiabaticity Pulses[a]

Pulse	$\omega_2(t)$	$\Delta\omega(t)$
Hyperbolic secant	$\omega_2(\max)\,\mathrm{sech}(\beta t)$	$\lambda\,\tanh(\beta t)$
Gaussian	$\omega_2(\max)\,\exp(-\beta t^2)$	$\lambda\,\mathrm{erf}(\beta t)$
Lorentzian	$\omega_2(\max)\,(1+\beta^2 t^2)^{-1}$	$\lambda\,[\tan^{-1}(\beta t) + \beta t(1+\beta^2 t^2)^{-1}]/2$
Cosine	$\omega_2(\max)\,\cos(\beta t)$	$\lambda\,[\beta t + \sin(\beta t)\cos(\beta t)]/2$
WURST-2	$\omega_2(\max)\,\cos^2(\beta t)$	$\lambda\,[12\beta t + 8\sin(2\beta t) + \sin(4\beta t)]/32$
Chirp	$\omega_2(\max)$ (constant)	$\lambda\beta t$
Arctan	$\omega_2(\max)\,(1+\beta^2 t^2)^{-1/2}$	$\lambda\,\tan^{-1}(\beta t)$

[a]Parameters β and λ depend on particular function.

5. APPLICATIONS

5.1. H–H Decoupling

Homonuclear decoupling during acquisition in the directly detected dimension is probably one of the most challenging applications of decoupling. The feasibility of band-selective time-shared H–H decoupling was first demonstrated for small molecules in I-BURP-2 decoupled 2D COSY (Kupče and Freeman, 1993a), G3 decoupled 2D ROESY (Hammarstrom and Otting, 1994), and I-SNOB-3 decoupled 2D homonuclear HMQC (Kupče *et al.*, 1995) experiments. These applications employed linearly polarized decoupling waveforms and relatively high decoupler duty cycle of up to 25%. Although the experiments demonstrated substantial increase in resolution and simplification of spectral patterns, no improvement in sensitivity could be obtained.

Up to 70% improvement in sensitivity in a medium-size protein (villin 14T, 126 residues) has been observed in HSQC-based experiments, in 2D ^{15}N–^{1}H HSQC and 3D NOESY-HSQC spectra (Kupče and Wagner, 1995) using adiabatic (WURST-2) decoupling. The consumption of RF power in these experiments is minimized by using 20- to 25-ms-long inversion pulses and a relatively low adiabaticity factor of 1.2. This not only minimizes the artifacts associated with homonuclear decoupling, but also allows considerably wider bandwidths to be covered.

A 3-ppm-wide decoupling band was centered at 4.47 ppm and covered 120 out of 126 villin 14T α-proton resonances. Because the length of the free induction decay in protein spectra is usually less than 0.2 s, short supercycles should be used. Under these circumstances, the five-step supercycle of Tycko *et al.* (1985) was found to give the best performance.

To obtain good-quality spectra, it is very important to achieve excellent water suppression. In our hands the HSQC proposed by Kay *et al.* (1992) provided extremely efficient water suppression, which is achieved by combined use of magnetic field gradients and RF field gradients (Messerli *et al.* 1989; Kay *et al.*, 1993).

Although it is relatively easy to achieve essentially maximum sensitivity and resolution gains in relatively small proteins (see Figs. 16 and 17), the H–H decoupling becomes rather inefficient in larger proteins (>15 kDa) and in any case when the transverse relaxation rate ($1/T_2$) becomes larger than the J coupling. This is shown in Fig. 18 for the 156-residue protein N-ada-17. Some of the resonances still experience a notable line narrowing and increase in intensity, but the overall improvement is not as dramatic.

As saturation of the water signal is inevitable in experiments of this kind, homo-decoupling involving observation of NH protons may also fail to produce

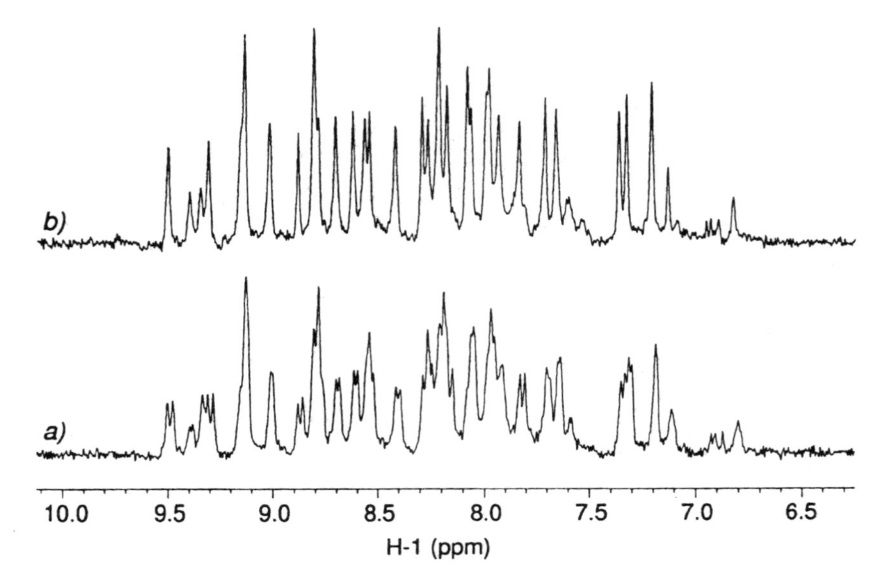

Figure 16. The NH region of the proton spectrum of the 38-residue protein agitoxin 2 (300 μM in 95% H_2O, 5% D_2O) recorded at 400 MHz on a Varian Unity Plus using the $^{15}N-^{1}H$ HSQC pulse sequence: (a) normal spectrum; (b) using time-shared homonuclear decoupling during acquisition. The WURST-2 waveform was applied to CH protons (3.0 to 6.0 ppm) at 5% duty cycle. Essentially all NH doublets collapse into singlets resulting in substantial increase in spectral resolution and signal intensities. The artifacts introduced by adiabatic homo-decoupling are hardly detectable.

any improvement in sensitivity because of exchange between NH protons and water.

Adiabatic decoupling (SESAM) based on hyperbolic secant pulses has also been applied to α protons (from 3.0 to 6.0 ppm) in a 20 mM solution (95% H_2O, 5% D_2O) of bovine pancreatic trypsin inhibitor (BPTI) in a 2D NOESY experiment (Weigelt *et al.*, 1996). Presaturation of the water signal was used in combination with the WATERGATE sequence to maximize the water suppression efficiency and minimize the artifacts introduced by the decoupling. Although the homonuclear decoupling was primarily used to suppress *J* cross-peaks, these peaks were partially reintroduced by the asynchronous (time-shifted) mode of decoupling. An additional 90° pulse applied just before the acquisition ensured complete suppression of the *J* cross-peaks. The failure to obtain sensitivity improvement in this experiment can be attributed to a relatively short length of the inversion pulse (6-ms hyperbolic secant) used for the decoupling, which, in turn, required use of excess power and (possibly) excess decoupler duty cycle (not quoted).

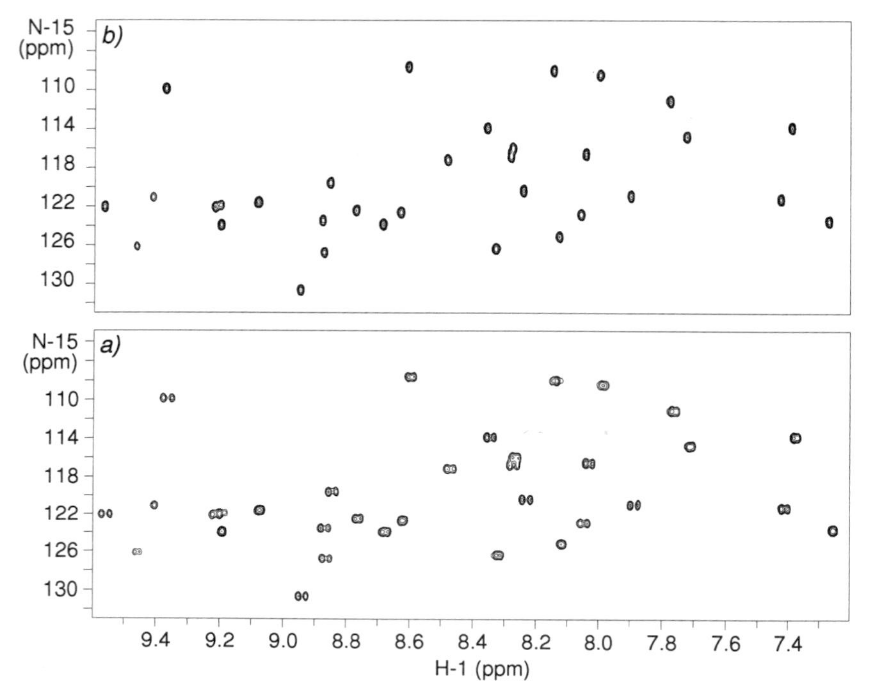

Figure 17. The 2D $^{15}N-^{1}H$ HSQC spectra of the 38-residue protein Agitoxin 2 (300 μM in 95% H_2O, 5% D_2O) recorded at 400 MHz: (a) normal spectrum; (b) homo-decoupled spectrum (for experimental details see Fig. 16). Both spectra were processed and plotted equally. Note the increased resolution and simplification in the spectrum (b).

5.2. C–C Decoupling

Spectra of ^{13}C-enriched proteins show large (30 to 60 Hz) carbon–carbon couplings that dramatically reduce both the sensitivity and resolution in many NMR experiments. Although constant-time experiments can be used to remove the C–C splittings (Santoro and King, 1992), they introduce a delay that leads to a significant loss of sensitivity related to efficient relaxation of protonated carbons.

The homonuclear C–C decoupling has become a standard part of many important experiments in biomolecular NMR, such as HNCA, HNCOCA (Kay *et al.*, 1990), or HCCH TOCSY (Bax *et al.*, 1990; Kay *et al.*, 1993). Most of these experiments employ decoupling of carbonyls as they are well separated from the other ^{13}C resonances in protein spectra and therefore are a relatively easy target for homonuclear decoupling. The SEDUCE-1 decoupling (McCoy and Mueller, 1992a) has been used extensively for these purposes (Yamazaki *et al.*, 1994a,b).

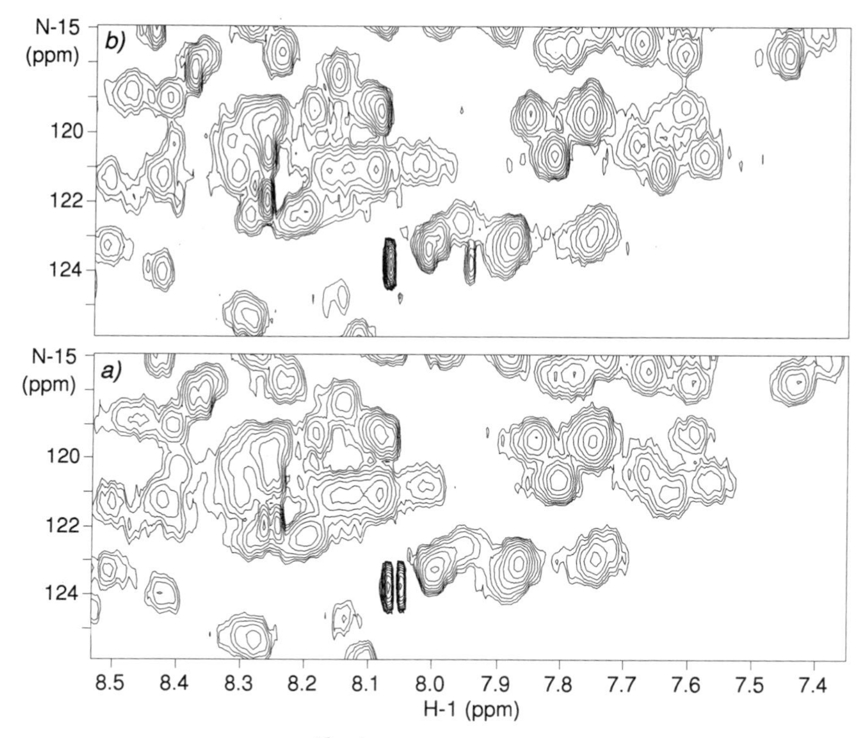

Figure 18. A crowded region of 2D ^{15}N–^{1}H HSQC spectra of the 17-kDa protein N-ada-17 (1 mM in 90% H_2O, 10% D_2O) recorded at 400 MHz: (a) normal spectrum; (b) homo-decoupled spectrum (for experimental details see Fig. 16). Although the resolution improvement is still notable, it is reaching its limits and is not as impressive as in the case of smaller proteins (see Fig. 17).

As discussed above, adiabatic decoupling waveforms can improve the quality of the decoupling and decrease the power dissipated in the sample. Further benefits from adiabatic homonuclear decoupling can be obtained by extending the decoupling to other regions, for instance to β-carbons, and by simultaneous decoupling of several bands.

5.2.1. ^{13}C HSQC

One of the simplest experiments to implement the C–C decoupling is the 2D ^{13}C HSQC (Kupče and Wagner, 1996). A simple gradient echo sequence employing a selective ^{13}C 180° r-SNOB pulse (Kupče *et al.*, 1995) is introduced into the standard pulse sequence (Kay *et al.*, 1992) to preserve only the bandwidth of interest (C_α region) and minimize the artifacts folding in from the decoupled regions (see Fig. 19).

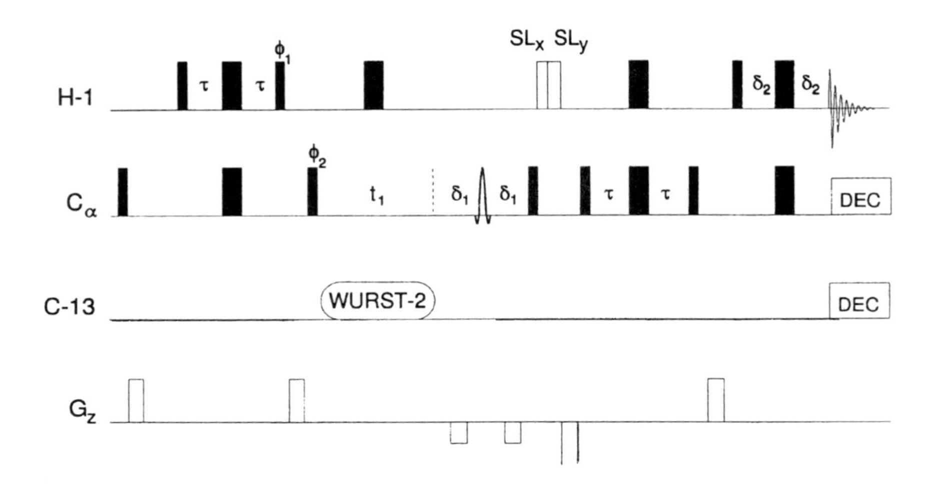

Figure 19. The HSQC pulse sequence used to obtain 2D homo-decoupled ^{13}C–^{1}H correlated spectra. The pulse sequence was adapted from Kay *et al.* (1992) and modified to include C–C decoupling and a selective refocussing (r-SNOB) pulse (Kupče *et al.*, 1995) to minimize the artifacts introduced by homo-decoupling. The phases are $\phi_1 = 90°, 270°$, $\phi_2 = 0°, 0°, 180°, 180°$, rec = 0°, 180°, 180°, 0°.

The efficiency of C–C decoupling is shown in Fig. 20 for the 126-residue protein villin 14T (Markus *et al.*, 1994). Apart from the C_α resonances of 14 glycine residues that cluster in the region between 39.6 and 42.6 ppm, all other α-carbons appear between 46 and 64 ppm. The conventional ^{13}C HSQC spectrum of this region revealing a severe overlap of cross-peaks is shown in Fig. 20a.

Decoupling of carbonyls that typically appear between 167 and 177 ppm is readily achieved using WURST-2 waveform. A 4-ms-long WURST-2 pulse superimposed with a five-step TPG-5 supercycle were centered at 172 ppm covering a 15-ppm-wide band. Relatively high adiabaticity of ca. 2 for the adiabatic inversion pulses can be afforded because the carbonyls are well separated from the α-carbons. This required an RMS RF amplitude of only 0.47 kHz providing a clean decoupling (see Fig. 20b) and the expected twofold increase in C_α signal intensities (Fig. 21b).

Further simplification of spectra and sensitivity gain can be obtained by applying a second decoupling field to β-carbons. Most of the β-carbon signals appear in the region between 23 and 43 ppm. The C_α–C_β couplings are somewhat smaller (around 40 Hz) than C_α–CO couplings (ca. 55 Hz) allowing use of longer decoupling pulses and, therefore, less RF power for the same bandwidth. In practice, it is convenient to use the same pulse length for both CO and C_β regions. Usually a 5-ms-long pulse is a reasonably good compromise. The spectrum recorded using two simultaneous WURST-2 decoupling waveforms (Fig. 20c) cen-

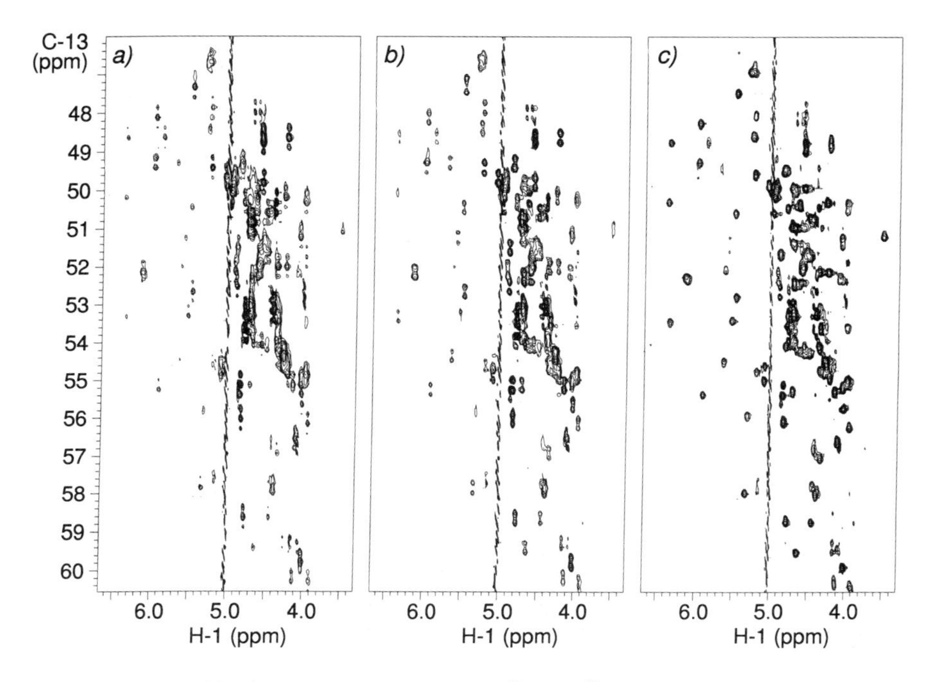

Figure 20. The 2D ^{13}C–^{1}H correlated HSQC spectra of ^{13}C- and ^{15}N-labeled 126-residue protein villin 14T (1 mM in 90% H_2O, 10% D_2O) recorded at 750 MHz (proton) frequency on a Unity Plus (Varian) spectrometer. Four transients were acquired per t_1 value with a total of 1024 increments. (a) The conventional spectrum, (b) CO-decoupled spectrum, and (c) CO- and C_β-decoupled spectrum. (Adapted from Kupče & Wagner, 1996).

tered on CO (172 ppm) and β-carbons (33 ppm) and covering bands 13 ppm (CO) and 20 ppm (C_β) wide is very similar to that recorded for proteins with a natural abundance of ^{13}C, except the sensitivity is considerably higher. In addition to the effect of ^{13}C labeling, a sensitivity gain by a factor of 4 (time saving by a factor of 16) is obtained using band-selective homonuclear decoupling at two frequency bands (see Fig. 21c).

Unfortunately, decoupling of all β-carbons is frequently impossible. Particularly, C_β of the serine peaks overlap with the region of α-carbons. The β-carbons of alanines appear at the high-field end of the carbon spectrum and attempts to extend the bandwidth of C_β decoupling beyond the 20-ppm bandwidth usually lead to unacceptable levels of modulation sidebands and loss of signal intensities. And finally, the β resonances of threonines appear at the low-field end of the aliphatic part of the carbon spectrum. These two groups of resonances can still be decoupled by applying additional RF fields centered at 13 ppm (alanines) and 67.5 ppm

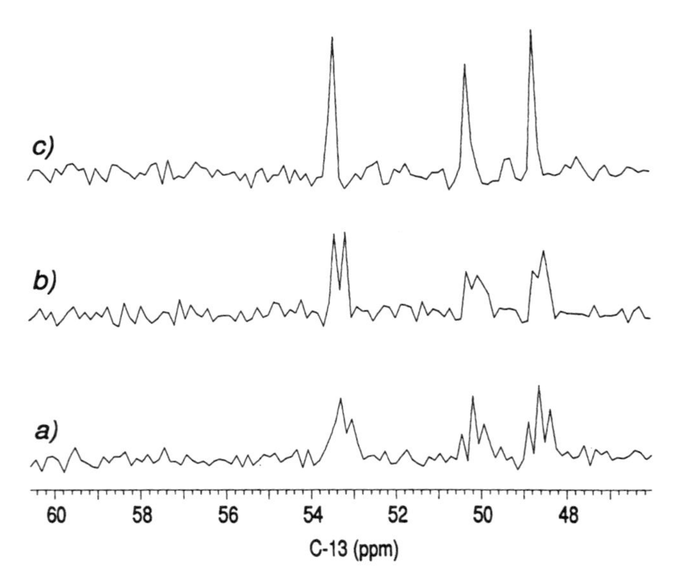

Figure 21. Traces extracted at $\delta_H = 6.36$ ppm from spectra shown in Fig. 20 demonstrate the sensitivity enhancement and spectral simplification obtained using homonuclear decoupling.

(threonines). Excluding empty spectral regions from the decoupling helps to keep the level of the decoupling field at a minimum. An example of three-band decoupling including the alanine resonances is shown in Fig. 22. A four-band decoupling including also the threonine resonances has been used in HNCA and HCCH TOCSY experiments (Matsuo *et al.*, 1996a,b), and will be discussed below. Such a gradual C–C decoupling may be helpful for signal assignment.

5.2.2. HNCA and HN(CO)CA Experiments

The HNCA and HN(CO)CA experiments (Kay *et al.*, 1990; Grzesiek and Bax, 1992a,b; Yamazaki *et al.*, 1994) are now standard techniques in biomolecular NMR. Because the pulse sequences contain a C_α evolution period, both CO and C_β decoupling can be used to improve sensitivity and resolution of these experiments. It has been demonstrated that the C_β decoupled (Cbd) versions of HNCA and HN(CO)CA experiments (Matsuo *et al.*, 1996a) are significantly more sensitive than corresponding constant-time experiments.

The Cbd-HNCA and Cbd-HN(CO)CA pulse sequences are shown in Fig. 23. These sequences employ the WATERGATE (Sklenar *et al.*, 1993) module instead of the sensitivity enhancement and coherence selection modules suggested originally (Yamazaki *et al.*, 1994a,b). In addition, two water-selective e-SNOB flip-back

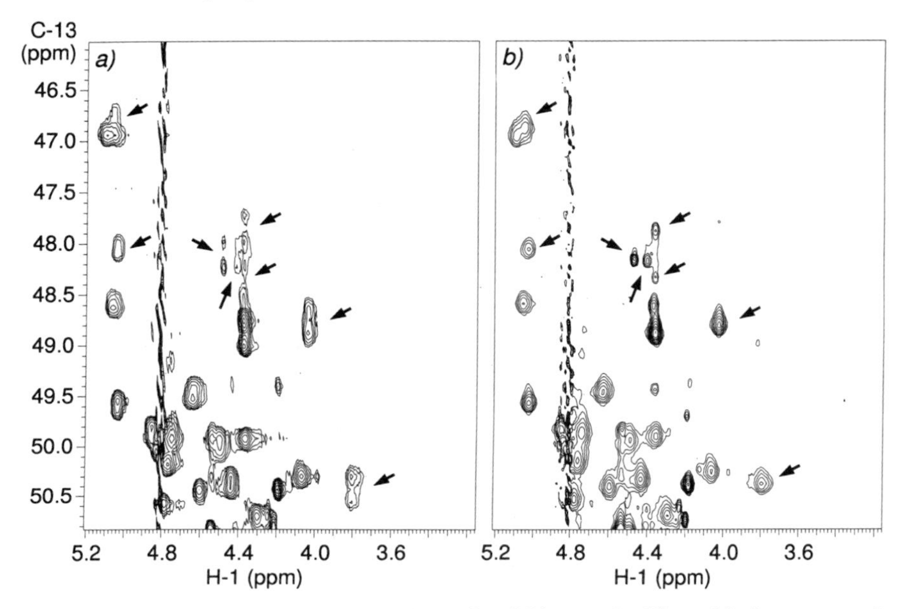

Figure 22. The effect of introducing the third decoupling field centered at 15 ppm (alanine resonances) on the spectrum of Fig. 20c. Only the alanine C_α region is shown. Several alanine C_α peaks indicated by arrows collapse into singlets.

pulses are incorporated into the sequences so as to keep water magnetization along the +*z* axis during the experiment and, in particular, prior to acquisition. Four simultaneous homonuclear WURST-2 decoupling fields are applied to carbonyls (167–177 ppm) and β-carbons (13.5–17.5, 20–40, and 65.5.–69.5 ppm) during the C_α evolution period leaving only the serine resonances undecoupled. The improvement in spectral resolution with β-carbon decoupling is demonstrated in Fig. 24 for the first plane of a Cbd-HNCA of ^{13}C- and ^{15}N-labeled lysozyme.

The sequences have been applied (Matsuo *et al.*, 1996a) both to deuterated and protonated ^{13}C- and ^{15}N-labeled eukaryotic translation initiation factor eIF-4E protein (213 residues) and to cyclin-dependent kinase inhibitor p16 (140 residues). Significant sensitivity improvements over the constant-time versions of these experiments have been observed. Figure 25 compare traces from Cbd-, CT-, and nondecoupled HNCA spectra of 100% deuterated eIF-4E (213 residues) at 1 mM concentration within a CHAPS micelle. Figure 26 compares the first ^1H–^{13}C planes from 3D HNCA spectra recorded with a 250 μM sample of the perdeuterated cyclin-dependent kinase inhibitor p16.

One drawback of the Cbd experiments is that glycine C_α signals partially overlap with the major decoupling band (20–40 ppm) and are usually strongly affected by the RF field. The problem usually is less severe at higher magnetic fields.

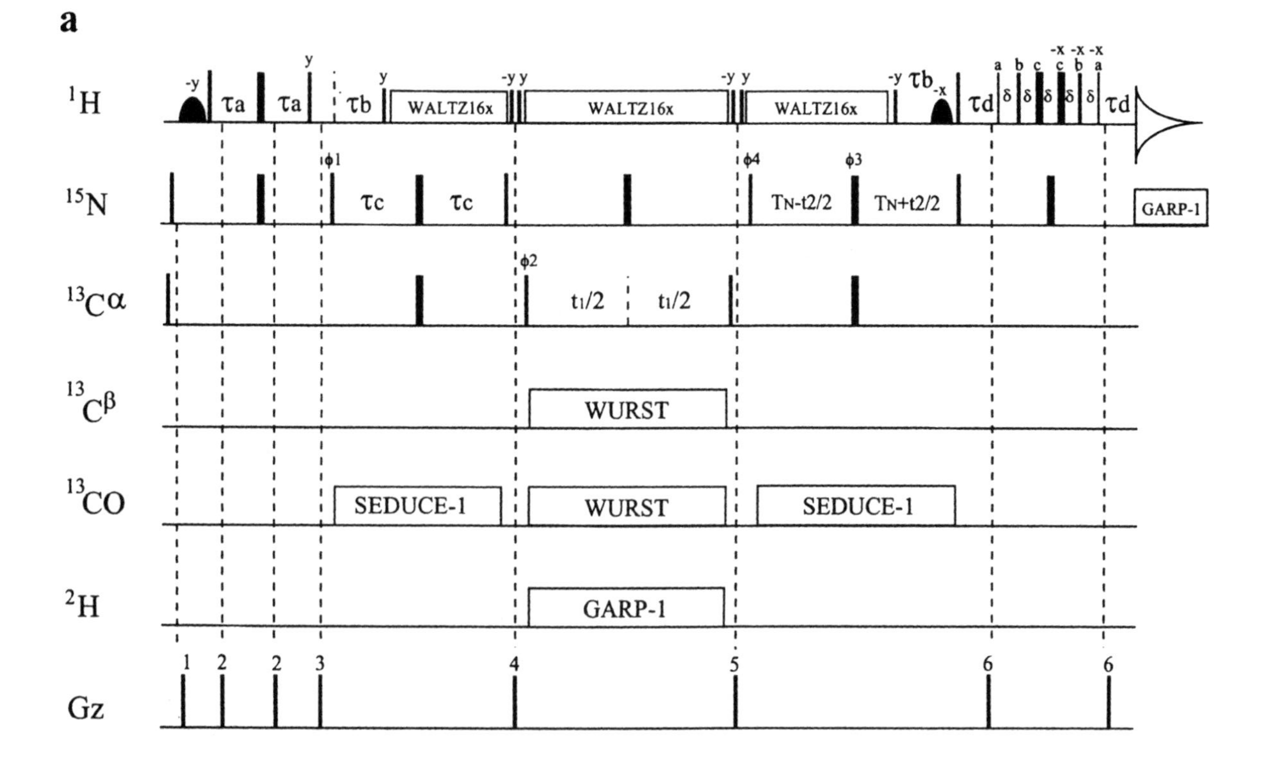
a
1H
15N
13Cα
13Cβ
13CO
2H
Gz
τa
τa
τb
WALTZ16x
WALTZ16x
WALTZ16x
τb
τd
δ
τd
φ1
τc
τc
φ4
φ3
TN-t2/2
TN+t2/2
GARP-1
φ2
t1/2
t1/2
WURST
SEDUCE-1
WURST
SEDUCE-1
GARP-1
1
2
2
3
4
5
6
6

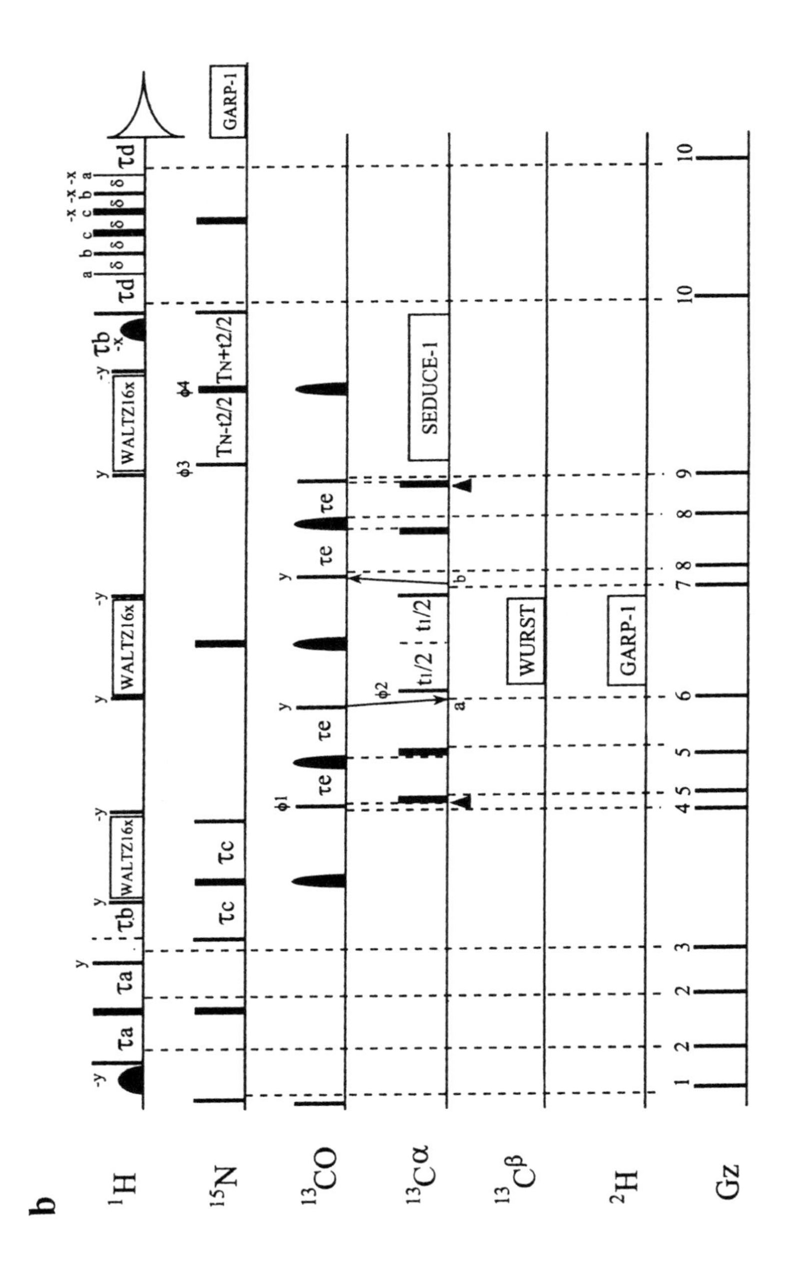
b
1H
15N
13CO
13Cα
13Cβ
2H
Gz
WALTZ16x
WALTZ16x
WALTZ16x
GARP-1
SEDUCE-1
WURST
GARP-1
τa
τa
τb
τc
τc
τe
τe
τe
τe
τb
τd
τd
δ
t1/2
t1/2
TN-t2/2
TN+t2/2
φ1
φ2
φ3
φ4

Figure 23. (a) Pulse sequences for (a) the Cbd-HNCA and (b) the Cbd-HN(CO)CA experiment (Matsuo *et al.*, 1996a). In both experiments, narrow and wide pulses represent 90° and 180° pulses, respectively. Two selective e-SNOB pulses (Kupče *et al.*, 1995) are applied to maintain water along the $+z$ axis. In the WATERGATE sequence (Sklenar *et al.*, 1993), a, b, and c correspond to 1/13, 9/13, and 19/13 * 2π pulse, respectively. Unless indicated otherwise, pulses are applied along the x axis. The ^{1}H, ^{15}N, and ^{13}C carriers are centered at 4.71 (water at 30°C), 118, and 52.5 ppm, respectively. (a) Proton pulses are applied using a 25-kHz field with the exception of the water-selective 90° pulses, which are applied as e-SNOB shaped pulses (Kupče *et al.*, 1995) generated with the Pbox software (Varian NMR Instruments). ^{1}H decoupling is achieved by WALTZ-16x with 90° flanking pulses and a 6.8-kHz field as described for other triple resonance experiments for backbone assignment (Yamazaki *et al.*, 1994a,b). The hard ^{15}N pulses were applied with a 5.2 kHz field, and a 827 Hz field was used for GARP-1 (Shaka *et al.*, 1985) decoupling during acquisition. The ^{13}Ca 90° pulse is applied using a 15.1 kHz field, and ^{13}CO decoupling is achieved by a 690 Hz field with SEDUCE-1 profile (McCoy and Mueller, 1992a, 1993). All decoupling periods are interrupted during the application of the gradient pulses (Kay, 1993). The ^{2}H decoupling is achieved using a 770 Hz field with GARP-1 profile (Shaka *et al.*, 1985) centered at 3 ppm. The C^β decoupling is achieved by using a WURST-2 profile (Kupče and Freeman, 1996a,b) generated with the Pbox tool. For the C^β of alanine and threonine, a 500-Hz bandwidth is decoupled centered at 15.5 and 67.5 ppm, respectively, and major C^β decoupling is achieved with a 2500-Hz bandwidth centered at 30 ppm. The CO decoupling is also applied in the C^α evolution period using the WURST-2 profile with a 2500-Hz bandwidth centered at 173 ppm. The rms B_1 field is 556 Hz to achieve the C^β and CO decoupling. The phase of the WURST decoupling is cycled $\pm x$ to suppress cycling side bands. The delays used are $\tau a = 2.3$ ms, $\tau b = 5.5$ ms, $\tau c = 12.4$ ms, $\tau d = (2\tau a - 3.37\delta)/2$, $\delta = 0.21$ ms, and $T_N = 12.4$ ms. The phase cycling is: $\phi 1 = (x, -x)$; $\phi 2 = 2(x), 2(-x)$; $\phi 3 = 4(x), 4(-x)$; rec = $(x, -x, -x, x)$. Quadrature detection in F1 and F2 are achieved by States–TPPI (Marion *et al.*, 1989), of $\phi 2$ and $\phi 4$, respectively. The durations and strengths of the gradients (rectangular) are g1 = (0.5 ms, 8 G/cm), g2 = (0.5 ms, 4 G/cm), g3 = (1.0 ms, 10 G/cm), g4 = (1.0 ms, −5 G/cm), g5 = (0.6 ms, 10 G/cm), g6 = (1.0 ms, 15 G/cm). (b) Cbd-HN(CO)CA experiment. The differences from the Cbd-HNCA are as follows. The ^{13}C carrier is first centered at 173 ppm, at point a the carrier is shifted to 52.5 ppm (C^α) and subsequently returned to 173 ppm at point b. All carbon 90° pulses are applied using a field strength of 3.90 kHz to minimize excitation of uninteresting spins (Marion *et al.*, 1989). All $^{13}C^\alpha$ 180° pulses (8.80 kHz) are applied in phase-modulated manner. ^{13}CO 180° pulses are applied using the SEDUCE-1 profile (McCoy and Mueller, 1992a, 1993) with a pulse length of 230 μs, and the ^{13}CO 180° pulses between a and b were applied in phase-modulated manner. The two arrowheads indicate the positions of the Bloch–Siegert compensation pulses (Grzesiek and Bax, 1992b). A SEDUCE-1 profile (McCoy and Mueller, 1992a, 1993) of 25-ppm bandwidth was used for ^{13}Ca decoupling during t_2. The C^β decoupling is achieved using a WURST-2 profile (Kupče and Freeman, 1996a,b) generated with the Pbox tool. For the C^β of alanine and threonine, 500-Hz bandwidths are used centered at 15.5 and 67.5 ppm, respectively, and major C^β decoupling is achieved with a 2500-Hz bandwidth centered at 30 ppm. The rms B_1 field is 424 Hz to achieve the C^β decoupling. The delays used are $\tau a = 2.3$ ms, $\tau b = 5.5$ ms, $\tau c = 12.4$ ms, $\tau d = (2\tau a - 3.37\delta)/2$, $\tau e = 4.1$ ms, $\delta = 0.21$ ms, and $T_N = 12.4$ ms. The phase cycling is: $\phi 1 = (x, -x)$; $\phi 2 = 2(x), 2(-x)$; $\phi 3 = x$; $\phi 4 = 4(x), 4(-x)$; rec = $(x, -x, -x, x)$. Quadrature detection in F1 and F2 are achieved by States–TPPI (Marion *et al.*, 1989), of $\phi 2$ and $\phi 3$, respectively. The durations and strengths of the gradients (rectangular) are g1 = (0.5 ms, 8 G/cm), g2 = (0.5 ms, 4 G/cm), g3 = (1.0 ms, 10 G/cm), g4 = (1.0 ms, 1 G/cm), g5 = (1.0 ms, 7 G/cm), g6 = (1.0 ms, −15 G/cm), g7 = (1.0 ms, 8 G/cm), g8 = (1.0 ms, 0.4 G/cm), g9 = (0.6 ms, 10 G/cm), g10 = (1.0 ms, 15 G/cm).

On the other hand, the lack of C_β coupling makes the glycine signals relatively intense and easy to observe in regular HNCA and HN(CO)CA experiments.

Another drawback of homonuclear decoupling is the Bloch–Siegert (ω_{BS}) shifts of the C_α resonances. This is only a minor problem, because ω_{BS} will be the same in both the Cbd-HNCA and Cbd-HNCOCA experiments, provided the same decoupling field is used. Therefore, the sequential assignment of spin systems by matching the C_α chemical shifts is straightforward. On the other hand, connecting sidechain spin systems with backbone resonances may require quantitative information about the Bloch–Siegert shifts. As discussed above, the ω_{BS} can be estimated using the Bloch equations. In Fig. 27 the Bloch–Siegert shifts measured from constant-time HN(CO)CA and Cbd-HN(CO)CA of eIF-4E are overlaid with the calculated curve. The measured Bloch–Siegert shifts are rather small, with a maximum of 0.25 ppm. In practice, such a calibration curve can easily be established by matching only a few well-resolved HNCA peaks of a simple and well-behaved protein with the calculated offset dependence of the Bloch–Siegert shift. The calibration curve will be valid for any other protein provided the same decoupling waveform and the same field strength are used.

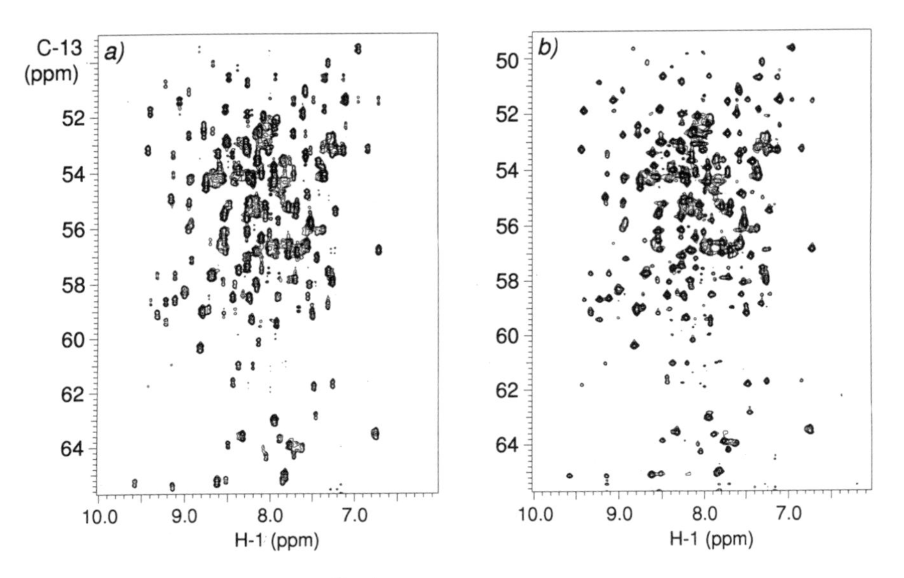

Figure 24. Comparison of the first ^{13}C–^{1}H planes from (a) regular and (b) C_β decoupled 3D HNCA experiments of 2 mM T4 lysozyme recorded at a (proton) frequency of 750 MHz on a Varian Unity Plus spectrometer.

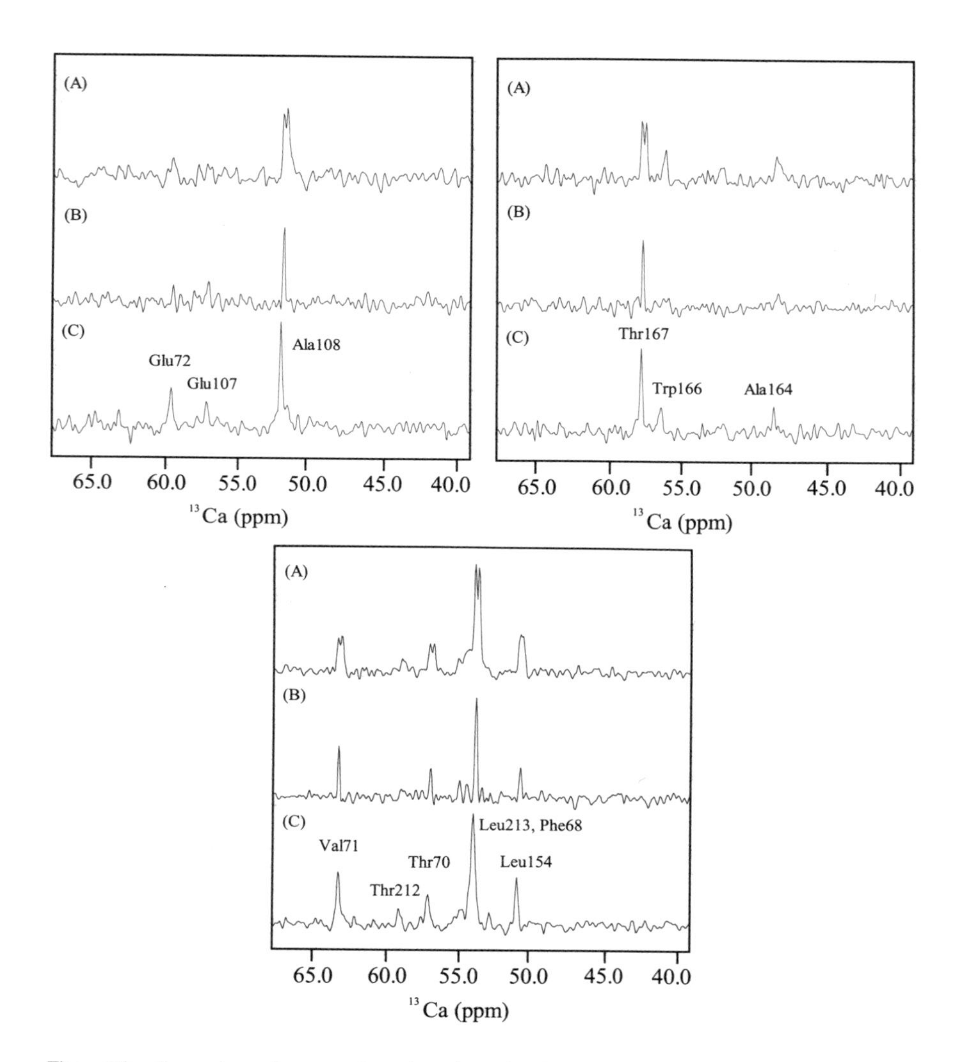

Figure 25. Comparison of cross sections along the carbon frequency through cross-peaks in 3D HNCA experiments recorded with a 1 mM sample of ^{13}C, ^{15}N and ^{2}H-labeled eIF-4E. The three panels show cross-peaks for different types of amino acid residues. The top trace (A) in each panel is from a regular undecoupled HNCA experiment. The middle trace (B) is from an experiment with constant-time C_α evolution. The bottom trace (c) is from the Cbd-HNCA recorded with the pulse sequence of Fig. 23a. Adapted from Matsuo *et al.* (1996a).

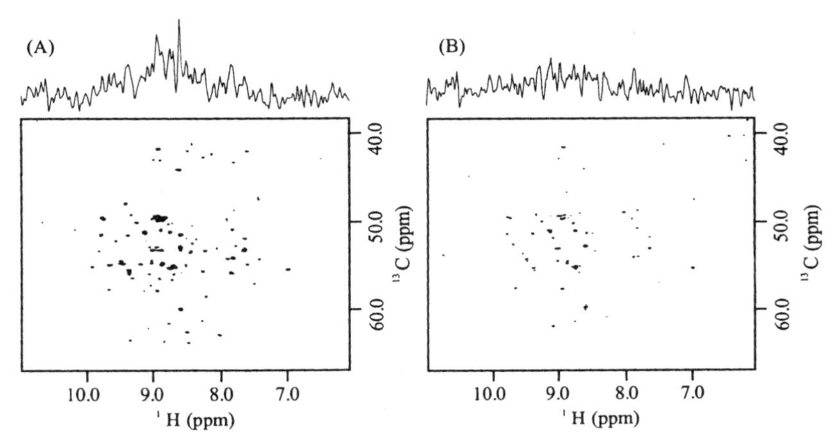

Figure 26. Comparison of the first ^{1}H–^{13}C planes from 3D HNCA experiments recorded with a 250 μM sample of the cyclin-dependent kinase inhibitor p16. (A) Spectrum recorded with the Cbd-HNCA of Fig. 23a with homonuclear carbon decoupling; (B) regular constant-time HNCA experiment. The traces on top are skyline projections.

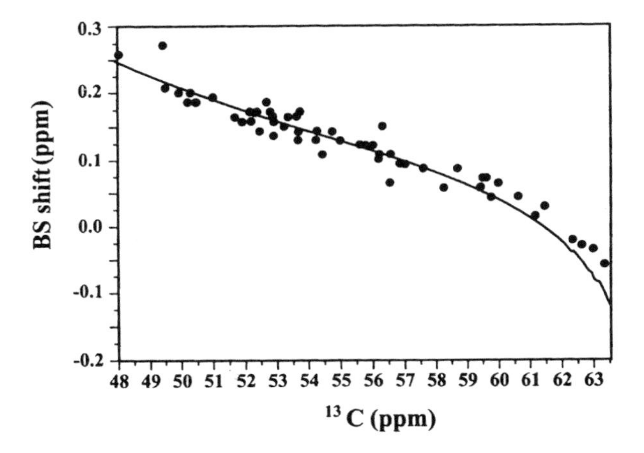

Figure 27. Calibration of the Bloch–Siegert shifts. The $^{13}C_{\alpha}$ chemical shift difference between the constant-time HN(CO)CA and the Cbd-HN(CO)CA spectra resulting from the Bloch–Siegert effect (Matsuo *et al.*, 1996a). The 2D ($^{13}C_{\alpha}$/HN) spectra of 100% deuterated eIF-4E protein were recorded using the pulse sequence of 3D CT-HN(CO)CA (Yamazaki *et al.*, 1994b) and the pulse sequence of Fig. 23b. The experiments were acquired as a 95×512 complex matrix with spectral widths of 3700 and 8000 Hz, respectively. The data were doubled in $^{13}C_{\alpha}$ dimension by linear prediction, apodized using a 90°-shifted squared sine-bell window function, zero filled to 4096 complex points followed by Fourier transformation. The digital resolution of the spectra in $^{13}C_{\alpha}$ dimension is 0.9 Hz (0.0072 ppm). A total of 63 well-resolved peaks were used for comparison. The solid curve represents a simulation of the Bloch–Siegert effects calculated on the basis of the known theory (Bloch and Siegert, 1940; Ramsey, 1955; McCoy and Mueller, 1992c).

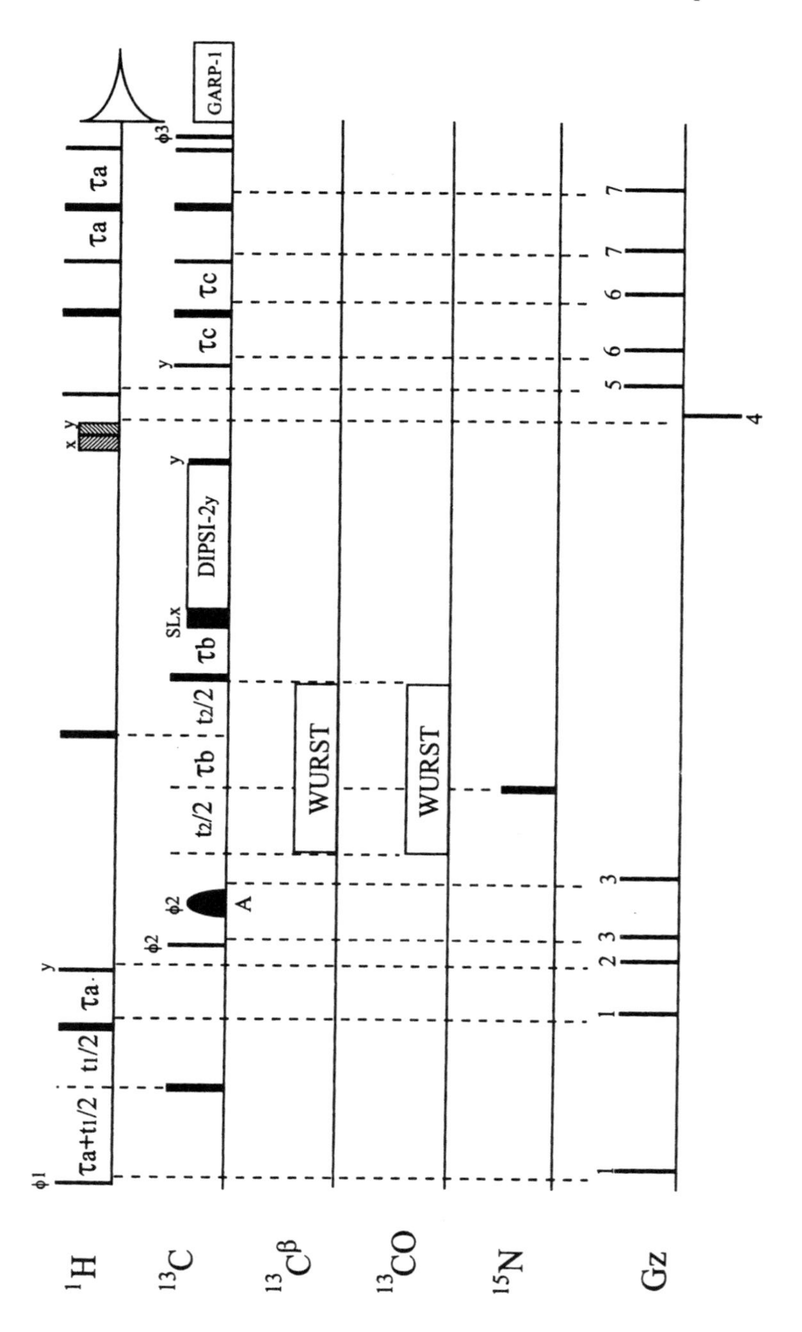

1H
13C
13Cβ
13CO
15N
Gz
τa+t1/2
t1/2
τa
t2/2
τb
SL,x
DIPSI-2y
τc
GARP-1
WURST
WURST
A
φ1
φ2
φ3
x
y
1 1 2 3 3 4 5 6 6 7 7

Figure 28. Pulse sequence of the Cbd-HCCH TOCSY experiment (Matsuo *et al.*, 1996). All narrow and wide pulses represent 90° and 180° pulses, respectively. The ^{1}H, ^{15}N, and ^{13}C carriers are centered at 4.71 (water at 30°C), 118, and 43.0 ppm, respectively. The C^{α}-selective 180° pulse (pw = 1.2 ms, bw = 1500 Hz) is applied with r-SNOB shape (Kupče *et al.*, 1995) as a frequency-shifted pulse at point A. All other ^{13}C pulses are applied at a field strength of 23 kHz with the exception of the 2-ms SLx pulse, the DIPSI-2 mixing sequence (Shaka *et al.*, 1985), the 90° pulse immediately following the DIPSI-2 scheme where RF field strengths of 4.7 kHz were employed, and the WURST-2 (Kupče and Wagner, 1995, 1996; Kupče and Freeman, 1996a,b) decoupling sequence. All of the pulses in the DIPSI-2 sequence are applied along $\pm y$. The phase of the WURST-2 decoupling is cycled independently (x,y) to suppress the spurious peaks introduced by homonuclear decoupling. The ^{1}H pulses are applied using a 44-kHz field with the exception of the SL pulses, which are applied as consecutive 7- and 4.4-ms pulses with an RF field strength of 10 kHz.Note that this water suppression method was proposed previously by Kay *et al.* (1993). The hard ^{15}N pulses were applied with a field strength of 5.0 kHz. Carbon decoupling during acquisition is achieved using the GARP-1 decoupling sequence (Shaka *et al.*, 1985) with a field strength of 4.2 kHz. The C^{β} decoupling is achieved by using the WURST-2 sequence (Kupče and Freeman, 1995; Kupče and Wagner, 1995) in the t_2 period. For the C^{β} of alanine and threonine, a 600-Hz bandwidth is decoupled, centered at 15.0 and 68.0 ppm, respectively, and major C^{β} decoupling is achieved with a 2400-Hz bandwidth centered at 30 ppm. The CO decoupling is also applied in the C^{α} evolution period by using the WURST-2 sequence with a 2000-Hz bandwidth centered at 173 ppm. A B_1(rms) field strength of 420 Hz is used to simultaneously achieve the C^{β} and CO decoupling. The C^{α}-selective 180° pulse, the WURST-2 decoupling sequence, and the DIPSI-2 sequence were implemented with the Pbox tool (Kupe and Freeman, 1993b). The delays used are $\tau a = 1.6$ ms, $\tau b = 1.2$ ms, $\tau c = 1.1$ ms. The phase cycles are $\phi 1 = (x, -x)$; $\phi 2 = 2(x), 2(-x)$; $\phi 3 = 4(x), 4(-x)$; rec = $(x, -x, -x, x)$. Quadrature detection in F1 and F2 are achieved by States–TPPI (Marion *et al.*, 1989), of $\phi 1$ and $\phi 2$, respectively. The durations and strengths of the gradients (rectangular) are g1 = (0.5 ms, 8 G/cm), g2 = (2.0 ms, 15 G/cm), g3 = (0.5 ms, 8 G/cm), g4 = (7.0 ms, −30 G/cm), g5 = (4.4 ms, 30 G/cm), g6 = (0.5 ms, 8 G/cm), g7 = (0.5 ms, 8 G/cm).

5.2.3. HCCH TOCSY

The HCCH TOCSY experiment (Bax *et al.*, 1990; Kay *et al.*, 1993) is a particularly valuable technique for sidechain resonance assignments. Unfortunately, for larger proteins extensive signal overlap makes the assignments difficult to obtain. The performance of the HCCH TOCSY experiment can be substantially improved by introducing C_β decoupling, which increases both resolution and sensitivity.

The Cbd-HCCH TOCSY pulse sequence (Matsuo *et al.*, 1996b) shown in Fig. 28 is based on a pulse sequence proposed by Kay *et al.* (1993). A gradient echo sequence based on the r-SNOB refocusing pulse is incorporated to select only the C_α region of the ^{13}C spectrum. This retains only coherence pathways from H_α via C_α to sidechain carbons and protons. The opposite pathway starting from the sidechain protons is suppressed. As a result, the Cbd-HCCH TOCSY experiment selects only cross-peaks among H_α (ω_1), C_α region (ω_2), and sidechain protons (ω_3), reducing the size of the data matrix and the experiment time dramatically as compared with the standard version of the experiment. The spectral width in $\omega_1(H_\alpha)$ is reduced by a factor of 2 (from 6 ppm to 3 ppm) and in $\omega_2(C_\alpha)$ by a factor of 3

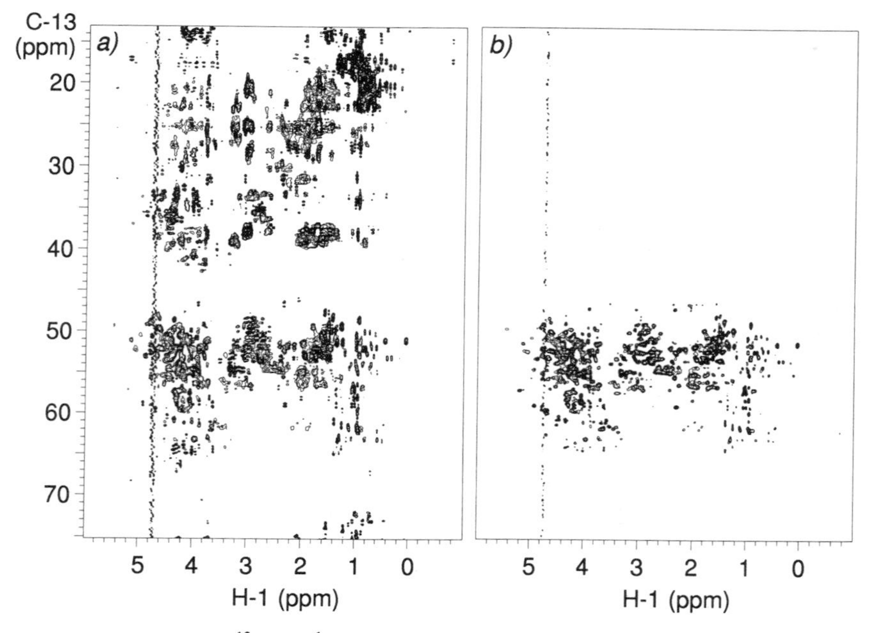

Figure 29. Comparison of $^{13}C(\omega_2)-^{1}H(\omega_3)$ planes from (a) regular and (b) Cbd-HCCH TOCSY spectra of 2 mM sample of T4 lysozyme recorded at a (proton) frequency of 400 MHz on a Varian Unity Plus spectrometer using pulse sequences of Fig. 28.

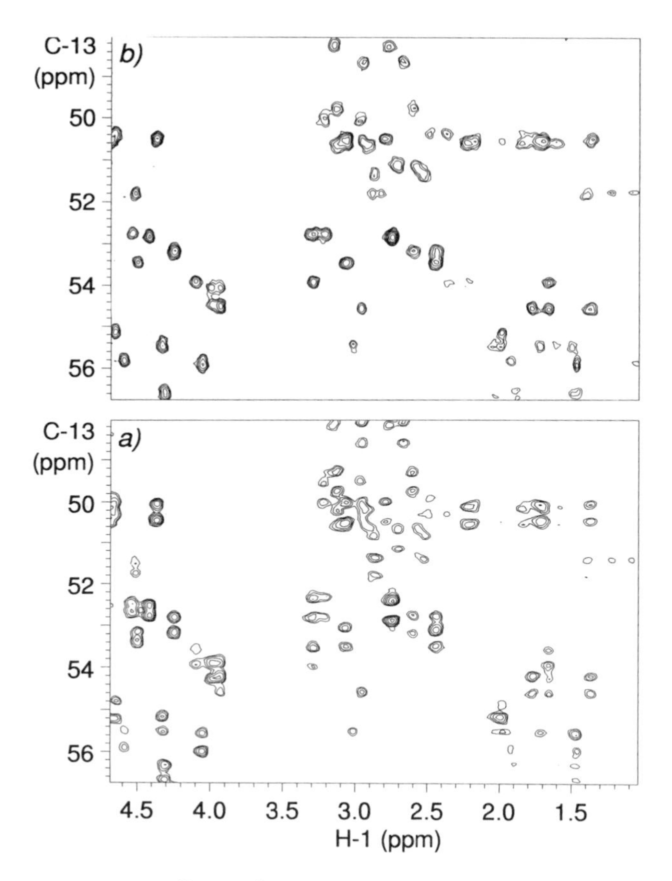

Figure 30. Expansions of the $^{13}C(\omega_2)$–$^{1}H(\omega_3)$ planes from (a) regular and (b) Cbd-HCCH TOCSY spectra of 300 μM 38-residue protein agitoxin 2 recorded at a (proton) frequency of 400 MHz.

(from about 60 ppm to about 20 ppm). Assuming that a minimum number of transients per experiment is required, the total experimental time can be reduced by a factor of 6. If folding is used in the carbon dimension of the traditional experiment, the carbon spectral width in ω_2 can be kept small. Thus, folding without C_β decoupling could reduce the experimental time by a factor of 3, but leaves the inconvenience of analyzing folded spectra.

The C_β decoupling is implemented in the same way as for the Cbd-HNCA and Cbd-HN(CO)CA. It is essential to use the same homonuclear decoupling waveform and field strength, as this leads to the same Bloch–Siegert shifts and facilitates connection with sidechain spin systems.

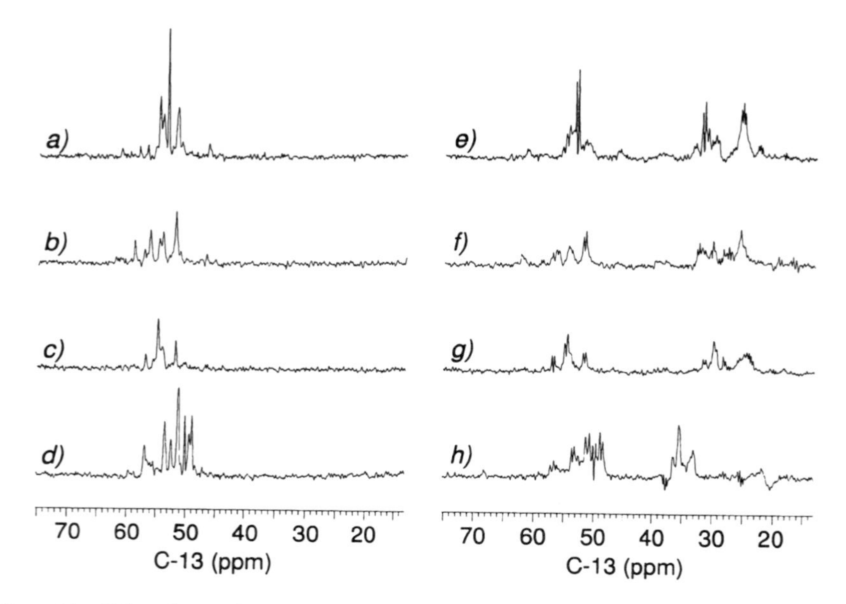

Figure 31. Selected traces from the HCCH TOCSY spectra of T4 lysozyme shown in Fig. 29. Traces (a)–(d) are from the conventional HCCH TOCSY spectrum extracted at $\delta_H = 2.89, 2.44, 2.24$, and 2.38 ppm, whereas the corresponding traces (e)–(h) are from the Cbd spectra. Note significant improvements in sensitivity and resolution in the Cbd experiment.

A comparison of $^{13}C(\omega_2)$–$^{1}H(\omega_3)$ 2D planes from a conventional HCCH TOCSY using CO decoupling and Cbd-HCCH TOCSY recorded for a sample of T4 lysozyme is shown in Fig. 29. In the Cbd experiment, the cross-peaks in the sidechain region are completely eliminated by the selective gradient echo at position A in the pulse sequence in Fig. 28. The same pulse achieves selection of H_α signals in the t_1 period. It should be noted that the eliminated regions do not contain any useful information. Although in the present example a full carbon spectral width of 6.25 kHz was recorded for demonstration purposes, in real applications reduced window sizes for both $\omega_1(H_\alpha)$ and $\omega_2(C_\alpha)$ can be used.

In Fig. 29 the C_β decoupling effect is clearly visible. There are a number of cross-peaks visible in the Cbd-HCCH TOCSY spectrum (Fig. 29b) that are missing in the conventional spectrum (Fig. 29a). A more detailed insight is provided by Fig. 30, which shows the same pulse sequences applied to a smaller protein, the 38-residue agitoxin, which gives less crowded spectra. Only two serine signals are not decoupled in this spectrum. The sensitivity gain is even more obvious from the cross sections shown in Fig. 31. The collapse of the multiplets and increased peak heights in the Cbd experiment clearly shows all of the benefits of the C_β decoupling.

6. CONCLUSIONS

Elimination of homonuclear splittings by means of adiabatic homonuclear decoupling can dramatically improve resolution and sensitivity of many important experiments, such as HSQC, HNCA, HN(CO)CA, and HCCH TOCSY used in biomolecular NMR as standard tools for establishing sequential connectivities and assignment of various spin systems.

The adiabatic decoupling proves to be the most economical in terms of power consumption and is the most efficient decoupling technique for homonuclear applications known to date. Low decoupling power levels allow band-selective decoupling in close proximity to resonances to be observed, creating minimal artifacts outside the decoupling region. This is crucial for decoupling of β-carbons in many C–C decoupled experiments [HNCA, HN(CO)CA, HCCH TOCSY] and for CH–NH decoupling in H–H decoupled experiments (^{15}N HSQC, NOESY, NOESY-HSQC). Decoupling schemes requiring low RF peak amplitude, such as WURST pulses, tend to provide a better overall performance.

Adiabatic decoupling is also very flexible making simultaneous multiple frequency decoupling straightforward. This significantly enhances the potential of homonuclear decoupling, especially in the case of C–C decoupled experiments.

The sensitivity of these experiments depends crucially on the number of increments used in the indirectly detected dimensions. The maximum evolution times should reflect the ratios of corresponding transverse relaxation times:

$$\frac{t_H(\max)}{t_X(\max)} = \frac{T_2(H)}{T_2(X)} \tag{35}$$

where X is usually ^{13}C or ^{15}N. In most cases, the transverse relaxation times of X are much longer than those of the attached protons. Hence, to obtain the best sensitivity and resolution, the X evolution periods should be sampled out further than the proton evolution. Typically, the ^{13}C sampling times should be at least 20–50 ms, depending on experiment type and size of the protein.

Alternatively, homonuclear decoupling can be achieved using constant-time experiments or selective 180° pulses. Although the first technique usually leads to considerable relaxation-related loss of sensitivity, the selective pulse decoupling is not free of artifacts (McCoy and Mueller, 1993) and is also more difficult to set up, especially if several bands must be decoupled.

The Boch–Siegert shifts introduced by adiabatic homonuclear decoupling are usually small (0.25 ppm or less) and can be easily corrected for. On the other hand, the correction may not even be necessary provided the same decoupling waveform and field strength have been used for all experiments.

REFERENCES

Abragam, A., 1961, *The Principles of Nuclear Magnetism,* Oxford University Press, London.
Allen, L., and Eberly, J. H., 1975, *Optical Resonance and Two-Level Atoms*, Wiley, New York.
Anderson, W. A., and Freeman, R., 1962, *J. Chem. Phys.* **37:**85.
Basus, V. J., Ellis, P. D., Hill, H. D. W., and Waugh, J. S., 1979, *J. Magn. Reson.* **35:**19.
Bauer, C., Freeman, R., Frenkiel, T., Keeler, J., and Shaka, A. J., 1984, *J. Magn. Reson.* **58:**442.
Baum, J., Tycko, R., and Pines, A., 1985, *Phys. Rev.* **A32:**3435.
Bax, A., and Grzesiek, S., 1993, *Acc. Chem. Res.* **26:**131.
Bax, A., Clore, G. M., and Gronenborn, A., 1990, *J. Magn. Reson.* **88:**425.
Bloch, F., 1954, *Phys. Rev.* **93:**944.
Bloch, F., and Siegert, A., 1940, *Phys. Rev.* **57:**522.
Bloom, A., and Shoolery, J. N., 1955, *Phys. Rev.* **97:**1261.
Bohlen, J.-M., and Bodenhausen, G., 1993, *J. Magn. Reson. Ser. A* **102:**293.
Chen, L., Wang, T.-C. L., Ricca, T. L., and Marshall, A. G., 1987, *Anal. Chem.* **59:**449.
Eggenberger, U., Schmidt, P., Sattler, M., Glaser, S. J., and Griesinger, C., 1992, *J. Magn. Reson.* **100:**604.
Emsley, L., and Bodenhausen, G., 1990, *Chem. Phys. Lett.* **168:**297.
Ernst, R. R., 1966, *J. Chem. Phys.* **45:**3845.
Fujiwara, T., and Nagayama, K., 1988, *J. Magn. Reson,* **77:**53.
Geen, H., and Freeman, R., 1991, *J. Magn. Reson.* **93:**93.
Geen, H., Wu, X. L., Xu, P., Friedrich, J., and Freeman, R., 1989, *J. Magn. Reson.* **81:**646.
Gibbs, A., and Morris, G. A., 1991, *J. Magn. Reson.* **91:**77.
Grzesiek, S., and Bax, A., 1992a, *J. Magn. Reson.* **96:**432.
Grzesiek, S., and Bax, A. 1992b, *J. Am. Chem. Soc.* **114:**6291.
Gutowsky, H. S., McCall, D. W., and Slichter, C. P., 1951, *Phys. Rev.* **84:**589.
Hahn, E. L., and Maxwell, D. E., 1951, *Phys. Rev.* **84:**1286.
Hammarstrom, A., and Otting, G., 1994, *J. Am. Chem. Soc.* **116:**8847.
Jesson, J. P., Meakin, P., and Kneissel, G., 1973, *J. Am. Chem. Soc.* **95:**618.
Kay, L. E., 1993, *J. Am. Chem. Soc.* **115:**2055.
Kay, L. E., Ikura, M., Tschudin, R., and Bax, A., 1990, *J. Magn. Reson.* **89:**496.
Kay, L. E., Keifer, P., and Saarinen, T., 1992, *J. Am. Chem. Soc.* **114:**10663.
Kay, L. E., Xu, G. Y., Singer, A. U., Muhandiram, D. R., and Forman-Kay, J. D., 1993, *J. Magn. Reson., Ser. B* **101:**333.
Kupče, Ē., and Freeman, R., 1993a, *J. Magn. Reson. Ser. A* **102:**364.
Kupče, Ē., and Freeman, R., 1993b, *J. Magn. Reson. Ser. A* **105:**234.
Kupče, Ē., and Freeman, R., 1995, *J. Magn. Reson. Ser. A* **115:**273.
Kupče, Ē., and Freeman, R., 1996a, *J. Magn. Reson. Ser. A* **117:**246.
Kupče, Ē., and Freeman, R., 1996b, *Chem. Phys. Lett.* **250:**523.
Kupče, Ē., and Wagner, G., 1995, *J. Magn. Reson. Ser. B* **109:**329.
Kupče, Ē., and Wagner, G., 1996, *J. Magn. Reson. Ser. B* **110:**309.
Kupče, Ē., Boyd, J., and Campbell, I. D., 1995, *J. Magn. Reson. Ser. B* **106:**300.
Kupče, Ē., Freeman, R. Wider, G., and Wuthrich, K., 1996, *J. Magn. Reson. Ser. A* **122:**81.
Levitt, M. H., and Freeman, R., 1982, *J. Magn. Reson.* **43:**502.
Levitt, M. H., Freeman, R., and Frenkiel, T., 1982, *J. Magn. Reson.* **47:**328.
Levitt, M. H., Freeman, R., and Frenkiel, T., 1983, *Adv. Magn. Reson.* **11:**47.
Marion, D., Ikura, M., Tschudin, R., and Bax, A., 1989, *J. Magn. Reson.* **85:**393.
Markus, M. A., Nakayama, T., Matsudira, P., and Wagner, G., 1994, *J. Biomol. NMR* **4:**553.
Matsuo, H., Kupče, Ē., Li, H., and Wagner, G., 1996a, *J. Magn. Reson. Ser. B* **113:**91.
Matsuo, H., Kupče, Ē., and Wagner, G., 1996b, *J. Magn. Reson. Ser. B* **113:**190.

McCall, S. L., and Hahn, E. L., 1969, *Phys. Rev.* **183:**457.
McCoy, M. A., and Mueller, L., 1992a, *J. Am. Chem. Soc.* **114:**2108.
McCoy, M. A., and Mueller, L., 1992b, *J. Magn. Reson.* **98:**674.
McCoy, M. A., and Mueller, L., 1992c, *J. Magn. Reson.* **99:**18.
McCoy, M. A., and Mueller, L., 1993, *J. Magn. Reson. Ser. A* **101:**122.
Messerli, B.A., Wider, G., Otting, G., Weber, C., and Wüthrich, K., 1989, *J. Magn. Reson.* **85:**608.
Morris, G. A., 1988, *J. Magn. Reson.* **80:**547.
Morris, G. A., and Freeman, R., 1978, *J. Magn. Reson.* **29:**433.
Patt, S. L., 1992, *J. Magn. Reson.* **96:**94.
Ramsey, N. F., 1955, *Phys. Rev.* **100:**1191.
Royden, V., 1954, *Phys. Rev.* **96:**543.
Santoro, J., and King, G. C., 1992, *J. Magn. Reson.* **97:**202.
Shaka, A. J., and Keeler, J., 1987, *Prog. NMR Spectrosc.* **19:**47.
Shaka, A. J., Keeler, J., Frenkiel T., and Freeman, R., 1983a, *J. Magn. Reson.* **52:**335.
Shaka, A. J., Keeler, J., and Freeman, R., 1983b, *J. Magn. Reson.* **53:**313.
Shaka, A. J., Barker, P. B., and Freeman, R., 1985, *J. Magn. Reson.* **64:**547.
Silver, M. S., Joseph, R. I., and Hoult, D. I., 1985, *Phys. Rev. A* **31:**2753.
Sklenar, V., Piotto, M., Leppik, R., and Saudec, V., 1993, *J. Magn. Reson. Ser. A* **102:**241.
Slichter, C. P., 1996, *Principles of Magnetic Resonance*, Springer, Berlin.
Tannus, A., and Garwood, M., 1996, *J. Magn. Reson. Ser. A* **120:**133.
Tomlinson, B. L., and Hill, H. D. W., 1973, *J. Chem. Phys.* **59:**1775.
Tycko, R., Pines, A., and Gluckenheimer, R., 1985, *J. Chem. Phys.* **83:**2775.
Walker, J. S., 1988, *Fourier Analysis,* Oxford University Press, London.
Waugh, J. S., 1982a, *J. Magn. Reson.* **49:**517.
Waugh, J. S., 1982b, *J. Magn. Reson.* **50:**30.
Weigelt, J., Hammarstrom, A., Bermel, W., and Otting, G., 1996, *J. Magn. Reson. Ser B* **110:**219.
Yamazaki, T., Lee, W., Revington, M., Mattiello, D. L., Dahlquist, F. W., Arrowsmith, C. H., and Kay, L. E., 1994a, *J. Am. Chem. Soc.* **116:**6464.
Yamazaki, T., Lee, W., Arrowsmith, C. H., Muhandiram, D. R., and Kay, L. E., 1994b, *J. Am. Chem. Soc.* **116:**11655.
Zhang, S., Wu, J., and Gorenstein, D. G., 1996, *J. Magn. Reson. Ser. A*, **123:**181.
Zuiderweg, E. R. P., and Fesik, S. L., 1991, *J. Magn. Reson.* **93:**653.
Zuiderweg, E. R. P., Zeng, L., Brutscher, B., and Morshauser, R. C., 1996, *J. Biomol. NMR* **8:**147.

6

Pulse Sequences for Measuring Coupling Constants

Geerten W. Vuister, Marco Tessari, Yasmin Karimi-Nejad, and Brian Whitehead

1. INTRODUCTION

It has been recognized since the early days of NMR that *J* coupling constants contain very useful information regarding molecular conformation (Karplus, 1959, 1963; Bystrov, 1976). For small (bio)molecules the magnitude of the *J* coupling constants can often be measured directly from the splitting of the resonances of interest. However, the accurate measurement of the magnitude of the *J* coupling constants has been problematic for larger biomolecules in which the measurements of in-phase or antiphase splittings failed. A large number of so-called "direct methods" have been proposed to overcome this problem (Oschkinat and Freeman, 1984; Kessler *et al.*, 1985; Neuhaus *et al.*, 1985; Kay and Bax, 1989; Titman and Keeler, 1990; Ludvigsen *et al.*, 1991; Smith *et al.*, 1991; Szyperski *et al.*, 1992).

Geerten W. Vuister,* Marco Tessari,* Yasmin Karimi-Nejad, and Brian Whitehead* • Department of NMR Spectroscopy, Bijvoet Center for Biomolecular Research, Utrecht University, 3584 CH Utrecht, The Netherlands. **Current address*: NSR Center, University of Nÿmegen, Tuernoviveld 1, 6525 ED Nÿmegen, The Netherlands.

Biological Magnetic Resonance, Volume 16: Modern Techniques in Protein NMR, edited by Krishna and Berliner. Kluwer Academic / Plenum Publishers, 1999.

With the recent advent of isotope labeling techniques for proteins and RNA and DNA molecules, interest in the measurement of J couplings has again surged. The usage of isotope labels has prompted the development of a series of experiments aimed at measuring a large array of both homo- and heteronuclear coupling constants. As a result of these developments, valuable structural information can now be obtained in a relatively straightforward way for medium-sized biomolecules. Moreover, the newly measured J coupling data have allowed the reparametrization of the Karplus curves describing the dependences of the ^3J values on the intervening torsion angles (Vuister and Bax, 1993a; Wang and Bax, 1995, 1996; Hu and Bax, 1996, 1997b). It is to be expected that the newly obtained curves will be of higher accuracy than those derived solely on the basis of small model compounds, in particular for the range of dihedral angles that were actually used in the parametrization.

These new techniques can be subdivided on the basis of the underlying principle for measuring the J coupling. In the exclusive correlation spectroscopy (E.COSY) methods (Griesinger *et al.*, 1985, 1986, 1987) (discussed in Section 3), two spins are correlated without disturbing the energy levels of a third spin, which is J coupled to both other spins. The resulting E.COSY pattern then allows the measurement of a small J coupling, provided that the second J coupling is large enough to allow separation of the multiplet components. E.COSY methods have now been used to measure a large variety of ^2J and ^3J coupling constants (Montelione *et al.*, 1989, Wider *et al.*, 1989; Sørensen, 1990; Delaglio *et al.*, 1991; Gemmecker and Fesik, 1991; Schmieder *et al.*, 1991; Eggenberger *et al.*, 1992; Emerson and Montelione, 1992; Griesinger and Eggenberger, 1992; Seip *et al.*, 1992; Vuister and Bax, 1992; Madsen *et al.*, 1993; Weisemann *et al.*, 1994a; Löhr and Rüterjans, 1995, 1997; Wang and Bax, 1995, 1996).

A second class of experiments (discussed in Section 4) aims to quantify the signal modulation or attenuation resulting from the active coupling and is referred to as quantitative J correlation experiments (Archer *et al.*, 1991; Billeter *et al.*, 1992; Blake *et al.*, 1992; Grzesiek *et al.*, 1992; Vuister and Bax, 1993a,b; Vuister *et al.*, 1993a,b, 1994; Bax *et al.*, 1994; Kuboniwa et al., 1994; Hu and Bax, 1996, 1997a,b; Hennig *et al.*, 1997; Hu *et al.*, 1997).

Alternative methods for measuring J couplings include the so-called P-FIDS or C′-FIDS methods (Schwalbe *et al.*, 1994; Rexroth *et al.*, 1995a) and the ZQ/DQ methods (Rexroth *et al.*, 1995b; Otting, 1997).

2. DIRECT METHODS FOR MEASURING J COUPLINGS

The direct methods attempt to measure the J couplings from in-phase or antiphase splittings of a particular resonance in 1D or 2D spectra. In cases where the magnitude of the J coupling is smaller than the natural linewidth, this typically requires strong resolution enhancement and/or iterative fitting programs to obtain accurate J couplings (Neuhaus *et al.*, 1985; Kay and Bax, 1990; Ludvigsen *et al.*, 1991; Smith *et al.*, 1991; Szyperski *et al.*, 1992). Related methods use deconvolution of a multiplet with a reference multiplet (Oschkinat and Freeman, 1984; Kessler *et al.*, 1985; Titman and Keeler, 1990).

Although the resolution and digitization in the acquisition dimension usually can be chosen sufficiently large to resolve J splittings in cross-peaks, the direct methods are fraught with problems. For example, the active coupling in the COSY experiments results in an antiphase pattern that tends to overestimate the coupling constant if the linewidth becomes comparable to the J coupling (Neuhaus *et al.*, 1985). The robustness of frequency domain fitting routines is highly dependent on the quality of the data, i.e., the signal-to-noise ratio and the line shape, and the number of parameters, i.e., the number of passive couplings, and often these methods are dogged by the well-known problem of false minima (Yang and Havel, 1994).

More sophisticated methods employ the occurrence of an active J coupling of interest in multiple regions of the spectrum or in multiple spectra. In the DISCO procedure (Oschkinat and Freeman, 1984; Kessler *et al.*, 1985) the (F_2) traces at a particular cross-peak position (Fig. 1A) are combined with the corresponding (F_2) diagonal traces, shifted 90° in phase (Fig. 1B). After appropriate scaling, addition yields one component of the multiplet (Fig. 1C), whereas subtraction yields the other component (cf. Fig. 1D). Comparison of the last two spectra allows a more accurate measurement of the coupling constant. Passive splittings are also more readily evaluated because the complexity of the multiplet is reduced. A related method employs both NOESY and COSY spectra (Ludvigsen *et al.*, 1991).

A second and more elegant approach using antiphase and in-phase patterns obtained from COSY and TOCSY spectra, respectively, was proposed by Titman and Keeler (1990). The corresponding traces from COSY and TOCSY spectra are inverse Fourier transformed, scaled, and the resulting time-domain signals are multiplied by $\cos(\pi J^* t)$ and $\sin(\pi J^* t)$, respectively, where J^* represents a trial coupling:

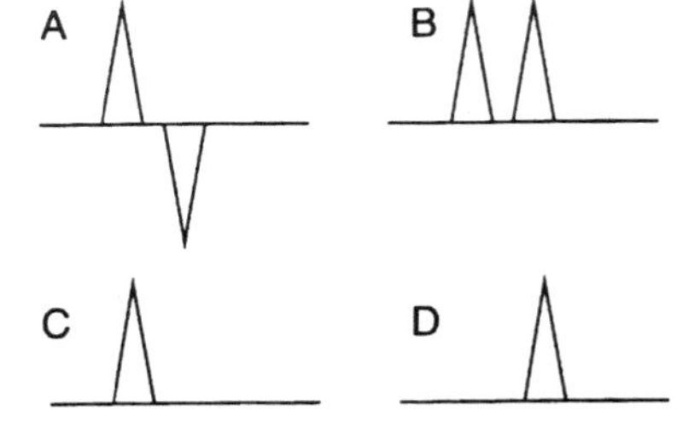

Figure 1. Schematic outline of the DISCO procedure (Oschkinat and Freeman, 1984) showing COSY antiphase cross-peak trace (A), in-phase diagonal trace (B), sum (C), and difference (D) spectra of traces A and B.

$$S^{COSY}(t) = A \exp(i\omega t) \sin(\pi J t) \cos(\pi J^* t)\, \Pi_i \cos(\pi J_i t) \tag{1a}$$

$$S^{TOCSY}(t) = B \exp(i\omega t) \cos(\pi J t) \sin(\pi J^* t)\, \Pi_i \cos(\pi J_i t) \tag{1b}$$

where ω denotes the angular frequency of the resonance of interest and $\Pi_i \cos(\pi J_i t)$ represents the modulation resulting from all other passive couplings. The Fourier transforms of $S^{COSY}(t)$ and $S^{TOCSY}(t)$ generate two quite different spectra unless $A = B$ and $J^* = J$. As A/B and J^* are varied in a 2D iterative search program, the sum of the squares of the differences between corresponding points in the two spectra passes through a minimum, which yields $J^* = J$. A related method is J deconvolution (Jones *et al.*, 1993) in which the J coupling is determined by J doubling, and the splitting is removed by a deconvolution procedure. The passive couplings are determined from the resulting multiplet, and all J couplings are optimized using the original data.

3. THE E.COSY METHODS

The E.COSY method was first proposed for homonuclear spin systems for the extraction of J coupling constants and is formally equivalent to a superposition of multiple-quantum filtered spectra with appropriate weighting (Griesinger *et al.*, 1985, 1986, 1987). For unlabeled proteins, yielding homonuclear 2D COSY-like spectra, the method has not gained widespread usage. However, the advent of ^{15}N and ^{13}C labeling in proteins has turned this method into a valuable tool for measuring J couplings.

As proposed by Wang and Bax (1995), the different E.COSY schemes in this chapter will be referred to by the experiment name from which they are derived, while the passive unperturbed spin will be indicated in square brackets. Thus, the HNCA E.COSY experiment (Schmieder *et al.*, 1991; Seip *et al.*, 1992, 1994; Madsen *et al.*, 1993) for measuring $^3J(H^NH^\alpha)$ (*vide infra*) will be denoted HNCA[H^α].

3.1. Explanation of the E.COSY Principle

The E.COSY method allows the measurement of a (partially) unresolved coupling between spins I and X, J_{IX}, if there is a third spin S that has a large, resolved J coupling with the X spin, J_{SX}. In addition, it must be possible to transfer magnetization between the I and S spins without disturbing the spin state of the X nucleus. The principle is most easily understood by considering two spins I and S, both heteronuclear to spin X (Fig. 2A). The $(F_1,F_2) = (\nu_S,\nu_I)$ cross-peak is shown in Fig. 3A in the absence of J interactions with the X-nucleus in t_1 and t_2. Suppose the large J_{SX} interaction is now present during t_1. Consequently, the $(F_1,F_2) =$

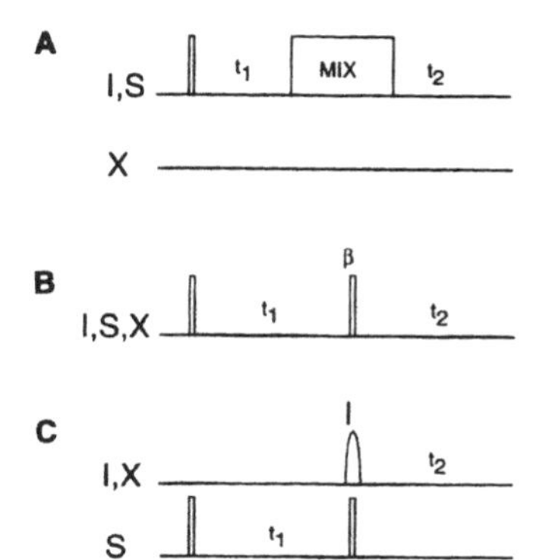

Figure 2. Schematic outline of three E.COSY schemes. In scheme A, both the I and S spins are heteronuclear to spin X. In scheme B, all three spins I, S, and X are homonuclear. In scheme C, spins I and X are heteronuclear to spin S. See text for further details.

(ν_S,ν_I) cross-peak will show an in-phase doublet splitting along the F_1 axis resulting from this interaction (Fig. 3B), the upfield component being associated with the X spin in the α state and the downfield-shifted component being associated with the X spin in the β state (assuming $J_{SX} > 0$). Likewise, interaction between the I spin and the X spin during t_2 would result in a doublet splitting along the F_2 axis (Fig. 3C), which may or may not be resolved. If, however, the J interactions J_{SX} and J_{IX} are active during t_1 and t_2, respectively, and the spin state of the X nucleus is preserved during the transfer of magnetization between the I and S spins, the

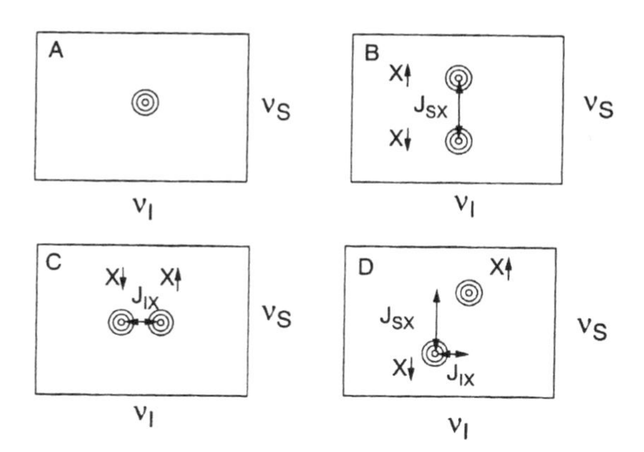

Figure 3. Schematic outline of the E.COSY principle. (A) $J(SX)$ interaction decoupled in t_1 and $J(IX)$ interaction decoupled in t_2. (B) $J(SX)$ interaction active in t_1 and $J(IX)$ interaction decoupled in t_2. (C) $J(SX)$ interaction decoupled in t_1 and $J(IX)$ interaction active in t_2. (D) $J(SX)$ interaction active in t_1 and $J(IX)$ interaction active in t_2, while assuming that the transfer of magnetization from S to I has not disrupted the spin state of X.

E.COSY pattern depicted in Fig. 3D will be generated because each of the two components of the multiplet is now displaced along both the F_1 and the F_2 axes.

An alternative explanation of the E.COSY pattern can be constructed in terms of the product operator formalism (Sørensen *et al.*, 1983) by realizing that the E.COSY pattern results from the superposition of a double in-phase multiplet with a double antiphase multiplet (Fig. 4). Taking again the simple case of spins *I* and *S*, both heteronuclear to spin *X*, and evaluating the effects of the pulse sequence depicted in Fig. 2A, the transverse *S* magnetization, generated by the first 90°(*I*,*S*) pulse, evolves under the influence of chemical shift and J_{SX} interaction:

$$S_y \xrightarrow{t_1} S_y\cos(\omega_S t_1)\cos(\pi J_{SX} t_1) - S_x\sin(\omega_S t_1)\cos(\pi J_{SX} t_1) - 2S_xX_z\cos(\omega_S t_1)\sin(\pi J_{SX} t_1) - 2S_yX_z\sin(\omega_S t_1)\sin(\pi J_{SX} t_1) \quad (2)$$

The mixing period selects one of the quadrature components of the magnetization terms of Eq. (2) and transfers magnetization from the *S* spin to the *I* spin without disturbing the *X*-spin polarization:

$$\text{Eq. (2)} \xrightarrow{\text{mix}} I_y\cos(\omega_S t_1)\cos(\pi J_{SX} t_1) - 2I_yX_z\sin(\omega_S t_1)\sin(\pi J_{SX} t_1) \quad (3)$$

The *I* spin now evolves during t_2 under the influence of chemical shift and J_{IX} interaction:

$$\text{Eq. (3)} \xrightarrow{t_2} \{I_y\cos(\omega_I t_2) - I_x\sin(\omega_I t_2)\}\cos(\pi J_{IX} t_2)\cos(\omega_S t_1)\cos(\pi J_{SX} t_1) - 2X_z\{I_x\cos(\omega_I t_2) + I_y\sin(\omega_I t_2)\}\sin(\pi J_{IX} t_2)\cos(\omega_S t_1)\cos(\pi J_{SX} t_1) - 2X_z\{I_y\cos(\omega_I t_2) - I_x\sin(\omega_I t_2)\}\cos(\pi J_{IX} t_2)\sin(\omega_S t_1)\sin(\pi J_{SX} t_1) + \{I_x\cos(\omega_I t_2) + I_y\sin(\omega_I t_2)\}\sin(\pi J_{IX} t_2)\sin(\omega_S t_1)\sin(\pi J_{SX} t_1) \quad (4)$$

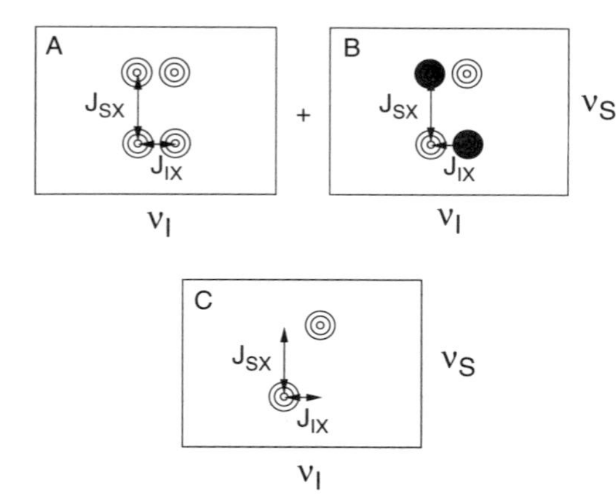

Figure 4. Schematic multiplet patterns illustrating the E.COSY principle in terms of product operators. The superposition of a double in-phase pattern (A) with a double antiphase pattern (B) results in the E.COSY pattern (C) via cancellation of half of the components (gray-shaded cross-peak components have a negative intensity).

Only the first and last terms of Eq. (4) contain observable operators which result in the double in-phase pattern shown in Fig. 4A and the double antiphase pattern shown in Fig. 4B, respectively. The superposition results in cancellation of half of the multiplet components (Fig. 4C), and the magnitude of the J_{IX} coupling can be determined from the displacement of the two multiplet components along the F_2 axis. Note that the presence of additional passive couplings to either spin I or spin S results in an additional in-phase splitting along the F_2 or F_1 axis, respectively.

The situation in which the X spin is heteronuclear to both I and S spins presents the simplest case of the E.COSY experiment, in that the X spin can be left unperturbed by simply applying no pulses to this nucleus. In case the X spin is homonuclear to both the I and S spins, as shown in Fig. 2B, the transition within the multiplet must be preserved by the application of a β pulse. The use of a β flip angle in the so-called β-COSY (Bax and Freeman, 1981a,b) represents a compromise between proper cancellation of the different multiplet patterns on the one hand and sensitivity on the other hand. Alternatively, the β pulse can be replaced by a multiple-quantum filter together with appropriate superposition as in the original E.COSY experiment (Griesinger *et al.*, 1985, 1986, 1987).

A third situation encountered often, in particular in the case of $^{13}C/^{15}N$-labeled proteins, is shown in Fig. 2C. Here, the S spin is heteronuclear to the I and X spins. In transferring the magnetization from I to S, the X spin has to remain invariant. Hence, pulses applied in this transfer sequence must be selective for the I spin. Nonselective sequences that effectively restore the original X state, such as the sensitivity-enhancement scheme (Palmer *et al.*, 1991), also fall in this category.

3.2. Extraction of the J Value and Accuracy of the Method

The relative displacement of the two components of the E.COSY multiplet allows for the determination of the value of the J coupling constant (Fig. 4C). To evaluate the magnitude of this displacement, several procedures can be applied. Eventually, the accuracy of all of these procedures depends on the accuracy by which peak positions in the spectrum can be determined. A manual evaluation of the E.COSY patterns, e.g., manually aligning the F_2 traces of the two components, is susceptible to bias and more accurate methods need to be evaluated.

The most simple alternative employs an (interactive) peak-picking procedure using peak interpolation. Commonly used software packages, e.g., PIPP (Garrett *et al.*, 1991) or REGINE (Kleywegt *et al.*, 1993), employ some form of parabolic interpolation around a local maximum that allows for an accurate determination of the peak maximum. Typically, for well-resolved peaks with a sufficient signal-to-noise ratio and well-defined line shape, interpolation can easily improve the precision by a factor of 5–10 as compared with the linewidth of the resonance (Vuister and Bax, 1994; Wang and Bax, 1996).

The second method employs a point-by-point trace-fitting procedure in the frequency domain (Schwalbe *et al.*, 1993). The major drawback of this procedure is that the accuracy of the extracted J value is determined by the digital resolution of the frequency-domain spectrum. By far the most elaborate, but probably also the most unbiased and accurate method, it employs a continuous alignment in the time domain (Karimi-Nejad *et al.*, 1994; Schmidt *et al.*, 1995), rather than a point-by-point alignment in the frequency domain. Thus, the accuracy by which the coupling constant can be determined becomes independent of the digital resolution.

Following Karimi-Nejad *et al.* (1994), this procedure amounts to a convolution of one of the traces of the multiplet by a frequency-shifting function, $\delta(\pi J)$, where J denotes the J coupling to be determined. Varying J allows the difference between the frequency-shifted trace and the unperturbed trace to be minimized. The resulting value, J_{min}, is taken as the coupling constant. Let $S^{\text{u}}(\omega)$ denote the upper trace through the multiplet and $S^{\text{l}}(\omega)$ the lower trace (as indicated in Fig. 5A), both of length n. The time-domain trace of the lower multiplet, $S^{\text{l}}(t)$, is obtained by inverse Fourier transformation of $S^{\text{l}}(\omega)$. Multiplication of the time-domain trace by the complex function $\exp(i\pi Jt)$ amounts to a shift of πJ in the frequency domain. Thus, the frequency-shifted trace can be expressed as

$$S^{\text{l}}(\omega + \pi J) = A \times FT\{S^{\text{l}}(t)\exp(i\pi Jt)\} \tag{5}$$

where A denotes a scaling factor. The χ^2 residual of the difference spectrum of the two traces:

$$\chi^2 = \sum_n \{S^{\text{u}}(\omega) - S^{\text{l}}(\omega + \pi J)\}^2 \tag{6}$$

has to be minimized by a non-linear fit adjusting the parameters A and J. The procedure is illustrated in Fig. 5. Panel A shows the Asn^{36} C$^{\alpha}$–H$^{\beta 2}$ and C$^{\alpha}$–H$^{\beta 3}$ F_2,F_3 multiplets in an HCCH-E.COSY spectrum of ribonuclease T1. The upper trace (solid) and lower trace (dashed) of one C$^{\alpha}$–H$^{\beta}$ multiplet are indicated and shown in detail in panel B. Minimization of Eq. (6) yields $J = 10.6 \pm 0.4$ Hz and the resulting superposition is shown in panel C. The experimental error of the J value was obtained by standard error propagation assuming that the noise of the individual spectral data points is uncorrelated and exhibits uniform variances around the mean-squared noise intensity, which was arbitrarily set to 1 in Eq. (6).

An evaluation of the random errors in the accuracy of the J value can be obtained by an intentional collapse of the multiplet along the frequency axis in which the J coupling is to be determined. By decoupling the passive spin along the corresponding time axis, the displacement between the two components is now expected to vanish. The resulting spectrum can be used for evaluation of the random errors resulting from the procedure to extract the J value. For example, in the case of the measurement of the ^2J(C$^{\alpha}$C′) from the CT-HSQC[C′], it was found that using

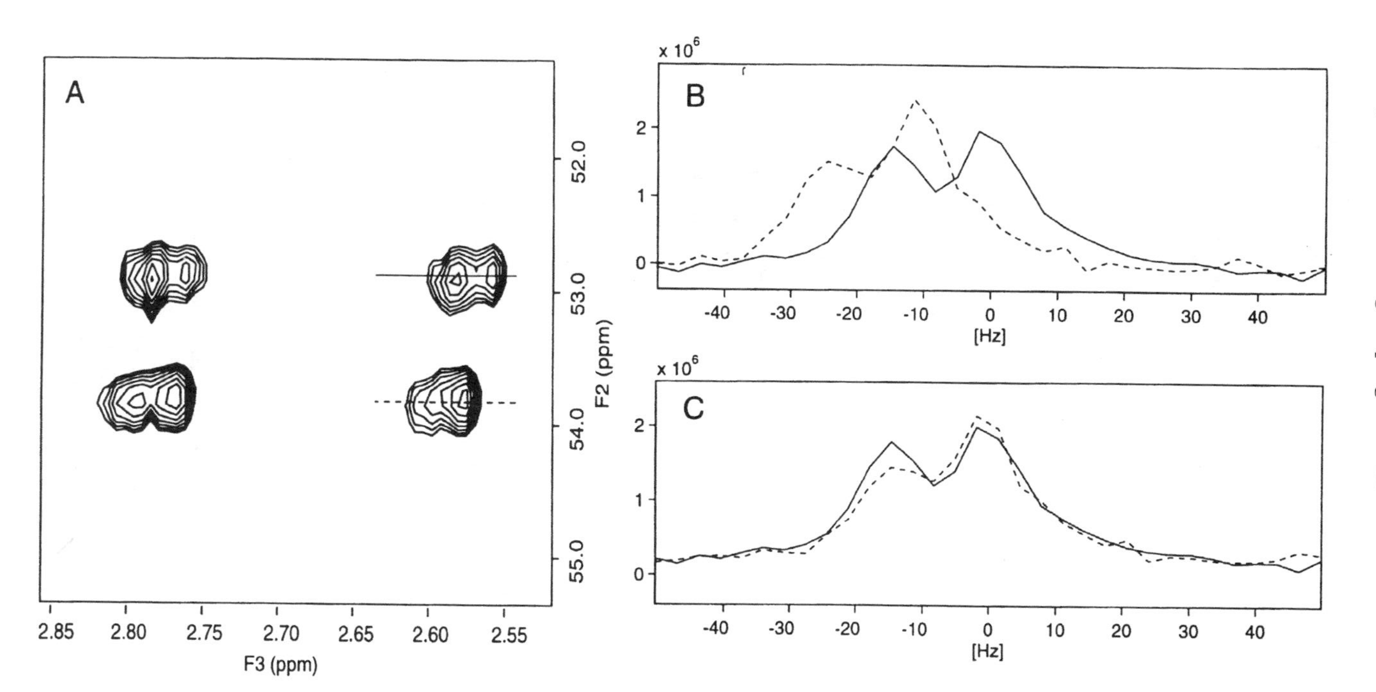

Figure 5. (A) Section of the F_2–F_3 slice from the HCCH-E.COSY spectrum of ribonuclease T_1 (data from Karimi-Nejad *et al.*, 1994), taken at the H^{α} F_1 frequency of Asn^{36}. Correlations to the β-protons are shown. Splitting along the F_2 axis results from the large $^1J(H^{\alpha}C^{\alpha})$ interaction active during t_2. Displacement of the multiplet components along the F_3 axis results from the $^3J(H^{\alpha}H^{\beta})$. (B) Traces through the low-field H^{β} cross-peaks (indicated in A). (C) Traces after a time-domain frequency shift, the value of which was determined by minimizing Eq. (6). The resulting J value is 10.6 ± 0.4 Hz. (Reproduced from Karimi-Nejad, 1996).

a double constraint peak-picking procedure the rms difference between the two components was –0.05 ± 1 Hz (Vuister and Bax, 1992).

Experimentally, the accuracy of the measured J coupling is determined by the degree to which the spin polarization of the X spin can be preserved during the experiment. Failure to do so results in (partial) collapse of the E.COSY multiplet, compromising the accuracy of the measurement. Thus, the time needed for transfer of magnetization between the I and S spins should be short relative to the $T_{1,\mathrm{sel}}$ of the X spin. For the measurement of $^3J(\mathrm{H^NH^\alpha})$ on ubiquitin using the HNCA[H^α] experiment (see Section 5.4) the effect of the H^α spin flips was estimated to result in a 0.8- to 1.0-Hz underestimate of the true J coupling (Wang and Bax, 1996). An interesting variation on the triple-resonance based E.COSY schemes in this respect is the H(N)CA,CO[H^α] experiment (Löhr and Rüterjans, 1997) for measuring $^3J(\mathrm{C'H^\alpha})$, in which the time needed for transfer of magnetization is reduced to the duration of the 90° pulse.

Differential relaxation of the different multiplet components (London, 1990; Harbison, 1993; Norwood, 1993; Norwood and Jones, 1993) also affects the accuracy of all E.COSY measurements (Norwood, 1995; Zhu *et al.*, 1995). As can be seen in Fig. 5, the high-field transitions of the H^α–H^β multiplets are of lower intensity, owing to cross-correlation effects involving the $\mathrm{H}^{\beta2}$–$\mathrm{H}^{\beta3}$ spins. Also, the multiplet components display differential linewidths. In general, the distortions in the E.COSY pattern will result in an underestimate of the measured J coupling (Rexroth *et al.*, 1995b).

4. THE QUANTITATIVE *J* CORRELATION METHODS

A suite of experimental techniques, commonly referred to as the quantitative J correlation method (QJ), for measuring a diverse array of J couplings has been developed in recent years (Archer *et al.*, 1991; Bax *et al.*, 1992, 1994; Grzesiek *et al.*, 1992, 1995; Vuister and Bax, 1993a,b; Vuister *et al.*, 1993a,b, 1994; Hu and Bax, 1996, 1997a,b; Hennig *et al.*, 1997; Hu *et al.*, 1997). In the QJ method, the magnetization is allowed to dephase under the influence of a J coupling for a constant period of duration T. The magnitude of the resulting in-phase and antiphase magnetization terms are determined, and the J coupling can be extracted from the ratio of these terms.

4.1. Spin-Echo-Based Quantitative *J* Correlation Schemes

Figure 6 depicts the three different variations in which the QJ method can be used to measure the J interaction between two J-coupled spins, I and S. The first variation is a spin-echo difference-based scheme (Fig. 6A). This scheme can be used when there is one spin, S, heteronuclear to the I spin with a J coupling, J_{IS}.

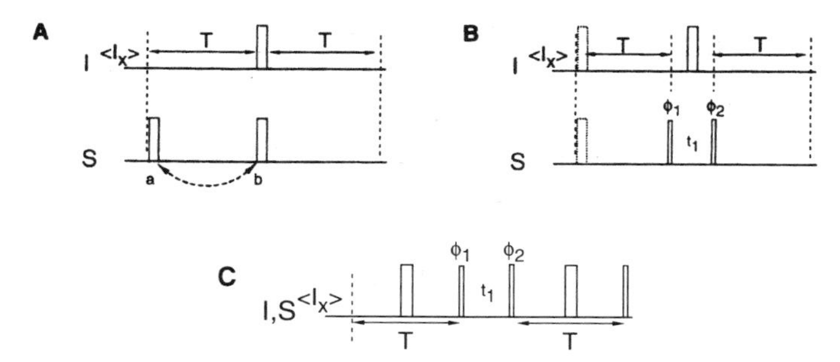

Figure 6. Three schemes of the quantitative J correlation method for determining J couplings. Refer to the text for a full explanation. The phases in schemes B and C assume the presence of an I_x operator at the start of the pulse sequence element. (A) QJ scheme based on a spin-echo difference experiment. The 180° S pulse is applied at position a for the reference experiment and position b for the experiment in which the J interaction is active. (B) QJ scheme based upon the HMQC experiment. The phases are $\phi_1 = x,-x$; $\phi_2 = 2x,2(-x)$; receiver $= x,2(-x),x$ or $4x$ (for the reference). The dashed 180°(I,S) pulses are optional and can be used for chemical shift labeling of the I spin. (C) QJ scheme based on the COSY experiment. The phases are $\phi_1 = x,-x$; $\phi_2 = 2x,2(-x)$; receiver $= 4x$.

Assuming transverse magnetization of the I spin, generated by some preparation sequence, the J_{IS} interaction is active for the full duration of $2T$ when the 180°(S) pulse is applied at position b:

$$I_x \xrightarrow{2T} I_x\cos(2\pi J_{IS}T) + 2I_yS_z\sin(2\pi J_{IS}T) \tag{7}$$

where I and S denote the spin operators in the product operator formalism (Sørensen *et al.*, 1983). Typically, the $2I_yS_z$ term does not result in observable magnetization or can be purged away. Application of the 180°(S) pulse in position a effectively decouples the IS interaction and a reference experiment is obtained. The magnetization detected in the reference experiment experiences identical relaxation and other losses, such as those resulting from passive couplings, as compared with the experiment with the 180°(S) applied at time b. Hence, the J coupling can be calculated from the normalized relative difference of the two spectra in a straightforward fashion:

$$(S_a - S_b)/S_a = 1 - \cos(2\pi J_{IS}T) = 2\sin^2(\pi J_{IS}T) \tag{8}$$

where S_a and S_b denote the signal obtained with the 180°(S) pulse in positions a and b, respectively. The presence of more spins J-coupled to the I spin can be accommodated for, provided that the magnitude of these J interactions is of known, and constant, magnitude (Grzesiek *et al.*, 1993). Note that chemical shift labeling of the

I spin can be easily implemented by moving the 180° pulses in concert in a constant-time fashion.

4.2. HMQC-Based Quantitative J Correlation Schemes

The second variation of the QJ method is based on the heteronuclear multiple-quantum correlation (HMQC) experiment and shown in Fig. 6B. This variation is most useful when there are multiple spins S_i, heteronuclear to spin I, with varying magnitude of the J couplings J_{ISi}. Considering first an isolated IS spin pair and assuming transverse I-spin magnetization generated by some preparation sequence, the J_{IS} coupling is active for the duration of T:

$$I_x \xrightarrow{T} I_x\cos(\pi J_{IS}T) + 2I_yS_z\sin(\pi J_{IS}T) \tag{9}$$

The subsequent $90°(S,\phi_1)$ pulse selects for the $2I_yS_z$ component and converts it into heteronuclear multiple-quantum terms. Chemical shift labeling with the S-spin frequency is obtained during t_1 in a pseudo-single-quantum manner:

$$2I_yS_z \xrightarrow{90°(S,\phi_1)} \xrightarrow{t_1} -2I_yS_y\cos(\omega_St_1) + 2I_yS_x\sin(\omega_St_1) \tag{10}$$

where the $\sin(\pi J_{IS}T)$ term of Eq. (9) has been temporarily omitted from Eq. (10). The subsequent $90°(S,\phi_2)$ pulse selects one of the quadrature components and converts it back into antiphase magnetization which is allowed to rephase for a period T:

$$-2I_yS_y\cos(\omega_St_1) \xrightarrow{90°(S,\phi_2)} \xrightarrow{T}$$

$$-2I_yS_z\cos(\omega_St_1)\cos(\pi J_{IS}T) + I_x\cos(\omega_St_1)\sin(\pi J_{IS}T) \tag{11}$$

Only the I_x operator results in observable magnetization, so that after Fourier transformation with respect to t_1, the volume of the cross-peak at the $F_1 = \omega_S$ frequency, V_S, can be written as

$$V_S = A\sin^2(\pi J_{IS}T) \tag{12}$$

where the previously omitted trigonometric term has been reintroduced and A denotes a proportionality constant that incorporates all experimental factors, constant losses related to relaxation and other passive couplings, and the effects of data processing. A reference experiment, selecting for the I_x operator of Eq. (9), is obtained by changing the receiver phase cycle (see caption of Fig. 6B). Obviously, the I_x operator does not evolve in t_1 and thus the dimensionality of the reference experiment is one lower, as compared with the first experiment. During the second period of duration T, the J_{IS} coupling is again active:

$$I_x \cos(\pi J_{IS}T) \xrightarrow{T} I_x \cos^2(\pi J_{IS}T) + 2I_yS_z \cos(\pi J_{IS}T) \sin(\pi J_{IS}T) \tag{13}$$

Only the I_x term results in observable magnetization so that the volume of the reference peak, V_I, becomes

$$V_I = A' \cos^2(\pi J_{IS}T) \tag{14}$$

where A' is a constant closely related to A. Assuming for simplicity momentarily identical transverse relaxation times of the in-phase and antiphase magnetization, the constants A and A' are related by a simple scaling factor resulting from the different Fourier transformations of both experiments (Vuister *et al.*, 1993b). Thus, after scaling the magnitude of the J coupling, J_{IS} can be easily calculated from the ratio of the cross-peak volumes:

$$V_S/V_I = \tan^2(\pi J_{IS}T) \tag{15}$$

In the case of several spins S, it is straightforward to show that the additional spins act as passive spins in both versions of the experiment. The development of a small amount of double antiphase magnetization involving spins S_i and S_j, $4I_xS_iS_j$, potentially may result in small errors. However, because of the $\sin(\pi J_{ISi}T)$ $\sin(\pi J_{ISj}T)$ dependence of this term, it can safely be neglected (Vuister *et al.*, 1993b). Hence, the expressions for the cross-peak volumes for the ith S spin, V_{Si}, and the reference, V_I, become

$$V_{Si} = A \sin^2(\pi J_{ISi}T) \prod_{j \neq i} \cos^2(\pi J_{ISj}T) \tag{16a}$$

$$V_I = A' \cos^2(\pi J_{ISi}T) \prod_{j \neq i} \cos^2(\pi J_{ISj}T) \tag{16b}$$

where J_{ISi} denotes the J coupling between the I spin and the ith S spin. From Eq. (16) it can be seen that the terms involving spin S_j cancel when computing the ratio of V_{Si} and V_I:

$$V_{Si}/V_I = \tan^2(\pi J_{ISi}T) \tag{17}$$

which is identical to Eq. (15). Chemical shift labeling of the I spin can be implemented in a straightforward fashion, by shifting the dashed $180°(I,S)$ pulses in a constant-time manner.

4.3. COSY-Based Quantitative J Correlation Schemes

The third variation of the QJ method is shown in Fig. 6C and based on a COSY-type experiment. It can be used in the case of one or more spins S_i, homonuclear to spin I, with varying J couplings, J_{ISi}. The scheme is highly analogous to the scheme discussed for the second variation of the QJ method (Fig. 6B), albeit that reference and cross peaks are present in one spectrum, thereby facilitating the analysis. Again, considering first only an isolated IS spin pair and starting with transverse I-spin magnetization, the J_{IS} coupling is active for a duration T:

$$I_x \xrightarrow{T} I_x \cos(\pi J_{IS}T) + 2I_yS_z \sin(\pi J_{IS}T) \tag{18}$$

The 90°(ϕ_1) pulse now results in a COSY-like transfer:

$$\text{Eq. (18)} \xrightarrow{90°(\phi_1)} I_x \cos(\pi J_{IS}T) - 2I_zS_y \sin(\pi J_{IS}T) \tag{19}$$

During the t_1 period, the maximum duration of which is short relative to $(J_{IS})^{-1}$, the I_x and $2I_zS_y$ operators precess with the ω_I and ω_S frequencies, respectively, resulting in

$$\text{Eq. (19)} \xrightarrow{t_1} I_x \cos(\omega_I t_1) + I_y \sin(\omega_I t_1) - 2I_zS_y \cos(\omega_S t_1) + 2I_zS_x \sin(\omega_S t_1) \tag{20}$$

where the earlier trigonometric terms have been dropped from Eq. (20). The 90°(ϕ_2) pulse selects one of the quadrature components and converts the antiphase S magnetization back to antiphase I magnetization. During the rephasing period of duration T, the J_{IS} interaction is again active resulting in a total of four terms:

$$\begin{aligned}\text{Eq. (20)} \xrightarrow{90°(\phi_2)} \xrightarrow{T} &\{I_x \cos(\pi J_{IS}T) - 2I_zS_y \sin(\pi J_{IS}T)\}\cos(\omega_I t_1) \\ &+ \{2I_yS_z \cos(\pi J_{IS}T) - I_x \sin(\pi J_{IS}T)\} \cos(\omega_S t_1)\end{aligned} \tag{21}$$

The last 90° pulse again converts the antiphase terms, and thus only the I_x terms contribute to the observed I-spin signal. After Fourier transformation with respect to the t_1 dimension, peaks at the ω_I and ω_S frequencies are obtained along this frequency axis, referred to as the "diagonal" and "cross-peak," respectively. Reintroducing the earlier omitted trigonometric terms and neglecting differential relaxation of in-phase and antiphase terms momentarily, the volumes of the cross-peak, V_S, and diagonal peak, V_I, are written as

$$V_S = -A \sin^2(\pi J_{IS}T) \tag{22a}$$

$$V_I = A \cos^2(\pi J_{IS} T) \tag{22b}$$

where A again denotes a proportionality constant that includes all experimental and other constant factors. As above for the HMQC-based QJ methods, the magnitude of the J_{IS} coupling can be extracted in a straightforward fashion from the cross-peak to diagonal-peak ratio:

$$V_S/V_I = -\tan^2(\pi J_{IS} T) \tag{23}$$

In this experiment, similar to the second variation of the QJ method, the presence of several spins S_i that are J coupled to spin I, results in additional passive coupling terms that cancel from the ratio V_{Si}/V_I. The J_{ISi} can therefore still be extracted from the cross-peak to diagonal-peak ratio using Eq. (23).

As the maximum duration of the t_1 period typically is kept short compared to the transverse relaxation time of the I and S spins, the line shape in the F_1 dimension is determined primarily by the apodization function. Consequently, the ratio of cross-peak intensity to diagonal-peak intensity yields almost identical results.

4.4. Sources of Systematic Errors

4.4.1. Effects of Relaxation

The assumption of identical relaxation rates of in-phase and antiphase magnetization is usually not valid as the in-phase magnetization relaxes at a slower rate than the antiphase magnetization (Bax *et al.*, 1990; London, 1990; Peng *et al.*, 1991). Consequently, the intensity of the peak resulting from the antiphase operator (Fig. 6, scheme B or C) or with the dephasing active (Fig. 6, scheme A), will be reduced relative to the peak resulting from the in-phase operator. Therefore, the J coupling will be systematically underestimated using the QJ method. As was shown before for the $\{^{13}C'\}$ spin-echo difference ^{13}C HSQC experiment (Grzesiek *et al.*, 1993) and the HNHA experiment (Vuister and Bax, 1993a), these errors can be compensated for if an estimate of the selective T_1 of the S spin is available.

To a first approximation (Bax *et al.*, 1990; London, 1990; Peng *et al.*, 1991), the apparent transverse relaxation time of the $2I_yS_z$ operator, $T_{app}(2I_yS_z)$, can be expressed in terms of the transverse relaxation time of the I spin, $T_2(I_x)$, and the selective T_1 of the S spin, $T_{1,sel}(S)$:

$$\frac{1}{T_{app}(2I_yS_z)} = \frac{1}{T_2(I_x)} + \frac{1}{T_{1,sel}(S)} \tag{24}$$

Using this approximation, the resulting systematic errors in the J values were estimated to be on the order of 1–2% for the $\{^{13}C'\}$ spin-echo difference ^{13}C HSQC experiment (Grzesiek *et al.*, 1993). For the HNHA experiment, the systematic errors are somewhat larger because of the relatively fast relaxation of the H^{α} spins, but can be easily compensated for, as explained below (Vuister and Bax, 1993a).

The time evolution of the I_x and $2I_yS_z$ operators under the influence of J coupling and relaxation can be written in the following matrix equation:

$$d/dt\begin{pmatrix} I_x \\ 2I_yS_z \end{pmatrix}=\begin{pmatrix} -\dfrac{1}{T_2(I_x)} & \pi J \\ -\pi J & -\dfrac{1}{T_{\text{app}}(2I_yS_z)} \end{pmatrix}\begin{pmatrix} I_x \\ 2I_yS_z \end{pmatrix} \quad (25)$$

which can be solved by simple numerical integration.

Figure 7 shows the apparent J values, calculated for the third QJ method (Fig. 6C) by numerical integration of Eq. (25) followed by application of Eq. (23) versus the true J values. The curves are shown assuming $T_{1,\text{sel}}(S) = 100$ ms and $T = T_2(I_x)$, $T_{1,\text{sel}}(S) = 100$ ms and $T = 2 \times T_2(I_x)$, and $T_{1,\text{sel}}(S) = 200$ ms and $T = T_2(I_x)$. The curves for $T_{1,\text{sel}}(S) = 100$ ms and 200 ms using $T = T_2$ are almost linear, allowing the correction of the calculated J coupling by a simple factor. In contrast, extending

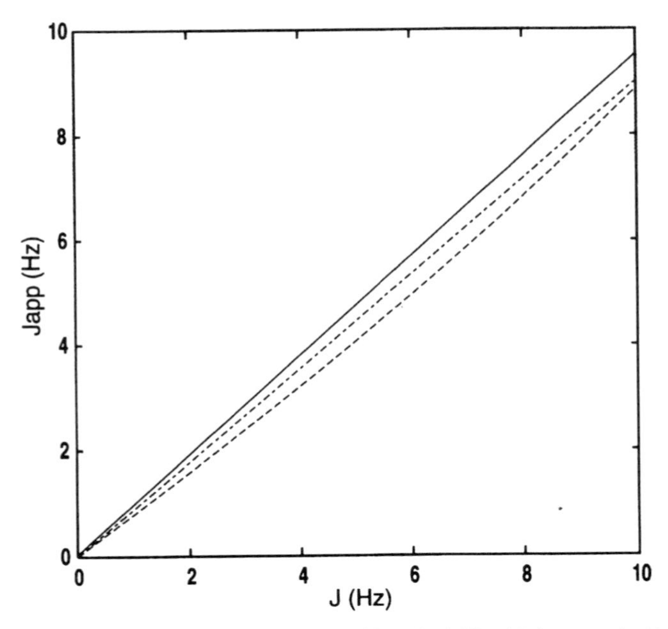

Figure 7. Apparent J values calculated for the third QJ method (Fig. 6C) by numerical integration of Eq. (25) followed by application of Eq. (23) versus the true J values. The curves were computed assuming $T_{1,\text{sel}}(S) = 100$ ms and $T = T_2(I_x)$ (dashed-dotted), $T_{1,\text{sel}}(S) = 100$ ms and $T = 2 \times T_2(I_x)$ (dashed), and $T_{1,\text{sel}}(S) = 200$ ms and $T = T_2(I_x)$ (solid).

the dephase/rephase period to $2 \times T_2(I_x)$ disrupts the linearity, in addition to giving rise to larger errors.

In the case of several spins S_i that are J coupled to spin I, i.e., in schemes B and C of Fig. 6, not only the "leakage" of the antiphase magnetization term, but also the effects of cross-relaxation between spins S_i and S_j ($S_{z,i}$–$S_{z,j}$ spin flips) have to be taken into account. The exchange of magnetization between the two antiphase terms, $2I_yS_{z,i}$ and $2I_yS_{z,j}$, results in an increase of the smaller antiphase term at the expense of the larger antiphase term. The magnitude of this effect increases with increasing dephase/rephase times and is also proportional to the cross-relaxation rate between the two spins S_i and S_j. For the LRCH experiment on selectively $^{13}C^{\delta}$-Leu labeled staphylococcal nuclease, these effects were investigated (Vuister *et al.*, 1993b). Values of the $^3J(C^{\delta2}H^{\beta2})$ and $^3J(C^{\delta2}H^{\beta3})$ coupling constants of Leu^{137}, derived from the LRCH data, are shown in Fig. 8 as a function of three dephase/rephase delays, T. Immediately apparent is the decrease of the larger of the two couplings, i.e., the $^3J(C^{\delta2}H^{\beta2})$, with increasing T. This decrease results from the generally faster relaxation of the antiphase magnetization as discussed above, and the loss related to spin flips with the second β-proton. In contrast, the measured values for $^3J(C^{\delta2}H^{\beta3})$ are rather constant, as the loss caused by the faster relaxation of the antiphase magnetization is compensated by the gain from the spin flips.

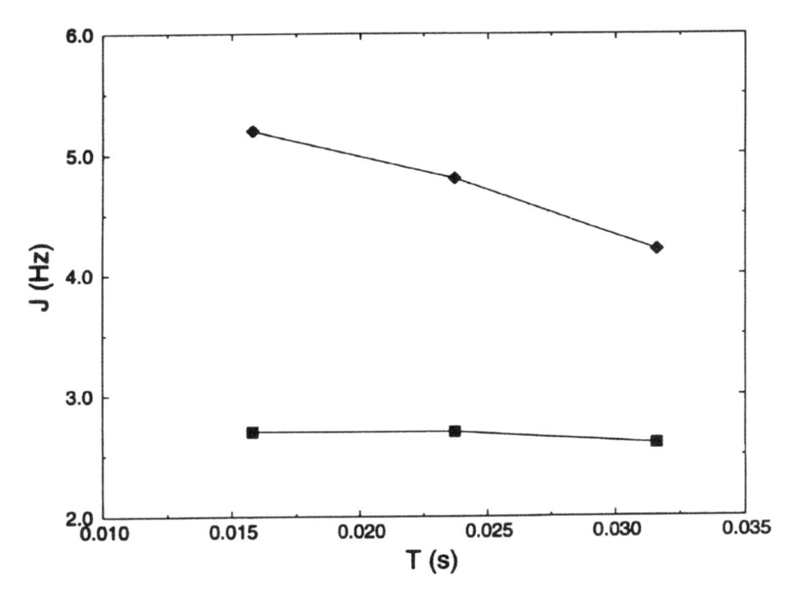

Figure 8. Values of the $^3J(C^{\delta2}H^{\beta2})$ (diamonds) and $^3J(C^{\delta2}H^{\beta2})$ (squares) coupling constants of Leu^{137} derived from the LRCH data of selectively $^{13}C^{\delta}$-Leu labeled staphylococcal nuclease (Vuister *et al.*, 1993b) as a function of three different dephase/rephase delays, T.

Extrapolation to $T = 0$ yields the true values of $^3J(C^{\delta2}H^{\beta2})$ and $^3J(C^{\delta2}H^{\beta3})$, which are 5.4 and 2.6 Hz, respectively.

4.4.2. Effects of RF Inhomogeneity

The proper inversion by a 180° pulse or an exact 90° pulse is of importance for the QJ methods, as the method relies on quantification of transferred and non-transferred magnetization pathways. Typically, the effects of RF inhomogeneity will systematically lower the measured J coupling. For example, if a fraction f of the spins is not inverted by the 180° (S) pulse (Fig. 6A), the difference spectrum is attenuated by an amount $(1 - f)$, resulting in a decrease of the measured coupling of $(1 - f/2)$. For the $\{^{13}C'\}$ spin-echo difference ^{13}C HSQC experiment on calmodulin, this effect was estimated to reduce the measured $^3J(C^{\gamma}C')$ coupling by 1–5% (Grzesiek *et al.*, 1993).

QJ schemes B and C of Fig. 6 suffer from the effects of RF inhomogeneity related to the 90° pulses surrounding the t_1 evolution period. Equations (10) and (19) show that the cross-peak intensity is affected by a factor $\sin^2(\alpha)$ and $\sin^4(\alpha)$ for schemes B and C, respectively, where α denotes the nominal flip angle of the 90° pulse. In scheme B, the nontransferred antiphase magnetization, proportional to $\cos^2(\alpha)$, attenuates the reference peak by a factor $\{1 - \cos^2(\alpha)\tan^2(2\pi JT)\}$, whereas in scheme C, the reference peak is attenuated by a factor $\{1 - \cos^4(\alpha)\tan^2(2\pi JT)\}$. Thus, the errors in the calculated coupling constants become dependent on the magnitude of J, but always will decrease the calculated J coupling.

4.4.3. Effects of Incomplete Labeling

Incomplete labeling affects those QJ experiments in which the dephasing spin, i.e., the S spin, is a non-^1H nucleus. If the random labeling of the S spin is L, a fraction $(1 - L)$ of the magnetization is not subject to dephasing. For scheme A, this results in a scaling of the relative difference [see Eq. (8)] by a factor L, yielding a measured J value that is approximately underestimated by a factor $L^{-1/2}$ (Grzesiek *et al.*, 1993). For schemes B and C, the diagonal-peak intensity is increased by an amount $(L^{-1} - 1)$ relative to the cross-peak intensity. As an example, for the HN(CO)CO experiment (Section 5.5), a labeling percentage $L = 97\%$ would result in an approximate 2% underestimate of the J coupling.

5. THE BACKBONE ANGLE ϕ

Knowledge about the backbone angles ϕ and ψ is of great importance in restricting the conformational space of the backbone and thereby helps to establish the global fold of the protein. Historically, the $^3J(H^NH^{\alpha})$ received the most attention because it can be measured in a relatively straightforward way in small proteins.

Therefore, information about the ϕ angle was used in NMR structure determination as early as 1984 (Pardi *et al.*, 1984).

5.1. *J* Couplings Related to ϕ

Figure 9 shows the Fisher projection of a peptide fragment around the backbone angle ϕ. From the figure, it is evident that the torsion angle ϕ is associated with up to six ^3J coupling constants. For nonglycine residues, these coupling constants are $^3J(H^NH^\alpha)$, $^3J(H^NC^\beta)$, $^3J(H^NC')$, $^3J(C'^{i-1}H^\alpha)$, $^3J(C'^{i-1}C^\beta)$, and $^3J(C'^{i-1}C')$. Table 1 lists the experiments aimed at measuring these six coupling constants.

As the magnitude of the $^3J(H^NH^\alpha)$ is also relatively easy to determine in smaller polypeptides, a parametrization of the Karplus curve based on protein data has been available for some time (Pardi *et al.*, 1984). A recent reparametrization on the basis of a larger number of observations obtained from newer experiments (Vuister and Bax, 1993a; Wang and Bax, 1996) has adjusted the magnitude of the *A*, *B*, and *C* parameters somewhat, but has not changed the overall shape of the Karplus curve for $^3J(H^NH^\alpha)$. The curves corresponding to three different parametrizations are shown in Fig. 10A. The most recent one was obtained by Wang and Bax (1996) for ubiquitin using optimized dihedral angles (see curve with small dashes in Fig. 10A). Thus, all curves show that the $^3J(H^NH^\alpha)$ value is approximately 10 Hz for β-sheet regions and in the range of 4–6 Hz for residues in α-helical conformation.

A second coupling constant related to the backbone angle ϕ is the $^3J(H^NC')$. It can be measured from the HNCA[C′] E.COSY experiment (Seip *et al.*, 1994; Weisemann *et al.*, 1994a; Wang and Bax, 1995). Values of this coupling constant recently determined for ubiquitin prompted also for a reparametrization of the Karplus curve which substantially differs from the parametrization obtained from the studies on small model compounds (Wang and Bax, 1995). The resulting curve is also shown in Fig. 10A, and illustrates the complementarity to the $^3J(H^NH^\alpha)$, but also the rather limited range of the $^3J(H^NC')$ coupling constant in proteins.

A third coupling constant yielding information about the backbone angle ϕ is the $^3J(H^NC^\beta)$. This coupling constant has been measured in an E.COSY fashion from 3D ^{15}N-separated NOESY[C] or 4D $^{13}C/^{15}N$-separated NOESY[C] experiments on a fully $^{13}C/^{15}N$ labeled protein (Seip *et al.*, 1994) or by using an

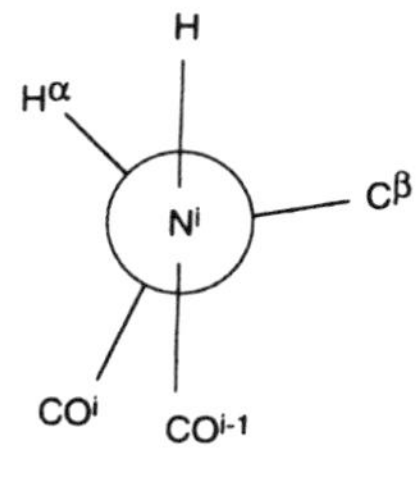

Figure 9. Fisher projection around the backbone torsion angle ϕ.

Table 1
Experiments for Measuring J Couplings Related to ϕ

Coupling	Experiment[a]	Reference	Labeling	Remark
$^3J(H^NH^\alpha)$	COSY	Pardi *et al.* (1984)	—	From splittings in COSY
	COSY/NOESY	Ludvigsen *et al.* (1991)	—	DISCO-like procedure
	^{15}N HMQC	Kay *et al.* (1989)	^{15}N	High-resolution HMQC
		Kay and Bax (1990)		
	HNCA[H^α]	Schmieder *et al.* (1991)	^{15}N/^{13}C	Different schemes for leaving H^α untouched
		Seip *et al.* (1992, 1994)		
		Madsen *et al.* (1993)		
		Görlach *et al.* (1993)		15N-$^1H^N$ MQ scheme
		Weisemann *et al.* (1994a)		
		Tessari *et al.* (1995)		
		Löhr and Rüterjans (1995)		
		Wang and Bax (1996)		Selective H^N pulses in revINEPT
	(H)NCAHA[H^N]	Löhr and Rüterjans (1995)	^{15}N/^{13}C	H^α detection in H_2O
	HSQC/HMQC-J	Neri *et al.* (1990)	^{15}N	
		Billeter *et al.* (1992)		
		Kuboniwa *et al.* (1994)		CT version
	HNHA	Vuister and Bax (1993a)	^{15}N	
		Kuboniwa *et al.* (1994)		
$^3J(H^NC')$	HNCA[C']	Weisemann *et al.* (1994a)	^{15}N/^{13}C	
		Seip *et al.* (1994)		
		Wang and Bax (1995)		
	(H)CANNH[C']	Löhr and Rüterjans (1995)	^{15}N/^{13}C	
$^3J(H^NC^\beta)$	3D/4D-separated NOESY[C^β]	Seip *et al.* (1994)	^{15}N/^{13}C	
	(H)CANNH[C^β]	Löhr and Rüterjans (1995)	^{15}N/^{13}C	Requires C^α selective pulses
	HNCA[$C\beta$]	Wang and Bax (1996)	^{15}N/^{13}C	Requires C^α selective pulses
$^3J(C'^{i-1}H^\alpha)$	(H)NCAHA[C']	Löhr and Rüterjans (1995)	^{15}N/^{13}C	
	HCAN[C']	Wang and Bax (1996)	^{15}N/^{13}C	Sample in D_2O preferable
	H(N)CA,CO[H^α]	Löhr and Rüterjans (1997)	^{15}N/^{13}C	
	HN(CO)HB	Grzesiek *et al.* (1992)	^{15}N/^{13}C	Separate reference experiment
$^3J(C'^{i-1}C')$	HN(CO)CO	Hu and Bax (1996)	^{15}N/^{13}C	
$^3J(C'^{i-1}C^\beta)$	HN(CO)C	Hu and Bax (1997b)	^{15}N/^{13}C	

[a]Experiments are named analogous to the nomenclature proposed for triple-resonance experiments (Ikura *et al.* 1990). The passive spin in the E.COSY-type experiments is indicated in square brackets (Wang and Bax, 1995).

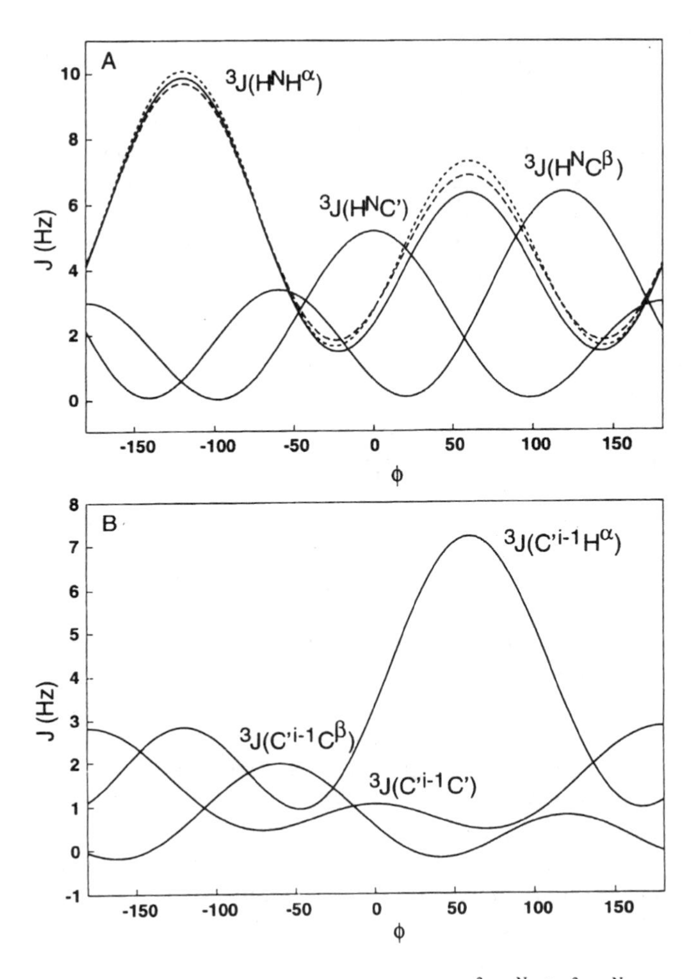

Figure 10. Karplus curves showing the dependences of the (A) $^3J(H^NH^\alpha)$, $^3J(H^NC')$, and $^3J(H^NC^\beta)$ and (B) $^3J(C'^{i-1}H^\alpha)$, $^3J(C'^{i-1}C^\beta)$, and $^3J(C'^{i-1}C')$ coupling constants on the torsion angle ϕ. Curves shown are: $^3J(H^NH^\alpha) = A\cos^2(\phi - 60°) + B\cos(\phi - 60°) + C$ with $A, B, C = 6.51, -1.76, 1.60$ (solid) (Vuister and Bax, 1993a); $A, B, C = 6.41, -1.46, 1.9$ (large dashes) (Pardi *et al.*, 1984); and $A, B, C = 6.98, -1.38, 1.72$ (small dashes) (Wang and Bax, 1996). $^3J(H^NC') = 4.0\cos^2(\phi \pm 180°) - 1.1\cos(\phi \pm 180°) + 0.1$ (Wang and Bax, 1995, 1996). $^3J(H^NC^\beta) = 4.7\cos^2(\phi + 60°) - 1.5\cos(\phi + 60°) + 0.2$ (Bystrov, 1976). $^3J(C'^{i-1}H^\alpha) = 3.75\cos^2(\phi + 120°) - 2.19\cos(\phi + 120°) + 1.28$ (Wang and Bax, 1996). $^3J(C'^{i-1}C^\beta) = 1.5\cos^2(\phi - 120°) - 0.6\cos(\phi - 120°) - 0.1$ (Bystrov, 1976). $^3J(C'^{i-1}C') = 1.33\cos^2(\phi) - 0.88\cos(\phi) + 0.62$ (Hu and Bax, 1996). N.B.: Unfortunately, Wang and Bax (1995, 1996) incorporate a 180° phase shift into their equations for $^3J(H^NC')$ and $^3J(C'^{i-1}H^\alpha)$ by defining the dihedral angle ϕ as H^{Ni}–N^i–$C^{\alpha i}$–C'^i instead of the IUPAC definition C'^{i-1}–N^i–$C^{\alpha i}$–C'^i (IUPAC, 1970).

HNCA-based E.COSY experiment employing selective C^{α} pulses (Löhr and Rüterjans, 1995; Wang and Bax, 1996). Reparametrization of the corresponding Karplus relationship on the basis of the values measured for ubiquitin (Wang and Bax, 1996) again substantially changed the earlier curve derived on the basis of FPT-INDO calculations, most notably in the region of positive ϕ values. However, as only two experimental data points were obtained with $^3J(H^NC^{\beta})$ values corresponding to $\phi > 0$, the maximum of the $^3J(H^NC^{\beta})$ Karplus curve (at $\phi = 120°$) may be ill-determined. In fact, reparametrization using refined backbone angles substantially changed this maximum. Unfortunately, the range of $^3J(H^NC^{\beta})$ values in proteins will be rather limited as the maximum value is found for ϕ angles that are conformationally disallowed in proteins. Therefore, the applicability will critically depend on the accuracy of the measurement of the $^3J(H^NC^{\beta})$ values.

The fourth coupling constant related to ϕ is the $^3J(C'^{i-1}C')$. This coupling constant can be measured from the HN(CO)CO experiment (Hu and Bax, 1996) in a straightforward fashion. Recent reparametrization of the Karplus curve for $^3J(C'^{i-1}C')$ was also performed using the values obtained for ubiquitin (Hu and Bax, 1996). The result is shown in Fig. 10B and indicates that this coupling constant is predominantly useful in restricting the backbone angle ϕ for values around $\pm 180°$. For β-sheet regions, however, it can provide useful additional information as the $^3J(H^NH^{\alpha})$ curve allows for two possibilities when the measured $^3J(H^NH^{\alpha})$ values are in the range of 8–9 Hz. The steep dependence of $^3J(C'C')$ on ϕ then allows for a discrimination between these two possibilities.

Values for the $^3J(C'^{i-1}H^{\alpha})$ coupling constant have recently been measured in proteins (Löhr and Rüterjans, 1995, 1997; Wang and Bax, 1996) and its Karplus-type dependence on the torsion angle ϕ has been reparametrized using the values measured in ubiquitin (Wang and Bax, 1996). The corresponding curve in Fig. 10B reveals that a substantially larger value is expected for $\phi = 60°$ as compared with the range of negative values of ϕ. Hence, it is a useful coupling constant for identification of residues in the positive region of the Ramanchandran plot.

The $^3J(C'^{i-1}C^{\beta})$ is the last coupling constant for which values measured in proteins have become available and its Karplus-type dependence on the torsion angle ϕ has also been reparametrized (Hu and Bax, 1997b).

5.2. The HNHA Experiment

The HNHA experiment (Vuister and Bax, 1993a; Kuboniwa *et al.*, 1994) is a QJ experiment for measuring the $^3J(H^NH^{\alpha})$ coupling constant. The pulse sequence for this experiment is shown in Fig. 11. The experiment belongs to the third class of QJ experiments (see Fig. 6C) for measuring J couplings involving two J correlated homonuclear spins, as discussed in Section 4.3. The general dephase/rephase scheme of this third variation of the QJ method is overlaid with a multiple-quantum-type correlation sequence, yielding a 3D experiment. The ^{15}N chemical

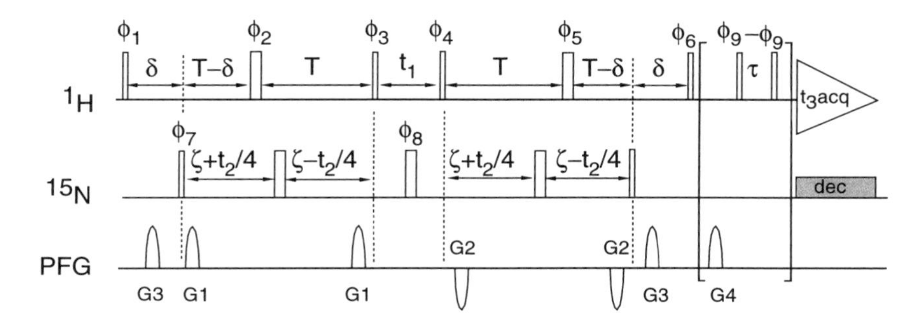

Figure 11. Pulse scheme of the 3D HNHA experiment (Vuister and Bax, 1993a). Narrow and wide bars denote 90° and 180° pulses, respectively. Unless indicated otherwise, all pulses are applied along the x-axis. The water flip-back modification of Kuboniwa *et al.* (1994) is shown in brackets. The following phase cycle is used: $\phi_1 = x$; $\phi_2 = 4x, 4(-x)$; $\phi_3 = x,-x$; $\phi_4 = 2x,2(-x)$; $\phi_5 = 4x,4(-x)$; $\phi_6 = y,-y$; $\phi_7 = x$; $\phi_8 = 8x,8y$; receiver = $8x$, $8(-x)$. For the water flip-back modification: $\phi_1 = \phi_2 = \phi_3 = 2x,2(-x)$; $\phi_4 = \phi_5 = \phi_6 = x$; $\phi_7 = x$; $\phi_8 = x,y$; $\phi_9 = 4x,4y$; receiver = $x,2(-x),x,y,2(-y),y$. Quadrature detection is obtained with the States–TPPI method, incrementing phases ϕ_7 for t_1 and ϕ_1, ϕ_2, and ϕ_3 for t_2. The delay δ is set to 5.4 ms and the delay T is set to 10–20 ms, the optimum usually found for $2T \approx T_2(\mathrm{H^N})$.

shift evolution is accomplished by displacement of the first and third 180° ^{15}N pulses. Note that no evolution of the ^1J(HN) coupling takes place during this time as the ^{15}N–^{1}H magnetization is present as multiple-quantum magnetization.

As discussed in Section 4.3, the J coupling is extracted from the cross-peak to diagonal-peak intensity ratio by application of Eq. (23). As the diagonal peaks have two $\mathrm{H^N}$ frequency coordinates, i.e., along the F_1 and F_3 axes, these peaks display a resolution comparable to a 2D correlation spectrum acquired with a maximum acquisition time of 4ζ in t_2. However, because of the constant-time nature of the evolution in this domain, the time-domain data are easily extended by mirror-image linear prediction (Zhu and Bax, 1992), effectively yielding a maximum acquisition time of 8ζ. As the $^1\mathrm{H^N}$–^{15}N correlation map is usually well dispersed, and 8ζ typically is in the range of 80–100 ms, the overlap does not present a serious problem.

Strip plots taken along the F_1 axis of the HNHA spectrum of the photoactive yellow protein (PYP) obtained at the $^1\mathrm{H^N}$–^{15}N resonance frequencies of Ala^{45}, Glu^{46}, Gly^{47}, His^{108}, and Met^{109} are shown in Fig. 12. The cross-peaks to the intraresidual H^α appear in the region of 4–6 ppm. They are of opposite sign with respect to the diagonal peaks located in the region of 7–10 ppm, in agreement with Eq. (22). The differences in intensities between the cross-peaks of the first three strips and the intensities observed in the last two strips are evident. Extracting the ^3J values by application of Eq. (23) yields 3.4, 4.6, 9.5, and 10.3 Hz for the Ala^{45}, Glu^{46}, His^{108}, and Met^{109} strips, respectively. The large values observed for His^{108} and Met^{109} are in complete agreement with the fact that these residues of PYP adopt

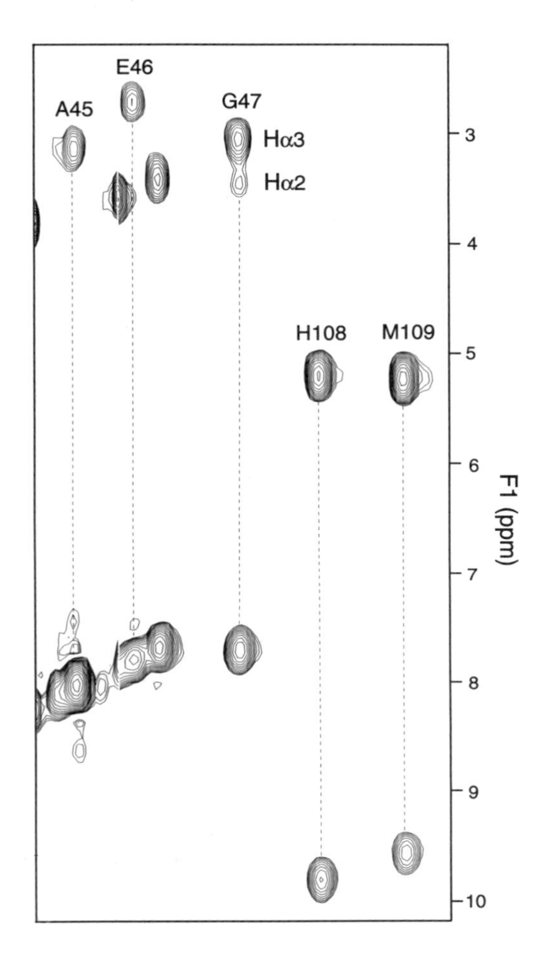

Figure 12. F_1 strips from the 3D HNHA spectrum of a 2 mM solution of uniformly ^{15}N labeled PYP in 95/5 v/v H_2O/D_2O solution at pH 5.8, recorded at 311 K with the pulse sequence of Fig. 11. The data consisted of $50\,(t_1, {}^1H) \times 59\,(t_2, {}^{15}N) \times 512\,(t_3, {}^1H)$ complex points with maximum acquisition times of 11.6 (t_1), 26.5 (t_2), and 51.3 (t_3) ms. The strips have been taken at the $^1H^N$–^{15}N resonance frequencies of Ala^{45}, Glu^{46}, Gly^{47}, His^{108}, and Met^{109}.

an extended conformation in solution (Düx *et al.*, 1997). In the case of Gly^{47}, two correlations to both α-protons are observed. The differences in the cross-peak intensities directly reflect the differences in the magnitude of both J couplings, which are calculated to be 6.3 and 4.0 Hz for the downfield- and upfield-shifted cross-peaks, respectively.

As previously discussed, the assumption that in-phase and antiphase magnetization relax at identical rates is invalid and results in a systematic underestimate of the J coupling. It was also shown that the calculated J coupling can be corrected for by multiplication with a constant factor if an estimated value of the $T_{1sel}(H^\alpha)$ is available, provided that the duration of the dephase/rephase periods does not exceed the T_2 of the amide protons. In the case of staphylococcal nuclease with $\tau_c \approx 9$ ns

(Vuister and Bax, 1993a) and Ca^{2+}-free calmodulin (Kuboniwa *et al.*, 1994) this factor was 1.1. For the present spectrum of PYP for which $\tau_c \approx 6.5$ ns (Düx *et al.*, 1998), the data were corrected by a factor 1.05. However, in some cases (Kuboniwa *et al.*, 1994) the HNHA data, even after correction, still showed a systematic difference relative to values obtained using the CT-HMQC-J experiment. This effect was only partially explained by RF inhomogeneity.

The HNHA experiment offers excellent sensitivity, in spite of the long dephasing and rephasing delays. For the experiment performed on staphylococcal nuclease, the lower limit on the $^3J(\mathrm{H^NH^\alpha})$ was 3 Hz (Vuister and Bax, 1993a), whereas in the case of PYP, values as low as 2.1 Hz were measured. Compared with other experiments for measuring the $^3J(\mathrm{H^NH^\alpha})$ (see Table 1), the HNHA experiment also has the advantage that J couplings to both α-protons of glycine residues can be determined, provided that these do not have degenerate chemical shifts.

5.3. The J-Modulated $\{^1\mathrm{H}–^{15}\mathrm{N}\}$-COSY Experiment

Analogous to the HNHA experiment is the J-modulated $\{^1\mathrm{H}–^{15}\mathrm{N}\}$-COSY experiment (Billeter *et al.*, 1992; Kuboniwa *et al.*, 1994). In this experiment the modulation resulting from evolution of the $^3J(\mathrm{H^NH^\alpha})$ is sampled in a pseudo-3D fashion: Two chemical-shift dimensions provide the resolution, i.e., the ^{15}N and ^{1}H, whereas J modulation is obtained in the third dimension of this experiment.

For the water-flipback constant-time implementation of this experiment (CT-HMQC-J) (Kuboniwa *et al.*, 1994), the sampling in the third dimension is achieved by recording a series of 2D spectra with varying values for the constant-time delay, T. Using an expression for the time dependence of the in-phase and antiphase magnetization analogous to Eq. (25), Kuboniwa *et al.* (1994) derived the following expression for the time dependence of the cross-peak intensity of an $^{15}\mathrm{N}–^1\mathrm{H}$ correlation:

$$S(T) = S(0)\,e^{-T/(T_{2,\mathrm{MQ}} + 2T_{1,\alpha})}\left\{\cos(\pi J^{\mathrm{r}} T) + \frac{\sin(\pi J^{\mathrm{r}} T)}{2\pi J^{\mathrm{r}} T_{1,\alpha}}\right\} \tag{26}$$

where

$$J^{\mathrm{r}} = \left(J_{\mathrm{HH}}^2 - \frac{1}{(2\pi T_{1,\alpha})^2}\right)^{1/2} \tag{27}$$

and J^{r} denotes the apparent J coupling derived from the fit of Eq. (26) to the experimentally obtained intensities, J_{HH} denotes the actual J coupling to be determined, and $T_{1,\alpha}$ is the selective T_1 of the H^α spin [see Eq. (24)]. As in the HNHA experiment, the results are corrected for by assuming a uniform value of $T_{1,\alpha}$.

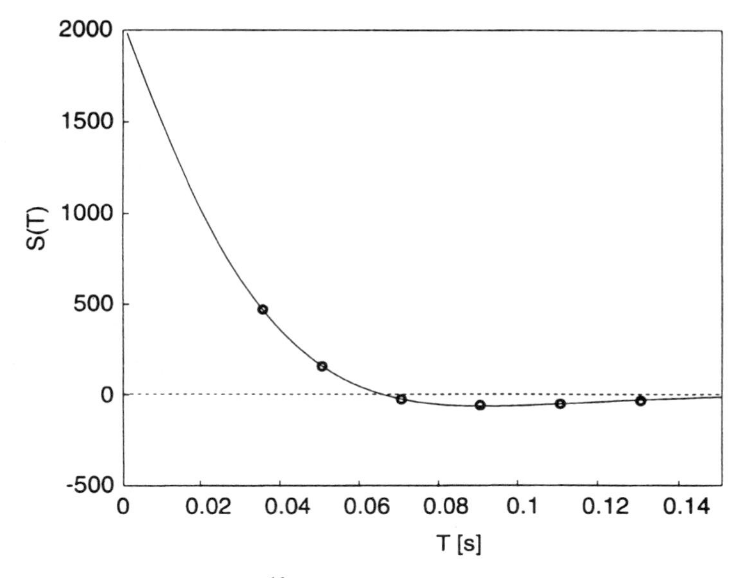

Figure 13. Signal intensity of the Ala[16] cross-peak in the CT-HMQC-J experiment (Kuboniwa *et al.*, 1994) recorded on a 2 mM solution of uniformly ^{15}N

Figure 13 shows the signal intensity, $S(T)$, obtained from the CT-HMQC-J experiment on ^{15}N-labeled PB92 for Ala[16] as a function of six different values for T. The optimized fit of Eq. (26) to these experimental intensities, assuming $T_{1\alpha}$ = 100 ms, is also plotted. Clearly, the data points are well-described by this curve which yields a $T_{2,MQ}$ = 35 ms and $^3J(H^NH^\alpha)$ = 8.6 Hz. In general, reliable fits can be obtained provided the signal intensities show at least one zero crossing, i.e., the $^3J(H^NH^\alpha)$ must be sufficiently large with respect to the $T_{2,MQ}$ so as to avoid overdamping. In the case of critically or overdamped signals, only a reliable upper estimate can be obtained (Billeter *et al.*, 1992). Note that this situation is equivalent to the lack of a cross-peak in the HNHA experiment. Also in this case, only an upper estimate of the coupling constant can be obtained.

5.4. The HNCA[H^α] Experiment

The HNCA[H^α] experiment employs the E.COSY principle for measuring $^3J(H^NH^\alpha)$. The experiment belongs to the third class of E.COSY experiments (Fig. 2C), in which the spin detected during acquisition, i.e., H^N, is homonuclear with respect to the spin state that must be preserved, i.e., H^α. Several schemes have been proposed for maintaining the H^α spin state, but most commonly a bilinear rotation is used (Sørensen, 1990; Seip *et al.*, 1992), optionally combined with the sensitiv-

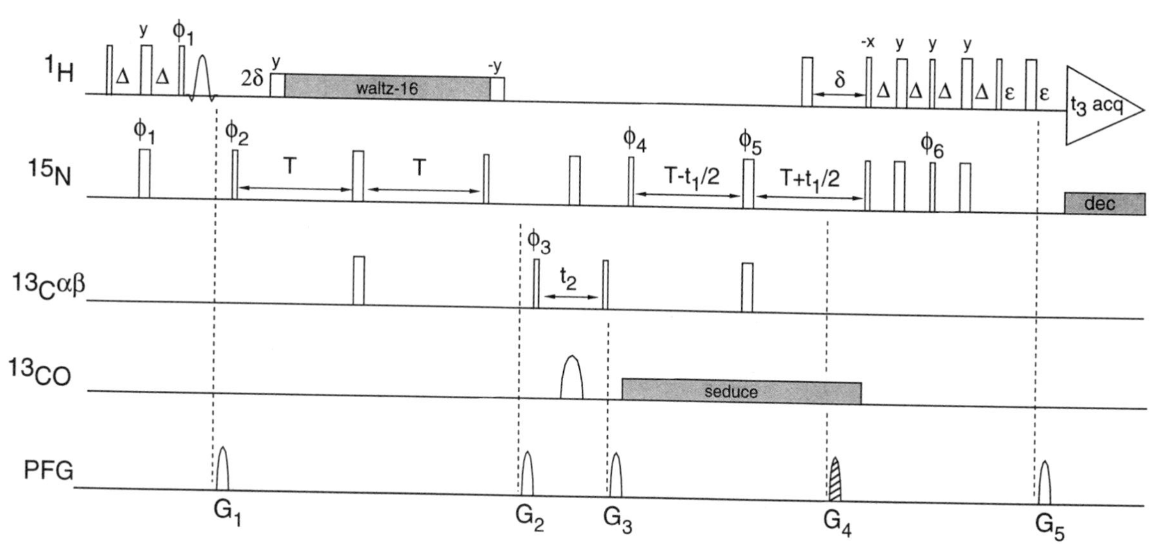

Figure 14. Pulse sequence of the HNCA[H^{α}] experiment adapted from sequences published before (Madsen *et al.*, 1993; Seip *et al.*, 1994; Weisemann *et al.*, 1994a). Narrow and wide bars denote 90° and 180° pulses, respectively. The shaped water-flip-back ^1H pulse was implemented as a 270° Gaussian pulse of 3.3-ms duration. The 180° pulses on the $C^{\alpha\beta}$ had a zero-excitation profile at the center of the C′ resonances (175 ppm). The phase cycle was: $\phi_1 = y,-y$; $\phi_2 = 2x,2(-x)$; $\phi_3 = 4x,4(-x)$; $\phi_4 = x$; $\phi_5 = 2x,2(-x)$; $\phi_6 = -y$; receiver $= x,2(-x),x,-x,2x,-x$. P/N gradient-coherence selection with sensitivity enhancement (Kay *et al.*, 1992) was accomplished inverting the sign of gradient G_4 in conjunction with inversion of phase ϕ_6. In addition, for every t_1 increment, phase ϕ_4 and the receiver phase were incremented by 180°. Quadrature detection during t_2 was obtained using the States–TPPI method, incrementing phase ϕ_3. The delays were as follows: $\Delta = 2.25$ ms; $\delta = 2.7$ ms; $T = 24.8$ ms; $\varepsilon = G_5 + 100\ \mu s$. Gradient durations and strengths: $G_1 = 1$ ms, -15 cm^{-1}; $G_2 = 1$ ms, 5 Gcm^{-1}; $G_3 = 1$ ms, 5 G cm^{-1}; $G_4 = 2$ ms, 30 G cm^{-1}; $G_5 = 1$ ms, 6.084 G cm^{-1}.

ity-enhancement scheme (Palmer *et al.*, 1991; Kay *et al.*, 1992). The latter implementation is shown in Fig. 14 and is based on earlier published sequences that have been described in terms of product operators on several previous occasions (Madsen *et al.*, 1993; Seip *et al.*, 1994; Weisemann *et al.*, 1994a).

Figure 15 shows the strips along the ^{13}C axis of the HNCA[H$^\alpha$] spectrum of uniformly ^{15}N/^{13}C labeled SH2 (Src homology domain 2) domain of c-Src taken at the ^{1}H resonance frequencies of Thr^{171}, Phe^{172}, Val^{174}, Glu^{176}, Tyr^{184}, Lys^{200}, Tyr^{202}, and Phe^{212}. For most traces, two cross-peaks are visible which originate from the intraresidual and sequential ^{15}N-to-^{13}C$^\alpha$ transfers. The delay T (see Fig. 14) is chosen to optimize the intraresidual connectivity. Both cross-peaks are split along the F_2 axis by the large one-bond $^1J(C^\alpha H^\alpha)$ coupling. The pulse sequence is designed to preserve the H$^\alpha$ spin state. Hence, an E.COSY pattern is observed resulting from the $^3J(H^N H^\alpha)$ and $^4J(H^N H^\alpha)$ splittings along the F_3 axis for the intraresidual and sequential connectivities, respectively.

Longitudinal relaxation of the H$^\alpha$ will disturb the E.COSY pattern and result in a systematic underestimate of the measured values. In fact, all of the values measured from the strips in Fig. 15 were systematically smaller than the values backcalculated from the dihedral angles ϕ of the refined NMR solution structure of the SH2 domain of c-Src (Xu *et al.*, 1995). This effect has been analyzed in detail

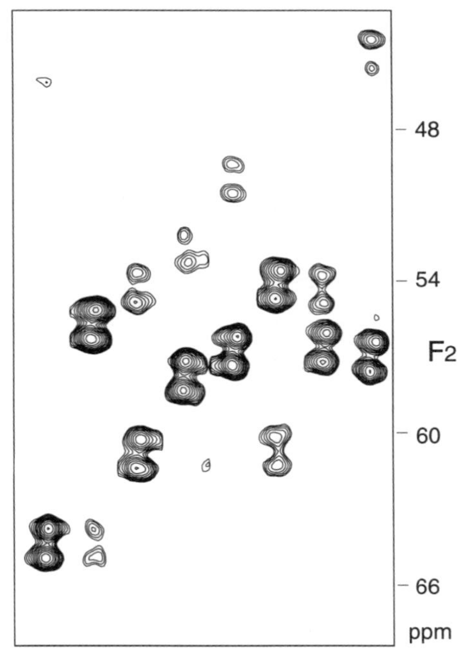

Figure 15. Strips along the ^{13}C axis of the HNCA[H$^\alpha$] spectrum of a 1 mM solution of uniformly ^{15}N/^{13}C labeled SH2 domain of c-Src in 95/5 v/v H_2O/D_2O taken at the ^{1}H resonance frequencies of Thr^{171}, Phe^{172}, Val^{174}, Glu^{176}, Tyr^{184}, Lys^{200}, Tyr^{202}, and Phe^{212}. The spectrum resulted from a data set with 37 (t_1, ^{15}N) × 40 (t_2, ^{13}C) × 512 (t_3, ^{1}H) complex points with maximum acquisition times of 24.7, 10.9, and 77 ms, respectively, recorded with the pulse sequence of Fig. 14.

for the $^3J(H^NH^\alpha)$ values measured for ubiquitin (Wang and Bax, 1996). In this study, simulated HNCA[H^α] E.COSY patterns were subjected to a data analysis procedure identical to the one used for the experimental data. The values derived from the simulated data were compared with the actual values used to generate these data. The observed systematic differences were then used to correct the experimental results.

Weisemann *et al.* (1994a) concluded that the coherence transfer between the α and β states of the H^α spin resulting from $^3J(H^\alpha H^\beta)$, $^3J(NH^\beta)$, or $^2J(NH^\alpha)$ couplings was negligible (< 5%). However, the ε–180°(H)–ε spin-echo sequence used for applying the refocusing gradient G_5, produces a nonnegligible phase difference along the F_3 axis between the two components of the E.COSY multiplet, resulting from the $^3J(H^NH^\alpha)$ evolution during 2ε. When extracting the J values using the procedure described in Section 3.2, this effect can be compensated for.

5.5. The HN(CO)CO Experiment

The HN(CO)CO experiment (Hu and Bax, 1996) is displayed in Fig. 16. An HNCO experiment relays the magnetization from the amide proton to the C'^{i-1} nucleus after which a pulse scheme belonging to the third class of the QJ method is used to generate C′–C′ correlations. The current scheme is modified slightly with respect to the original implementation by introduction of a SEDUCE composite pulse sequence (McCoy and Müller, 1992) for decoupling of the C^α spins, and gradient-coherence selection with sensitivity enhancement (Kay *et al.*, 1992) in the final reverse INEPT.

Both diagonal and cross-peaks can be observed in this experiment, allowing for direct calculation of the coupling constant using Eq. (23). For each C′ nucleus, both sequential C′ nuclei are separated by three intervening bonds. In addition, for Asp and Asn residues, the sidechain γ-carbonyl is separated by three bonds from the backbone C′. Thus, one can observe up to three correlations along the F_2 axis, or even four correlations in the case of Asp or Asn residues.

Examples of F_2 strips taken from the HN(CO)CO spectrum of $^{13}C/^{15}N$-labeled SH2 of c-Src are shown in Fig. 17. In the strip taken at the ^{15}N–^1H resonance position of Phe^{172}, a strong correlation to the C′ of Thr^{171} can be seen. This peak corresponds to the so-called "diagonal peak" [see Eq. (22b)]. In addition, a correlation to the C′ of Phe^{172} is observed, resulting from the out-and-back transfer caused by the $^3J(C'C')$ coupling between Thr^{171} and Phe^{172}. Using the ratio of peak intensities, the $^3J(C'C')$ can be calculated to be 2.4 Hz by application of Eq. (23). The next strip, taken at the ^{15}N–^1H resonance frequency of Leu^{173}, shows a diagonal peak to Phe^{172}. Naturally, based on the inherent symmetry of the experiment a correlation to Thr^{171} would be expected. Unfortunately, this correlation overlaps with the other sequential connectivity to Leu^{173}. In general, the dispersion of the carbonyl resonances is rather limited (ca. 10 ppm) and because the maximum t_2

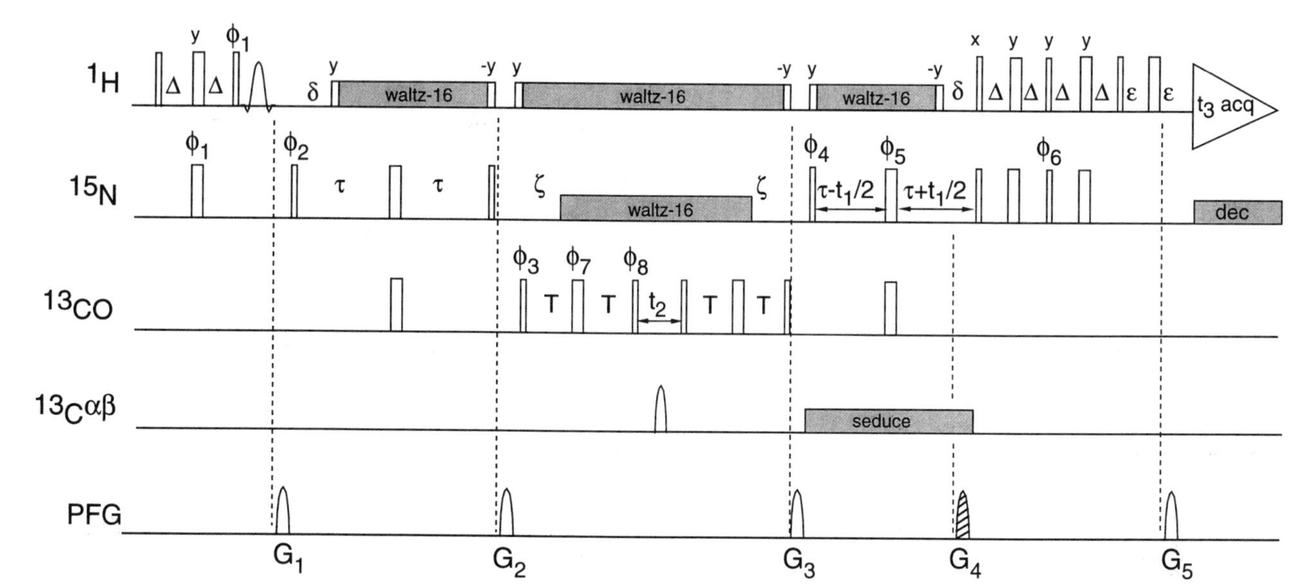

Figure 16. Pulse sequence of the HN(CO)CO experiment adapted from Hu and Bax (1996). Narrow and wide bars denote 90° and 180° pulses, respectively. C′ pulses had a zero excitation in the $C^{\alpha\beta}$ region. The phase cycle was: $\phi_1 = y,-y$; $\phi_2 = 2x,2(-x)$; $\phi_3 = 4x,4(-x)$; $\phi_4 = x$; $\phi_5 = 2x,2(-x)$; $\phi_6 = -y$; $\phi_7 = x$; $\phi_8 = 8x,8(-x)$; receiver $= x,2(-x),x,-x,2x,-x$. Quadrature detection during t_2 was obtained using the States–TPPI method, incrementing phases ϕ_3, ϕ_7, and ϕ_8. The delays were as follows: $\Delta = 2.25$ ms; $\delta = 5.4$ ms; $\tau = 24.8$ ms; $\varepsilon = G_5 + 100\ \mu s$, $T = 32.5$ ms (*N.B.*: in the original sequence of Hu and Bax (1996), T was taken as 45–55 ms]; $\zeta = 32$ ms. Other experimental parameters were as described in the caption of Fig. 14.

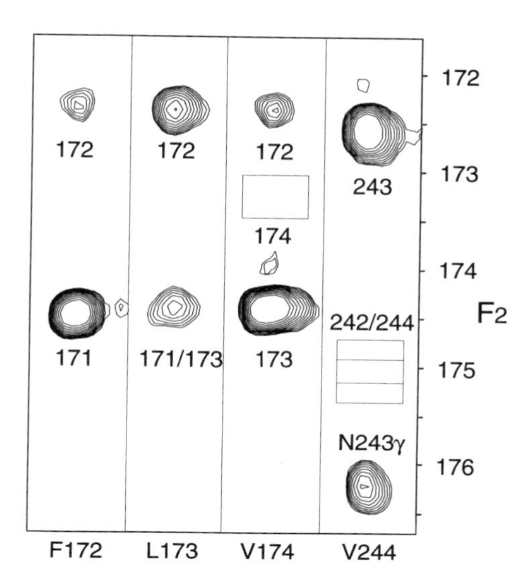

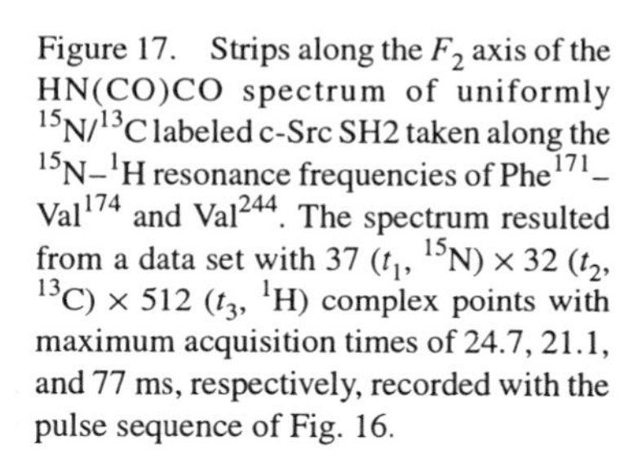
Figure 17. Strips along the F_2 axis of the HN(CO)CO spectrum of uniformly $^{15}N/^{13}C$ labeled c-Src SH2 taken along the $^{15}N-^{1}H$ resonance frequencies of Phe^{171}–Val^{174} and Val^{244}. The spectrum resulted from a data set with 37 (t_1, ^{15}N) × 32 (t_2, ^{13}C) × 512 (t_3, ^{1}H) complex points with maximum acquisition times of 24.7, 21.1, and 77 ms, respectively, recorded with the pulse sequence of Fig. 16.

acquisition time is kept relatively short, the overlap of "diagonal" and cross-peaks presents a nonnegligible problem. In the case of Src, in 35 out of 111 potential F_2 traces, overlap between two or more correlations will occur. This effect becomes even more pronounced for Asp and Asn residues which potentially also display correlations to the sidechain C′.

Correlations to the sidechain C′ of Asn and Asp residues carry valuable information regarding the sidechain dihedral angle χ_1. An example is given in the last strip of Fig. 17. Asn^{243} shows no correlations to the C′ spins of Thr^{242} or Leu^{244}. At the resonance position of Asn^{243} C^{γ}, an intense cross-peak is observed, yielding a $^{3}J(C'C'^{\gamma})$ coupling constant of 2.9 Hz, in agreement with the g^{-} conformation of this residue around its χ_1 angle.

6. THE BACKBONE ANGLE ψ

The backbone torsion angle ψ is far less amenable to analysis using J couplings (Table 2). The Fisher projection is shown in Fig. 18. In fact, as the $^{3}J(N^{i}N^{i+1})$ coupling constants are of vanishingly small magnitude (Bystrov, 1976), only the $^{3}J(H^{\alpha i}N^{i+1})$ and $^{3}J(C^{\beta i}N^{i+1})$ coupling constants potentially provide information about the torsion angle ψ. E.COSY-type experiments employing either the H^{α} as the passive spin (Seip *et al.*, 1994; Weisemann *et al.*, 1994a) or the nitrogen as a passive spin (Wang and Bax, 1995) have been used for measuring $^{3}J(H^{\alpha i}N^{i+1})$. Alternatively, the HSQC-NOESY[N] (Wider *et al.*, 1989) (see Section 7.1) or HNHB experiment (Archer *et al.*, 1991; Madsen *et al.*, 1993; Düx *et al.*, 1997) (see

Table 2
Experiments for Measuring J Couplings Related to ψ

Coupling	Experiment[a]	References	Labeling	Remark
$^3J(H^{\alpha}N^{i+1})$	NOESY[N]	Montelione *et al.* (1989)	^{15}N	E.COSY on $d_{\alpha N}$
	HSQC-NOESY[N]	Wider *et al.* (1989)	^{15}N	E.COSY on $d_{\alpha N}$
	HNHB	Archer *et al.* (1991)	^{15}N	
		Madsen *et al.* (1993)		
		Düx *et al.* (1997)		
	HN(CO)CA[H^{α}]	Seip *et al.* (1994)	$^{15}N/^{13}C$	
	HCACO[N]	Wang and Bax (1996)	$^{15}N/^{13}C$	Sample in D_2O

[a]See note *a* of Table 1.

Section 7.2) provides the correlation to the sequential H^{α} from which the $^3J(H^{\alpha i}N^{i+1})$ can be derived. Although in the latter experiment the ^3J couplings to the H^{β} act as passive couplings that attenuate the sequential H^{α} correlation, for residues with $-150° < \psi < 0°$ it is quite feasible to measure $^3J(H^{\alpha i}N^{i+1})$ in well-behaved proteins.

Recently, values of $^3J(H^{\alpha i}N^{i+1})$ coupling constants in ubiquitin measured in HCACO[N] E.COSY experiments have been used to reparametrize the Karplus curve describing the dependence of the $^3J(H^{\alpha i}N^{i+1})$ on ψ (Wang and Bax, 1995). The resulting curve is shown in Fig. 19. Also dotted are the experimental $^3J(H^{\alpha i}N^{i+1})$ values measured for PYP using a modified HNHB experiment (Düx *et al.*, 1997). The experimental ^3J values match the shape of the curve reasonably well, exhibiting the largest (negative) values around $\psi \approx -60°$. However, the data also suggest a somewhat lower minimum, in agreement with results obtained before for ribonuclease T1 (Karimi-Nejad, 1996). Substituent effects (Haasnoot *et al.*, 1980) could also be responsible for this effect as these have been deemed (partially) responsible for the scatter observed between theoretical and experimental values (Wang and Bax, 1995).

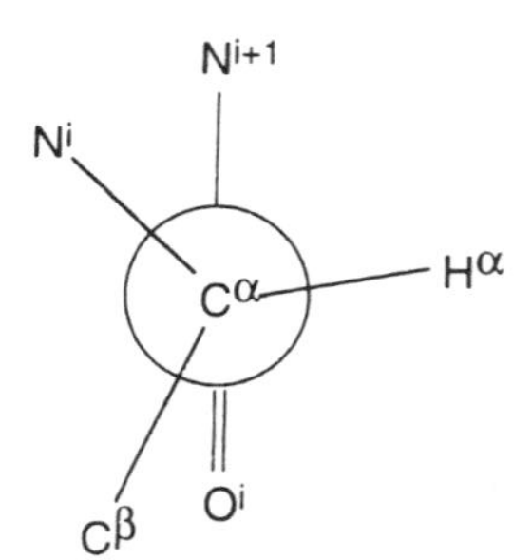

Figure 18. Fisher projection of the backbone torsion angle ψ.

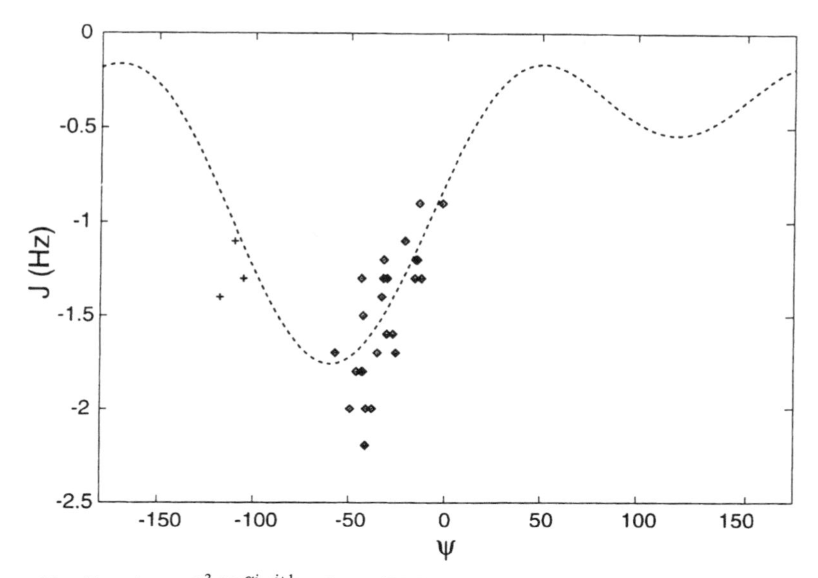

Figure 19. Experimental $^{3}J(\mathrm{H}^{\alpha i}\mathrm{N}^{i+1})$ values of PYP measured using the modified HNHB experiment (Düx *et al.*, 1997) versus ψ values of the refined average structure (Düx *et al.*, 1998). Glycine $^{3}J(\mathrm{H}^{\alpha 3}\mathrm{N}^{i+1})$ values are indicated by "+" and are offset in ψ by $-120°$. Karplus curve describing the dependence of the $^{3}J(\mathrm{H}^{\alpha i}\mathrm{N}^{i+1})$ on ψ, using parameters taken from Wang and Bax (1995): $^{3}J(\mathrm{H}^{\alpha i}\mathrm{N}^{i+1}) = -0.88 \cos^2(\psi - 120) - 0.61 \cos(\psi - 120) - 0.27$. Note that in the original parametrization the authors used a 180° phase shift in their definition of the dihedral angle ψ with respect to the IUPAC definition (IUPAC, 1970) which has been used here.

The detection limit of the correlations to the sequential H^{α} determines the lower limit of the $^{3}J(\mathrm{H}^{\alpha i}\mathrm{N}^{i+1})$ couplings that can be obtained from the HNHB experiment. Thus, for PYP, correlations corresponding to coupling constants < 0.7 Hz were not observed, i.e., no data were obtained for $\psi > 0°$ or $\psi < -150°$. In principle, the E.COSY HSQC-NOESY[N] experiment will allow for the measurement of smaller couplings, although the accuracy of the method depends on the intensity of the sequential $\mathrm{H}^{\alpha i}$–$\mathrm{H}^{\mathrm{N}i+1}$ cross-peak.

Regarding the small $^{3}J(\mathrm{C}^{\gamma}\mathrm{N})$ values measured in proteins to date (Vuister *et al.*, 1993a; Hu *et al.*, 1997; Hu and Bax, 1997a), the $^{3}J(\mathrm{C}^{\beta i}\mathrm{N}^{i+1})$ values across the peptide bond are expected to be small and no experimental data have as yet been reported for proteins.

7. THE SIDECHAIN ANGLES χ_1 and χ_2

Information on the χ_1 rotameric state and stereospecific assignment of β-methylene protons is of great importance to the structure determination process. Figure 20 shows the Fisher projection around the sidechain angle χ_1, which

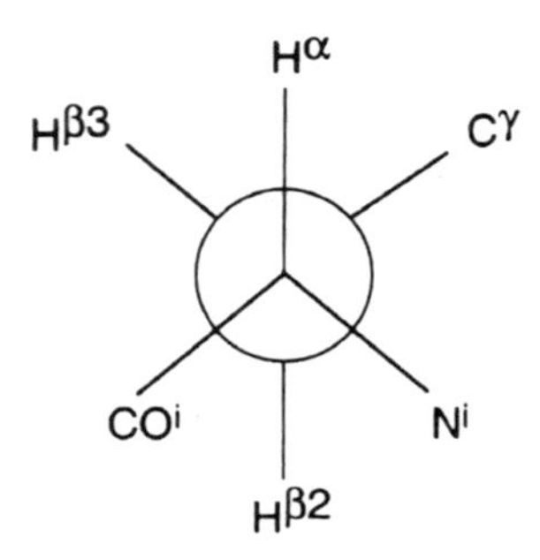

Figure 20. Fisher projection of the sidechain torsion angle χ_1 for residues Asp, Asn, Leu, Lys, Arg, Glu, Met, Pro, His, Phe, Tyr, and Trp. For Ala C^{γ} is substituted by $H^{\beta 1}$, for Ser C^{γ} is substituted by O^{γ}, and for Cys C^{γ} is substituted by S^{γ}. For Val C^{γ} is substituted by $C^{\gamma 1}$, $H^{\beta 2}$ by $C^{\gamma 2}$, and $H^{\beta 3}$ by H^{β}. For Thr C^{γ} is substituted by $C^{\gamma 2}$, $H^{\beta 2}$ by $O^{\gamma 1}$, and $H^{\beta 3}$ by H^{β}. Finally, for Ile C^{γ} is substituted by $C^{\gamma 2}$, $H^{\beta 2}$ by $C^{\gamma 1}$, and $H^{\beta 3}$ by H^{β}. Note that in accordance with IUPAC rules (IUPAC, 1970), the χ_1 torsion angle is defined by the orientation of the highest-ranking C^{β} substituent relative to the N nucleus, and therefore differs for different residue types, e.g., Leu versus Thr.

illustrates that, for the majority of amino acids, this information can be obtained from as many as nine different ^3J coupling constants. In practice, however, not all of these can be measured straightforwardly and a summary of the experiments aimed at measuring a selection of these ^3J coupling constants is presented in Table 3.

The Karplus curves describing the dependences of $^3J(H^{\alpha}H^{\beta 2})$, $^3J(H^{\alpha}H^{\beta 3})$, $^3J(H^{\alpha}C^{\gamma})$, $^3J(C'H^{\beta 2})$, $^3J(C'H^{\beta 3})$, $^3J(NH^{\beta 2})$, and $^3J(NH^{\beta 3})$ upon the torsion angle χ_1 are shown in Fig. 21. Parametrization of all of these curves is based on studies on small model compounds and/or *ab initio* calculations, as no parametrizations using protein data are available. For example, two curves for $^3J(H^{\alpha}H^{\beta 2})$ are shown (Bystrov, 1976; DeMarco *et al.*, 1978a), which differ by almost 2 Hz in their maximum value observed for $\chi_1 = -60°$. In analogy to the large changes in the ϕ-dependent Karplus curves, observed on reparametrizations using protein data (Section 5), it can be expected that the χ_1-dependent Karplus curves will also substantially change once accurate protein data become available.

For some ^3J coupling constants, such as $^3J(NC^{\gamma})$ or $^3J(C'C^{\gamma})$, no parametrizations are available. Values measured in proteins, however, suggest narrow ranges for these two coupling constants, being ~2.5 and < 0.5 Hz for *trans* and *gauche* $^3J(NC^{\gamma})$, respectively (Vuister *et al.*, 1993a; Hu *et al.*, 1997, Hu and Bax, 1997a) and ~3.0–4.0 and < 1.0 Hz for *trans* and *gauche* $^3J(C'C^{\gamma})$, respectively (Grzesiek *et al.*, 1993; Karimi-Nejad *et al.*, 1994; Hu and Bax, 1997a; Hu *et al.*, 1997).

Experiments aimed at measuring ^3J coupling constants related to χ_2 are listed in Table 4. For most residues, only $^3J(HH)$, $^3J(CH)$, and $^3J(CC)$ coupling constants provide information about the χ_2 angle, although for selected residues, coupling constants involving other nuclei could be envisioned, e.g., the sidechain carbonyl C^{δ} nucleus for Glu and Gln residues or the N^{δ} nucleus for His residues.

7.1. The HSQC-NOESY[N] Experiment

The 3D HSQC-NOESY[N] experiment (Wider *et al.*, 1989) for the measurement of $^3J(NH^{\beta})$ couplings is shown in Fig. 22. The experiment belongs to the first

Table 3
Experiments for Measuring J Couplings Related to χ_1

Coupling	Experiment[a]	References	Labeling	Remark
$^3J(H^\alpha H^\beta)$	E.COSY	Griesinger *et al.* (1986)	—	Original E.COSY
	HCCH COSY[H^β]	Gemmecker and Fesik (1991), Griesinger and Eggenberger (1992)	^{13}C	
	HCCH-TOCSY[H^β]	Gemmecker and Fesik (1991), Emerson and Montelione (1992)	^{13}C	
	HXYH[H^β]	Tessari *et al.* (1995)	^{13}C	
	HAHB	Vuister *et al.* (1994)	^{12}C-reverse	
	HACAHB	Grzesiek *et al.* (1995)	^{13}C	
	CT-HACAHB	Tessari *et al.* unpublished	^{13}C	
$^3J(C'H^\beta)$	HN(CO)HB	Grzesiek *et al.* (1992), Bax *et al.* (1994)	$^{15}N/^{13}C$	
	HNCACH-TOCSY[C′]	Weisemann *et al.* (1994b)	$^{15}N/^{13}C$	
$^3J(NH^\beta)$	NOESY[N]	Montelione *et al.* (1989)	^{15}N	
	HSQC-NOESY[N]	Wider *et al.* (1989)	^{15}N	
		Chary *et al.* (1991)		
	HNHB	Archer *et al.* (1991)	^{15}N	
		Madsen *et al.* (1993)		
		Düx *et al.* (1997)		Modified HNHB
$^3J(H^\alpha C^\gamma)$	HMBC	Bax *et al.* (1988)	^{13}C	
	TOCSY	Zuiderweg and Fesik (1991)	^{13}C	Band-selective decoupling
	HMQC-TOCSY[C]	Edison *et al.* (1992)	Partial ^{13}C	
	LRCH	Vuister and Bax (1993b)	^{13}C	Val, Thr, Ile $C^\gamma H_3$
$^3J(NC^\gamma)$	{^{15}N} sed[b] ^{13}C CT-HSQC	Vuister *et al.* (1993b)	$^{15}N/^{13}C$	Val, Thr, Ile $C^\gamma H_3$
	{C^γ} sed ^{15}N CT-HSQC	Hu *et al.* (1997)	$^{15}N/^{13}C$	Tyr, Phe, His, Trp
	HNCG	Hu and Bax (1997a)	$^{15}N/^{13}C$	
$^3J(C'C^\gamma)$	{C′} sed ^{13}C CT-HSQC	Grzesiek *et al.* (1993)	$^{15}N/^{13}C$	Val, Thr, Ile $C^\gamma H_3$
	HCCCH-COSY[C′]	Schwalbe *et al.* (1993)	^{13}C	
	{C^γ} sed HNCO	Hu *et al.* (1997)	$^{15}N/^{13}C$	Tyr, Phe, His, Trp
$^3J(C'C'^\gamma)$	HN(CO)CO	Hu and Bax (1996)	$^{15}N/^{13}C$	Asp and Asn

[a]See note *a* of Table 1.

[b]sed: spin-echo difference.

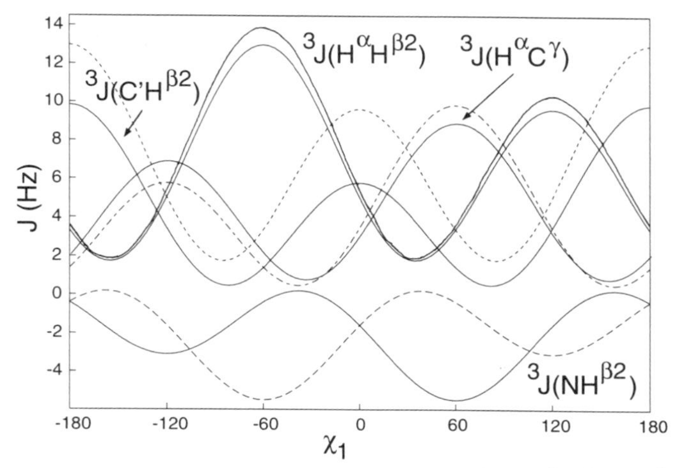

Figure 21. Karplus curves showing the dependences of the $^3J(H^{\alpha}H^{\beta2})$, $^3J(H^{\alpha}H^{\beta3})$, $^3J(H^{\alpha}C^{\gamma})$, $^3J(C'H^{\beta2})$, $^3J(C'H^{\beta3})$, $^3J(NH^{\beta2})$, and $^3J(NH^{\beta3})$ on the torsion angle χ_1. The curves shown are: $^3J(H^{\alpha}H^{\beta2}) = 9.5\cos^2(\chi_1 - 120°) - 1.68\cos(\chi_1 - 120°) + 1.8$ (DeMarco *et al.*, 1978a) and $^3J(H^{\alpha}H^{\beta2}) = 10.2\cos^2(\chi_1 - 120°) - 1.8\cos(\chi_1 - 120°) + 1.9$ (Bystrov, 1976) (solid thin line), $^3J(H^{\alpha}H^{\beta3}) = 9.5\cos^2(\chi_1) - 1.68\cos(\chi_1) + 1.8$ (DeMarco *et al.*, 1978a) (short dashes), $^3J(H^{\alpha}C^{\gamma}) = 7.1\cos^2(\chi_1 + 120°) - 1.0\cos(\chi_1 + 120°) + 0.7$ (Schmidt, 1997 and references therein), $^3J(C'H^{\beta2}) = 7.2\cos^2(\chi_1) - 2.04\cos(\chi_1) + 0.6$ (Fischman *et al.*, 1980), $^3J(C'H^{\beta3}) = 7.2\cos^2(\chi_1 + 120°) - 2.04\cos(\chi_1 + 120°) + 0.6$ (Fischman *et al.*, 1980) (short –long dashes), $^3J(NH^{\beta2}) = -4.4\cos^2(\chi_1 + 120°) + 1.2\cos(\chi_1 + 120°) + 0.1$ (DeMarco *et al.*, 1978b), and $^3J(NH^{\beta3}) = -4.4\cos^2(\chi_1 - 120°) + 1.2\cos(\chi_1 - 120°) + 0.1$ (DeMarco *et al.*, 1978b) (long dashes).

Table 4
Experiments for Measuring *J* Couplings Related to χ_2

Coupling	Experiment[a]	References	Labeling	Remark
^3J(HH)	see Table 3 under $^3J(H^{\alpha}H^{\beta})$			
^3J(CH)	HMBC	Bax *et al.* (1988)	^{13}C	
	HMQC-TOCSY[C]	Edison *et al.* (1992)	Partial ^{13}C	
	HSQC-TOCSY[C]	Sattler *et al.* (1992)	—	
	LRCH	Vuister *et al.* (1993b)	^{13}C-selective	
		Vuister and Bax (1993b)	^{13}C	Ile, Leu $C^{\delta}H_3$
^3J(CC)	LRCC	Bax *et al.* (1992, 1994)	^{13}C	Ile, Leu $C^{\delta}H_3$
	$HN(CO)CAC_{ali}$	Hennig *et al.* (1997)	$^{15}N/^{13}C/^{2}H$	

[a]See note *a* of Table 1.

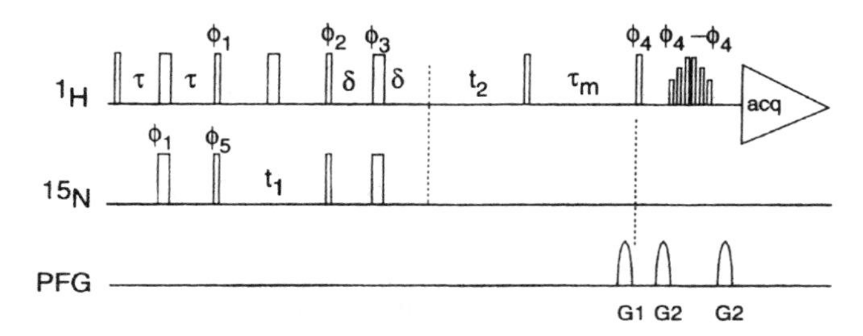

Figure 22. Pulse sequence of the HSQC-NOESY[N] experiment adapted from Wider *et al.* (1989) by inclusion of pulsed-field gradients (Bax and Pochapsky, 1992) and a WATERGATE (Saudek *et al.*, 1994) detection scheme. Narrow and wide bars denote 90° and 180° pulses, respectively. The phase cycle is as follows: $\phi_1 = y,-y$; $\phi_2 = 2(45°),2(225°)$; $\phi_3 = 8x,8(-x)$; $\phi_4 = 8x,8y$; $\phi_5 = 4x,4(-x)$; receiver = $x,2(-x),x,-x,2x,-x,y,2(-y),y,-y,2y,-y$. Quadrature detection was obtained using the States–TPPI method, incrementing phases ϕ_5 for t_1 and ϕ_2 and ϕ_3 for t_2.The delays were as follows: $\tau = 2.25$ ms; $\delta = 2.7$ ms; $\tau_m = 80$ ms. Gradient strengths and durations: $G_1 = 3$ ms, -7 G cm^{-1}; $G_2 = 1$ ms, 30 G cm^{-1}.

category of E.COSY experiments (Fig. 2A) in which the passive spin, i.e., the ^{15}N, is heteronuclear to spins *I* and *S*, i.e., proton spins in this case.

Figure 23 shows a part of the F_2–F_3 cross section from the 3D HSQC-NOESY[N] spectrum of PYP. The H^N resonance frequencies of residues Lys^{104}, Phe^{62}, and Leu^{40} are indicated. A large doublet splitting along the F_2 axis is visible, resulting from the ^1J(NH^N) coupling active during t_2. As the ^{15}N spin state remains

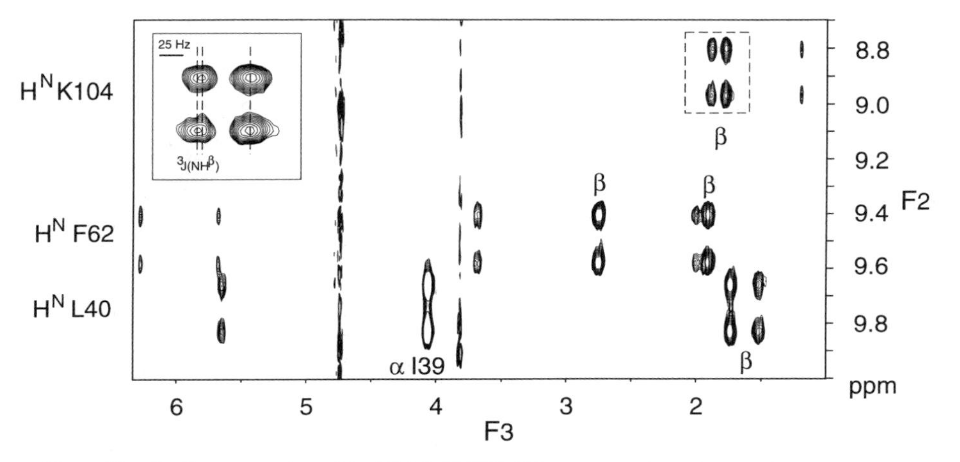

Figure 23. F_2–F_3 cross section of the HSQC-NOESY[N] spectrum of a 2 mM solution of uniformly ^{15}N labeled PYP, recorded with the pulse sequence of Fig. 22. The spectrum resulted from a data set with 47 (t_1, ^{15}N) × 96 (t_2, ^{1}H) × 1024 (t_3, ^{1}H) complex points with maximum acquisition times of 29.4, 31.7, and 154 ms, respectively.

undisturbed by the mixing and detection part of the pulse sequence, an E.COSY-type pattern is observed as a result of the $^1J(\mathrm{NH^N})$ and long-range $^3J(\mathrm{NH}^i)$ couplings.

The $^3J(\mathrm{NH}^\beta)$ coupling constants are of particular interest as these values provide information about the χ_1 dihedral angle (Fig. 20) and can also aid in the stereospecific assignment of β-methylene protons. The $\mathrm{H^N}$–H^β cross-peaks of Lys[104] are shown enlarged in the inset of Fig. 23. Indicated are the displacements of the two components along the F_3 axis, determined to be –4.0 and –0.5 Hz.

In addition to the $\mathrm{H^N}$–H^β intraresidual cross-peaks, typically cross-peaks with intraresidual and sequential H^α are observed. Examples of these are also shown in Fig. 23. These cross-peaks allow extraction of the values of $^2J(\mathrm{NH}^\alpha)$ and $^3J(\mathrm{NH}^{\alpha i-1})$ coupling constants. In particular, the latter is of considerable importance as its magnitude is correlated with the backbone angle ψ (Bystrov, 1976; Wang and Bax, 1995).

7.2. The HNHB Experiment

The original HNHB experiment (Archer *et al.*, 1991; Madsen *et al.*, 1993) belongs to the second class of QJ methods (Fig. 6B) and requires a separate 2D reference experiment for quantification of the *J* coupling (Bax *et al.*, 1994). The different dimensionalities of the two spectra turn this into a somewhat cumbersome procedure. Figure 24 shows a modified HNHB pulse sequence (Düx *et al.*, 1997) that allows the magnitude of the *J* couplings to be determined directly from the ratio of a "cross-peak" to a "diagonal peak" in a single spectrum, similar to the third class of QJ experiments (Fig. 6C). The known and rather constant value of the $^1J(\mathrm{NH})$

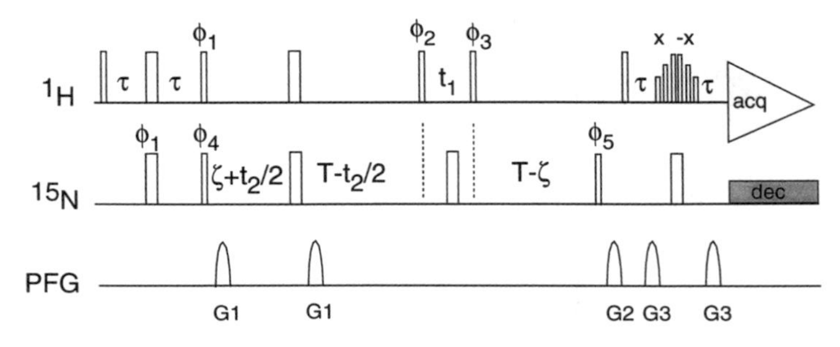

Figure 24. Pulse sequence for the modified HNHB experiment (Düx *et al.*, 1997). Narrow and wide bars denote 90° and 180° pulses, respectively, whereas the final reverse INEPT was implemented using a WATERGATE sequence (Saudek *et al.*, 1994). The phases are as follows: $\phi_1 = y,-y$; $\phi_2 = 2x,2(-x)$; $\phi_3 = 4x,4(-x)$; $\phi_4 = 4x,4(-x)$; $\phi_5 = 8x,-8(-x)$; receiver = 2(*x*,–*x*,–*x*,*x*),2(–*x*,*x*,*x*,–*x*). Duration and strength of the gradients were as follows: G_1 = 1 ms, –7 G cm^{-1}; G_2 = 1 ms, –20 G cm^{-1}; G_3 = 1 ms, 30 G cm^{-1}. The delays were as follows: τ = 2.5 ms; T = 32.6 ms; ζ = 2.7 ms. Note that setting $\zeta = 0$ and $T = (2\times{}^1J(\mathrm{NH^N}))^{-1}$ yields the HNHB version proposed by Madsen *et al.* (1993). (Reproduced from Düx *et al.*, 1997).

in proteins is used to yield a cross-peak of known intensity which can be used as a reference peak. Small variations in the magnitude of the $^{1}J(\mathrm{NH^N})$ (ca. ±2 Hz) do not result in significant variations in the intensities of the reference peak if the de- and rephasing times have different durations. Thus, it has been shown that the systematic errors in the $^{3}J(\mathrm{NH})$ values resulting from this scheme remained well below 3% when using Eq. (23) for the calculation of the coupling constants (Düx *et al.*, 1997).

The overlay of the de- and rephase schemes with the ^{15}N evolution affords an excellent resolution in this dimension. Moreover, as the ^{15}N–^{1}H correlation is usually well dispersed, the overlap of the "diagonal" resonances is in general only a minor problem and quantitative information can usually be extracted straightforwardly. Furthermore, even for those residues with overlapping ^{15}N–^{1}H correlations, the intensity of the cross-peak still provides a qualitative indication of the magnitude of the J coupling.

Two strips along the F_1 axis of the HNHB spectrum of PYP are shown in Fig. 25. Both strips show correlations to two H^{β} protons the intensities of which are directly proportional to the magnitude of the $^{3}J(\mathrm{NH}^{\beta})$ couplings. The coupling constants were calculated to be 2.4 and 4.8 Hz for Asp^{36} and 1.4 and 3.9 Hz for Gln^{41} by application of Eq. (23). Previously measured values for FKBP (Xu *et al.*, 1992) or ribonuclease T_1 (Karimi-Nejad *et al.*, 1994) fell in the range 0.5–5.5 Hz. Thus, the present values are indicative of rotameric averaging. As discussed in Section 4.4.1, the effects of $\mathrm{H}^{\beta2}$–$\mathrm{H}^{\beta3}$ cross-relaxation will also tend to equalize the measured J values. This effect is expected to be small in general, and in particular for PYP, which has a rotational correlation time of 6.4 ns (Düx *et al.*, 1998). Indeed, for several residues values in the range of 5–6 Hz, with the second J coupling being 0–1 Hz, were measured corresponding to a staggered rotameric conformation of the sidechain of these residues.

7.3. The {^{13}C′} and {^{15}N} Spin-Echo Difference ^{13}C CT-HSQC Experiments

The pulse schemes of the {^{15}N} spin-echo difference ^{13}C CT-HSQC (Vuister *et al.*, 1993b) and {^{13}C′} spin-echo difference ^{13}C CT-HSQC (Grzesiek *et al.*, 1993) experiments are shown in Fig. 26. The pulse schemes are based on the first class of QJ methods, discussed in Section 4.1. When the 180°(^{13}C′) or 180°(^{15}N) pulses are applied at position *b*, signal attenuation results from dephasing related to the $J(\mathrm{CC'})$ or $J(\mathrm{CN})$ couplings, respectively. Applying both 180° pulses at position *a* yields the reference experiment. Reference and attenuated spectra can be used to calculate the J coupling constants, using Eq. (8).

The experiments can be most easily applied to CH_3 groups which display a reduced heteronuclear dipolar broadening as a result of rapid rotation around the threefold symmetry axis and also have a highly nonexponential transverse relaxa-

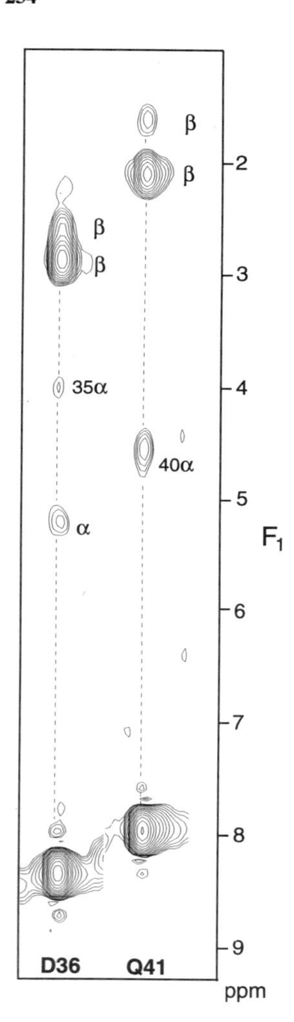

Figure 25. Two F_1 strips from the HNHB spectrum of PYP, taken at the $(F_2, F_3) = (^{15}N, {}^{1}H^{N})$ resonance frequencies of residues Asp^{36} and Gln^{41}. Diagonal peaks between 7 and 9 ppm are of sign opposite compared to cross-peaks. The spectrum was recorded with the pulse sequence of Fig. 24 on a 2 mM solution of uniformly ^{15}N labeled PYP. The data consisted of 47 $(t_1, {}^1H) \times$ 96 $(t_2, {}^{15}N) \times 512$ $(t_3, {}^1H)$ complex points with maximum acquisition times of 9.6 (t_1), 60.1 (t_2), and 76.8 (t_3) ms. (Reproduced from Düx *et al.*, 1997).

tion behavior. Thus, the slowly relaxing component of the transverse magnetization has a width of only a few hertz, which allows the magnetization to be transverse for the required 58 ms. The experiments are therefore particularly useful for obtaining stereospecific assignments of valine γ-methyl groups, as well as providing information about the χ_1 rotamer for valine, threonine, and isoleucine residues.

Figure 27 displays the methyl region of the ^{13}C CT-HSQC reference spectrum, and $\{^{15}N\}$-attenuated and $\{^{13}C'\}$-attenuated ^{13}C CT-HSQC spin-echo difference

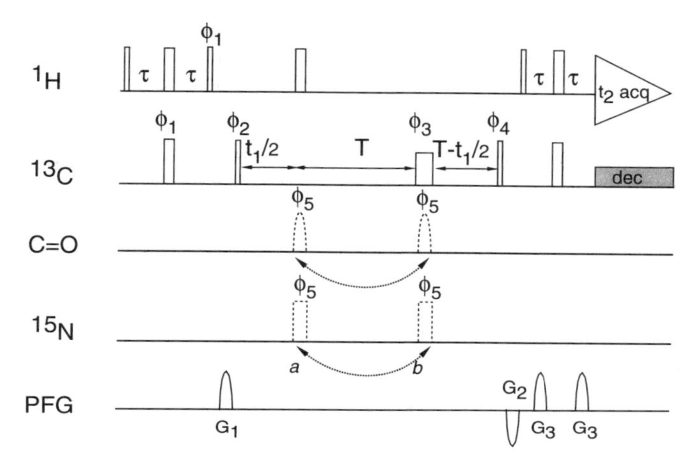

Figure 26. Pulse schemes of the {^{15}N} spin-echo difference ^{13}C CT-HSQC experiment (Vuister et al., 1993a) and {^{13}C′} spin-echo difference experiment ^{13}C CT-HSQC (Grzesiek *et al.*, 1993). Narrow and wide bars denote 90° and 180° pulses, respectively. Setting the 180°(^{15}N) at position *b* and the 180°(^{13}CO) at position *a* yields the experiment in which the *J*(CN) couplings are active. Setting the 180°(^{15}N) at position *a* and the 180°(^{13}CO) at position *b* yields the experiment in which the *J*(CC′) couplings are active, whereas setting both the 180°(^{15}N) and the 180°(^{13}CO) at position *a* yields the reference experiment. Unless indicated otherwise, all pulses are applied along the *x*–axis. The following phase cycle is used: $\phi_1 = y,-y$; $\phi_2 = x$; $\phi_3 = 2x,2(-x),2y,2(-y)$; $\phi_4 = 8x,8(-x)$; receiver = $2(x,-x),4(-x,x),2(x,-x)$. Quadrature detection in t_1 was obtained by the States–TPPI method, incrementing phase ϕ_2. The delay τ is set to 1.7 ms and the delay *T* is set to 28.6 or 57.2 ms.

spectra of BB10010. The data can be used for stereospecific assignment of the valine residues. For example, Val^{38} displays a cross-peak to one of its two C^{γ} methyl groups in the {^{15}N}-attenuated spectrum (panel B) [corresponding to $^{3}J(C^{\gamma}N) = 1.9$ Hz] and no cross-peak to its second C^{γ} methyl [boxed, corresponding to $^{3}J(C^{\gamma}N) <$ 0.9 Hz], whereas the reverse occurs in the {^{13}C}-attenuated spectrum (panel C, corresponding to couplings of <1.0 and 4.0 Hz, respectively). These data are consistent with a χ_1 angle of 180° and the stereospecific assignment as indicated in the figure. In addition to valine residues, isoleucine C^{γ} methyl and threonine C^{γ} methyl display cross-peaks in the difference spectra, corresponding to *trans* conformations with respect to the N and C′ nuclei.

Editing of ^{13}C–^{1}H correlation spectra is also obtained for Pro $C^{\delta}H_2$, Lys $C^{\varepsilon}H_2$, Arg $C^{\delta}H_2$, Gln $C^{\gamma}H_2$, and Asn $C^{\beta}H_2$ resonances as a result of substantial ^{1}J(CN) and ^{2}J(CN) couplings (Vuister *et al.*, 1993) or Glu $C^{\gamma}H_2$ and Asp $C^{\beta}H_2$ resonances using ^{1}J(CC′) (Grzesiek *et al.*, 1993).

Alternative experiments for measuring $^{3}J(C^{\gamma}N)$ and $^{3}J(C^{\gamma}C')$ are listed in Table 3.

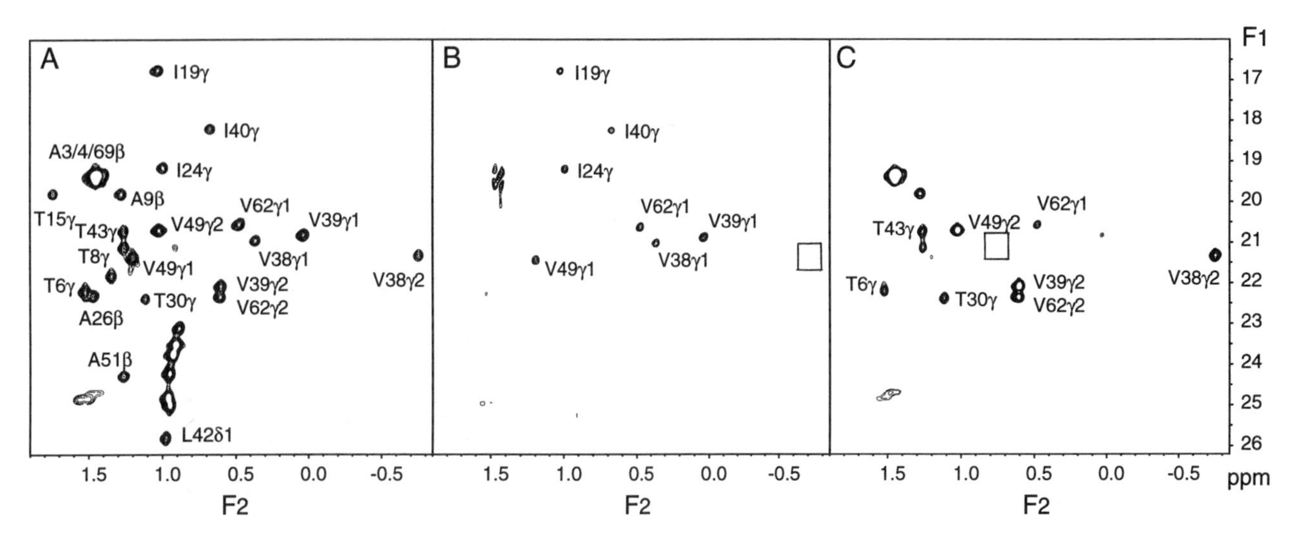

Figure 27. Methyl region of the ^{13}C CT-HSQC reference spectrum (A), and {^{15}N} (B) and {$^{13}C'$} (C) spin-echo difference ^{13}C CT-HSQC spectra of a 2 mM solution of uniformly $^{15}N/^{13}C$ labeled BB10010 in H_2O, recorded with the pulse sequence of Fig. 26.

7.4. The HACAHB Experiment

The HACAHB experiment (Grzesiek *et al.*, 1995; Tessari *et al.*, unpublished) is a QJ experiment for measuring the ${}^3J(\mathrm{H}^\alpha\mathrm{H}^\beta)$, which belongs to the third class of QJ experiments (Section 4.3) and provides a direct measure of the J coupling from the cross-peak to diagonal-peak intensity ratio.

The previously proposed 2D HAHB experiment (Vuister *et al.*, 1994) also yielded the ${}^3J(\mathrm{H}^\alpha\mathrm{H}^\beta)$ values from the cross-peak to diagonal-peak ratio, albeit only for the non-^{13}C-labeled phenylalanine residues of the Phe reverse-labeled heat-shock transcription factor DNA-binding domain. For larger unlabeled proteins the overlap of diagonal peak resonances in the 2D HAHB spectrum impairs the extraction of the ${}^3J$ values. This problem is overcome in ^{13}C-labeled proteins by correlating the proton spins with their directly attached ^{13}C nuclei.

The pulse scheme of the constant-time (CT) implementation of the HACAHB experiment (CT-HACAHB) (Tessari *et al.*, unpublished) is shown in Fig. 28. It comprises a heteronuclear correlation experiment in a constant-time HMQC-type fashion overlaid with de- and rephase periods to generate the required in-phase and antiphase proton terms. This concept is highly similar to the HNHA experiment (Vuister and Bax, 1993a) in which the dephasing and rephasing for the ${}^3J(\mathrm{H}^\mathrm{N}\mathrm{H}^\alpha)$ coupling constant were overlaid on the ^{15}N HMQC-type evolution period.

The simultaneous displacement of the first and third 180° ^{13}C pulses during the period of duration $4T$ causes chemical shift labeling of the ^{13}C spin. The evolution related to the homonuclear ${}^1J(\mathrm{C}^\alpha\mathrm{C}^\beta)$ scalar coupling is exactly refocused at the end of this period by choosing $T = \{4 \times {}^1J(\mathrm{C}^\alpha\mathrm{C}^\beta)\}^{-1}$. The dephasing of the zero- and double-quantum terms related to heteronuclear couplings other than ${}^1J(\mathrm{H}^\alpha\mathrm{C}^\alpha)$, e.g., ${}^3J(\mathrm{H}^\gamma\mathrm{C}^\alpha)$ or ${}^2J(\mathrm{H}^\beta\mathrm{C}^\alpha)$, is suppressed by moving the 180° proton pulses in concert with the 180° ^{13}C pulses.

The detection limit of the correlations in the CT-HACAHB spectrum depends critically on the T_2 relaxation behavior of the ${}^{13}\mathrm{C}^\alpha$ spin. It is well known, however, that in the slow tumbling limit the usually large one-bond ${}^{13}\mathrm{C}$–^{1}H dipolar interaction does not, to a first approximation, affect the decay of ${}^{13}\mathrm{C}$–^{1}H two-spin coherence-order (Griffey and Redfield, 1987). In the CT-HACAHB pulse scheme, the ^{13}C magnetization is present as multiple-quantum coherence for substantial periods, thus improving the overall sensitivity of the experiment.

A full operator treatment shows that the signal observed in t_3 is given by:

$$S(t_1,t_2,t_3) = A\ \{[\cos^2(\rho)\cos(\omega_{\mathrm{H}\alpha}t_1) - \sin^2(\rho)\cos(\omega_{\mathrm{H}\beta}t_1)]\cos(\omega_{\mathrm{C}\alpha}t_2)\exp(i\omega_{\mathrm{H}\alpha}t_3) + [\cos(\rho)\sin(\rho)\cos(\omega_{\mathrm{H}\alpha}t_1) - \cos(\rho)\sin(\rho)\cos(\omega_{\mathrm{H}\beta}t_1)]\cos(\omega_{\mathrm{C}\alpha}t_2)\, i\exp(i\omega_{\mathrm{H}\beta}t_3)\} \tag{28}$$

where $\rho = \{2\pi\, {}^3J(\mathrm{H}^\alpha\mathrm{H}^\beta)\,(T+\Delta)\}$ and A includes the trigonometric factors resulting from the de- and rephasing related to the one-bond heteronuclear coupling,

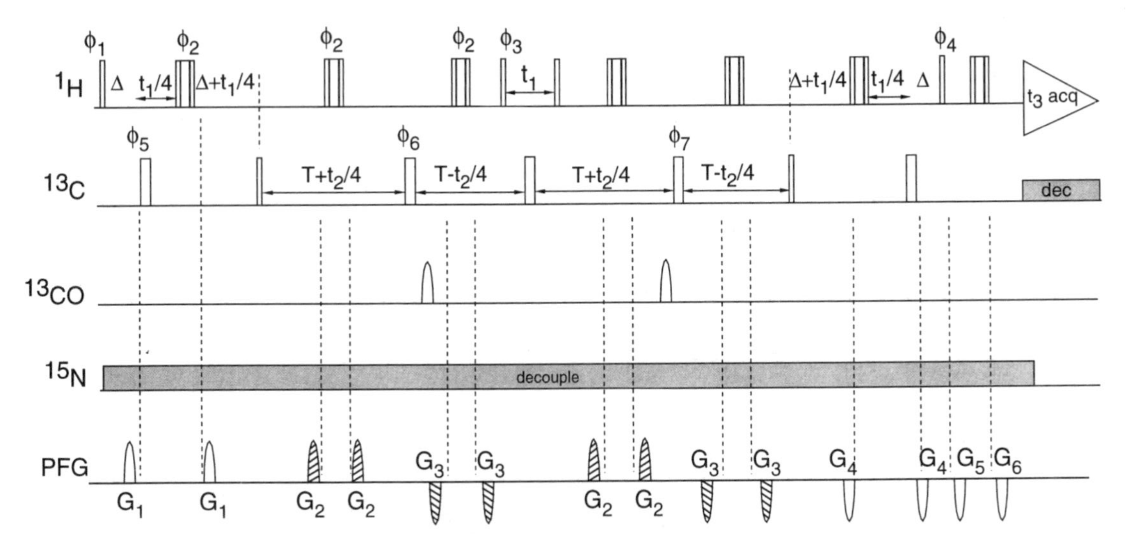

Figure 28. Pulse sequence of the CT-HACAHB experiment (Tessari *et al.*, unpublished). Narrow and wide bars denote 90° and 180° pulses, respectively. ^{1}H 180° pulses are implemented as $90_y180_x90_y$ composite pulses. 180° pulses on the ^{13}C channel applied during the $4T$ constant-time period have a zero-excitation profile at the center of the carbonyl frequency and 180° C′ pulses are implemented as off-resonance Gaussian pulses with a duration of 307 μs. The ^{13}C carrier is placed in the center of the aliphatic region (43 ppm). All pulses are applied along the x-axis unless indicated otherwise. The phases are as follows: $\phi_1 = x,-x$; $\phi_2 = x$; $\phi_3 = 2x,2(-x)$; $\phi_4 = y$; $\phi_5 = 4x,4(-x)$; $\phi_6 = x,-x$; $\phi_7 = 8x,8(-x)$;$\phi_8 = 8x,8(-x)$; receiver = $2(x,-x),4(-x,x),2(x,-x)$. Duration and strength of the pulsed-field gradients (PFG) were as follows: $G_1 = 0.5$ ms, 4 G cm^{-1}, $G_2 = 0.4$ ms, 22 G cm^{-1}; $G_3 = 0.4$ ms, -22 G cm^{-1}; $G_4 = 1.0$ ms, -7 G cm^{-1}; $G_5 = 0.8$ ms, -8 G cm^{-1}; and $G_6 = 0.8$ ms, -30 G cm^{-1}. PFGs G_1 and G_4 are used to remove imperfections resulting from the 180° pulses. PFGs G_2, G_3, G_5, and G_6 serve a similar purpose but in addition are also used for gradient coherence selection of multiple-quantum coherences (Ruiz Cabello *et al.*, 1992). Gradients G_2 and G_3 were alternated in sign on successive FIDs so as to record P- and N-type spectra. Quadrature detection in t_1 was obtained using the States–TPPI method incrementing phases ϕ_1, ϕ_2, and ϕ_3. The delays Δ and T are set to 1.7 and 7.15 ms, respectively.

${}^1J(\mathrm{H}^{\alpha}\mathrm{C}^{\alpha})$, and from passive homonuclear couplings during the period $4(T+\Delta)$. The first two terms result in purely in-phase magnetization detected at the H^{α} frequency during t_3. The following two terms result in dispersive antiphase magnetization which rephases during t_3 and will be detected at the H^{β} frequency. Note that because the experiment is inherently symmetric, in principle also antiphase H^{α} magnetization originating from H^{β} will be present. However, these undesired components will resonate in the carbon dimension at the C^{β} frequency and thus, provided the C^{α} and C^{β} resonances do not overlap, they will be separated.

Fourier transformation of the signal $S(t_1,t_2,t_3)$ with respect to the three axes yields spectra with a diagonal peak at $(\omega_{\mathrm{H}\alpha},\omega_{\mathrm{C}\alpha},\omega_{\mathrm{H}\alpha})$, and cross-peaks at $(\omega_{\mathrm{H}\beta},\omega_{\mathrm{C}\alpha},\omega_{\mathrm{H}\alpha})$. Equation (28) also shows that the ${}^3J(\mathrm{H}^{\alpha}\mathrm{H}^{\beta})$ can be calculated from the ratio of these two peak volumes using Eq. (23). The presence of additional passive J couplings results in additional trigonometric terms which cancel from the ratio of cross-peak to diagonal-peak volumes, as discussed in Section 4.3.

Figure 29 shows strips along the F_1 axis taken from the 3D CT-HACAHB spectrum of PB92 from *Bacillus alcalophilus* in $\mathrm{H_2O}$ solution at the ${}^{13}\mathrm{C}^{\alpha},{}^{1}\mathrm{H}^{\alpha}$ resonance frequencies of Asn^{179}, Asn^{85}, Ser^{126}, and Tyr^{208}. The H^{α}, H^{β}, and H^{N} regions are indicated and diagonal peaks in the H^{α} region have an opposite sign relative to the cross-peaks in the H^{β} and H^{N} regions, in accordance with Eq. (28). The ${}^3J$ values obtained for the various cross-peaks are also indicated in Fig. 29. For the resonances not showing a correlation in the CT-HACAHB spectrum, an upper

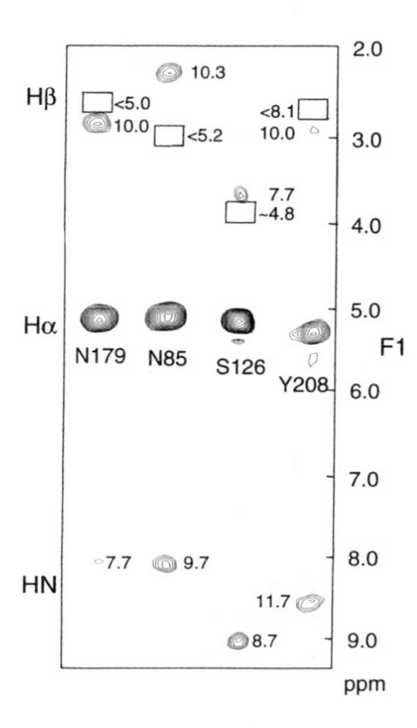

Figure 29. F_1 strips taken from the 3D CT-HACAHB spectrum of a 2 mM solution of uniformly ${}^{15}\mathrm{N}/{}^{13}\mathrm{C}$ labeled PB92 (269 amino acids, 27 kDa) in $\mathrm{H_2O}$ solution taken at the ${}^{13}\mathrm{C}^{\alpha},{}^{1}\mathrm{H}^{\alpha}$ resonance frequencies of Asn^{179}, Asn^{85}, Ser^{126}, and Tyr^{208}. The spectrum was acquired using $48 \times 54 \times 384$ complex datapoints with acquisition times of 9.6, 13.25, and 77 ms in t_1, t_2, and t_3, respectively. Boxes indicate the H^{β} resonance frequencies for which no cross-peaks or cross-peaks at the noise level were observed.

estimate of the value of $^3J(H^{\alpha}H^{\beta})$ can be obtained from the noise level of the experiment and the intensity of the diagonal peak. For example, the β-protons of Asn^{179} resonate at 2.81 and 2.65 ppm. At 2.81 ppm a strong cross-peak is found yielding a $^3J(H^{\alpha}H^{\beta})$ coupling constant of 10.0 Hz. At the resonance frequency of the other β-proton, 2.65 ppm, no cross-peak above the noise level is observed, indicating that this $^3J(H^{\alpha}H^{\beta})$ coupling constant should be smaller than 5.0 Hz. This is in agreement with the crystal (van der Laan *et al.*, 1992) and solution structure of PB92 (Martin *et al.*, 1997) in which the χ_1 angle of Asn^{179} has a g^+ conformation ($\chi_1 = -83°$). In general, the values obtained for the different coupling constants correlate well with the dihedral angles observed for PB92. In fact, the more intermediate $^3J(H^{\alpha}H^{\beta})$ values observed for Ser^{126} are indicative of some rotameric averaging, in support of the conformational flexibility of this residue in solution (Martin *et al.*, 1997).

Recording the spectrum in H_2O solution also enables the measurement of the $^3J(H^NH^{\alpha})$ coupling constants. However, the correlations to the H^{β} are attenuated by the presence of a passive $^3J(H^NH^{\alpha})$ coupling and hence, when sensitivity is the limiting factor, the CT-HACAHB experiment is preferentially recorded in D_2O solution.

As discussed in Section 4.4.1, the accuracy of the extracted coupling constants depends on the differential relaxation rates of the in-phase and antiphase magnetization terms (Vuister and Bax, 1993a). Similarly to the HNHA experiment, the systematic underestimation of the measured coupling constants can be compensated by applying a uniform relative correction factor. The cross-relaxation in the case of the β-methylene proton provides an additional source of systematic errors (see Section 4.4.1). This effect depends on the duration of the de- and rephase periods and is estimated to yield an approximate 10–20% underestimate for conformations in which one of the H^{β} is in a *trans* conformation with respect to the H^{α} (Grzesiek *et al.*, 1995).

7.5. The LRCC and LRCH Experiments

As was the case for the $\{^{15}N\}$ and $\{^{13}C\}$ spin-echo difference ^{13}C CT-HSQC experiments discussed in Section 7.3, the LRCC (Bax *et al.*, 1992) and LRCH (Vuister *et al.*, 1993b; Vuister and Bax, 1993b) experiments also employ the favorable relaxation properties of the methyl groups. Hence, the LRCC experiment yields information regarding $^3J(C^{\delta}C^{\alpha})$ in leucine and isoleucine residues whereas the LRCH experiment provides information regarding $^3J(C^{\delta}H^{\beta})$ in leucine and isoleucine residues and $^3J(C^{\gamma}H^{\alpha})$ in valine, isoleucine, and threonine residues. The HN(CO)CAC$_{ali}$ experiment proposed recently (Hennig *et al.*, 1997) also employs the QJ principle for measuring $^3J(C^{\alpha}C^{\delta})$ couplings, albeit that the favorable relaxation properties of a fully deuterated C^{α} nucleus are used to allow for the substantial de- and rephasing periods.

Slightly modified versions of the LRCC and LRCH experiments are shown in Figs. 30A and 30B, respectively. The LRCC experiment belongs to the third class of QJ experiments (Section 4.3), whereas the LRCH belongs to the second class (Section 4.2), requiring a separate reference experiment. The original LRCC experiment was implemented as a 2D experiment. In practice, however, the 3D variation shown in Fig. 30A is to be preferred. Although the original experiments were recorded in D_2O solution, recording the experiments in H_2O solution does not present a problem in the current implementation.

Three strips from the LRCC spectrum of $^{15}N/^{13}C$-labeled BB10010 taken along the F_1 axis are shown in Fig. 31. The strips display a large cross-peak at the $C^{\delta}H^{\delta}$ resonance frequencies of Ile^{24}, Ile^{19}, and $Leu^{\delta 2}$, respectively. In addition, correla-

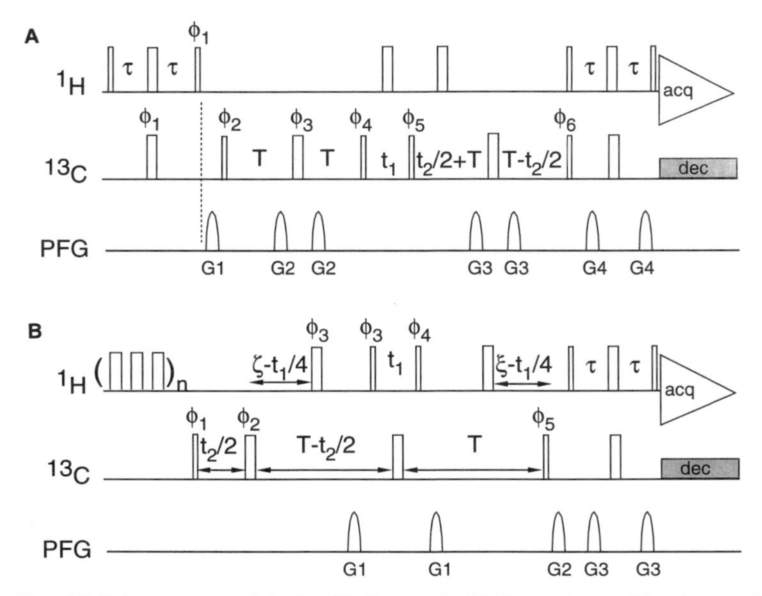

Figure 30. (A) Pulse sequence of the LRCC (Bax *et al.*, 1992) experiment. The phase cycle is as follows: $\phi_1 = y,-y$; $\phi_2 = 2x,2(-x)$; $\phi_3 = x$; $\phi_4 = 4y,4(-y)$; $\phi_5 = 8y,8(-y)$; $\phi_6 = 4x,4(-x)$; receiver = $x,2(-x),x,-x,2x,-x$. Quadrature detection is obtained using the States–TPPI method incrementing phases ϕ_{2-4} for t_1 and phases ϕ_{2-5} for t_2. The delays were set to: $\tau = 1.75$ ms, $T = 14$ ms. Gradient strengths and durations: $G_1 = 1$ ms, 10 G cm^{-1}; $G_2 = 1$ ms, -18 G cm^{-1}; $G_3 = 1$ ms, 20 G cm^{-1}; $G_4 = 1$ ms, 13 G cm^{-1}. (B) Pulse sequence of the LRCH (Vuister and Bax, 1993b) experiment. The phase cycle is as follows: $\phi_1 = x$; $\phi_2 = x,-x$; $\phi_3 = 2x,2(-x)$; $\phi_5 = 8y,8(-y)$; receiver = $P,2(-P),P$ with $P = x,2(-x),x$ for the 3D version or $P = 4x$ for the reference experiment. Quadrature detection is obtained using the States–TPPI method incrementing phases ϕ_1 and ϕ_3 for t_1 and t_2, respectively. The delays were set to: $\tau = 1.75$ ms, $T = 29.6$ ms, $\zeta = 3$ ms, $\xi = 4$ ms. Gradient strengths and durations: $G_1 = 1$ ms, 5 G cm^{-1}; $G_2 = 1$ ms, -10 G cm^{-1}; $G_3 = 1$ ms, 13 G cm^{-1}. Note that the time between the ^{13}C 90°(ϕ_1) and 90°(ϕ_5) pulses has to remain constant at $2T$ when incrementing t_1.

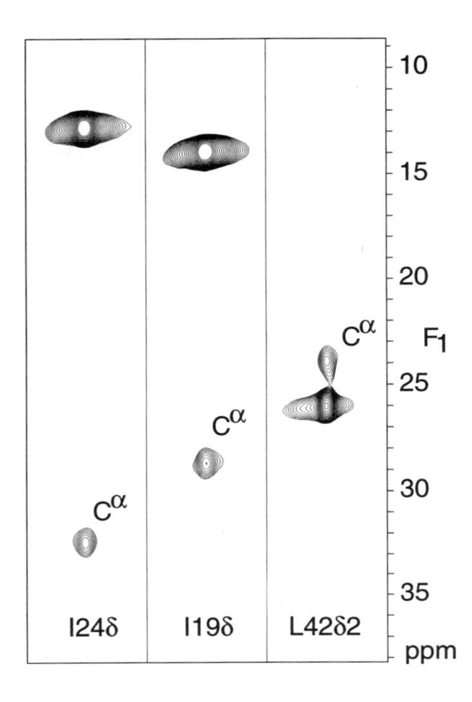

Figure 31. F_1 strips from the LRCC spectrum of BB10010 taken at the $C^{\delta}H^{\delta}$ resonance positions of Ile^{24}, Ile^{19}, and Leu^{42}. Correlations to the C^{α} are folded once in the F_1 dimension. The spectrum resulted from a data set recorded with the sequence of Fig. 30A as $37 \times 72 \times 512$ complex datapoints with acquisition times of 9.9, 23.8, and 77 ms in t_1, t_2, and t_3, respectively.

tions to the C^{α} nuclei are indicated (folded once in the F_1 dimension). Application of Eq. (23) yields coupling constants of 3.0, 2.7, and 3.8 Hz, respectively. Values of $^3J(CC)$ around 3 Hz are indicative of a *trans* conformation and similar values have been observed before for the calmodulin–M13 complex (Bax *et al.*, 1992). The combined usage of the $^3J(C^{\delta}C^{\alpha})$ and $^3J(C^{\delta}H^{\beta})$ couplings, obtained from the LRCC and LRCH experiments, allows the stereospecific assignment of C^{δ} groups in leucine residues, provided that no rotameric averaging around the χ_2 angle takes place, the two prochiral β-protons have nondegenerate chemical shifts, and the stereospecific assignment of these protons is known (Vuister *et al.*, 1993b). However, the latter information can be obtained from the values of the J couplings related to the torsion angle χ_1 (Table 3).

8. USAGE OF *J* COUPLINGS

The magnitude of the ^3J coupling constants can be correlated with the intervening dihedral angle through a Karplus equation (Karplus, 1959, 1963). Thus, the J coupling can be translated into a structural restraint. The multivalued nature of the Karplus equation, however, often requires information about a second J coupling that is dependent on the same torsion angle to resolve potential ambiguity. In the case of rotameric averaging or prochiral groups, it is an absolute necessity to obtain reliable information about several J coupling constants. Alternatively, NOE information in conjunction with J couplings can be used to establish stereospecific assignments and determine a rotameric state. Rather than employing a rotatorial model for translating the J value into a dihedral angle (see Section 8.2), J couplings can be used in direct-refinement protocols with either a single-conformer J-value-dependent potential or time- or ensemble-averaged J-value-dependent potentials (Kim and Prestegard, 1990; Mierke *et al.*, 1992,1994; Torda *et al.*, 1993; Garrett *et al.*, 1994).

8.1. Parametrization of Karplus Curves

The accuracy of the appropriate Karplus curve is crucial to a meaningful interpretation of a measured J value. The dependence of the magnitude of a ^3J coupling constant on the intervening torsion angle θ is generally described by a Karplus dependence (Karplus, 1959, 1963):

$$^3J = A\cos^2(\theta) + B\cos(\theta) + C \tag{29}$$

The adjustable parameters A, B, and C are typically determined by *ab initio* calculations or by studies on conformationally restricted small molecules (Bystrov, 1976). Alternatively, ^3J values measured in proteins are compared with the dihedral angles found in the corresponding crystal structures (Pardi *et al.*, 1984; Ludvigsen *et al.*, 1991; Vuister and Bax, 1993a; Wang and Bax, 1995,1996; Hu *et al.*, 1997; Hu and Bax, 1997b) or NMR-derived solution structures (Wang and Bax, 1996; Hu *et al.*, 1997). As proper parametrization is of major importance for the interpretation of the ^3J values (see Karimi-Nejad *et al.*, 1994, for an example), it is useful to discuss some common sources of errors in the parametrization procedures.

The parametrization using small molecules is hampered by both the usually low number of data points and the differences in the electronic structure of the model compounds relative to proteins. Consequently, in several instances parametrization of Karplus curves obtained from studies on small molecules yielded results that could not be directly used for the analysis of ^3J values in proteins (Wang and Bax, 1995, 1996; Hu and Bax, 1996; Hu *et al.*, 1997).

Parametrization using ^3J values measured in proteins is also complicated by serious difficulties, some of which are not easily overcome. The most common sources of errors are: (1) the error in the measured ^3J values; (2) the errors in the values of the dihedral angle θ; (3) when using dihedral angles derived from crystal structures: the genuine differences between crystal and solution structures, such as those caused by internal mobility in solution or those caused by distortions resulting from crystal contacts; (4) the limited range of accessible θ angles; and (5) the limited number of data points.

The experimental errors in the measured ^3J values translate into errors in the parameters A, B, and C (point 1 above). Generally, there will be a range of values for A, B, and C which are in accordance with the experimental data. Some insight into this range can be obtained by a so-called jackknife procedure. The A, B, and C parameters are usually determined by a least-squares fit of the measured ^3J values versus θ (using some error estimate for the ^3J value). If, however, arbitrarily 10% of the measured values are omitted and the A, B, and C parameters are determined again, slightly different values are obtained. The remaining 10% of the data can be used to assess the error. Repetition of this procedure for different randomly chosen sets of 10% deletion yields a range of accessible parameters. An example of this is shown in Fig. 32 for the parametrization of the $^3J(H^NH^\alpha)$ coupling constant on the basis of the $^3J(H^NH^\alpha)$ values measured for staphylococcal nuclease using the HNHA experiment (Vuister and Bax, 1993a).

Errors resulting from point 2 above could be minimized by using only high-resolution crystal structures, preferentially in several crystal forms. Differences between solution and NMR structures (point 3 above) can be assessed by careful comparison. The problem associated with point 4 needs particular attention if ^3J values are analyzed that significantly fall outside the range used for parametrization. Also, conclusions pertaining to dihedral angle values that have not been sampled in the parametrization process should be evaluated with great care. The rapidly expanding database of ^3J values should improve the errors resulting from point 5.

8.2. Analysis of *J* Couplings in Proteins

When analyzing ^3J coupling constants in proteins, their translation into the corresponding dihedral angles is hampered by two problems. First, the Karplus equation is not a single-valued function in a mathematical sense, yielding up to four dihedral angle values for a given ^3J value. Therefore, to obtain an unambiguous answer it is necessary to determine a set of coupling constants related to the same torsion angle in the protein. Second, conformational mobility, especially encountered in protein side chains, may lead to coupling constants that are averages over multiple conformations (Pople, 1958; Hoch *et al.*, 1985). In these cases, the observed time-averaged coupling constant is given by

$$\langle J \rangle = \int_{-\pi}^{\pi} J(\theta)\, P(\theta) d\theta \tag{30}$$

where $P(\theta)$ is the probability of observing the value $J(\theta)$ for the torsion angle θ at a given temperature T according to

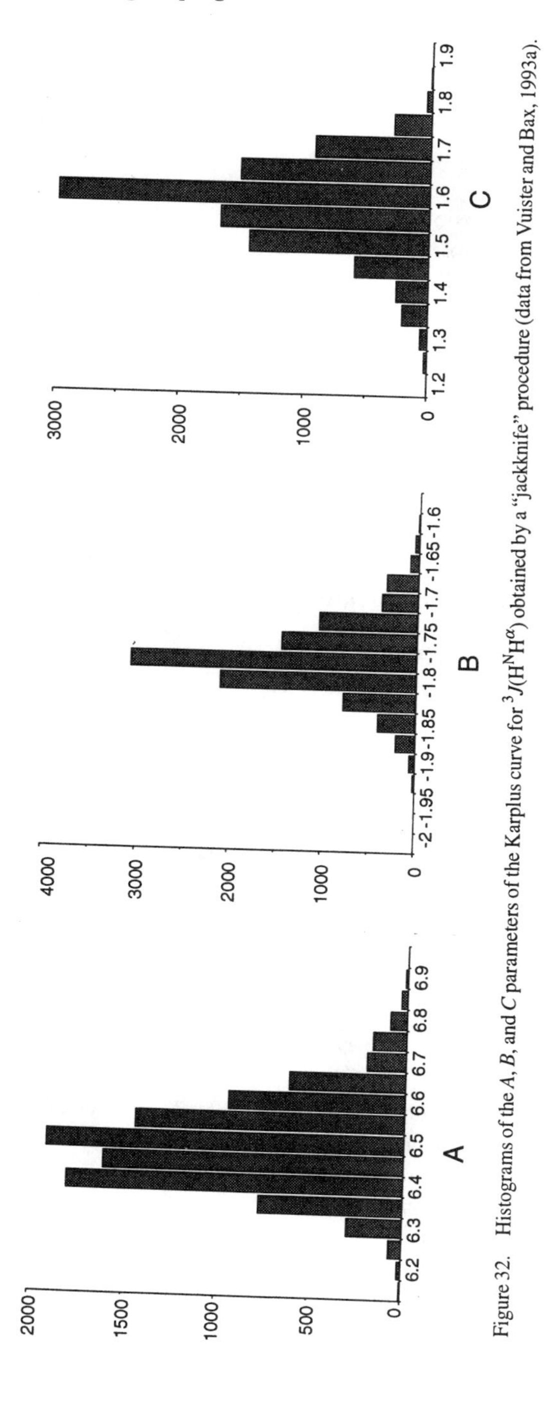

Figure 32. Histograms of the *A*, *B*, and *C* parameters of the Karplus curve for $^3J(H^NH^\alpha)$ obtained by a "jackknife" procedure (data from Vuister and Bax, 1993a).

$$P(\theta) = \frac{e^{-E(\theta)/kT}}{\int_{-\pi}^{\pi} e^{-E(\theta)/kT} d\theta} \tag{31}$$

and $E(\theta)$ defines the torsional potential. If all possible values of θ have the same probability, i.e., if complete rotational averaging occurs, $P(\theta) = 1/2\pi$ and one obtains according to Eqs. (29) and (30) an average coupling constant from the Karplus relationship

$$\langle {}^3J \rangle = A/2 + C \tag{32}$$

Only if $E(\theta)$, which depends on the intrinsic dihedral potential as well as on the environment (e.g., the solvent or the protein matrix) of the torsion angle under consideration, were known could rotameric probability distributions as well as coupling constant values be predicted. As this is usually not the case, one has to set up suitable models for the description of rotatorial mobility.

8.2.1. Models of Rotatorial Mobility

In the simple single-site model (Fig. 33A), the experimental value, $J^{\exp}$, is explained by assuming a fixed dihedral angle, θ:

$$J^{\exp} = J^{\text{calc}}(\theta) \tag{33}$$

where $J^{\text{calc}}(\theta)$ is derived from the appropriate Karplus curve. The model has only one adjustable parameter, i.e., θ.

The second model assumes a unimodal Gaussian distribution, where the angle θ is allowed to fluctuate about an expectation value, $\langle\theta\rangle$, with a given width, σ_θ (Fig. 33B), resulting in a two-parameter model according to

$$J^{\exp} = J^{\text{calc}}(\langle\theta\rangle, \sigma_\theta) = \frac{1}{\sigma_\theta\sqrt{2\pi}} \int_{-\pi}^{\pi} J^{\text{calc}}(\theta) e^{-1/2\,([\theta - \langle\theta\rangle]/\sigma_\theta)^2} d\theta \tag{34}$$

Traditionally, and especially in the case of peptide sidechains, the Pachler analysis (Pachler, 1963, 1964) (Fig. 33C) has found widespread use for the interpretation of motionally averaged coupling constants, assuming conformational equilibria between discrete staggered rotamers, e.g., the sidechain angle χ_1 (Kessler *et al.*, 1987). The populations of these staggered rotamers, p_{I}, p_{II}, and p_{III}, of the sites with $\theta = 180°$, $-60°$, and $+60°$, respectively, are allowed to vary to obtain the closest correspondence between experimentally derived J couplings and values back-calculated from the model. The resulting model is described by two independent adjustable parameters:

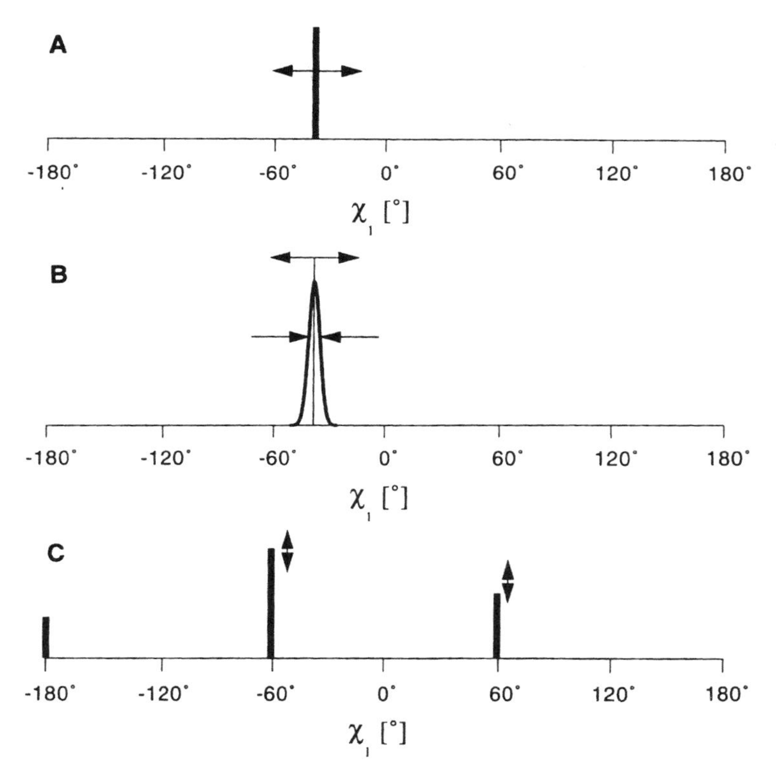

Figure 33. Models of rotatorial mobility (adapted from Schmidt, 1997): single-site model (A), Gaussian distribution (B), and staggered rotamers (C).

$$J^{\mathrm{exp}} = J^{\mathrm{calc}}\,(p_{\mathrm{I}}, p_{\mathrm{II}}, p_{\mathrm{III}})$$
$$= p_{\mathrm{I}}J^{\mathrm{calc}}(\theta = 180°) + p_{\mathrm{II}}J^{\mathrm{calc}}(\theta = -60°) + p_{\mathrm{III}}J^{\mathrm{calc}}(\theta = 60°) \tag{35a}$$

$$p_{\mathrm{I}} + p_{\mathrm{II}} + p_{\mathrm{III}} = 1 \tag{35b}$$

Possible limitations of this model arise from the assumption of a threefold torsion potential with minima at $\theta = -60°$, $+60°$, and $180°$ that might not always be correct (Gelin and Karplus, 1979; Fischman *et al.*, 1980; Schrauber *et al.*, 1993).

Continuous dihedral-angle distributions have also been used to describe the rotatorial dynamics of protein sidechains, both to explain theoretical coupling constant data derived from solvated MD simulations (Hoch *et al.*, 1985; Brüschweiler and Case, 1994) as well as to fit experimental data (Dzakula *et al.*, 1992 a,b; Karimi-Nejad *et al.*, 1994).

8.2.2. Application to Ribonuclease T1

A thorough analysis of coupling constants in the presence of internal motions should ideally apply different models for the characterization of rotameric averaging, and then compare the agreement of the corresponding theoretical coupling constants back-calculated from the model with the experimentally observed values.

Such an analysis has been carried out for the sidechains of the 104-amino-acid protein ribonuclease T_1, for which a total of 357 sidechain coupling constants, including ${}^3J(\mathrm{H}^\alpha\mathrm{H}^\beta)$, ${}^3J(\mathrm{C'H}^\beta)$, and ${}^3J(\mathrm{NH}^\beta)$, have been determined from various 2D and 3D heteronuclear E.COSY spectra (Karimi-Nejad *et al.*, 1994; Karimi-Nejad, 1996). The three different models, as discussed above, for internal dynamics about the torsion angle χ_1 were used to fit these experimental coupling constants. The parameters of these models were adjusted in a least-squares optimization applying the SIMPLEX algorithm (Nelder and Mead, 1965) to minimize the error function:

$$\chi^2 = \sum_{i=1}^{n} \left(\frac{J_i^{\mathrm{exp}} - J_i^{\mathrm{calc}}}{\sigma_{J_i}} \right)^2 \tag{36}$$

where the index i runs over all experimental coupling constants and σ_{J_i} are the experimental errors of the ${}^3J_i$ values determined by the J convolution method discussed in Section 3.2. In the case of sidechains containing diastereotopic methylene protons, the fit was carried out for both possible stereospecific assignments.

To compare the results obtained from the various models, their statistical significance can be evaluated from a computation of their χ^2 probability function Q (Press *et al.*, 1989):

$$Q(\chi^2, \upsilon) = \Gamma\left(\frac{\chi^2}{2}, \frac{\upsilon}{2}\right) / \Gamma\left(\frac{\upsilon}{2}\right) \tag{37}$$

Thereby, the fact that the applied models might vary in the number of adjustable parameters and hence also in the number of degrees of freedom, υ, is properly taken into account.

An example taken from the analysis of the coupling constant data of ribonuclease T1 (Karimi-Nejad *et al.*, 1994; Karimi-Nejad, 1996) may illustrate this procedure. The experimentally derived sidechain ${}^3J$ values for residues Asn^{43}, Asn^{44}, and Val^{78} are listed in Table 5, which also contains the optimized values of the adjustable parameters of the three models described above, the χ^2 values and statistical significances, Q, and the ${}^3J$ values back-calculated on the basis of each model. A comparison of the Q values computed for the various models shows that the sidechain conformation of Asn^{43} is best described by a static conformation with a χ_1 value of –84°. For Asn^{44} a Gaussian distribution centered around $\langle\chi_1\rangle = 96°$

with a width of 17° fulfills the experimental data best. The results obtained for Val^{78} show that this residue undergoes extensive rotameric averaging, which is best described by a three-site equilibrium between all three staggered rotamers.

More complicated models describing rotatorial mobility, e.g., multimodal Gaussian distributions or jumps between sites with arbitrary dihedral angle values, will increase the number of adjustable parameters, and therefore also require more observables to achieve the same statistical significance. Such models, however, have been used to fit combinations of J coupling and cross-relaxation data in proteins and carbohydrates (Dzakula *et al.*, 1992a,b; Poppe, 1993).

8.3. Application to the Photoactive Yellow Protein

PYP contains 13 glycine residues out of a total of 125 amino acids. Stereospecific assignment of the H^{α} resonances of these residues is of great importance to the structure determination process, as these residues are involved in only a limited number of NOE interactions. Thus, they contribute significantly to the observed conformational variability of the NMR-derived structures. Moreover, preliminary structure calculations indicated that several of these glycine residues adopted conformations with positive ϕ angles.

${}^3J(H^NH^{\alpha})$, ${}^3J(NH^{\beta})$, and ${}^3J(NH^{\alpha i-1})$ coupling constants were measured on a uniformly ^{15}N labeled sample of PYP using the HNHA (Vuister and Bax, 1993a; Kuboniwa *et al.*, 1994) and modified HNHB experiments (Düx *et al.*, 1997). Overall, 120 ${}^3J(H^NH^{\alpha})$ couplings were extracted from the HNHA spectrum, whereas 117 ${}^3J(NH^{\beta})$ and 29 ${}^3J(H^{\alpha}N)$ couplings could be extracted from the HNHB spectrum. Figure 34 shows two F_1 strips from the 3D HNHB spectrum of PYP taken at the $(F_2, F_3) = (^{15}N, {}^1H^N)$ resonance frequencies of residues Val^{83} and Asn^{87}, which directly follow residues Gly^{82} and Gly^{86}, respectively. F_1 strips taken from the HNHA spectrum at the $(F_2, F_3) = (^{15}N, {}^1H^N)$ resonance frequencies of residues Gly^{82} and Gly^{86} are also plotted. The correlations to the α-protons of the glycine residues are indicated in all strips. In both strips of the HNHB spectrum, relatively large cross-peaks are observed to one of the sequential α-protons. The coupling constants for Gly^{82} and Gly^{86} were calculated to be 2.0 and 1.2 Hz, respectively. The boxes indicate the resonance positions of the second sequential α-proton. The thermal noise level sets an upper limit for the coupling constants involving these protons (<0.7 Hz in both cases). ${}^3J(NH^{\alpha i-1})$ values of 1.2–2.0 Hz observed for one of the α-protons are consistent with a *trans* conformation of this proton with respect to the sequential nitrogen nucleus (Wang and Bax, 1995).

The strips of the HNHA spectrum show correlations to both α-protons. Application of Eq. (23) yields the ${}^3J(H^NH^{\alpha})$ coupling constants. For Gly^{86} we thus obtain 7.3 and 5.2 Hz for the upfield and downfield shifted resonances, respectively. The ${}^3J(H^NH^{\alpha})$ and ${}^3J(NH^{\alpha i-1})$ coupling constants for the 13 glycine residues are listed in Table 6. Although the ${}^3J(H^NH^{\alpha})$ coupling constants are dependent on the

Table 5
Examples of Sidechain Conformations in Ribonuclease T1 as Determined from ^3J Coupling Constants*

		$^3J_{\alpha\beta2}$ (Hz)[a]	$^3J_{C'\beta2}$ (Hz)[a]	$^3J_{N'\beta2}$ (Hz)[a]	$^3J_{\alpha\beta3}$ (Hz)[a]	$^3J_{C'\beta3}$ (Hz)[a]	$^3J_{N'\beta3}$ (Hz)[a]	χ^2	Q (%)
Asn[43]									
Experimental		10.1±0.6	0.4±0.3	−2.2±0.5	4±1.1	3.3±0.4	nd[b]		
χ_1 (°)	**−84**	**11.19**	**0.47**	**−1.81**	1.74	**3.66**	—	3.7	**45**
$<\chi_1> \pm \sigma\chi_1$ (°)	−84±1	11.15	0.47	−1.83	1.74	3.69	—	3.7	29
p_I, p_{II}, p_{III} (%)[c]	0,77,23	10.67	1.38	−1.59	3.38	3.36	—	23.5	0
Asn[44]									
Experimental		8.0±0.5	1.6±0.3	nd	1.8±0.9	6.9±0.7	−1.9±0.5		
χ_1 (°)	98	8.48	1.02	—	2.21	6.68	−2.58	7.0	14
$<\chi_1> \pm \sigma\chi_1$ (°)	**96±17**	7.79	**1.43**	—	**2.79**	**6.75**	−2.28	**2.4**	**49**
p_I, p_{II}, p_{III} [%][c]	0,41,59	7.31	1.38	—	3.38	6.34	−2.51	7.9	5
Val[78]									
Experimental		5.2±0.5	2.6±0.6	−3.8±0.4					
χ_1 (°)	123	5.44	3.05	−3.09				4.1	13
$<\chi_1> \pm \sigma\chi_1$ (°)	−55±41	5.22	2.68	−3.83				0.1	82
p_I, p_{II}, p_{III} [%][c]	**19,66,15**	5.20	**2.60**	**−3.79**				0.0	**97**

*Data from Karimi-Nejad, 1996.

[a] ^3J coupling constants listed under "Experimental" are the experimentally observed values, whereas the values in the following three rows have been back–calculated from the corresponding model.

[b] Experimentally not determined (nd) because of spectral overlap or lack of NOE cross-peak.

[c] Staggered rotamer populations (I = 180°, II = −60°, III = 60°).

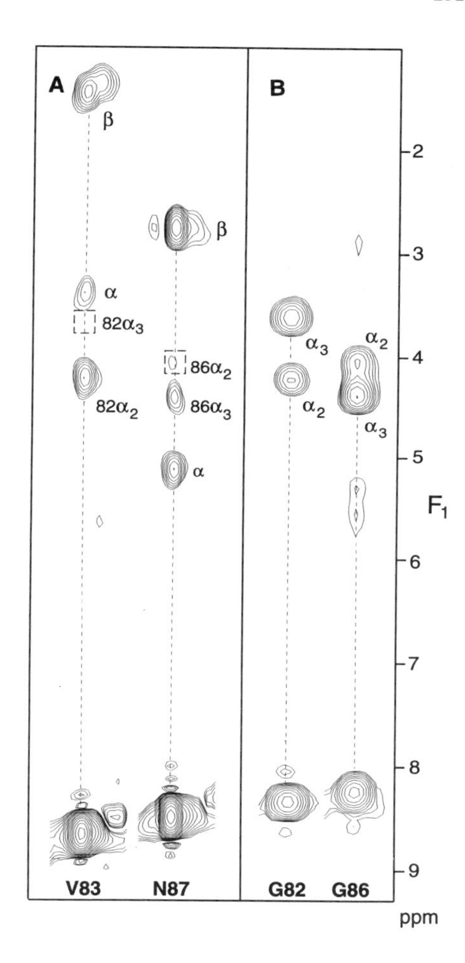

Figure 34. Strips from the HNHB (A) and HNHA (B) spectra of PYP, taken at the $(F_2, F_3) = (^{15}N, {}^{1}H^{N})$ resonance frequencies of residues Val^{83} and Asn^{87} (A), and Gly^{82} and Gly^{86} (B). Diagonal peaks between 7 and 9 ppm are of opposite sign compared with cross-peaks. The data have been acquired as described in the captions of Figs. 12 and 25 for the HNHA and HNHB experiments, respectively. (Reproduced from Düx *et al.*, 1997).

backbone angle ϕ (see Section 5.1), their usage in conjunction with the $^3J(NH^{\alpha i-1})$, which are dependent on the backbone angle ψ, allows the stereospecific assignment and the extraction of dihedral restraints, as discussed below.

The *trans* configuration of one of the α-protons with respect to the sequential nitrogen, inferred from the large value of the $^3J(NH^{\alpha i-1})$ coupling constant, is consistent with two possibilities of the backbone angle ψ, e.g., $\psi \approx -60°$ or $\psi \approx +60°$, corresponding to the two choices of the stereospecific assignment. A similar situation applies to the $^3J(H^NH^\alpha)$ coupling constants: Using the previously parametrized Karplus curve for $^3J(H^NH^\alpha)$ (Pardi *et al.*, 1984; Vuister and Bax, 1993a; Wang and Bax, 1996), both $\phi \approx +90°$ and $\phi \approx -90°$ yield the measured ^3J values, where each conformation is again associated with one choice of the stereospecific assignment. However, the α-proton associated with $^3J(H^NH^\alpha)$ = 7.3 Hz has $^3J(NH^{\alpha i-1})$ = 1.2 Hz. This excludes two, out of the total of four, possible solutions and only ϕ, $\psi \approx$ +90°, +60° or $\phi,\psi \approx$ –90°, –60° remain as possible conformations. The last two options are structurally considerably different and can easily be distinguished. For PYP, Gly^{86} adopts a conformation with its backbone ϕ,ψ angles in the upper-right quadrant of the Ramanchandran plot (Düx *et al.*, 1997). Thus, its H^α resonances can be unambiguously assigned, and dihedral constraints can be derived from the *J* coupling data. For the second example shown in Fig. 34, Gly^{82}, the reverse situation is encountered with $\phi,\psi \approx$ –60°, –60°. Overall, for all eight glycine residues that were not subject to conformational averaging and had nondegenerated H^αresonance frequencies, stereospecific assignments were obtained. Reliable and precise ϕ,ψ dihedral restraints were also derived for these residues from the *J* coupling data. The resulting assignments and dihedral restraints are also listed in Table 6. Application of these data to the structure calculation of PYP reduced the pairwise backbone RMSD from 0.83 Å to 0.73 Å (Düx *et al.*, 1998).

9. CONCLUDING REMARKS

The advent of ^{15}N and $^{15}N/^{13}C$ labeling in proteins has prompted the development of a large array of experiments for measuring almost any homo- or heteronuclear ^3J coupling constant in proteins. Many of these modern techniques for measuring *J* couplings can now be applied routinely to most proteins, providing valuable additional structural information.

The two most commonly used schemes, E.COSY and QJ methods, have specific advantages and disadvantages. E.COSY methods are suited equally well for the extraction of small and large couplings, in addition to providing information about the relative signs of the coupling constants. The accuracy of its results is dependent on the signal-to-noise ratio and the line shape of the multiplet, and the extent to which the spin state of the passive nucleus can be preserved. Sometimes the spin topology prohibits the E.COSY measurement and also overlap of different multiplets can present a problem. Moreover, the actual accurate determination of the coupling is somewhat more cumbersome compared with the QJ methods.

The extraction of the ^3J values is simple for the QJ-based methods. In addition, in those cases where the spin topology prohibits the E.COSY measurement, the QJ methods sometimes can yield the ^3J values. The lower limit of the coupling constants that can be determined by the QJ-based methods is set by the signal-to-noise ratio of the experiment, i.e., the size of the ^3J value relative to the effective transverse relaxation time. If the relaxation is too fast relative to the magnitude of the *J* coupling, a correlation is not observed and only an upper estimate of the *J* value can be obtained. In addition, no information about the sign of the ^3J value can be derived from the QJ experiments.

Table 6
$^3J(H^NH^\alpha)$ and $^3J(H^{\alpha i-1}N)$ Values, Stereospecific Assignments, and Torsion Angle Restraints for the Glycine Residues in PYP*

Residue	Shift (ppm)	$^3J(H^NH^\alpha)^a$ (Hz)	$^3J(H^\alpha N^{i+1})^{a,b}$ (Hz)	Assignment	Restraint[c] ϕ	ψ
G7	2.96	nd	<1.0	nd[d]		
	3.59	nd	<1.0	nd		
G21	4.06	5.3–7.0	0.9–1.5	ov[e]		
	4.14	5.3–7.0	0.9–1.5	ov		
G25	3.85	5.5	1.4	$H^{\alpha 3}$	$-100 < \phi < -80$	
	4.40	8.5	<1.3	$H^{\alpha 2}$	$20 < \psi < 35$ or $85 < \psi < 100$	
G29	4.50	4.2–6.4	<1.0	ov		
	4.68	4.2–6.4	<1.0	ov		
G35	3.99	4.0–6.1	0–1.1	ov		
	3.99	4.0–6.1	0–1.1	ov		
G37	3.66	5.1	<1.0	$H^{\alpha 2}$	$80 < \phi < 100$	
	4.58	8.2	1.1	$H^{\alpha 3}$		$10 < \psi < 25$
G47	4.16	6.3	<0.6	$H^{\alpha 3}$	$-70 < \phi < -50$	
	4.41	4.0	1.4	$H^{\alpha 2}$		$-35 < \psi < -15$
G51	4.00	5.6	1.3	$H^{\alpha 2}$	$80 < \phi < 100$	
	4.27	6.9	ov	$H^{\alpha 3}$		$-20 < \psi < -5$
G59	3.74	5.1	1.3	$H^{\alpha 2}$	$80 < \phi < 100$	
	4.41	7.5	ov	$H^{\alpha 3}$		$-30 < \psi < -10$
G77	3.54	3.3	2.0	$H^{\alpha 2}$	$-60 < \phi < -40$	
	4.14	5.9	ov	$H^{\alpha 3}$		$-75 < \psi < -40$
G82	3.62	6.5	<0.7	$H^{\alpha 3}$	$-70 < \phi < -60$	
	4.17	5.1	2.0	$H^{\alpha 2}$		$-75 < \psi < -40$
G86	4.04	5.2	<0.7	$H^{\alpha 2}$	$80 < \phi < 100$	
	4.36	7.3	1.2	$H^{\alpha 3}$		$15 < \psi < 30$
G115	3.73	6.0	<0.8	nd	averaged	
	4.20	6.3	<0.8	nd		averaged

*Reproduced from Düx *et al.*, 1997.

[a]All values have been corrected by 1.05 to account for the faster relaxation antiphase magnetization with respect to in-phase magnetization (see Section 4.4.1).

[b]Absolute value because sign cannot be determined from the data.

[c]Restraint ranges were derived assuming the absence of rotamer averaging and used NOE data to exclude certain ϕ,ψ combinations (see text).

[d]nd: not determined.

[e] ov: (partial) overlap.

Provided the *J* coupling data are handled with prudence, for static conformations they can be translated into accurate restraints that will greatly improve the accuracy of the NMR-derived structures. Moreover, the possibility of obtaining structural information from several sources opens the prospects for an independent validation of the NMR-derived structure; i.e., the presence of several *J* couplings pertaining to one torsion angle allows restraints, derived from one or a few *J* values, to be used in the structure determination process, whereas other *J* values, not translated into restraints, are used to validate the results. Alternatively, as the *J* coupling presents a time-averaged quantity, it also contains valuable information regarding conformational flexibility which can be appropriately modeled. In all, the measurement of *J* couplings constants has become a valuable source of information regarding biomolecular conformation and dynamics.

ACKNOWLEDGMENTS. G.W.V. thanks Dr. Ad Bax for collaboration on the topic of this chapter. The authors thank Petra Düx for sharing results on PYP prior to publication, and Professor Rolf Boelens and Professor Rob Kaptein for interest and support. G.W.V. has been financially supported by the Royal Netherlands Academy of Arts and Sciences. This work was also supported by the Netherlands Foundation for Chemical Research (SON) with financial assistance from the Netherlands Organization for Scientific Research (NWO). The authors also thank Professor Klaas J. Hellingwerf for providing the ^{15}N-labeled PYP, Drs. Rolf Boelens and Dick Schipper for the use of the ^{15}N/^{13}C-labeled PB92 and sharing of data, Drs. Linda Nicholson and Lisa Gentile for providing the ^{15}N/^{13}C-labeled Src, and Professor John Walto and Dr. Lee Higgins for the use of the ^{15}N/^{13}C-labeled BB10010 sample.

REFERENCES

Archer, S. J., Ikura, M., Torchia, D. A., and Bax, A., 1991, *J. Magn. Reson.* **95:**636.

Bax, A., and Freeman, R., 1981a, *J. Magn. Reson.* **44:**542.

Bax, A., and Freeman, R., 1981b, *J. Magn. Reson.* **45:**177.

Bax, A., and Pochapsky, S. S., 1992, *J. Magn. Reson.* **99:**638.

Bax, A., Sparks, S.W., and Torchia, D. A., 1988, *J. Am. Chem. Soc.* **110:**7926.

Bax, A., Ikura, M., Kay, L. E., Torchia, D. A., and Tschudin, R., 1990, *J. Magn. Reson.* **86:**304.

Bax, A., Max, D., and Zax, D., 1992, *J. Am. Chem. Soc.* **114:**6924.

Bax, A., Vuister, G. W., Grzesiek, S., Delaglio, F., Wang, A. C., Tschudin, R., and Zhu, G., 1994, *Methods Enzymol.* **239:**79.

Billeter, M., Neri, D., Otting, G., Qiu Qian, Y., and Wüthrich, K., 1992, *J. Biomol. NMR* **2:**257.

Blake, P. R., Summers, M. F., Adams, M. W. W., Park, J.-B., Zhou, Z. H., and Bax, A., 1992, *J. Biomol. NMR* **2:**527.

Brüschweiler, R., and Case, D., 1994, *J. Am. Chem. Soc.* **116:**11199.

Bystrov, V. F., 1976, *Prog. NMR Spectrosc.* **10:**41.

Chary, K. V. R., Otting, G., and Wüthrich, K., 1991, *J. Magn. Reson.* **93:**218.

Delaglio, F., Torchia, D. A., and Bax, A., 1991, *J. Biomol. NMR* **1:**439.

DeMarco, A., Llinas, M., and Wüthrich, K., 1978a, *Biopolymers* **17:**617.

DeMarco, A., Llinas, M., and Wüthrich, K., 1978b, *Biopolymers* **17:**2727.

Dzakula, Z., Westler, W. M., Edison, A. S., and Markley, J. L., 1992a, *J. Am. Chem. Soc.* **114:**6195.

Dzakula, Z., Edison, A. S., Westler, W. M., and Markley, J. L., 1992b, *J. Am. Chem. Soc.* **114:**6200.

Düx, P., Whitehead, B., Boelens, R., Kaptein, R., and Vuister, G. W., 1997, *J. Biomol. NMR* **10:**301.

Düx, P., Rubinstenn, G., Vuister, G. W., Boelens, R., Mulder, F. A. A., Hård, K., Hoff, W. D., Kroon, A., Hellingwerf, K. J., and Kaptein, R., 1998, *Biochemistry*, **37:**12689.

Edison, A. S., Westler, W. M., and Markley, J. L., 1992, *J. Magn. Reson.* **92:**434.

Eggenberger, U., Karimi-Nejad, Y., Thüning, H., Rüterjans, H., and Griesinger, C., 1992, *J. Biomol. NMR* **2:**583.

Emerson, S. D., and Montelione, G. T., 1992, *J. Magn. Reson.* **99:**413.

Fischman, A. J., Live, D. H., Wyssbrod, H. R., Agosta, W. C., and Cowburn, D., 1980, *J. Am. Chem. Soc.* **102:**2533.

Garrett, D. S., Powers, R., Gronenborn, A. M., and Clore, G. M., 1991, *J. Magn. Reson.* **95:**214.

Garrett, D. S., Kuszewski, J., Hancock, T. J., Lodi, P. J., Vuister, G. W., Gronenborn, A. M., and Clore, G. M., 1994, *J. Magn. Reson. B* **104:**99.

Gelin, B. R., and Karplus, M., 1979, *Biochemistry* **18:**1256.

Gemmecker, G., and Fesik, S. W., 1991, *J. Magn. Reson.* **95:**208.

Görlach, M., Wittekind, M., Farmer, B. T. II, Kay, L. E., and Mueller, L., 1993, *J. Magn. Reson. B* **101:**194.

Griesinger, C., and Eggenberger, U., 1992, *J. Magn. Reson.* **97:**426.

Griesinger, C., Sørensen, O. W., and Ernst, R. R., 1985, *J. Am. Chem. Soc.* **107:**6394.

Griesinger, C., Sørensen, O. W., and Ernst, R. R., 1986, *J. Chem. Phys.* **85:**6837.

Griesinger, C., Sørensen, O. W., and Ernst, R. R., 1987, *J. Magn. Reson.* **75:**474.

Griffey, R. H., and Redfield, A. G., 1987, *Q. Rev. Biophys* **19:**51.

Grzesiek, S., Ikura, M., Clore, G. M., Gronenborn, A. M., and Bax, A., 1992, *J. Magn. Reson.* **96:** 215.

Grzesiek, S., Vuister, G. W., and Bax, A., 1993, *J. Biomol. NMR* **3:**487.

Grzesiek, S., Kuboniwa, H., Hinck, A. P., and Bax, A., 1995, *J. Am. Chem. Soc.* **117:**5312.

Haasnoot, C. A. G., de Leeuw, F. A. A. M., and Altona, C., 1980, *Tetrahedron Lett.* **36:**2783.

Harbison, G. S., 1993, *J. Am. Chem. Soc.* **115:**3026.

Hennig, M., Ott, D., Schulte, P., Löwe, R., Krebs, J., Vorherr, T., Bermel, W., Schwalbe, H., and Griesinger, C., 1997, *J. Am. Chem. Soc.* **119:**5055.

Hoch, J. C., Dobson, C. M., and Karplus, M., 1985, *Biochemistry* **24:**3831.

Hu, J.-S., and Bax, A., 1996, *J. Am. Chem. Soc.* **118:**8170.

Hu, J.-S., and Bax, A., 1997a, *J. Biomol. NMR* **9:**323.

Hu, J.-S., and Bax, A., 1997b, *J. Am. Chem. Soc.* **119:**6360.

Hu, J.-S., Grzesiek, S., and Bax, A., 1997, *J. Am. Chem. Soc.* **119:**1803.

Ikura, M., Kay, L. E., and Bax, A., 1990, *Biochemistry* **29:**2577.

IUPAC, 1970, *Biochemistry* **9:**3471.

Jones, J. A., Grainger, D. S., Hore, P. J., and Daniell, G. J., 1993, *J. Magn. Reson. A* **101:**162.

Karimi-Nejad, Y., 1996, NMR-spektroskopische Untersuchungen zur Struktur der Ribonuclease T_1 und ihrer Komplexe mit 2′- und 3′- Guanosinmonophosphat, Thesis, Universität Köln.

Karimi-Nejad, Y., Schmidt, J. M., Rüterjans, H., Schwalbe, H., and Griesinger, C., 1994, *Biochemistry* **33**:5481.

Karplus, M., 1959, *J. Chem. Phys.* **30:**11.

Karplus, M., 1963, *J. Am. Chem. Soc.* **85:**2870.

Kay, L. E., and Bax, A., 1990, *J. Magn. Reson.* **86:**110.

Kay, L. E., Brooks, B., Sparks, S. W., Torchia, D. A., and Bax, A., 1989, *J. Am. Chem. Soc.* **111:**5488.

Kay, L. E., Keifer, P., and Saarinen, T., 1992, *J. Am. Chem. Soc.* **114:**10663.

Kessler, H., Müller, A., and Oschkinat, H., 1985, *Magn. Reson. Chem.* **23:**844.

Kessler, H., Griesinger, C., and Wagner, K., 1987, *J. Am. Chem. Soc.* **109:**6927.

Kim, Y., and Prestegard, J. H., 1990, *Proteins Struct. Funct. Genet.* **8:**377.
Kleywegt, G. J., Vuister, G. W., Padilla, A., Knechtel, R. M., Boelens, R., and Kaptein, R., 1993, *J. Magn. Reson.* **102:**166.
Kuboniwa, H., Grzesiek, S., Delaglio, F., and Bax, A., 1994, *J. Biomol. NMR* **4:**871.
Löhr, F., and Rüterjans, H., 1995, *J. Biomol. NMR* **5:**25.
Löhr, F., and Rüterjans, H., 1997, *J. Am. Chem. Soc.* **119:**1468.
London, R. E., 1990, *J. Magn. Reson.* **86:**410.
Ludvigsen, S., Andersen, K. V., and Poulsen, F. M., 1991, *J. Mol. Biol.* **217:**731.
Madsen, J. C., Sørensen, O.W., Sørensen, P., and Poulsen, F. M., 1993, *J. Biomol. NMR* **3:**239.
Martin, J. R., Mulder, F. A. A., Karimi-Nejad, Y., Van der Zwan, J., Mariani, M., Schipper, D., and Boelens, R., 1997, *Structure* **5:**521.
McCoy, M. A., and Müller, L., 1992, *J. Am. Me.d Soc.* **114:**2108.
Mierke, D. F., Grdadolnik, S. G., and Kessler, H., 1992, *J. Am. Chem. Soc.* **114:**8283.
Mierke, D. F., Huber, T., and Kessler, H., 1994, *J. Comp. Aided Mol. Design* **8:**29.
Montelione, G. T., Winkler, M. E., Rauenbuehler, P., and Wagner, G., 1989, *J. Magn. Reson.* **82:**198.
Nelder, J.A., and Mead, R., 1965, *Computer J.* **7:**308.
Neri, D., Otting, G., and Wüthrich, K., 1990, *J. Am. Chem. Soc.* **112:**3663.
Neuhaus, D., Wagner, G., Vasak, M., Kägi, J. H., and Wüthrich, K., 1985, *Eur. J. Biochem.* **151:**257.
Norwood, T. J., 1993, *J. Magn. Reson. A* **101:**109.
Norwood, T. J., 1995, *J. Magn. Reson. A* **114:**92.
Norwood, T. J., and Jones, K., 1993, *J. Magn. Reson. A* **104:**106.
Oschkinat, H., and Freeman, R., 1984, *J. Magn. Reson.* **60:**164.
Otting, G., 1997, *J. Magn. Reson.* **124:**503.
Pachler, K. G. R., 1963, *Spectrochim. Acta* **19:**2085.
Pachler, K. G. R., 1964, *Spectrochim. Acta* **20:**581.
Palmer, A. G., Cavanagh, J., Wright, P. E., and Rance, M., 1991, *J. Magn. Reson.* **93:**151.
Pardi, A., Billeter, M., and Wüthrich, K., 1984, *J. Mol. Biol.* **180:**741.
Peng, J., Thanabal, V., and Wagner, G., 1991, *J. Magn. Reson.* **95:**421.
Pople, J. A., 1958, *Mol. Phys.* **1:**3.
Poppe, L., 1993, *J. Am. Chem. Soc.* **115:**8421.
Press, W. H., Flannery, B. P., Teukolsky, S. A., and Vetterling W. T., 1989, *Numerical Recipes*, Cambridge University Press, London.
Rexroth, A., Szalma, S., Weisemann, R., Bermel, W., Schwalbe, H., and Griesinger, C., 1995a, *J. Biomol. NMR* **6:**237.
Rexroth, A., Schmidt, P., Szalma, S., Geppert, T., Schwalbe, H., and Griesinger, C., 1995b, *J. Am. Chem. Soc.* **117:**10389.
Ruiz-Cabello, J., Vuister, G. W., Moonen, C. T., van Gelderen, P., Cohen, J. S., and van Zijl,
P. C. M., 1992, *J. Magn. Reson.* **93:**151.
Sattler, M., Schwalbe, H., and Griesinger, C., 1992, *J. Am. Chem. Soc.* **114:**1126.
Saudek, V., Piotto, M., and Sklenar, V., 1994, *Bruker Rep.* **140/94:**6.
Schmidt, J., 1997, *J. Magn. Reson.* **124:**310.
Schmidt, J. M., Ernst, R. R., Aimoto, S., and Kainosho, M., 1995, *J. Biomol. NMR* **6:**95.
Schmieder, P., Thanabal, V., McIntosh, L. P., Dahlquist, F. W., and Wagner, G., 1991, *J. Am. Chem. Soc.* **113:**6323.
Schrauber, H., Eisenhaber, F., and Argos, F., 1993, *J. Mol. Biol.* **230:**592.
Schwalbe, H., Rexroth, A., Eggenberger, U., Geppert, T., and Griesinger, C., 1993, *J. Am. Chem. Soc.* **115:**7878.
Schwalbe, H., Marino, J. P., King, G. C., Wechselberger, R., Bermel, W., and Griesinger, C., 1994, *J. Biomol. NMR* **4:**631.
Seip, S., Balbach, J., and Kessler, H., 1992, *Angew. Chem. Int. Ed. Engl.* **31:**1609.

Seip, S., Balbach, J., and Kessler, H., 1994, *J. Magn. Reson. B* **104:**172.
Smith, L. J., Sutcliffe, M. J., Redfield, C., and Dobson, C. M., 1991, *Biochemistry* **30:**986.
Sørensen, O. W., 1990, *J. Magn. Reson.* **90:**433.
Sørensen, O. W., Eich, G. W., Levitt, M. H., Bodenhausen, G., and Ernst, R. R., 1983, *Prog. NMR Spectrosc.* **16:**163.
Szyperski, T., Güntert, P., Otting, G., and Wüthrich, K., 1992, *J. Magn. Reson.* **99:**552.
Tessari, M., Mariani, M., Boelens, R., and Kaptein, R., 1995, *J. Magn. Reson. B* **108:**89.
Titman, T., and Keeler, J. H., 1990, *J. Magn. Reson.* **89:**640.
Torda, A. E., Brunne, R. M., Huber, T., Kessler, H., and van Gunsteren, W. F., 1993, *J. Biomol. NMR* **3:**55.
van der Laan, J. M., Teplyakov, A. V., Kelders, H., Kalk, K. H., Misset, O., Mulleners, L. J., and Dijkstra, B. W., 1992, *Protein Eng.* **5:**405.
Vuister, G. W., and Bax, A., 1992, *J. Biomol. NMR* **2:**401.
Vuister, G. W., and Bax, A., 1993a, *J. Am. Chem. Soc.* **115:**7772.
Vuister, G. W., and Bax, A., 1993b, *J. Magn. Reson. B* **102:**228.
Vuister, G. W., and Bax, A., 1994, *J. Biomol. NMR* **4:**193.
Vuister, G. W., Wang, A. C., and Bax, A., 1993a, *J. Am. Chem. Soc.* **115:**5334.
Vuister, G. W., Yamazaki, T., Torchia, D. A., and Bax, A., 1993b, *J. Biomol. NMR* **3:**297.
Vuister, G. W., Kim, S.-J., Wu, C., and Bax, A., 1994, *J. Am. Chem. Soc.* **116:**9206.
Wang, A. C., and Bax, A., 1995, *J. Am. Chem. Soc.* **117:**1810.
Wang, A. C., and Bax, A., 1996, *J. Am. Chem. Soc.* **118:**2483.
Weisemann, R., Rüterjans, H., Schwalbe, H., Schleucher, J., Bermel, W., and Griesinger, C., 1994a, *J. Biomol. NMR* **4:**231.
Weisemann, R., Löhr, F., and Rüterjans, H., 1994b, *J. Biomol. NMR* **4:**587.
Wider, G., Neri, D., Otting, G., and Wüthrich, K., 1989, *J. Magn. Reson.* **85:**426.
Xu, R. X., Olejniczak, E. T., and Fesik, S. W., 1992, *FEBS Lett.* **305:**137.
Xu, R. X., Word, N., Davis, D. G., Rink, M. J., Willard, D. H., Jr., and Gampe, R. T., Jr., 1995, *Biochemistry* **34:**2107.
Yang, J.-X., and Havel, T., 1994, *J. Biomol. NMR* **4:**807.
Zhu, G., and Bax, A., 1992, *J. Magn. Reson.* **100:**202.
Zhu, L., Reid, B. R., and Drobny, G., 1995, *J. Magn. Reson A* **115:**206.
Zuiderweg, E. R. P., and Fesik, S. W., 1991, *J. Magn. Reson.* **93:**653.

7

Methods for the Determination of Torsion Angle Restraints in Biomacromolecules

C. Griesinger, M. Hennig, J. P. Marino, B. Reif, C. Richter, and H. Schwalbe

1. INTRODUCTION

The NMR spectra of isotopically labeled biomacromolecules provide a wealth of information about interatomic distances and angular geometries that can be used as conformational restraints in structure determination. NMR observables and restraints derived thereof can be categorized according to the underlying physical principles that define the interactions (Fig. 1). In general, the strength of these interactions depends on the relative orientation and/or distance between the interacting fields. The angular dependence of the interactions can be derived from quantum mechanical calculations or calibrated using appropriate model systems.

This review will focus on the determination of torsion angle restraints from coupling constants and from cross-correlated relaxation of multiple-quantum coherence. The determination of coupling constants is of particular importance in

---

C. Griesinger, M. Hennig, B. Reif, C. Richter, and H. Schwalbe • Institut für Organische Chemie, Johann Wolfgang Goethe Universität Frankfurt, D-60439 Frankfurt am Main, Germany **J. P. Marino** • Center for Advanced Research in Biotechnology, Rockville, Maryland 20850.

Biological Magnetic Resonance, Volume 16: Modern Techniques in Protein NMR, edited by Krishna and Berliner. Kluwer Academic / Plenum Press, 1999.

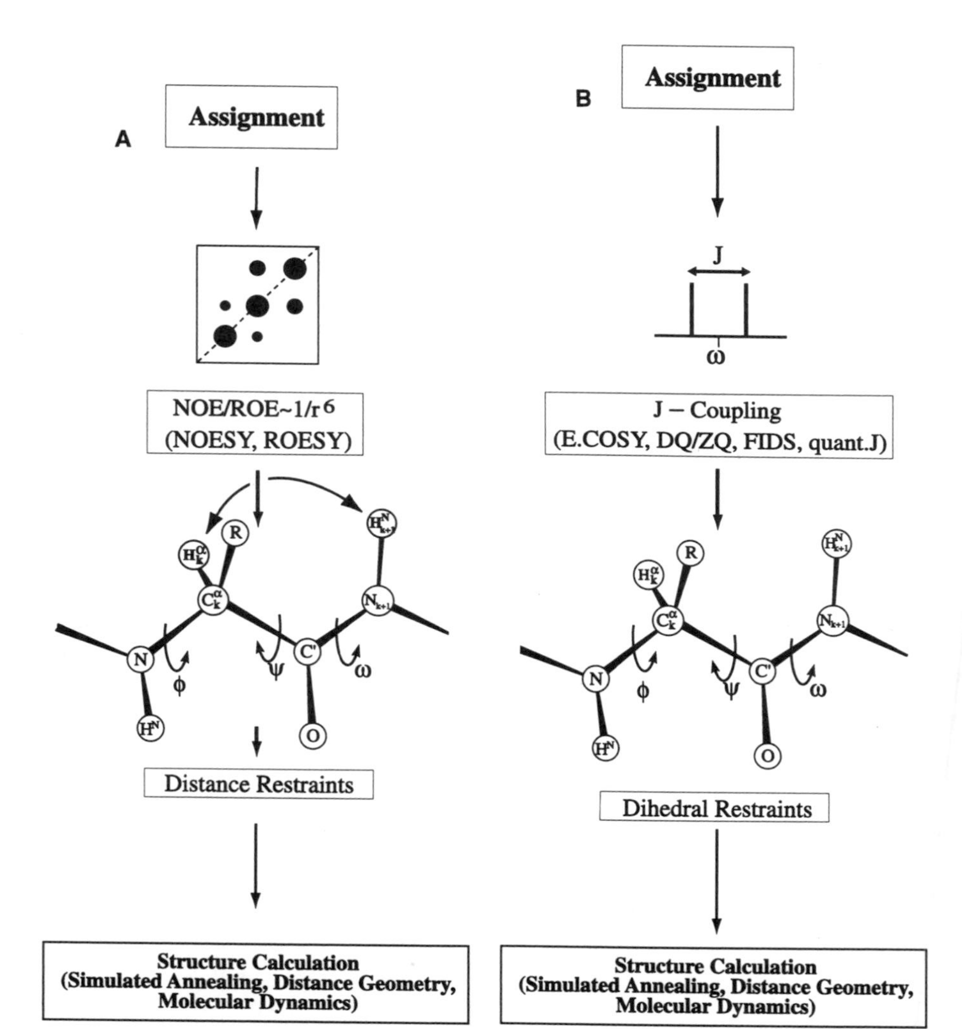
A
Assignment
NOE/ROE~1/r 6
(NOESY, ROESY)
R
N
C'
O
ϕ
ψ
ω
Distance Restraints
Structure Calculation
(Simulated Annealing, Distance Geometry, Molecular Dynamics)
B
Assignment
J
ω
J – Coupling
(E.COSY, DQ/ZQ, FIDS, quant.J)
R
N
C'
O
ϕ
ψ
ω
Dihedral Restraints
Structure Calculation
(Simulated Annealing, Distance Geometry, Molecular Dynamics)

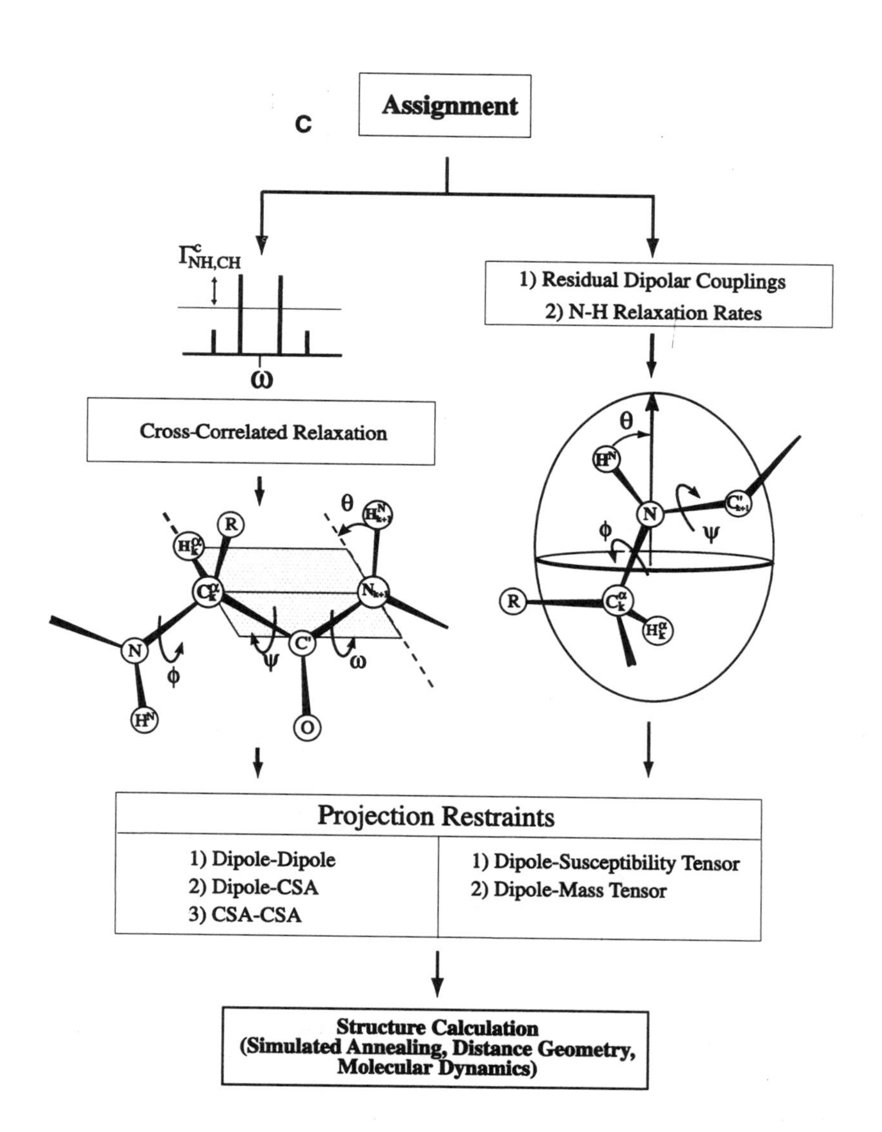
C
Assignment
$\Gamma^c_{NH,CH}$
ω
Cross-Correlated Relaxation
1) Residual Dipolar Couplings
2) N-H Relaxation Rates
θ
ϕ
ψ
ω
R
N
O
Projection Restraints
1) Dipole-Dipole
2) Dipole-CSA
3) CSA-CSA
1) Dipole-Susceptibility Tensor
2) Dipole-Mass Tensor
Structure Calculation
(Simulated Annealing, Distance Geometry,
Molecular Dynamics)

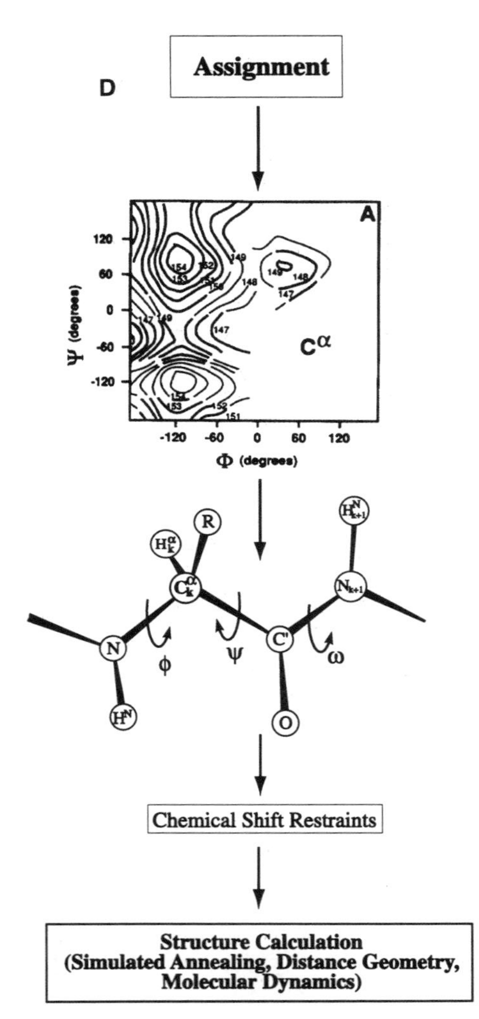

Figure 1. Structure determination protocol. In addition to the distance parameter, the NOE, all other parameters, *J* couplings, chemical shifts, residual dipolar couplings, anisotropic NH-relaxation analysis, and cross-correlated relaxation of multiple quantum coherence yield angular information that can be incorporated as restraints into structure calculation.

macromolecular systems that are insufficiently defined by NOE data alone because of either complex conformational averaging processes or the absence of protons. Although perdeuterated proteins are a typical example where the number of observable NOEs is too small to precisely define the overall protein fold, RNA molecules can also present the same problem of ill definition related to the absence of NOEs and complex conformational averaging processes in non-canonical secondary structure elements.

1.1. Angular Dependence of NMR Observables

1.1.1. Coupling Constants

The magnitude and sign of homo- and heteronuclear ${}^{n}J$ coupling constants reflect the local bonded geometry (Bystrov, 1976). To good approximation, the magnitude of ${}^{3}J$ coupling constants depends only on the torsion angle γ between the two spins under consideration. The functional dependence can be expressed using Karplus equations (Karplus, 1959, 1963) of the form ${}^{3}J(\gamma) = A \cdot \cos^2(\gamma) + B \cdot \cos(\gamma) + C$, for which A, B, and C are parameters in Hz that either are calculated from quantum-mechanical calculations or are parametrized empirically. In the last few years, the majority of pulse sequences designed to measure coupling constants have focused on the determination of vicinal coupling constants because their torsion angle dependence is best understood. Recently, however, new interest in the measurement of both geminal (${}^{2}J$) and direct (${}^{1}J$) coupling constants has arisen primarily for two reasons: (1) Heteronuclear ${}^{2}J$ coupling constants can provide valuable information for making stereospecific assignment (Schwarcz *et al.*, 1975; Cyr *et al.*, 1978; Hines *et al.*, 1993, 1994; Marino *et al.*, 1996) and (2) field dependence of ${}^{1}J$(N,H) or ${}^{1}J$(C,H) coupling constants caused by orientation related either to magnetic anisotropic susceptibility (Lohman and MacLean, 1978a,b, 1979; Bastiaan *et al.*, 1987; Tolman *et al.*, 1995; Kung *et al.*, 1995; Tjandra *et al.*, 1996, 1997; Tolman and Prestegard, 1996; Tjandra and Bax, 1997a) or to anisotropic reorientation in liquid crystals (Tjandra and Bax, 1997b) of the macromolecules relative to the static B_0 field can be exploited to derive new long-range structural restraints. There are five basic principles that underlie NMR methodologies currently available for the determination of torsion angle restraints in biomolecules: E.COSY, DQ/ZQ, FIDS/*J* modulation, quantitative *J*, and cross-correlated relaxation. These principles will be introduced and then recent applications to ^{13}C,^{15}N- and ^{13}C,^{15}N,^{2}H-labeled proteins and ^{13}C,^{15}N-labeled RNA will be discussed.

1.1.1.1. E.COSY Principle. E.COSY experiments (Griesinger *et al.*, 1985, 1986, 1987; Schwalbe *et al.*, 1995) rely on the *e*xclusive *c*orrelation *s*pectroscop*y* of spin states. In describing the principle of the E.COSY experiment, we will consider a spin system of three mutually coupled spins A, B, and C. If A and B are correlated such that the polarization of the third spin C remains unaffected (or is inverted), then the AB cross-peak consists of only two of four possible submultiplets, for which C is either in the α state $C^{\alpha}=1/2(1+2C_z)$ or in the β state $C^{\beta} = 1/2(1-2C_z)$. In the 2D experiment, the two pairs of frequencies $\Omega_A + \pi J(A,C)$ in ω_1, $\Omega_B + \pi J(B,C)$ in ω_2, and $\Omega_A - \pi J(A,C)$ in ω_1, $\Omega_B - \pi J(B,C)$ in ω_2, are correlated. The 2D frequency shift of the submultiplets is given by the displacement vector $\mathbf{J}_C$ [C for the passive spin C with unchanged (or inverted) spin states] with the components J(A,C) in ω_1 and J(B,C) in ω_2. As can be inferred from Fig. 2 and

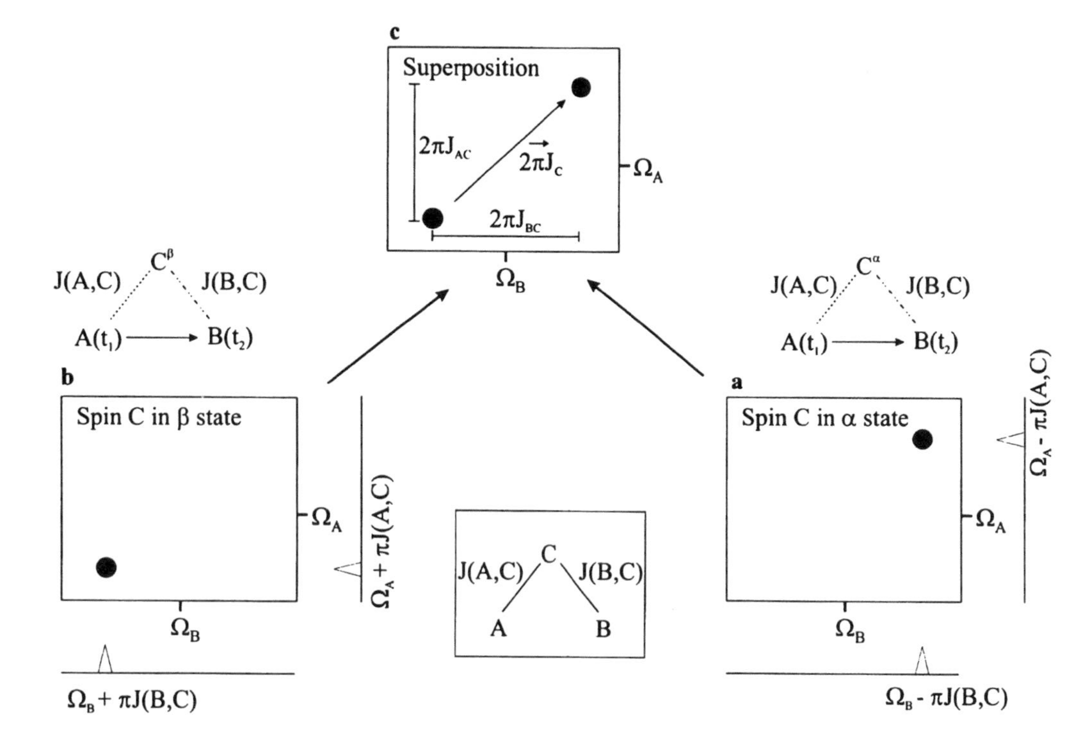

Figure 2. Correlation of a spin A that evolves during t_1 and B that evolves during t_2 without touching a third spin C. The multiplet pattern in (c) can be understood as a superposition of two spectra originating from molecules with C in the α state (a) and C in the β state (b). The displacement vector between the two submultiplets consists of the J(A,C) coupling in ω_1 and the J(B,C) coupling in ω_2.

neglecting for the moment relaxation effects, the size of a small coupling J(B,C) can be determined accurately provided that the *associated* coupling constant J(A,C) is larger than the resolution in ω_1. Furthermore, the relative signs of the two couplings can be determined from the orientation of the displacement vector $\mathbf{J}_C$. The $\mathbf{J}_C$ vector will point to the upper right (upper left) or lower left (lower right), if the signs of J(A,C) and J(B,C) couplings are the same (opposite).

For the design of E.COSY experiments a graphical representation of the three essential spins in an E.COSY triangle is helpful. In the corners of the triangle are the three essential spins that give rise to the cross-peak pattern shown in Fig. 2. Spins A and B are correlated by an arbitrary mixing process and need not be scalar coupled. The spin states of C must either be the same or inverted during the evolution periods t_1 and t_2.

1.1.1.2. DQ/ZQ Principle. Relaxation effects caused by differential relaxation of in-phase and antiphase operators, which depend on the geometry of the spin system, can lead to systematic errors in the interpretation of the cross-peak multiplets observed in an E.COSY-type experiment. The effects of so-called differential relaxation (Abragam, 1961; Harbison, 1993; Norwood, 1993) depend on the longitudinal relaxation rate of the passive spin C and are usually small when C is a heterospin. Differential relaxation can also be caused by so-called self-decoupling of the passive spin C related to exchange processes, for example. When the passive spin C is a proton, differential relaxation can alter the multiplet pattern such that a rigorous interpretation requires simulations that take both autocorrelated and cross-correlated relaxation pathways into account (Carlomagno *et al.*, in preparation). From the formula derived to quantify the effect of differential relaxation on J and its Taylor series expansion one obtains for $J >> \Delta\rho/2\pi$:

$$J^{\text{eff}} = \sqrt{J^2 - \left(\frac{\Delta\rho}{2\pi}\right)^2} = J - \frac{\Delta\rho^2}{(2\pi)^2(2J)} \tag{1}$$

where $\Delta\rho$ is the difference of the relaxation times for antiphase and in-phase transverse coherences. As the absolute error of J, $(J - J^{\text{eff}})$ is inversely proportional to the magnitude of J, the larger the value of J, the smaller the error introduced by differential relaxation. To obtain accurate measurement of small coupling constants in systems affected by the phenomenon of differential relaxation, it would therefore be advantageous to measure the small couplings as a function of the modulation of a much larger associated coupling. In this respect, the measurement of coupling constants as a function of the modulation of a large coupling associated with double and zero quantum coherence excited in an evolution period is an attractive alternative to the E.COSY experiment, whenever the passive spin C is a proton. Double and zero quantum coherence (called DQ/ZQ coherence) (Rexroth *et al.*, 1995) (Fig. 3) between A and B can be excited in a spin system A,B,C where there is a large associated coupling J(A,C) and a small coupling J(B,C). DQ coherence between A

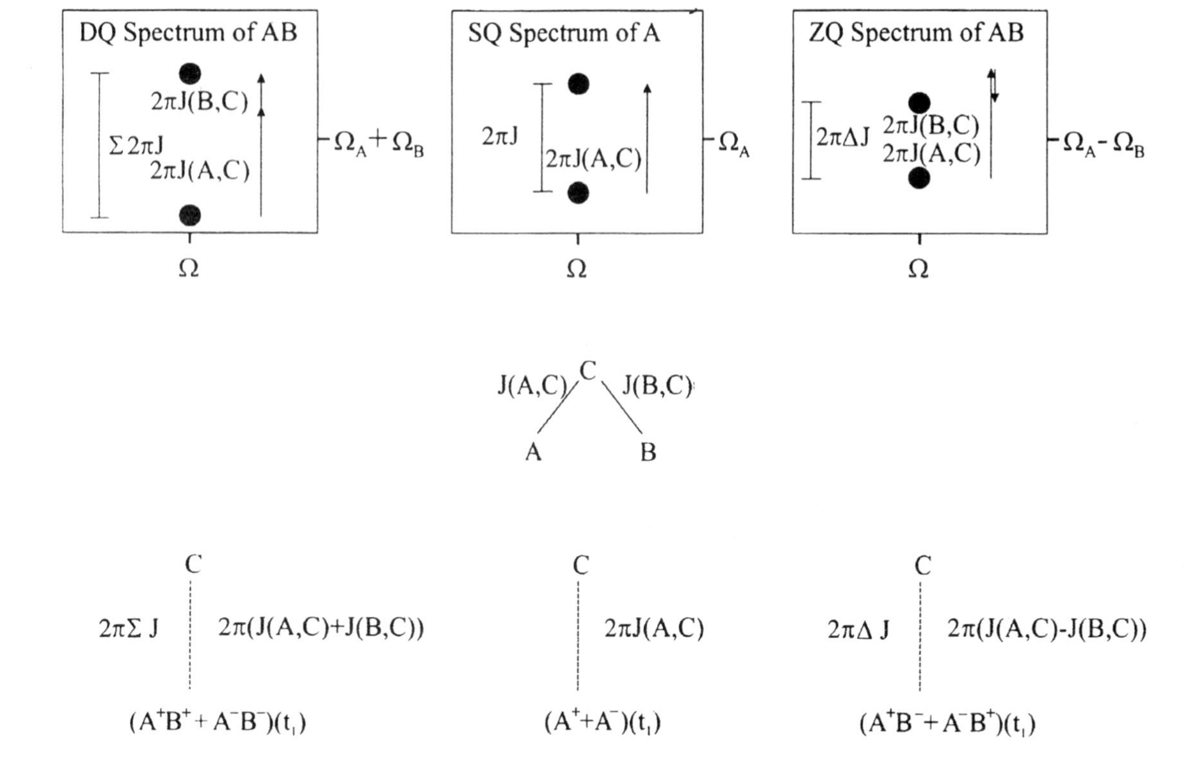

Figure 3. Determination of coupling constants from DQ, SQ, and ZQ spectra. Double quantum and zero quantum coherences of A and B are modulated by the coupling to a third spin C according to *J*(A,C)+*J*(B,C) and *J*(A,C)–*J*(B,C), respectively. Single quantum coherence of spin A evolves *J*(A,C). Recording of two of the spectra yields the desired coupling constants *J*(B,C) provided *J*(A,C) is resolved. Normally, J(A,C) will be a ^{1}J coupling. The detected nucleus is not specified as it is most often neither A nor B nor C.

and B evolves coupling to a third spin C as the sum of the two couplings J(A,C) + J(B,C), whereas ZQ coherence evolves as the difference of the two couplings J(A,C) – J(B,C). Adding or subtracting the splittings observed in the DQ and ZQ spectra yields either $2 \cdot J$(A,C) or $2 \cdot J$(B,C), respectively. Alternatively, DQ or ZQ coherence can be compared with SQ coherence. Information about the relative sign of the two couplings J(A,C) and J(B,C) is encoded in the relative size of the splitting in the DQ and in the ZQ spectrum, respectively if the splitting in the DQ spectrum is larger (smaller), the relative sign of J(A,C) and J(B,C) is the same (opposite). The DQ/ZQ principle is a 1D procedure for determining couplings, which offers advantages whenever both A and B are heterospins and cannot be detected for sensitivity reasons.

1.1.1.3. The FIDS Principle. Experiments based on the E.COSY and DQ/ZQ principles both rely on spin systems in which spin A is coupled to spin C via a large coupling that can be resolved in an evolution period. This yields in a three-spin system two well-resolved submultiplets from which couplings can easily be read off. These methods fail, however, if couplings to nuclei that are not coupled via a large coupling to another NMR-active nucleus are to be determined, e.g., quaternary carbons or phosphorus. To address the problem of determining coupling constants in these particular spin systems the FIDS (*f*itting of *d*oublets from *s*inglets) (Schwalbe *et al.*, 1993) procedure has been developed. The basic principle of this experiment is outlined in Fig. 4a. In a B,C spin system the multiplet of spin B contains the J(B,C) coupling constant among other couplings. This coupling can be determined by comparison of two spectra: a coupled spectrum in which the J(B,C) coupling evolves and a decoupled reference experiment in which the J(B,C) is removed. The procedure requires that each detected spin B be coupled at maximum to one unique spin C that is affected by the decoupling sequence. If this is the case, the spectra obtained in the coupled and the reference experiment should be identical except for the splitting induced by the coupling of interest, J(B,C). In the FIDS experiments, the coupled and decoupled experiment are connected via spectral convolution. Convolution of the decoupled reference spectra with a trial coupling $J(\mathrm{B,C})^{\mathrm{trial}}$ and minimizing the difference between coupled and convoluted reference spectrum yields J(B,C) in a fitting procedure that has only one variable. This procedure is also applicable if the isotopic abundance deviates from 100%, provided that the abundance is known.

A conceptually similar method to FIDS, which can also be used to measure couplings in these types of spin topologies, is the so-called J-modulated CT-HSQC experiment (Vuister *et al.*, 1993a, 1994). In the J-modulated CT-HSQC approach (Fig. 4b), two constant-time experiments are recorded: one in which the coupling of interest evolves for the constant-time period τ and the other in which the coupling is refocused. Integration and comparison of cross-peak intensities in the coupled and the decoupled experiment yields J(B,C). In both the FIDS and J-modulated methods, the transverse relaxation time of spin B does not need to be fitted and the

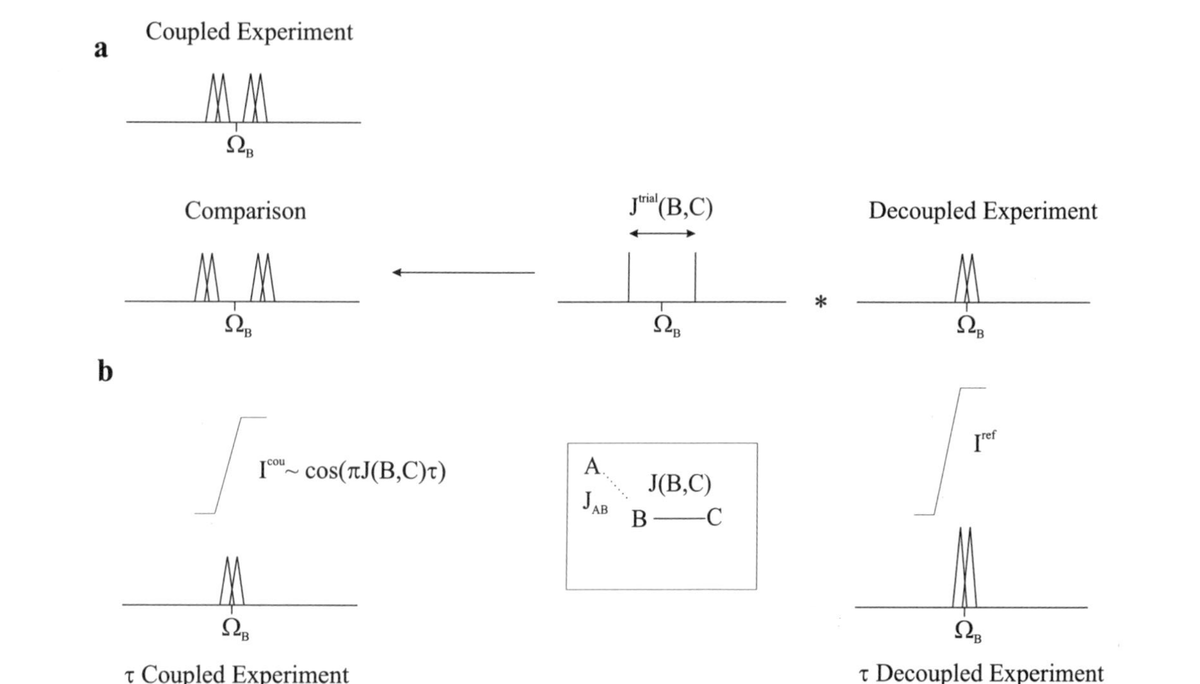

Figure 4. (a) FIDS method. The FIDS principle requires a spin B coupled to a spin C that can be selectively decoupled. Any other spin A must not be affected by decoupling of C. Recording of a spectrum of B with and without decoupling of C then provides the reference and the coupled spectrum. Convolution of the reference multiplet with an in-phase doublet related to a trial coupling *J*(B,C) and fitting to the coupled spectrum yields the desired coupling. (b) Echo-difference method—the constant-time version of (a). In the reference experiment, the coupling is refocused during a constant time τ. In the coupled experiment, the desired coupling *J*(B,C) evolves during τ. The intensities and integrals then correspond to $I^{cou}/I^{ref} = \cos(\pi J(B,C)\tau)$.

requirements in terms of spectral resolution are the same. The main difference between the FIDS and *J*-modulated methods is that the FIDS method measures the coupling of interest during a free evolution period, whereas the *J*-modulated method measures the coupling during a constant time period. The FIDS method therefore enjoys the advantage of allowing the determination of J(H,X) coupling constants in the acquisition period of a multidimensional experiment by the simple use of on- or off-resonance decoupling of the X nucleus.

1.1.1.4. Quantitative J Correlation Spectroscopy. In quantitative J correlations (Fig. 5), a coupling of interest is extracted from quantitative evaluation of the ratio of cross-peak to diagonal peak intensities in an out-and-back correlation experiment between B and C (Bax *et al.*, 1992; Blake *et al.*, 1992). In the preparation period of the out-and-back correlation experiment, in-phase coherence B_x evolves to antiphase $2B_yC_z$ during τ, a subsequent $90°_x$(B,C) pulse converts these two operators to B_x and $2B_zC_y$ which evolve and give rise to a reference peak derived from B_x and a cross-peak ($2B_zC_y$) in ω_1. After the evolution period, coherences are transferred back to B and are finally detected on B or a coupled spin. The transfer amplitude of the reference peak depends on $\cos^2(\pi J(A,B)\tau)$, and the transfer amplitude of the cross-peak, on $\sin^2(\pi J(A,B)\tau)$. Further passive couplings to B contribute equally to both the reference peak and the cross-peak. The coupling of interest J(B,C) can then be determined from the ratio of the reference (I_{ref}) and cross-peak (I_{CP}) intensity. Quantitative J experiments have been developed that are slightly different depending on whether B and C are homo- (Fig. 5a) or heteronuclear (Fig. 5b).

For the FIDS methods, the *J*-modulation and the quantitative *J*-correlation spectroscopy differential relaxation related to longitudinal eigenrelaxation of spin C introduces a systematic error in the measured J(B,C) coupling. The error is small for $\pi J(B,C) > (T_{1C})^{-1}$. This requirement is difficult to fulfill for protons as passive spins C but easy for any heteronuclear spin whose T_1 relaxation time increases with increasing molecular weight.

1.1.2. Cross-Correlated Relaxation for the Measurement of Angles between Tensors

Angular information on molecules is not only encoded in through-bond scalar couplings, but can also be derived from two tensorial interactions via cross correlated relaxation of double and zero quantium (DQ/ZQ) coherences (Reif *et al.*, 1997b). Common tensorial interactions are dipole couplings and chemical shift anisotropy. The strength of the interaction depends on the projection angle of two tensors onto each other provided these tensors are related in a defined way to the molecular frame. The dipole tensor between two spins A^1 and A^2 (B^1 and B^2) is axially symmetric with the axis of symmetry collinear to the internuclear bond vector $\mathbf{A^1A^2}$. The characterization of the angular dependence of the interaction of

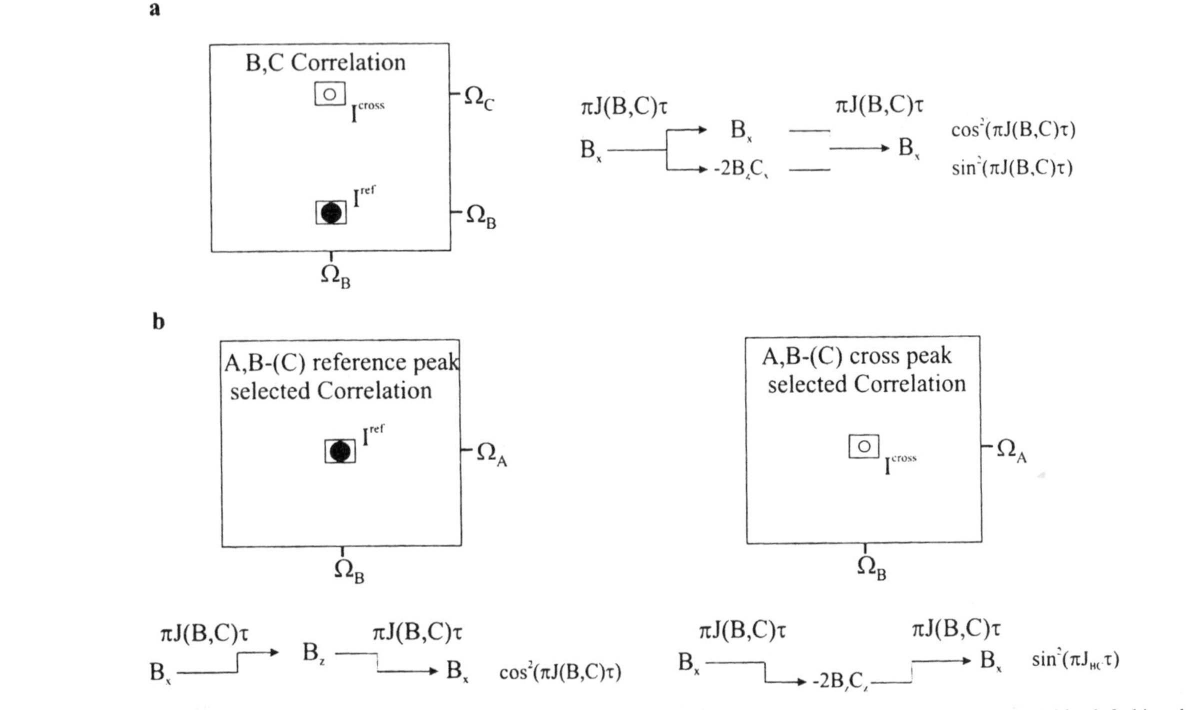

Figure 5. Quantitative *J*-correlation. In this experiment the desired coupling is used for coherence transfer from B to C in an out-and-back fashion during two constant time periods τ. The method is applicable for any spin topology. (a) B and C are of the same kind. The reference peak (B active in t_1) and the cross-correlation peak (C active during t_1) are modulated with $\cos^2(\pi JJ(B,C)\tau)$ and $\sin^2(\pi J(B,C)\tau)$, respectively. Coupling of B to other spins enters both transfer amplitudes in a similar manner and can be disregarded. (b) B and C are not of the same kind. In an A,B,C correlation experiment, the A,B reference peak (left) or the A,B,C cross-correlation (right) peak can be selected with appropriate phase cycling.

two dipole tensors $\mathbf{A}^1\mathbf{A}^2$ and $\mathbf{B}^1\mathbf{B}^2$ is therefore straightforward. The chemical shift anisotropy (CSA) tensor, which also can cause cross-correlated relaxation, is not known *a priori* and therefore needs to be determined experimentally or by quantum chemical calculations. The projections of tensors have been determined and interpreted in structural terms in solid-state *local field separated* (Hester *et al.*, 1976), spin diffusion (Dabbagh *et al.*, 1994), and multiple-quantum NMR spectroscopy (Schmidt-Rohr, 1996a,b; Feng *et al.*, 1996, 1997; Hong *et al.*, 1997). From the sideband pattern in solid-state spectra, the orientation of the two tensors with respect to each other can be derived.

In two pairs of nuclei (A^1–A^2 and B^1–B^2), projection-angle-dependent cross-correlated relaxation rates related to two dipolar couplings $\Gamma^{c}_{A^1A^2,B^1B^2}$ of double and zero quantum coherences between nuclei A^1 and B^1 can be measured provided the following three requirements are fulfilled: (1) The desired double and zero quantum coherence between nuclei A^1 and B^1 can be excited. (2) There is a resolvable

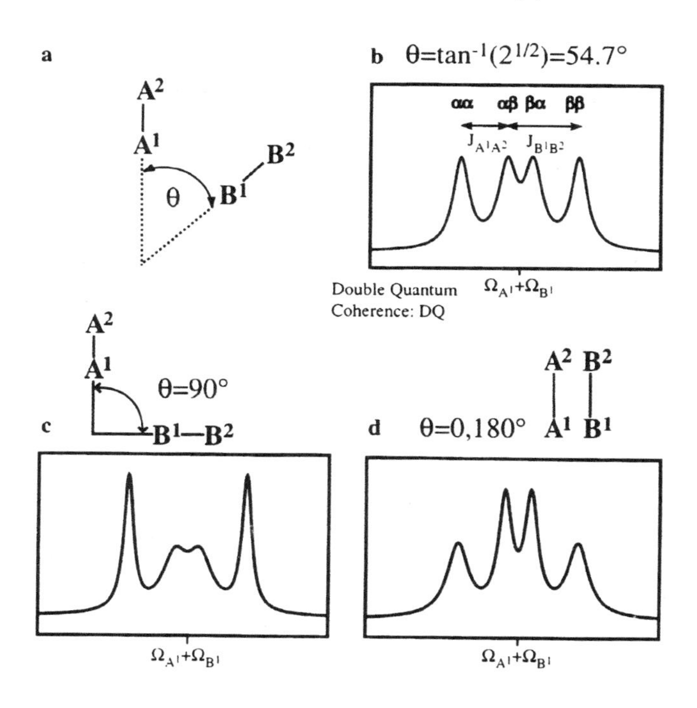

Figure 6. (a) Cross-correlated relaxation of double and zero quantum coherences. The requirement imposed on the spin system is that there are two pairs A^1–A^2 and B^1–B^2. The couplings $J(A^1,A^2)$ and $J(B^1,B^2)$ must be resolved. In addition, the dipolar relaxation A^1-A^2 and B^1–B^2 should be the main source of T_2 relaxation of A^1 and B^1, respectively. (b–d) Schematic multiplet pattern observed for intervectorial angles (θ) of 54.7°, 90°, and 180°. The cross-correlation of relaxation leads to different linewidths for the multiplet components.

coupling between A^1 and A^2 as well as between B^1 and B^2. (3) The main relaxation source for single quantum coherence of A^1 (SQC) is the dipolar coupling to A^2 and the main relaxation source for SQC of B^1 is the dipolar coupling to B^2.

To illustrate how cross-correlated relaxation can be used to measure the angle between two bond vectors, we will use the example of the generation of double and zero quantum coherence between spins A^1 and B^1 and call the angle between the A^1–A^2 and B^1–B^2 vectors, θ (Fig. 6).

In the experiment, double and zero quantum coherence between A^1 and B^1 evolve chemical shift $\Omega_{A^1}+\Omega_{B^1}$ and $\Omega_{A^1}-\Omega_{B^1}$ in an indirect detected period t_1 or t_2. A doublet of doublets of lines is generated with splittings related to scalar coupling of ${}^1J(A^1,A^2)$ and ${}^1J(B^1,B^2)$ if neither A^2 nor B^2 is decoupled during this indirect evolution period. In the absence of cross-correlated relaxation, all four multiplet components would have the same linewidth and intensities (Fig. 6b). α and β denote the polarization of A^2 and B^2, respectively. However, if one considers the effects of cross-correlated relaxation, the relative intensities of the lines will differ depending on the relative orientation of the two vectors. If the two vectors A^1–A^2 and B^1–B^2 are oriented orthogonal to each other, the inner two lines will be broader than the outer lines (Fig. 6c). The opposite is true for parallel orientation of the two vectors (Fig. 6d). Equal intensity for all four lines is also obtained if the two vectors span the *magic angle* $\theta = \text{arctg}(\sqrt{2}) \sim 54.7°$ (Fig. 6b).

2. DETERMINATION OF TORSION ANGLE RESTRAINTS IN RNA

The incorporation of ^{13}C and ^{15}N isotopes into RNA and DNA oligonucleotides has recently facilitated the development of a set of triple-resonance (^{1}H, ^{13}C, ^{15}N and ^{1}H, ^{13}C, ^{31}P) experiments designed for sequential through-bond assignment and structure determination (Batey *et al.*, 1992, 1995; Nikonowicz and Pardi, 1992; Nikonowicz *et al.*, 1992; Michnicka *et al.*, 1993; Quant *et al.*, 1994; Zimmer and Crothers, 1995; Agrofoglio *et al.*, 1997; Kainosho, 1997 and references therein; Smith *et al.*, 1997). The application of heteronuclear-based NMR experiments to oligonucleotides has provided many dramatic advantages that had previously been realized with isotopical labeling of protein samples. Uniform isotopic labeling of oligonucleotides now allows the unambiguous assignment of resonances using through-bond correlated NMR experiments, like the HCP (Heus *et al.*, 1994; Marino *et al.*, 1994a; Varani *et al.*, 1995), HCN (Sklenar *et al.*, 1993, 1994, 1996; Farmer *et al.*, 1993, 1994; Tate *et al.*, 1993; Simorre *et al.*, 1995, 1996; Marino *et al.*, 1997), and HCP-CCH-TOCSY (Marino *et al.*, 1994b; Wijmenga *et al.*, 1995), and limits reliance on the more ambiguous NOE-based assignment of sequential connectivity. In addition, the extraction of NOE-derived distance restraints is facilitated by the increased resolution obtained by combining NOESY experiments with evolution periods of heteronuclear chemical shifts (Aboul-ela and Varani,

1995; Varani *et al.*, 1996; Allain and Varani, 1997). Despite the high promise of isotopic labeling in RNA oligonucleotides, the dispersion of resonances obtained is usually poor for residues in canonical regions of secondary structure. Because of poor resolution, the gain in number of obtainable restraints from heteronuclear filtered NOESY experiments is typically smaller than for proteins. In fact, even for well-resolved residues the number of NOE-derived distance restraints is lower in oligonucleotides than in proteins of comparable size because of the low proton spin density. At the same time, the number of backbone degrees of freedom is larger. The incorporation of isotopic labels, however, also affords the possibility of measuring additional homo- and heteronuclear coupling constants and heteronuclear chemical shifts. The additional torsional restraints that can be derived from these measurements can help to define both the structural and dynamical properties of an oligonucleotide. Thus, the ambiguities caused by the lack of NOE-derived distances can be overcome to a large extent by the determination of torsion angles.

In this section, methods for the determination of coupling constants that rely on isotopic labeling of RNA and DNA oligonucleotides will be described. The focus of the review will be on the methodology of the experiments as the structural interpretation of the determined parameters has been reviewed elsewhere (van de Ven and Hilbers, 1988; Wijmenga *et al.*, 1994; Ippel *et al.*, 1996; Ramos *et al.*,

Table 1
Parametrization of Coupling Constants from the Literature

$$^3J(\mathrm{H,H}) = 13.24\cos^2\phi_{kl} - 0.91\cos\phi_{kl} + \Sigma_i\Delta\chi_i\,\{0.53 - 2.41\cos^2(\xi_i\phi_{kl} + 15.51\,|\Delta\chi_i|)\}$$

$$^3J(\mathrm{H,P}) = A\cos^2\psi(\mathrm{H,P}) + B\cos\psi(\mathrm{H,P}) + C$$

$$^3J(\mathrm{C,P}) = A\cos^2\psi(\mathrm{C,P}) + B\cos\psi(\mathrm{C,P}) + C$$

Coupling constant	Karplus parametrization	Reference
^3J(H1′,H2′)	$\Delta\chi_1 = 0.6980$, $\xi_1 = 1$	Haasnoot *et al.* (1981)
	$\Delta\chi_2 = 1.2240$, $\xi_2 = -1$	
	$\Delta\chi_3 = 1.3000$, $\xi_3 = 1$	
	$\Delta\chi_4 = 0.1625$, $\xi_4 = -1$	
^3J(H2′,H3′)	$\Delta\chi_1 = 0.0085$, $\xi_1 = 1$	
	$\Delta\chi_2 = 1.3000$, $\xi_2 = -1$	
	$\Delta\chi_3 = 1.3095$, $\xi_3 = 1$	
	$\Delta\chi_4 = 0.0770$, $\xi_4 = -1$	
^3J(H3′,H4′)	$\Delta\chi_1 = 0.0770$, $\xi_1 = 1$	
	$\Delta\chi_2 = 1.3095$, $\xi_2 = -1$	
	$\Delta\chi_3 = 1.2240$, $\xi_3 = 1$	
	$\Delta\chi_4 = 0.1530$, $\xi_4 = -1$	
^3J(H,P)	$A = 15.3$; $B = -6.1$; $C = 1.6$	Lankhorst *et al.* (1984)
^3J(C,P)	$A = 6.9$; $B = -3.4$; $C = 0.7$	Lankhorst *et al.* (1984)
^3J(C,P)	$A = 8.0$; $B = -3.4$; $C = 0.5$	Mooren *et al.* (1994)
^3J(C,P)	$A = 9.1$; $B = -1.9$; $C = 0.8$	Plavec and Chattopadhyaya (1995)

Table 2
Experiments for the Determination of Oligonucleotide Torsion Angles

Angle	Experiment	Described in section
α		
β	PFIDS	2.3.3
	Quantitative HCP	
γ	HCCH-E.COSY	2.2.1
	Forward-directed HCC-TOCSY-CCH-E.COSY	2.2.2
	C5′,H5′-selective HSQC	2.3.1
$\delta = \nu_3$	Forward-directed HCC-TOCSY-CCH-E.COSY	2.2.2
ε	PFIDS	2.3.3
	Quantitative HCP	
ζ		
χ	Refocused HMBC	2.3.2
ϕ_{12}	HCCH-E.COSY	2.2.1
$\phi_{12}, \phi_{23}, \phi_{34}$	Forward-directed HCC-TOCSY-CCH-E.COSY	2.2.2

1997; Marino *et al.*, in preparation). Experiments for the determination of homo- and heteronuclear coupling constants in oligonucleotides that define the sugar ring puckering, the conformation of the phosphodiester backbone, and the glycosidic angle χ will be presented. These experiments are especially useful for the characterization of noncanonical regions of RNA, in which fast conformational averaging is sometimes observed and the linear averaging of the coupling constants complements NOEs that reflect distances as $[1/<r^6>]^{-6}$.

2.1. Description of Conformations of Oligonucleotides

In oligonucleotides, the nucleotide phosphodiester backbone is defined by six backbone torsion angles ($\alpha, \beta, \gamma, \delta, \varepsilon, \zeta$) and the conformation around the glycosidic bond is defined by the torsion angle χ (Fig. 7) (Sänger, 1988).

The sugar pucker is defined by two parameters: the pseudorotation phase angle, P, and the pucker amplitude, ν_{max} [Kilpatrick *et al.*, 1947; Altona *et al.*, 1968; Altona and Sundaralingam, 1972; Haasnoot *et al.*, 1981; nomenclature according to *Eur. J. Biochem.* **131**:9–15 (1983)]. P and ν_{max} can be defined using ^3J(H,H) and ^3J(H,C) coupling constants (Fig. 8). ^3J(H,H) and ^3J(H,C) coupling constants can also be used to define the backbone angle γ and the glycosidic bond angle χ, while ^3J(H,P) and ^3J(C,P) coupling constants can be used to define the phosphodiester backbone angles, ε and β.

A rigorous interpretation of the measured coupling constants that involve H5′,H5″ protons in RNA and DNA requires that these diastereotopic protons first be stereospecifically assigned. Stereospecific assignment can be made based on the

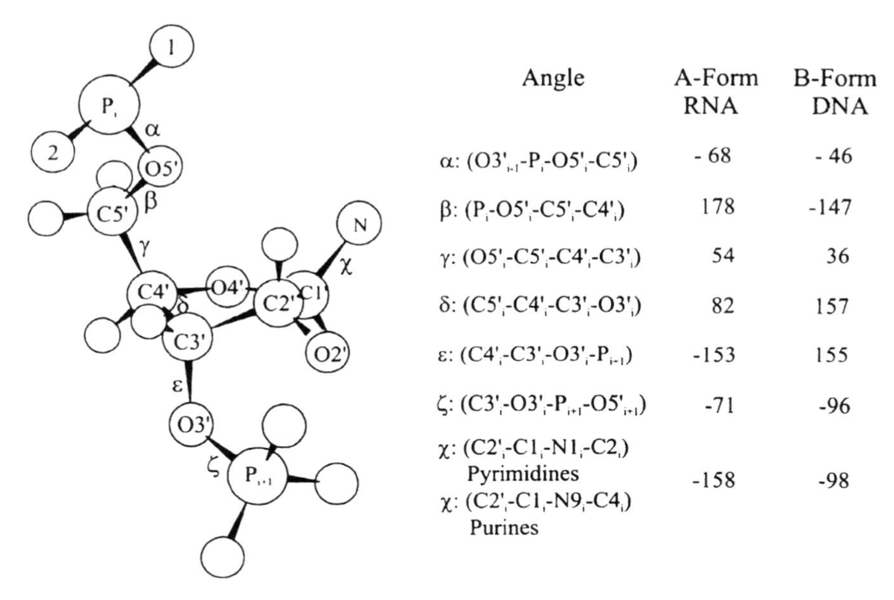

Angle	A-Form RNA	B-Form DNA
α: ($O3'_{i-1}$-P_i-$O5'_i$-$C5'_i$)	- 68	- 46
β: (P_i-$O5'_i$-$C5'_i$-$C4'_i$)	178	-147
γ: ($O5'_i$-$C5'_i$-$C4'_i$-$C3'_i$)	54	36
δ: ($C5'_i$-$C4'_i$-$C3'_i$-$O3'_i$)	82	157
ε: ($C4'_i$-$C3'_i$-$O3'_i$-P_{i+1})	-153	155
ζ: ($C3'_i$-$O3'_i$-P_{i+1}-$O5'_{i+1}$)	-71	-96
χ: ($C2'_i$-$C1_i$-$N1_i$-$C2_i$) Pyrimidines χ: ($C2'_i$-$C1_i$-$N9_i$-$C4_i$) Purines	-158	-98

Figure 7. Definition of exocyclic torsion angles. The conformation of the sugar phosphodiester backbone is defined by torsion angles α, β, γ, δ, ε, ζ, and χ in alphabetical order along the atoms of the backbone running from the 5′-end of the oligonucleotide to the 3′-end. The conformation around the glycosyl bond is defined by the torsion angle χ.

sign and magnitude of ^{2}J(H,C) coupling constants. In DNA, H2′,H2″ protons must also be stereospecifically assigned for a complete interpretation of coupling information that involves these proton spins. However, in most cases the stereospecific assignment of H2′,H2″ protons can simply be made based on chemical shift arguments.

2.2. Determination of Pseudorotation Phase P and Amplitude ν^{max} from ^{3}J(H,H) Couplings

2.2.1. The HCCH-E.COSY Experiment (Griesinger and Eggenberger, 1992; Schwalbe *et al.*, 1994)

In the HCCH-E.COSY experiment (Fig. 9a), ^{3}J(H,H) coupling constants can be determined in a spin system such as, ^{1}H1′–^{13}C1′–^{13}C2′–^{1}H2′, i.e., coupling constants in methine–methylene or methine-methine spin systems. The underlying experiment is a 2D (H)C(C)H-COSY experiment, in which coherence is transferred

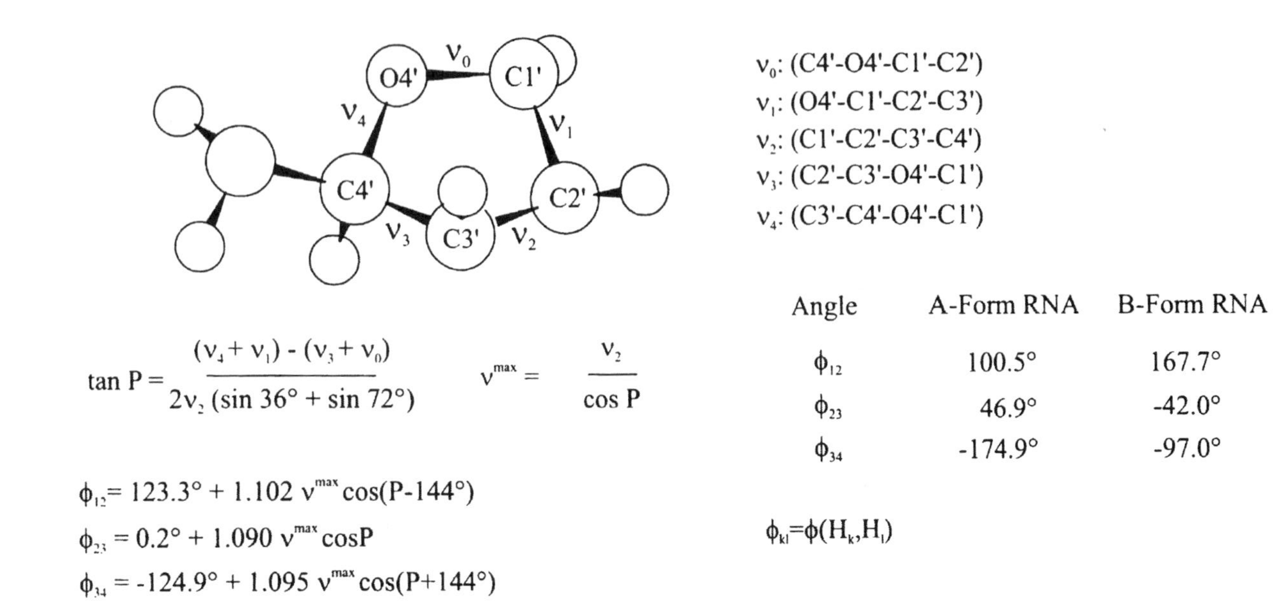

Figure 8. Definition of endocyclic torsion angles ν_0–ν_4. The conformation of the furanose ring can be described by two parameters, the pseudorotation amplitude ν^{max} and the pseudorotation phase *P*. *P* is defined via torsion angles between heavy atoms, while the conformation of the furanose ring can be assessed from ^3J(H,H) coupling constants by NMR. The ^3J(H,H) coupling constants can be translated into proton–proton torsion angles (ϕ_{kl} between H_k and H_l using the parametrization given in Table 1 and pseudorotation phase and amplitude are related to the proton–proton torsion angles as given in the figure (Haasnoot *et al.*, 1981).

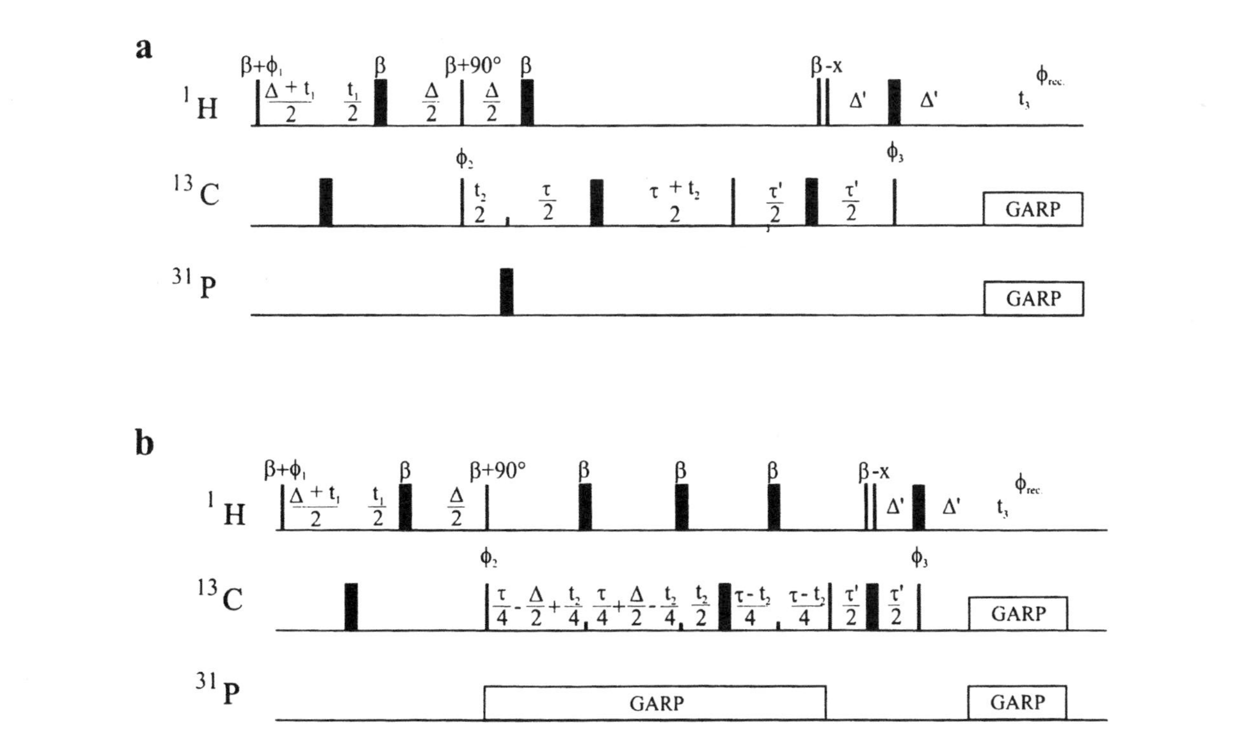

Figure 9. (a) Pulse sequence of the 3D HCCH-E.COSY. In all pulse sequences throughout this chapter, narrow and wide bars represent pulses of 90° and 180°, respectively. Filled half-egg-shaped pulses are selective 90° pulses, open half-egg-shaped pulses are selective 180° pulses. The default phase for a pulse is x. $\Delta = (2\,^1J(\mathrm{C,H}))^{-1}$, $\tau = (2\,^1J(\mathrm{C,C}))^{-1}$ for carbons with only one carbon bound, $\tau = 3/(4\,^1J(\mathrm{C,C}))$ for carbons with two carbons bound, $\tau' = (2\,^1J(\mathrm{C,C}))^{-1}$ or $\tau' = 1/(4\,^1J(\mathrm{C,C}))$ for carbons with one or two carbons bound, respectively. Δ and Δ' should be adjusted for varying $^1J(\mathrm{C,H})$ coupling constants in the ribose ring. $\phi_1 = x, -x$, $\phi_2 = x,x{-}x,-x,-x,-x$, $\phi_{rec} = x,-x,x,-x,x,x,-x$. (b) Same as for (a) but with suppression of C,H-dipole, CCSA cross-correlated relaxation.

sequentially for example from H1′ to H2′ in three steps that rely on the relatively large ^{1}J(C1′,H1′), ^{1}J(C1′,C2′) and ^{1}J(C2′,H2′) couplings, respectively. In the HCCH-E.COSY experiment, C1′ coherence created from H1′ coherence via a refocused INEPT transfer step evolves chemical shift and heteronuclear coupling to H1′ in t_1. The C1′ coherence is then defocused during τ to antiphase coherence $2C1'_xC2'_z$. Long-range ^{n}J(C,H) will be observed as in-phase splittings in t_1 and ^{n}J(C,C) couplings will attenuate the transfer efficiency each by $\cos(\pi\, {}^{n}J(C,C)\tau)$. Both interactions, however, have small effects during the short constant time period and will be neglected in the following discussion. τ is set to 1/(4J(C1′,C2′)) or 3/(4J(C1′,C2′)), which limits the resolution in t_1 to be $R = 1/t_1^{\max} = 1/\tau$ (see, however, Fig. 11). In the following relay step, antiphase coherence $2C1'_zC2'_y$ is refocused to $C2'_x$ and heteronuclear ^{1}J(C2′,H2) evolves to form $2C2'_yH2'_z$.

The ^{3}J(H1′,H2′) coupling constants can be determined from the C1′,H2′ (Fig. 11b or c) or C2′,H1′ generated cross-peak (Fig. 11b or d). In this experiment, as can be inferred from the E.COSY triangle (see Fig. 9), the starting proton spin (H1′ or H2′, respectively) serves as passive spin. The ratio of the connected transitions, in which the spin states of H1′ remain unaltered, to the nonconnected transitions depends on the flip angle β of the proton transfer pulse in the second DEPT transfer step (C2′ $\rightarrow$ H2′) (Olsen *et al.*, 1993). For an angle $\beta = 36°$, the ratio equals 10% (see also discussion in next paragraph). The transfer amplitude for the antiecho and echo transfer giving rise to the C1′,H2′ cross-peak is given by

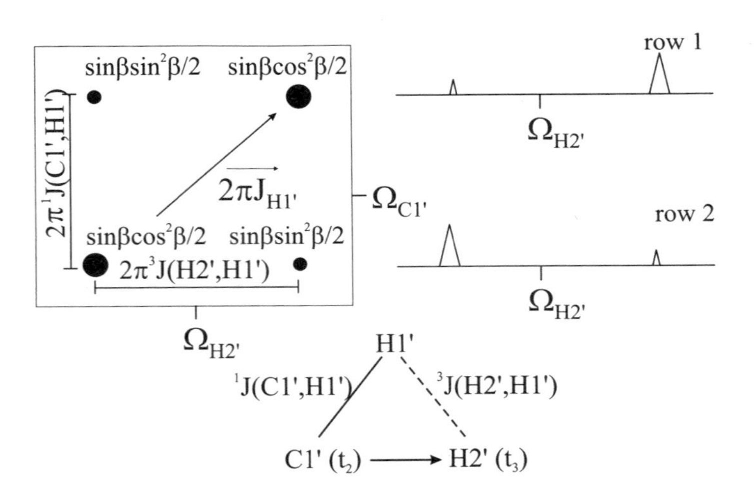

Figure 10. Schematic cross-peak pattern in the HCCH-E.COSY experiment. The coupling of interest can be determined from the relative displacement of multiplets in rows 1 and 2. As described in the text, the contributions of the nonconnected transitions depend on the angle β and can largely be suppressed during the process of coupling constant determination.

$$\sin^2 (\pi^1 J\,(\mathrm{H1',C1'})\,\Delta)\,\sin(\pi^1 J\,(\mathrm{C1',C2'})\,\tau)\,\sin(\pi^1 J\,(\mathrm{C2',C1'})\,\tau')$$
$$\cos(\pi^1 J\,(\mathrm{C2',C3'})\,\tau')\,\sin^2(\pi^1 J\,(\mathrm{H2',C2'})\,\Delta')\,\exp(\pm\, \mathrm{i}\Omega \mathrm{C1'} t_1)\exp(\mathrm{i}\Omega \mathrm{H2'} t_2)$$
$$\mathbf{\cos(\pm\,\pi^1 J\,(C1',H1')\,t_1 + \pi^3 J\,(H1',H2')\,(t_2 + \Delta'))} \tag{2}$$

with the portion of the equation that is relevant for the E.COSY displacement in bold.

The requirement in E.COSY to leave the spin states of the starting proton spin unaffected can also be implemented using a weighted summation of experiments with different flip angles 45° and 135° as an alternative to using a single small flip angle $\beta = 36°$. In the single β flip angle version of the experiment, the ratio of transfer efficiency between nonconnected and connected transitions equals at maximum $\tan^2\beta/2$. The appearance of nonconnected transitions introduces unsymmetrical signal components that systematically shift the submultiplets together, which leads to a systematic reduction in the size of the coupling constant. There are two ways to solve this problem. The first method requires the recording of two experiments with flip angles of 44.2° and 135.8° and weights of 6 and −1, respectively. The second method records a single experiment with $\beta = 36°$ and then utilizes a postacquisition processing procedure to remove the undesired component related to coherence transfer between nonconnected transitions. This postacquisition method multiplies row 1 (row 2) by $\tan^2\beta/2=0.11$ and subtracts the resulting row from row 2 (row 1). This scales the undesired multiplet component to $\tan^4\beta/2 = 0.01$ and therefore considerably reduces the contributions from the nonconnected transitions (see Fig. 10).

For the extraction of precise coupling constants without introducing biased selection criteria, the following procedure is applied. ω_1-summation to increase the signal-to-noise yields 1D rows that are displaced in ω_2 because of the ^3J(H,H) coupling. After inverse Fourier transformation and zero filling to increase the digitization, row 2 is shifted by an incremented frequency shift and squared after subtraction from row 1. The power integral over this difference spectrum versus the shift in hertz is plotted for corresponding cross-peaks (C1′,H2′ and C2′,H1′, for example). Because the coupling of interest evolves for $\cos(\pi^1 J(\mathrm{C1',H1'})t_1 + \pi^3 J(\mathrm{H1',H2'})(t_2 + \Delta'))$ during the experiment, there is an inherent phase distortion in the E.COSY multiplets. The acquired phase difference between the upper and lower trace $\Delta\phi$ depends on the size of the desired ^3J(H,H) coupling and the refocusing delay as follows: $\Delta\phi = 360° \cdot {}^3J(\mathrm{H1',H2'})\,\Delta$, which is taken into account in the coupling constant determination routine (Schwalbe *et al.*, 1994). The rms deviation of the ^3J(H,H) coupling is determined in the following manner: The integration region of the power difference spectrum is varied. The rms of the noise is then added to the minimum of the error integral. Varying the integration region over the power difference spectrum provides a noise-dependent measure for the minimum determination. The resulting error bar on the value of the minimum is

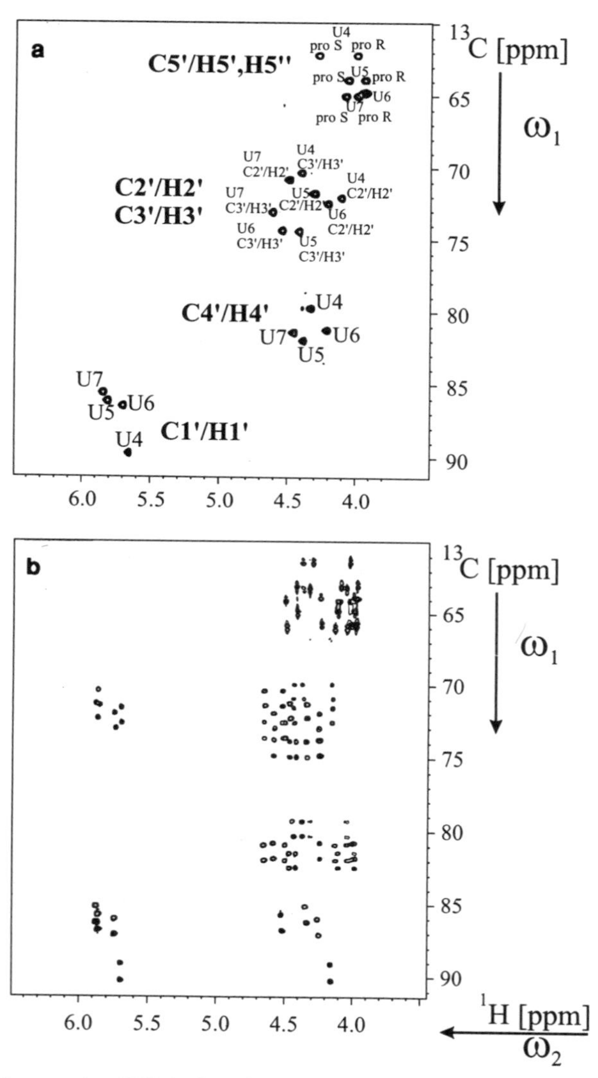

Figure 11. (a) Constant-time HSQC of a 1.5 mM RNA sample 5′-CGCUUUUGCG-3′ in which the four uridine residues are labeled with ^{13}C in the ribose ring. (b–d) 2D (H)C(C)H-E.COSY experiments with (b) $\tau = 3/(4\,^1J(C,C))$, $\tau' = 1/(4\,^1J(C,C))$ showing all correlations

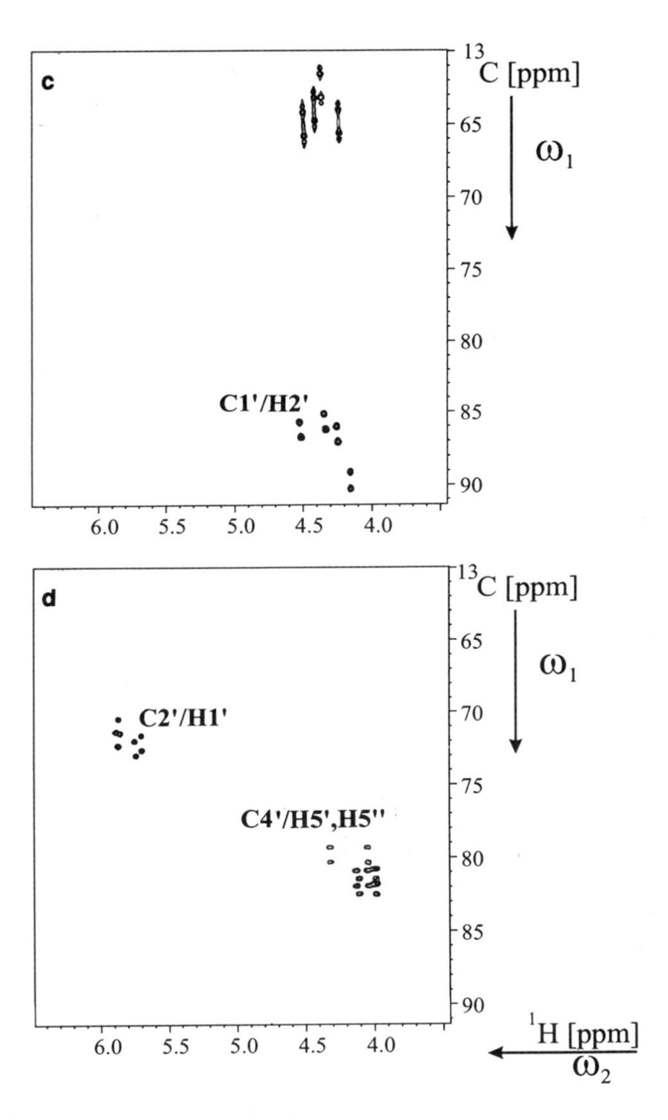

Figure 11. (c) $\tau = 1/(2^1J(C,C))$, $\tau' = 1/(4^1J(C,C))$ showing C1′,H2′ correlations; and (d) $\tau = 3/(4^1J(C,C))$, $\tau' = 1/(2^1J(C,C))$ showing C2′,H1′and C4′,H5′/H5″ correlations. Each 2D HCCH-E.COSY was recorded for 4 h.

translated into an error in the coupling constant determination by horizontal extrapolation.

Figure 11b–d shows the application of the HCCH-E.COSY experiment to a 10mer RNA 5′-GCGUUUUCGC-3′, in which the carbon atoms of the ribosyl ring in the four uridine residues are uniformly ^{13}C labeled.

All ^3J(H1′,H2′) coupling constants can be determined from inspection of the C1′,H2′ and the C2′,H1′ cross-peak region (Fig. 11). The resolution in this cross-peak region is also fair in larger RNAs. Quantification of the ^3J(H1′,H2′) coupling constants yields the assignment of C3′-*endo* puckering for U4, and C2′-*endo* puckering for U5 and U7. The ^3J(H1′,H2′) coupling constant for U6, however, is not in agreement with either canonical puckering mode. To determine a non-canonical conformation or conformational equilibria, all ^3J(H,H) couplings in the ribose ring need to be determined (see Section 2.2.2). A modified and improved version of the 2D and 3D HCCH-E.COSY experiment can also be applied to DNA oligonucleotides (Zimmer *et al.*, 1996). By applying 180° pulses during the constant-time evolution period t_1, the effects of seemingly large cross correlated dipole–CSA relaxation in the DNA H1′–C1′–C2′–H2′/H2″ spin system can be suppressed (see Fig. 9b).

2.2.2. The *Forward-Directed* HCC-TOCSY-CCH-E.COSY Experiment for the Measurement of a Complete Set of ^3J(H,H) Coupling Constants in RNA (Schwalbe *et al.*, 1995b; Marino *et al.*, 1996; Glaser *et al.*, 1996)

The measurement of a complete set of ^3J(H,H) coupling constants even in ^{13}C-labeled oligonucleotides from 2D or 3D HCCH-E.COSY is difficult because of the severe chemical shift overlap of the ribose H2′, H3′, H4′, H5′, and H5″ protons. As H1′ resonances are the best resolved in RNA, creation of a 3D experiment by correlation of H1′ chemical shifts in an HCC-TOCSY step prior to the 2D HCCH-E.COSY experiment (discussed in Section 2.2.1) should help alleviate the chemical shift overlap problem. However, in practice it was observed that simple application of an optimal TOCSY mixing time of 24 ms derived for the assignment of all carbons in a linear chain of five atoms did not yield spectra with well-defined cross-peaks as shown in Fig. 12a. The C2′,H2′ autocorrelation peaks, for example, overlap with the C2′,H3′ cross-peaks.

However, by judicious choice of the mixing period in the TOCSY and the constant time period in the CCH-E.COSY part of the experiment, the transfer amplitudes in the linear spin system C1′–C2′–C3′–C4′–C5′ with equal ^1J(C,C) can be tuned such that in the CCH-E.COSY plane only a restricted set of *forward-directed* cross-peaks originating from operators C_n',H_{n+1}' can be observed, while autocorrelation peaks (C_n',H_n') and *backward directed* peaks (C_n',H_{n-1}') are largely suppressed. The pulse sequence is shown in Fig. 13 and the schematic cross-peak

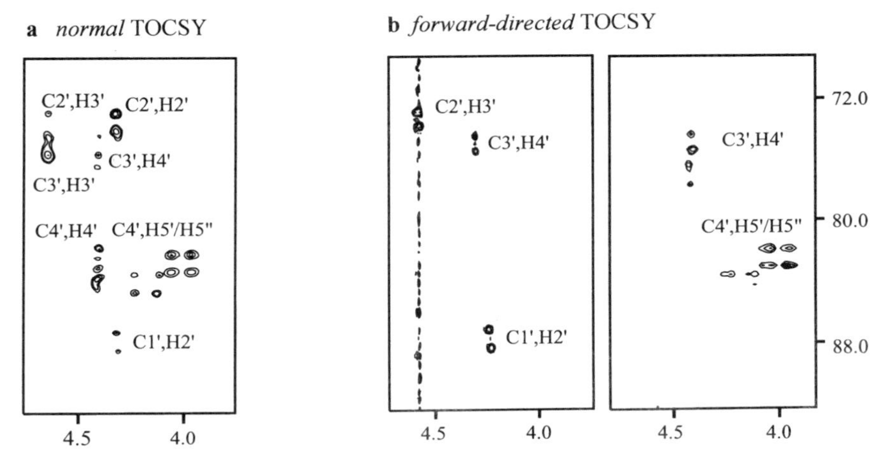

Figure 12. Comparison of the resolution and sensitivity of (a) a normal 3D HCC-TOCSY-CCH-E.COSY and (b) a forward-directed 3D HCC-TOCSY-CCH-E.COSY experiment. $\omega_1 = \Omega(^1\mathrm{H})$ 2D slices are shown.

pattern observed in the forward-directed TOCSY as compared with the nondirected TOCSY are presented in Fig. 14a,b. In Fig. 14a, the maximum number of possible cross-peaks is shown. The reduction in the number of cross-peaks in the forward-directed TOCSY simplifies the spectra without loss of information. This is because the autocorrelation peaks (C_n',H_n') do not contain any information on vicinal coupling constants, while the forward (C_n',H_{n+1}')- and backward (C_{n+1}',H_n')-directed cross-peaks have the same information content. The evolution of operators leading to forward- and backward-directed as well as autocorrelation peaks is shown in Fig. 15. The transfer amplitude during the latter CCH-E.COSY part of the sequence is essentially identical to the HCCH-E.COSY [Eq. (2)].

The mixing times in the forward-directed HCC-TOCSY-CCH-E.COSY can be optimized either to yield maximal sensitivity for the C1′,H2′, C2′,H3′, and C3′,H4′ forward-directed peaks (τ_I = 9.2 ms, τ_L = CT = 8.3 ms) for the determination of ^3J(H1′,H2′), ^3J(H2′,H3′), and ^3J(H3′,H4′), or to yield maximal sensitivity for the C3′,H4′ and C4′,H5′/H5″ peaks (τ_I = 13.5 ms, τ_L = CT = 7.6 ms) for the determination of ^3J(H3′,H4′) and ^3J(H4′,H5′/H5″).

Using this method, strikingly restrictive coherence transfer can be observed in linear spin systems with almost identical direct coupling constants as for example in ribose ring systems [^1J(C,C) = 40 Hz] or in amino acids with a linear spin system such as lysine, arginine, proline, or leucine side chains [^1J(C,C) = 35 Hz] (Eaton *et al.*, 1990; Glaser and Quant, 1996). Application of this experiment to the 10mer RNA allowed the determination of the ^3J(H1′,H2′), ^3J(H2′,H3′), and ^3J(H3′,H4′) coupling constants as summarized in Table 3.

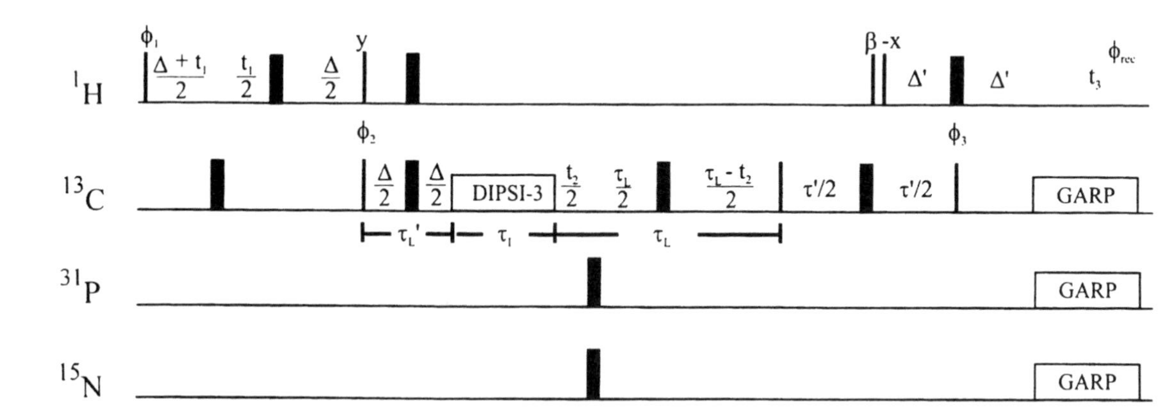

Figure 13. Pulse sequence of the forward-directed 3D HCC-TOCSY-CCH-E.COSY. $\beta = 45°$, $\Delta = (2^1J(\text{C,H}))^{-1}$, $\tau_I = 9.2$ ms, $\tau_L = 8.3$ ms to obtain C1′,H2′, C2′,H3′, and C3′,H4′ cross-peaks with maximum sensitivity, and $\tau_I = 13.5$ ms, $\tau_L = 7.6$ ms for the C3′,H4′ and C4′,H5′/H5″ cross-peaks for a uniform ^1J(C,C) coupling constant of 40 Hz, $\tau' = 1/(4^1J(\text{C,C}))$. $\phi_1 = x,-x$, $\phi_2 = x,x,-x,-x$, $\phi_3 = x,x,x,x,-x,-x,-x,-x$, $\phi_{rec} = x,-x,-x,x,-x,x,x,-x$.

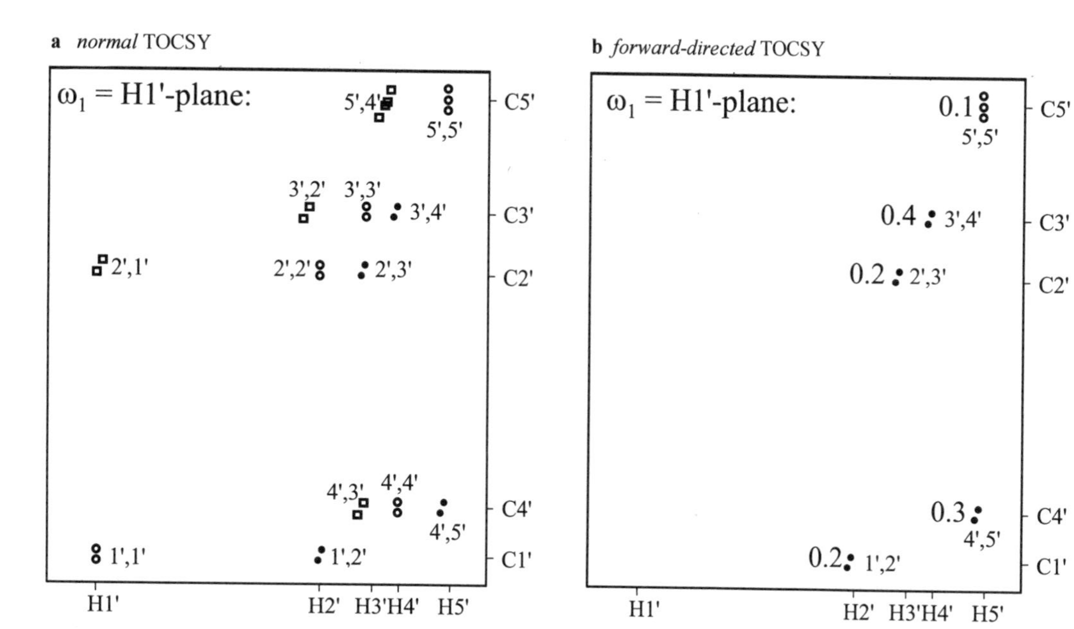

Figure 14. Schematic cross-peak pattern found in the $\omega_1 = \Omega(^1H)$ 2D slice of an RNA sugar moiety in (a) the normal 3D HCC-TOCSY-CCH-E.COSY spectrum (maximal number of possible cross-peaks) and (b) in the 3D forward-directed HCC-TOCSY-CCH-E.COSY experiment. Autocorrelation peaks are shown as open circles, filled circles represent forward-directed cross-peaks, and open squares represent backward-directed cross-peaks.

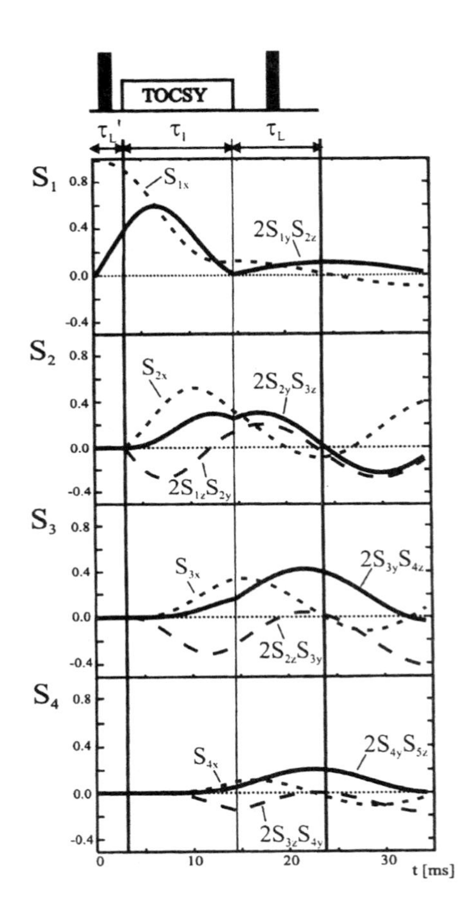

Figure 15. The evolution of coherences of interest in the forward-directed HCC-TOCSY-CCH-E.COSY as a function of the constant-time delay τ_L at an optimal TOCSY mixing time $\tau_I = 13.5$ ms and $\tau_L' = 3.2$ ms. Spins S_1–S_5 represent the linear spin system C1′–C5′ in the ribosyl ring of RNA with uniform ^1J(C,C) coupling constants of 40 Hz. Transfer amplitudes of forward-directed coherences ($2S_{iy}$, S_{i+1z}) are drawn as solid lines, while those for backward-directed ($2S_{i-1y}$,S_{iz}) and in-phase coherences (S_{ix}) are drawn as short and long dashed lines, respectively. S_i, represents the nucleus C_1' .

2.3. Determination of Exocyclic Torsion Angles

2.3.1. Determination of ^2J(C4′,H5′/H5″) for the Conformation around γ and the Stereospecific Assignment of Diastereotopic H5′/H5″ Protons

The H5′,H5″ protons in RNA oligonucleotides are diastereotopic. To determine the stereospecific assignment of the diastereotopic protons and the torsion angle γ, four coupling constants showing a differential dependence on the torsion angle γ must be determined. Newman projections of the three staggered rotamers around γ together with the coupling constant signature from ^3J(H4′,H5′/H5″) and ^2J(C4′,H5′/H5″) are shown in Fig. 17.

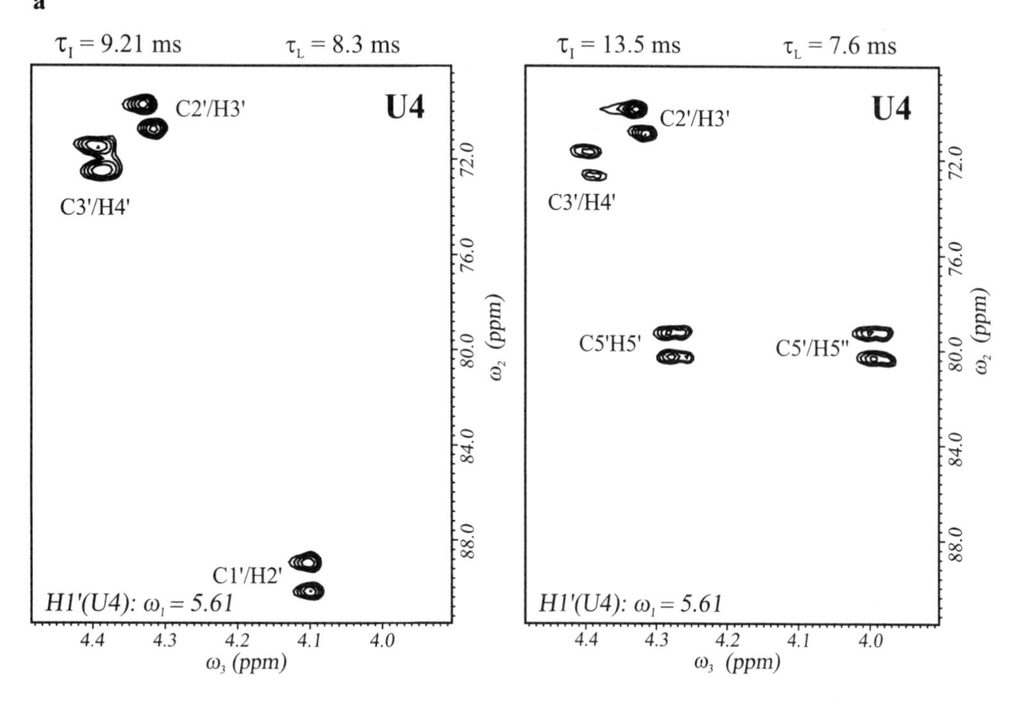

Figure 16. Forward-directed HCC-TOCSY-CCH-E.COSY of a 1.5 mM RNA sample 5′-CGCUUUUGCG-3′ in which the four uridine residues are labeled with ^{13}C in the ribose ring. U4 (a) and U7 (b) $\tau_I = 9.2$ ms, $\tau_L = 8.3$ ms to enhance the C1′,H2′, C2′,H3′, and C3′,H4′ cross-peaks (left panels), and $\tau_I = 13.5$ ms, $\tau_L = 7.6$ ms to enhance the C3′,H,4′ and C4′,H5′/H5″ cross-peaks (right panels). Each 3D spectrum was recorded for 48 h.

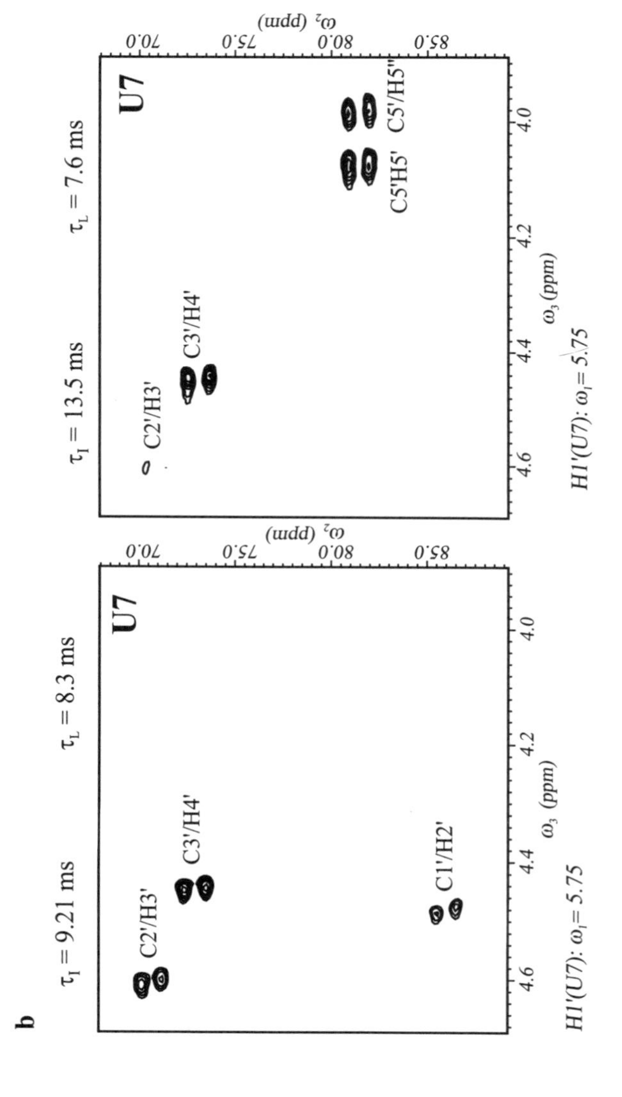

Figure 16. Continued

Table 3
Coupling Constants and Sugar Pucker as Derived from H,H Coupling Constants in 5′-CGCUUUUGCG-3′

	U4		U5		U6		U7	
^nJ(H,H) ^nJ(C,H)	Exp.	Pred.[a]	Exp.	Pred.	Exp.	Pred.	Exp.	Pred.
^3J(H1′,H2′)[b]	2.6 ± 0.3[c]	*2.29*	8.7 ± 0.1	*8.21*	6.8 ± 0.1	*7.97*	8.1	*8.24*
^3J(H2′,H3′)	5.2 ± 0.3	*5.53*	5.5 ± 0.1	*5.13*	5.4 ± 0.1	*6.67*	5.4 ± 0.3	*5.71*
^3J(H3′,H4′)	8.9 ± 0.2	*9.47*	1.6 ± 0.2	*1.97*	4.7 ± 0.1	*4.12*	3.1 ± 0.2	*2.87*
P^d; $\nu^{\max\, e}$; rmsJ[f]	44 44	0.42	144 43	0.42	123 42	1.07	134 44	0.24
% N[g]; rmsJ	93%	0.18	0%	0.42	32%	0.32	10%	0.40

[a]Predicted using parameters given in Haasnoot *et al.* (1981).
[b]Measured in Hertz.
[c]Couplings are the average of two independent measurements in different submultiplets.
[d]Pseudorotation phase.
[e]Pseudorotation amplitude.
[f]$\mathrm{rmsJ} = [\Sigma(J_{\mathrm{pred}} - J_{\mathrm{exp}})^2]^{1/2}/n^{1/2}$ measured in Hertz; n number of couplings.
[g]Two-state equilibrium with data for U4 and U5 representing pure N and S conformers, respectively. $\nu^{\max}$ is assumed to be 44°.

For the diastereotopic assignment of H5′,H5″ in RNA oligonucleotides, ^3J(H4′,H5′/H5″) coupling constants alone are insufficient for the stereospecific assignment in the $\gamma = -60°$ and $\gamma = +180°$ conformations as both have one large and one small coupling. However, as has been observed for ^2J(C,H) coupling constants

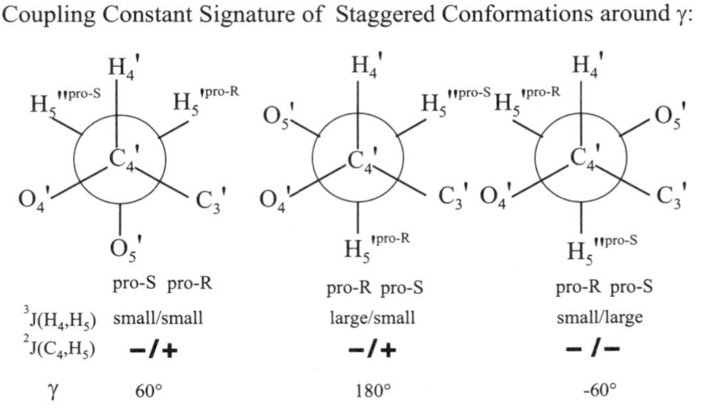

Figure 17. Newman projection around the backbone angle γ showing the three staggered rotamers with the predicted signs [^2J(C4′,H5′/H5″] and relative sizes [^2J(H4′,H5′/H5″] of coupling constants for the three rotameric states.

in model systems (Schwarcz *et al.*, 1975; Bock and Pederson, 1977; Cyr and Perlin, 1979) and by Hines *et al.* (1993, 1994) for the ^2J(C4′,H5′/H5″) coupling constants in RNA oligonucleotides, the sign dependence of the geminal ^2J(C4′,H5′/H5″) coupling constants can be used to assign the H5′,H5″ protons stereospecifically.

^2J(C,H) coupling constants have been determined successfully by Hines *et al.* (1993, 1994) in a Montelione–Wagner (Kessler *et al.*, 1988; Montelione *et al.*, 1989; Kurz *et al.*, 1991; Schmieder *et al.*, 1991; Zuiderweg and Fesik, 1991; Sattler *et al.*, 1992; Schleucher *et al.*, 1992) type E.COSY experiment that relies on a homonuclear correlation of H4′ with H5′,H5″, e.g., in a NOESY spectrum without touching the spin states of the C4′. A complete interpretation of the experiment is hampered by resonance overlap in the H4′,H5′/H5″ region. Complete stereospecific

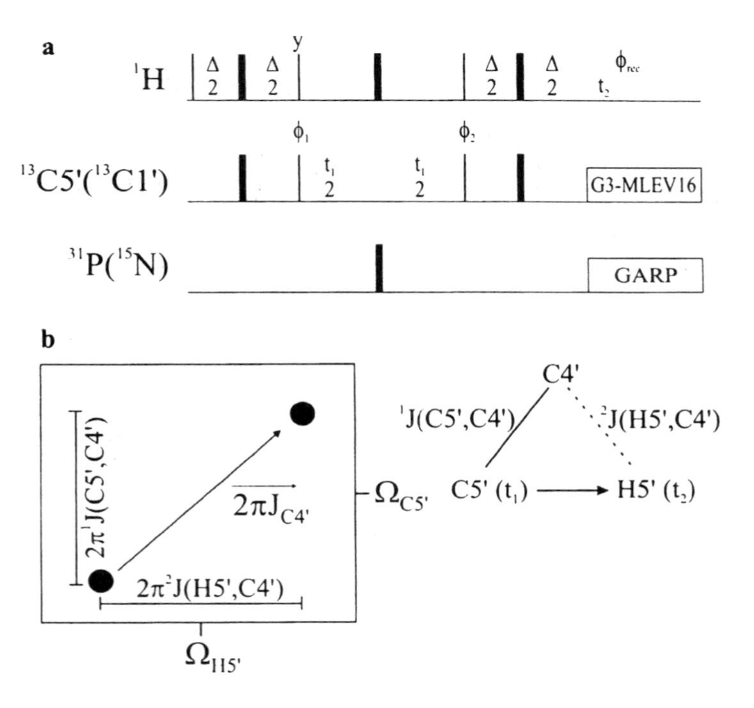

Figure 18. (a) Pulse sequence of the C5′,H5′ (Cl′,H1′) selective 2D ^{1}H, ^{13}C HSQC. The 90° pulses are selective for C5′(C1′) carbons. G4 and G4tr pulses as the first and second carbon 90° pulse had a length of 3.2 ms. The G3 pulses of 4.096 ms expanded according to MLEV16 have been used to selectively decouple C5′(C1′) without disturbing the spin states of C4′(C2′). $\Delta = (2^1J(\mathrm{C,H}))^{-1}$, $\phi_1 = x$, $\phi_2 = x,x,-x-x$, $\phi_{rec} = x,-x,-x,x$. (b) Schematic cross-peak pattern in the C5′ selective 2D ^{1}H, ^{13}C HSQC experiment.

assignments for all H5′,H5″ pairs in a 19mer RNA have been obtained from a set of 2D and 3D HSQC experiments using selective pulses on the C5′ resonances that leave the C4′ spin states untouched (Marino *et al.*, 1996). The basic experiment is a C5′,H5′ selective non-constant-time ^{1}H,^{13}C HSQC. G4 and time reverse G4 pulses (Emsley and Bodenhausen, 1989, 1990) are used as excitation pulses in the selective HSQC. The large ^1J(C5′,C4′) coupling constants of 40 Hz and smaller long-range J(C5′,C_i′) couplings evolve in ω_1 and lead to a splitting of the C5′,H5′ cross-peaks. Selective C5′ decoupling using a train of G3 pulses (Emsley and Bodenhausen, 1989, 1990) expanded with the MLEV16 (Levitt *et al.*, 1982) sequence for decoupling (Eggenberger *et al.*, 1992) decouple the C5′ region but leave the C4′ region untouched. The pulse sequence and the schematic cross-peak pattern are shown in Fig. 18.

The transfer function for the antiecho and echo part of the H5′,C5′ cross-peaks in the selective H5′,C5′ HSQC is given by

$$\sin^2(\pi {}^1J(\text{H5}',\text{C5}')\Delta \cos(\pi {}^2J(\text{H5}',\text{H5}'')\Delta) \cos(\pi {}^3J(\text{H5}',\text{H4}')\,\Delta)\, \Sigma_{i=1-4} \cos(\pi^{|i-n|}J(\text{C5}',\text{C}i')\, t_1 \cos(\pi {}^2J(\text{H5}',\text{H5}'')(\Delta + t_2)) \cos(\pi {}^3J(\text{H5}',\text{H4}')(\Delta + t_2)) \exp(\pm i\Omega_{\text{C5}'}t_1) \exp(i\Omega_{\text{H5}'}t_2) \cos(\pm \pi {}^1J(\text{C5}',\text{C4}')t_1 + \pi {}^3J(\text{C4}',\text{H5}')t_2) \quad (3)$$

In Fig. 19b, stereospecific assignments for U4, U5, and U7 are marked in the C5′,H5′ region of a C5′,H5′ selective HSQC. From the spread in chemical shifts in RNA oligonucleotides (Fig. 11a), it is apparent that selective correlation experiments based on the description above can also be achieved for the C1′,H1′ spin system, as it is another terminal carbon atom with only one directly bound carbon and the separation from the C2′ region is large enough to apply reasonably short selective pulses that cover the complete C1′ spectral width. From this experiment, ^2J(C2′,H1′) couplings can be extracted (Fig. 19a), which, however, do not strongly vary between C2′-*endo* and C3′-*endo* conformation.

2.3.2. Determination of ^3J(H1′,C8) and ^3J(H1′,C4) in Purine Nucleotides and ^3J(H1′,C6) and ^3J(H1′,C2) in Pyrimidine Nucleotides for the Conformation Around χ

For the determination of the local conformation around the glycosidic angle χ in oligonucleotides, ^3J(H1′,C) must be measured (Lemieux *et al.*, 1972; Davies *et al.*, 1985). In purines, both C8 and C4 are quaternary carbon atoms, and in pyrimidines, C2 is a quaternary carbon atom so that E.COSY methods cannot be applied to either nucleobase spin system because of the absence of a coupling partner with a large associated coupling constant. However, a quantitative J method can be applied to determine these couplings (Section 1.1.1.4). In a refocused HMBC (Bermel *et al.*, 1989) experiment, H1′ coherence can be correlated with the aromatic carbon atoms and quantification of cross-peak intensities allows the determination

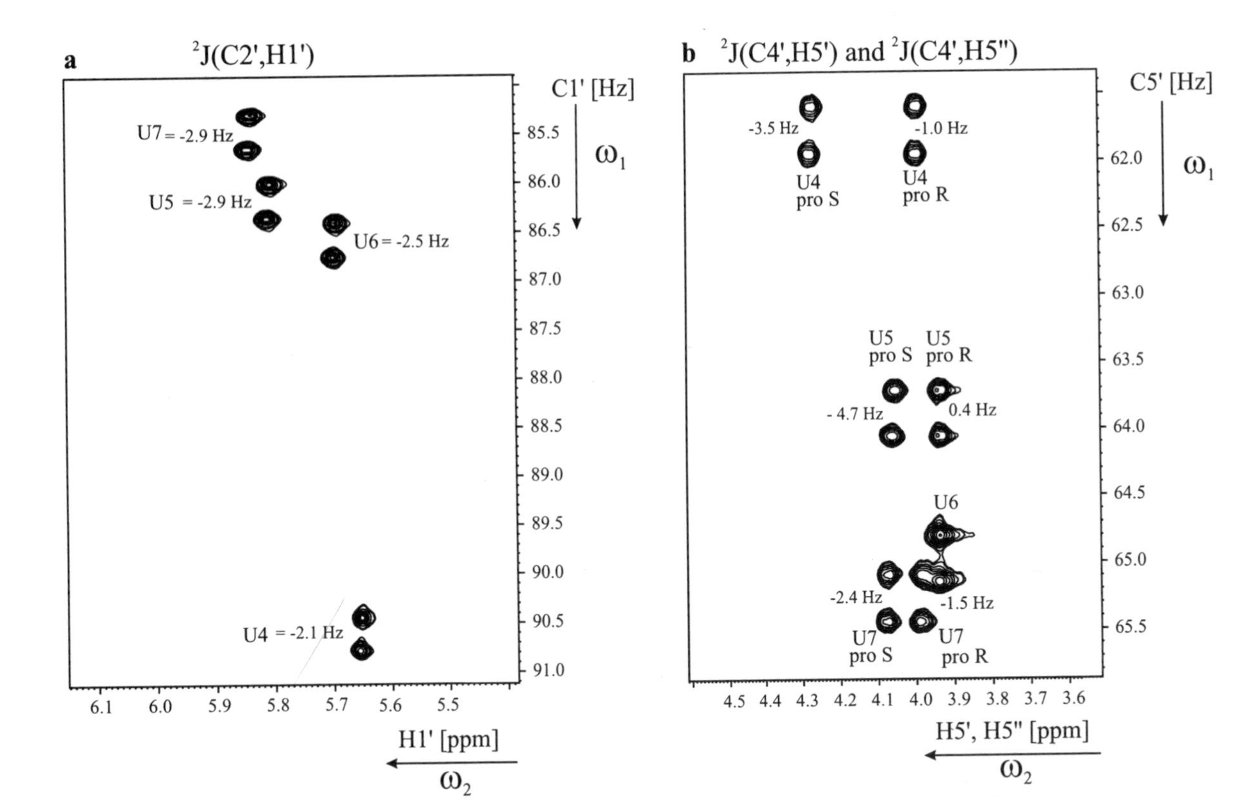

Figure 19. (a) C1′,H1′ selective 2D ^{1}H, ^{13}C HSQC and (b) C5′,H5′ selective 2D ^{1}H,^{13}C HSQC of a 1.5 mM RNA sample 5′-CGCUUUUGCG-3′ in which the four uridine residues are labeled with ^{13}C in the ribose ring. While the ^2J(C2′,H1′) are uniform for all four uridine residues and cannot be used for further conformational analysis, the stereospecific assignment of the H5′, H5″ can be derived from analysis of ^2J(C4′,H5′/H5″) together with the ^3J(H4′,H5′/H5″) coupling constants. Each 2D spectrum was recorded for 2 h.

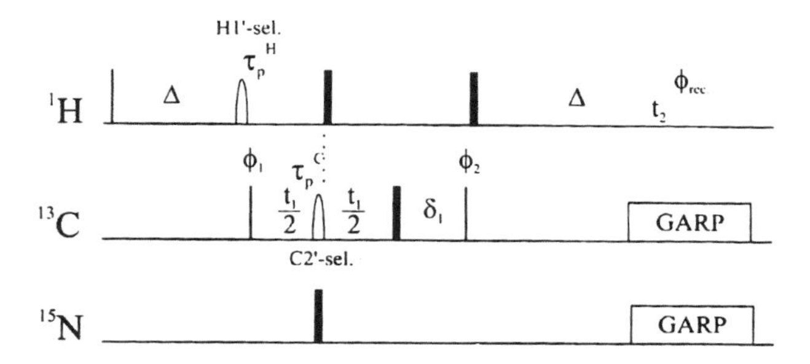

Figure 20. Pulse sequence for the refocused HMBC experiment. Δ = 25 ms. Open half-egg-shaped pulses denote selective pulses. $\delta_1 = \tau_p^{sel}(C2')$, $\phi_1 = x,-x$, $\phi_2 = x,x,-x,-x$, $\phi_{rec} = x,-x,-x,x$.

of the coupling constants rather accurately (Schwalbe *et al.*, 1994; Zhu *et al.*, 1994). An application has been shown for a ^{13}C,^{15}N-labeled DNA oligonucleotide (Zimmer *et al.*, 1996). The pulse sequence of this HMBC is shown in Fig. 20.

The selective pulses in the pulse sequence are applied to refocus the evolution of ^3J(H1′,H2′/H2″) (selective 180° pulse on H1′ protons) and to refocus the evolution of ^1J(C1′,C2′) during t_1 (selective 180° pulse on C2′ protons). The heteronuclear long-range coupling giving rise to the desired cross-peaks evolves for 2Δ, proton chemical shift Ω_H and carbon chemical shift is refocused for $t_1(0)$. Integration of cross-peaks H1′,C1′ and H1′,C8 yields the cross-peak intensity H1′,C8 that is proportional to $\sin^2(\pi {}^3J(\mathrm{H1'},\mathrm{C8})\Delta)$ and cross-peak intensity H1′,C1′ proportional to $\cos^2(\pi {}^3J(\mathrm{H1'},\mathrm{C8})\Delta)$ as is characteristic for quantitative J correlations:

$$^3J(\mathrm{H1'},\,\mathrm{C8}) = \frac{1}{\pi\Delta}\,\mathrm{arctg}\sqrt{\frac{I_{\mathrm{CP}}}{I_{\mathrm{ref}}}} \tag{4}$$

2.3.3. Determination of ^3J(C2′,P$_{i+1}$), ^3J(C4′,P$_{i+1}$), and ^3J(H3′,P$_{i+1}$) for the Determination of Conformation around the Backbone Angle ε and ^3J(H5′,P$_i$), ^3J(H5″,P$_i$), and ^3J(C4′,P$_i$) for the Conformation around β (Schwalbe *et al.*, 1993, 1994)

The conformation of the angles β and ε of the phosphodiester backbone in RNA oligonucleotides can be determined from measurement of ^3J(H,P) and ^3J(C,P) coupling constants as shown in Fig. 21a. As explained in Section 1.1.1.3, it is difficult to obtain coupling constants to ^{31}P using E.COSY techniques because ^{31}P has no NMR-active bonded substituent and the ^2J(C3′$_i$,P$_{i+1}$) and ^2J(C5′$_i$,P$_i$) coupling

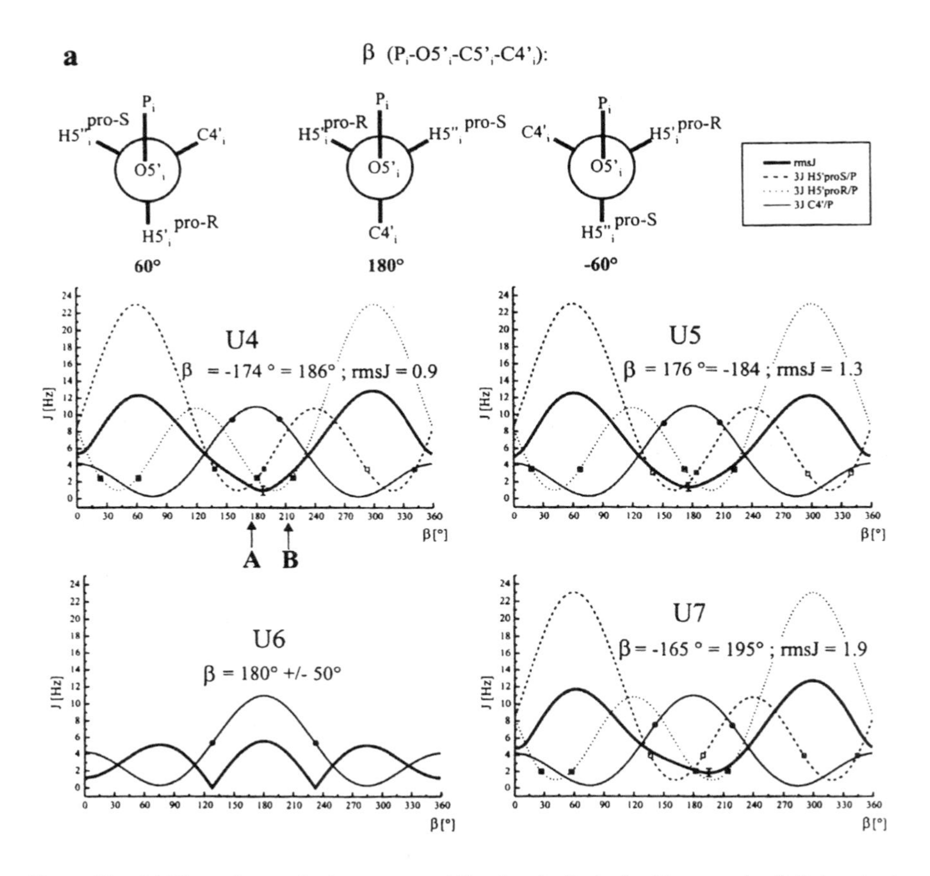

Figure 21. (a) Three staggered rotamers around the phosphodiester backbone angles β. Determined coupling constants for residues U4–U7 of a 1.5 mM RNA sample 5′CGCUUUUGCG-3′ in which the four uridine residues are labeled with ^{13}C in the ribose ring. The rmsJ between predicted and experimental coupling constants for a single conformation is shown as a thick line for each residue. Also shown are canonical values for β in A- and B-form helices. (b) Pulse sequence for the P-FIDS experiment. $\Delta = (2^1J(\text{C,H}))^{-1}$, $\tau = n/(^1J(\text{C,C}))$, with n = 1,2,3 depending on the T_2 relaxation times of carbons in the sample of interest. $\phi_1 = x,-x$, $\phi_2 = x,x,-x,-x$, $\phi_{rec} = x,-x,-x,x$. (c) Schematic cross-peak pattern in the P-FIDS experiment.

constants of 5 Hz are too small to be used as associated couplings (see, however, Schmieder *et al.*, 1992). On the other hand, the FIDS methodology can be applied to measure unique ^nJ(H,P) and ^nJ(C,P) couplings in oligonucleotides with the exception of ^nJ(C4′,P). For the latter a sum of two ^{31}P–^{13}C couples is measured in the FIDS experiment because there are two ^{31}P spins that couple to a given C4′. To

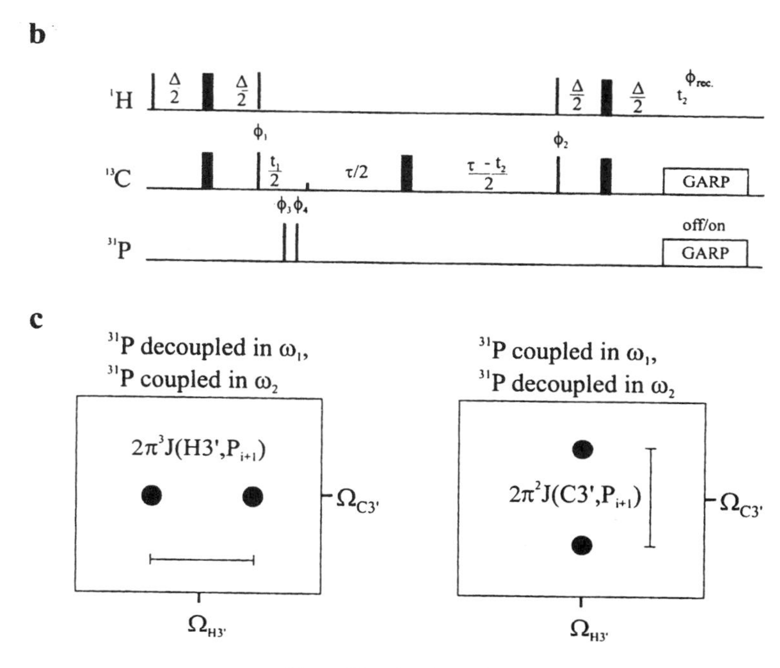

Figure 21. (continued).

demonstrate the FIDS technique, consider an H3′,C3′ cross-peak in an ^{1}H,^{13}C HSQC with alternating decoupling of ^{31}P in t_1 and t_2 as shown in Fig. 21b. In the first experiment, decoupling of ^{31}P by the 180°–(^{31}P) pulse ($\phi_3 = \phi_4$) in t_1 and no decoupling during t_2 (off-resonance broadband decoupling) yields an H3′$_i$,C3′$_i$ cross-peak that is a doublet related to the ^3J(H3′$_i$,P$_{i+1}$) coupling constant in ω_2. The second experiment employs decoupling of ^{31}P in t_2 (by on-resonance broadband decoupling) and no decoupling in t_1 ($\phi_3 = -\phi_4$). The resultant H3′$_i$,C3′$_i$ cross-peak is a doublet related to the ^2J(C3′$_i$,P$_{i+1}$) coupling constant in ω_1. For larger oligonucleotides, the coupling will not be resolved and the coupling is apparent only as a broadening of the coupled multiplet relative to the decoupled multiplet (see Fig. 22a). As described in Section 1.1.1.3., the unknown coupling can be obtained from the reference decoupled multiplet by deconvolution of the decoupled multiplet with a trial coupling J^{trial} and finding, in a one-parameter fitting routine, the minimum of the power difference spectrum between the deconvoluted reference multiplet and the experimental coupled multiplet (Fig. 22b).

Alternatively, a J-modulation ^{1}H,^{13}C HSQC can be used to measure J(C,P) couplings. This method requires that two constant-time ^{1}H,^{13}C HSQCs be recorded: one in which the J(C,P) evolves during the constant-time period CT, and the other

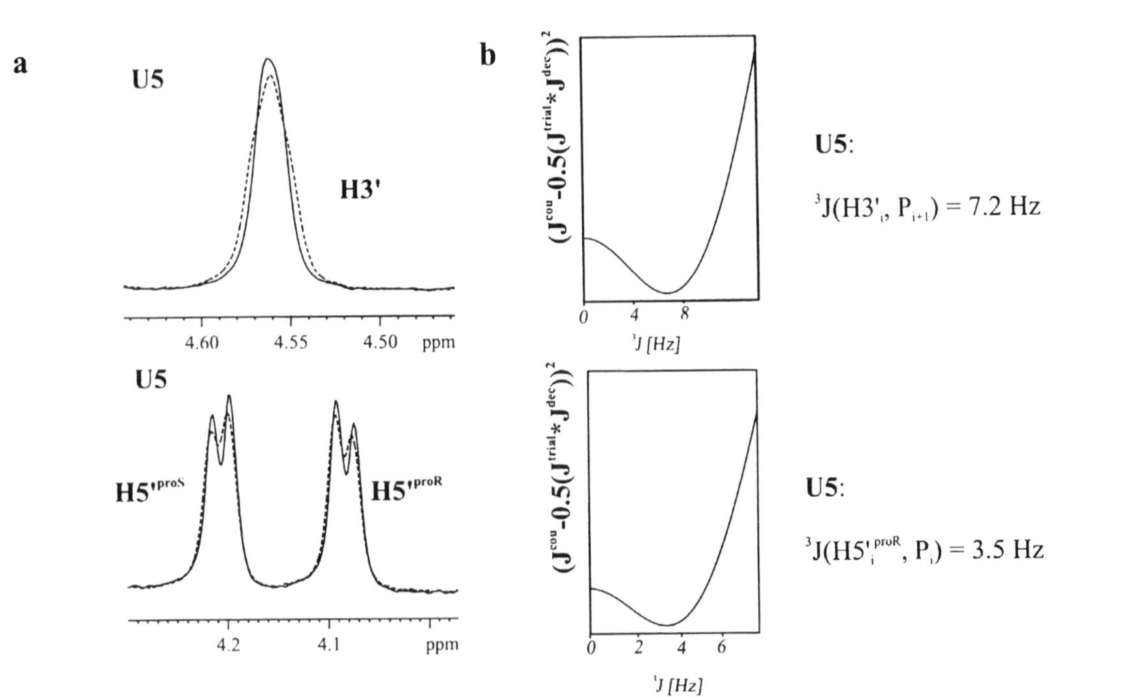

Figure 22. (a) Experimental 1D rows from the P-FIDS experiment with (solid line) and without (dashed line) ^{31}P decoupling during acquisition for H3′ and H5′/H5″ resonances for residue U5 of a 1.5 mM RNA sample 5′-CGCUUUUGCG-3′ in which the four uridine residues are labeled with ^{13}C in the ribose ring. The line broadening in the coupled (dashed) multiplets is clearly visible. (b) Fitting of the coupling of interest in a one-parameter fit gives a coupling of 7.2 Hz for $^3J(\mathrm{H3}'_i,P_{i+1})$ and 3.5 Hz for $^3J(\mathrm{H5}'_i,\mathrm{P}_i)$. In our application, $\tau = 2/^1J(\mathrm{C,C}) = 50$ ms (see Fig. 21a) and total measurement time was 16 h for both experiments. The experiments with different decoupling schemes were recorded in an interleaved manner.

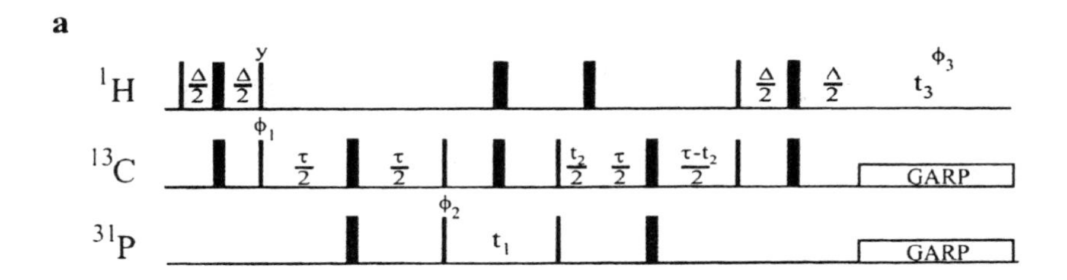

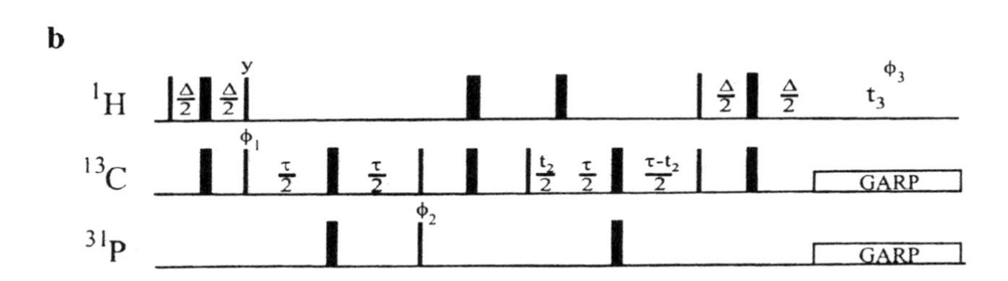

Figure 23. Pulse sequences for the quantitative 3D HCP experiment (a) and a 2D reference experiment (b). $\Delta = (2^1J(\text{C,H}))^{-1}$, $\tau = 1/^1J(\text{C,C}) = 25$ ms. In (a) $\phi_1 = x,-x$, $\phi_2 = x,x,-x-x$, $\phi_{rec} = x,-x,-x,x$; in (b), $\phi_1 = x,-x$, $\phi_2 = x,x,-x-x$, $\phi_{rec} = x,-x,-x,x$.

in which the J(C,P) coupling is refocused (Legault *et al.*, 1995). This yields an amplitude modulation (as shown in Fig. 5b) and comparison of the cross-peak intensities yields the size of the coupling constant. The evolution of the coupling constant can only be applied in the indirect carbon time domain. The duration of the CT period is restricted not by the size of the coupling of interest but by the size of the passive homonuclear J(C,C) coupling constants of the active carbon spin and by the T_2 relaxation of carbon. Because for carbon the ^1J(C,C) coupling constant is a factor of 10 larger than long-range ^nJ(C,C) couplings, the CT period can be tuned to $n/^1J$(C,C) in an ^{1}H,^{13}C HSQC.

As previously mentioned, the $C4'_i$ is coupled to two P_i and P_{i+1} with large couplings to each of ~9 Hz if β and ε are around 180°. For ^{13}C resonances that couple to two ^{31}P as the C4′, P-FIDS and J-modulated CT-HSQC fail to provide individual J(C4′,P) coupling constants because the two ^{31}P atoms cannot be selectively decoupled and therefore the coupling constants cannot be disentangled from the cross-peaks. However, the coupling constants can be obtained via the quantitative J-methodology (Section 1.1.1.4) from a 3D quantitative HCP. The cross-peak intensity of the $H4'_i,C4'_i,P_i$ cross-peak in the 3D HCP $I_{CP}(i,i)$ is proportional to $\sin^2(\pi^3J(C4'_i,P_i)\tau)\cos^2(\pi^3J(C4'_i,P_{i+1})\tau)$ and the cross-peak intensity of the

$H4'_i,C4'_i,P_{i+1}$ cross-peak $I_{CP}(i,i+1)$ is proportional to $\cos^2(\pi\, ^3J(C4'_i,P_i)\tau)\sin^2(\pi\, ^3J(C4'_i,P_{i+1})\tau)$. The intensity of the H4′,C4′ cross-peak $I_{ref}(i)$ in a 2D HC-(P) reference experiment is proportional to $\cos^2(\pi\, ^3J(C4'_i,P_i)\tau)\cos^2(\pi\, ^3J(C4'_i,P_{i+1})\tau)$. This is an application of the heteronuclear case described in Fig. 1.5b. The two desired ^3J coupling constants can be derived from

$$J(C4'_i, P_i) = \frac{1}{\pi\tau} \operatorname{arctg} \left(\frac{I_{CP}(i, i)}{I_{ref}(i)} \right)^{1/2}$$

$$J(C4'_i, P_{i+1}) = \frac{1}{\pi\tau} \operatorname{arctg} \left(\frac{I_{CP}(i, i+1)}{I_{ref}(i)} \right)^{1/2} \tag{5}$$

Additionally, ^nJ(C,P) coupling constants for carbons that are coupled to a single ^{31}P can be obtained from the quantitative evaluation of the cross-peaks in a 2D or 3D HCP experiment comparing the cross-peak intensities and cross-peaks in a 2D reference experiment as well (Fig. 24). For a ^{13}C resonance that couples only to a single ^{31}P, the cross-peak intensity I_{CP} is proportional to $\sin^2(\pi J(C,P)\tau)$ in the HCP experiment and $I_{ref} = \cos^2(\pi J(C,P)\tau)$ in the 2D reference experiment and the coupling is given by essentially the same formula:

$$J(C, P) = \frac{1}{\pi\tau_{CT}} \operatorname{arctg} \left(\frac{I_{CP}}{I_{ref}} \right)^{1/2} \tag{6}$$

Coupling constants determined in the P-FIDS experiment and in the quantitative HCP experiment are listed in Table 4.

2.4. ^{19}F-Labeled RNA Oligonucleotides

Because of its favorable NMR properties, 100% natural abundance, high sensitivity, and large chemical shift dispersion (Kalinowski *et al.*, 1994), ^{19}F labels have served as localized probes in a number of NMR studies (Chu and Horowitz, 1991; Parisot *et al.*, 1991; Chu *et al.*, 1992; Rastinejad and Lu, 1993). One of the prime target sites for ^{19}F substitution in RNA oligonucleotides is the 2′-OH position because oligoribonucleotides and their 2′-deoxy-2′-fluoro analogues have been shown to resemble each other at least at specific sites of the molecule (Reif *et al.*, 1997a). 2′-Deoxy-2′-fluoro RNA can serve as a suitable probe for conformational analysis bearing a number of advantages from an NMR spectroscopic point of view. 2′-Fluorination increases the chemical shift dispersion of proton and carbon resonances (see Fig. 24) and introduces another well-resolved nucleus allowing heteronuclear isotope filtering (Otting *et al.*, 1986; Otting and Wüthrich, 1989, 1990) to be performed. ^{19}F spins can also be used in E.COSY-type experiments to yield both homo- and heteronuclear coupling constants.

Table 4
Coupling Constants and Torsion Angles as Determined for 5′-CGCUUUUGCG-3′

	Couplings	U4	U5	U6	U7
β	$^3J(\text{H5}'^{\text{proS}}_i,\text{P}_i)^{a,b}$	3.6	3.0	*c*	4.0
	$^3J(\text{H5}'^{\text{proR}}_i,\text{P}_i)$	2.5	3.5	*c*	2.1
	$^3J(\text{C4}'_i,\text{P}_i)$	9.8	8.8	7.4	7.1
γ	$^3J(\text{H5}'^{\text{proS}}_i,\text{H4}'_i)^{d,e}$	2.6	2.7	*c*	3.3
	$^3J(\text{H5}'^{\text{proR}}_i,\text{H4}'_i)$	6.6	6.2	*c*	2.7
	$^2J(\text{H5}'^{\text{proS}}_i,\text{C4}'_i)$	−3.5	−4.7	*c*	−2.4
	$^2J(\text{H5}'^{\text{proR}}_i,\text{C4}'_i)$	−1.0	0.4	*c*	−1.5
ε	$^3J(\text{H3}'_i,\text{P}_{i+1})$	8.1	7.2	7.4	7.1
	$^3J(\text{C2}'_i,\text{P}_{i+1})$	2.1 (2.7)^f	6.0 (6.6)	5.1 (6.0)	3.8 (3.7)
	$^3J(\text{C4}'_i,\text{P}_{i+1})$	11.5	7.4	7.5	4.0
	$^2J(\text{C3}'_i,\text{P}_{i+1})$	4.8 (5.2)	5.0 (5.1)	5.3 (5.0)	4.5 (5.0)
	$^2J(\text{C5}'_i,\text{P}_i)$	4.6 (5.1)	4.4 (5.3)	5.3 (^{c})	4.9 (5.6)
	Backbone angle				
	β, rmsJ^g	−174°	176° 1.4	~180°^{h}	−162° 22
	ε, rmsJ^g	−154°	128° 1.7	127° 1/6	123° 0.5

^{a}Measured in Hertz.

^{b}Measured in PFIDS experiment in ω_2.

^{c}Overlapped.

^{d}Measured in HCC-TOCSY-CCH-E.COSY (Schwalbe *et al.*, 1995b).

^{e}Stereospecific assignments from C5′,H5′-selective HSQC (Marino *et al.*, 1996).

^{f}Values in parentheses from PFIDS experiment in ω_1.

^{g}rmsJ = $[\Sigma(J_{\text{pred}} - J_{\text{exp}})^2]^{1/2}/n^{1/2}$ measured in Hertz, *n* number of couplings, theoretical *J* calculation for *J*(C,P) and for *J*(H,P) parametrizations (Lankhorst *et al.*, 1984).

^{h}Based on the observation of only $^3J(\text{C4}'_i,\text{P}_i)$.

2.4.1. E.COSY Experiments for the Determination of ^nJ(H,F) and ^nJ(C,F) Coupling Constants

In ^{19}F-labeled molecules, correlation experiments like a NOESY or a ^{13}C,^{1}H HSQC can be performed to yield ^nJ(H,F) coupling constants. Using the E.COSY triangle as mnemonics shows that either an ^{1}H,^{1}H NOESY or an ^{1}H,^{13}C HSQC experiment in which the ^{19}F spin is left untouched can be used to measure *J*(H,F) coupling constants. From the H2′,H1′ cross-peaks in a NOESY or TOCSY experiment, ^3J(H1′,F2′) couplings can be determined because this cross-peak region is reasonably well-resolved because of ^{19}F labeling. Both the NOE transfer in a C3′-*endo* conformation or TOCSY transfer in a C2′-*endo* are efficient and the

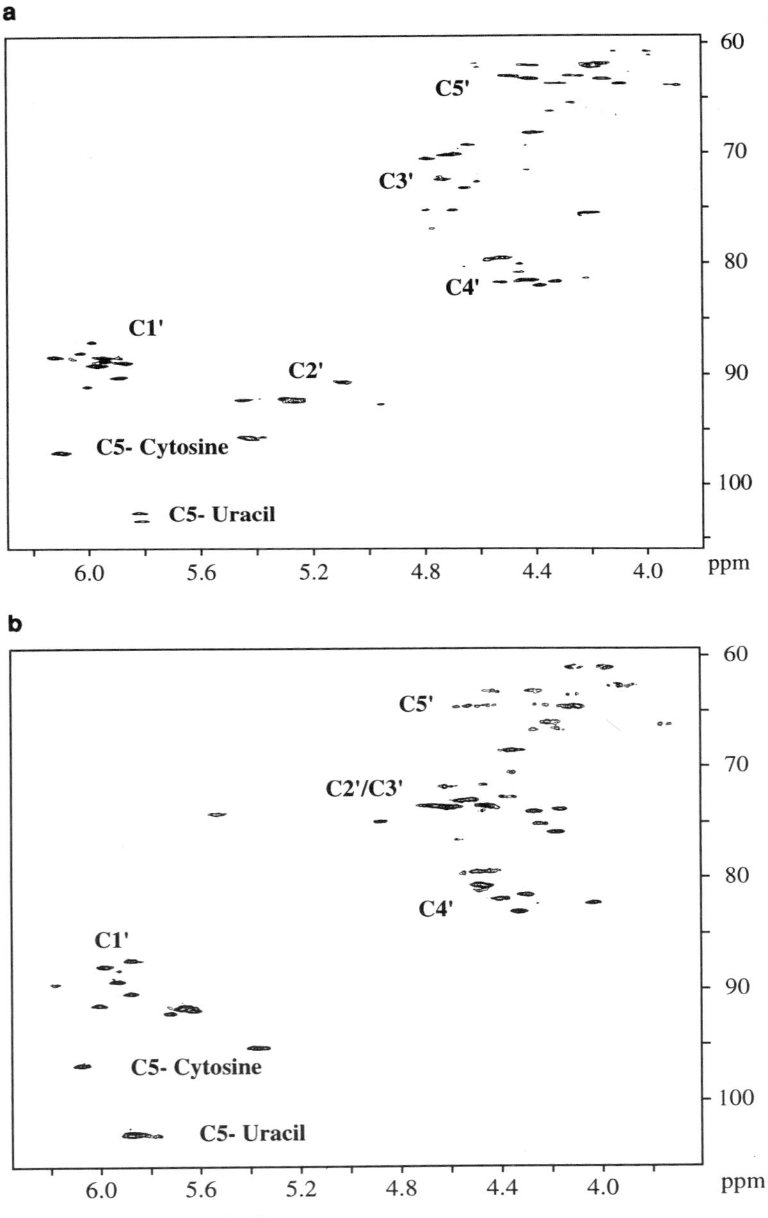

Figure 24. Comparison of 2D $^{1}H,^{13}C$ HSQC of a 2′-fluorinated RNA 5′-$C_fGC_fU_fU_fC_fGG\ C_fG$-3′, in which the 2′-OH group in pyrimidine nucleotides was substituted by ^{19}F in (a) and an unlabeled 5′-CGCUUCGGCG-3′ RNA in (b). The resolution enhancement especially on the C2′,H2′ cross-peak region, which does not overlap with the C3′,H3′ of because of the ^{19}F labeling, is clearly visible.

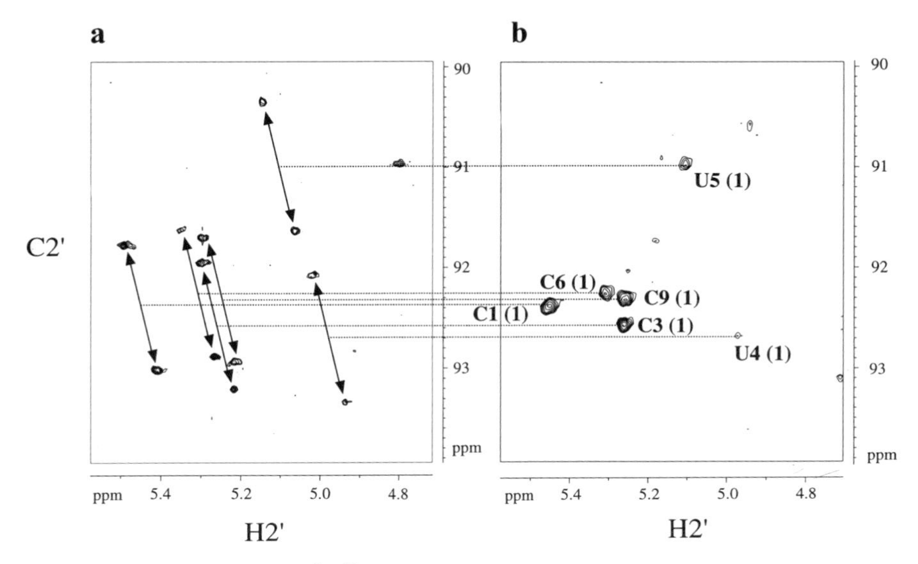

Figure 25. Comparison of 2D ^{1}H,^{13}C HSQC of a 2′-fluorinated RNA 5′ $C_fGC_fU_fU_fC_fGGC_fG$-3′ without (a) and with (b) ^{19}F decoupling. Because in (a) the spin states of ^{19}F remain untouched during the experiment, the C2′,H2′ is split as a result of ^{1}J(C2′,F2) coupling in ω_1 and the ^{2}J(H2′,F2) in ω_2. The fluorinated sample shows an equilibrium between a duplex (1) and a hairpin (2) conformation. At the contour level chosen in this plot, only cross-peaks of the major component (1) are visible.

associated coupling ^{2}J(H2′,F) is large (−51 Hz). As shown in Fig. 25, the two submultiplets of the H2′,C2′ cross-peak in the ^{13}C,^{1}H HSQC are displaced by a ^{1}J(C2′,F) of −191 Hz in ω_1 and ^{2}J(H2′,F) in ω_2.

2.4.2. ^{19}F,^{1}H HSQC-E.COSY Experiment, Determination of ^{3}J(H1′,H2′) Coupling Constants in 2′-Deoxy-2′-fluoro RNA

In unlabeled oligoribonucleotides, relative intensities of the moderately well resolved H1′,H2′ cross-peaks in a DQF-COSY experiment can be used to qualitatively determine ^{3}J(H1′,H2′) coupling constants. In 2′-deoxy-2′-fluoro RNA, ^{3}J(H1′,H2′) coupling constants can be determined from spin topology filtering in an E.COSY type experiment (Weisemann *et al.*, 1994). The experiment relies on the observation that the ^{2}J(H2′,F′) coupling constant does not depend on the sugar conformation and is approximately four times larger than the vicinal ^{3}J(H1′,F2′) and ^{3}J(H3′,F2′) coupling constants. Therefore, the H2′–^{19}F2′ pair resembles the H–^{15}N pair in a labeled protein or the H-^{13}C pair in a molecule with natural abundance of carbon. The pulse sequence for the determination of ^{3}J(H,H) in 2′-deoxy-2′-fluoro RNA is shown in Fig. 26. The experiment is essentially a gradient-enhanced ^{19}F,^{1}H HSQC for correlations via ^{3}J(H1′,F2′) and ^{3}J(H3′,F2′) coupling constants. The back transfer ensures that the passive H2′ is not touched between the evolution period t_1 and the detection period. This is achieved by tuning the delay Δ′ for the large ^{2}J(H2′,F2′) coupling, resulting in an E.COSY pattern much the same as in the HNCA-E.COSY experiment (Weisemann *et al.*, 1994). Neglecting homonuclear couplings, which are small compared with the heteronuclear coupling, H2′ behaves as an isolated spin. The $90_x 180_x 90_y 180_x 90_x 180_x$ pulse sandwich leaves *z*-magnetization unrotated yielding the desired behavior for the passive spin in an E.COSY experiment. The schematic cross-peak pattern of the E.COSY-type experiment is shown in Fig. 27: For the (F2′,H1′) cross-peak shown, the two submultiplet patterns are displaced by the displacement vector $\mathbf{J_F}$ with the components ^{2}J(H2′,F2′) as associated coupling in ω_1 and ^{3}J(H1′,H2′) as coupling of interest in ω_2.

a

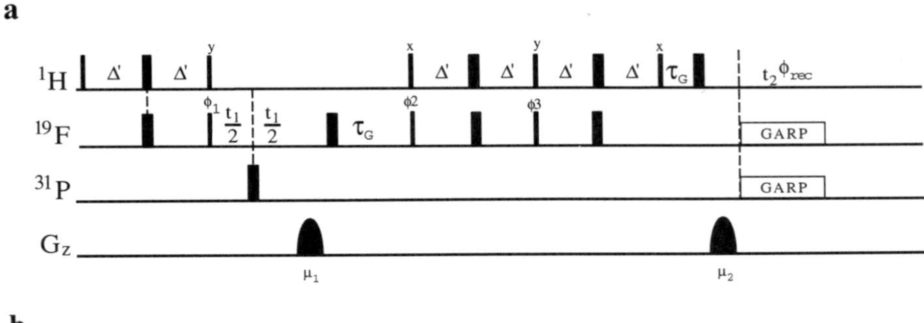

b

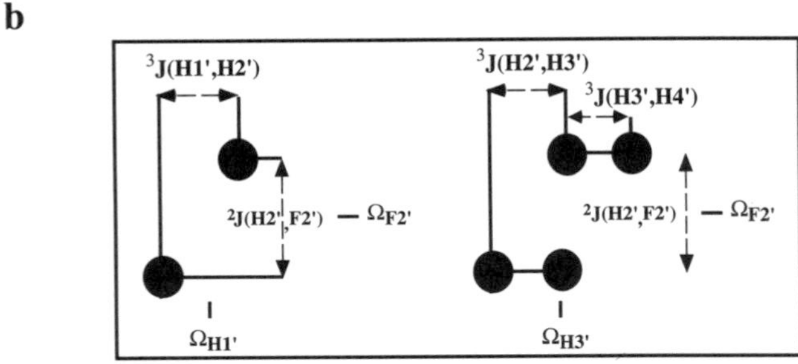

Figure 26. Pulse sequence of the ^{1}H,^{19}F HSQC-E.COSY experiment for the measurement of ^{3}J(H1′,H2′) coupling constants in fluorinated RNA oligomers. Narrow and wide bars are 90° and 180° pulses, respectively. Default phase is *x*. $\Delta' = (2^2J(\mathrm{F,H}))^{-1}$, $\tau_g = 1.5$ ms. $\phi_1 = x,-x$, $\phi_2 = x,x,-x,-x$, $\phi_3 = y,y,-y,-y$, $\phi_{rec} = x,-x,-x,x$.

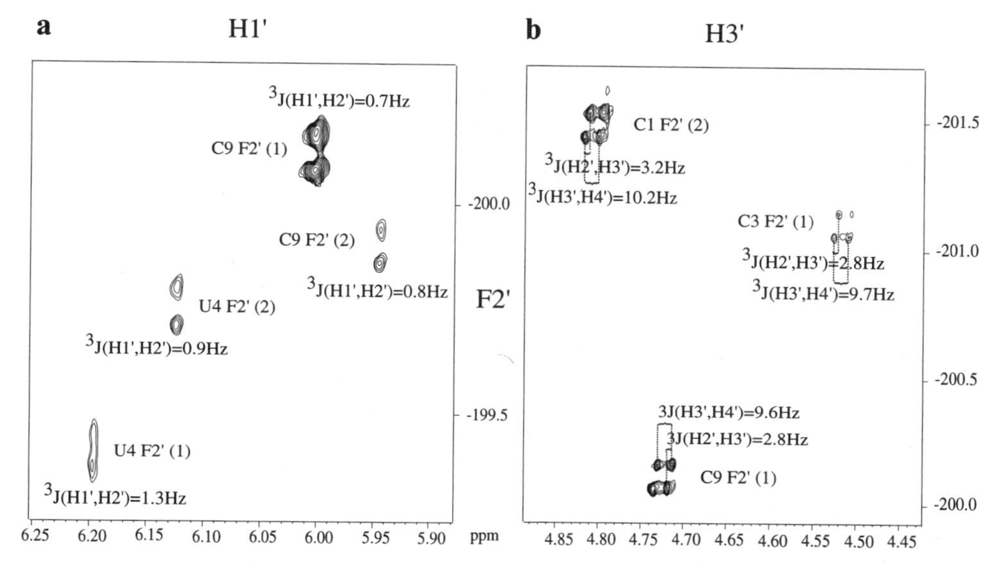

Figure 27. 2D ^{1}H,^{19}F HSQC-E.COSY experiment of a 2′-fluorinated RNA 5′-$C_fGC_fU_fU_fC_fGGC_fG$-3′ showing the H1′,F2′ and the H3′,F2′ cross-peak region.

Table 5
Measured Scalar Couplings in Fluorinated Nucleotides for the Hairpin (a) and Duplex Confomation (b)

	Couplings (Hz)	C1	C3	U4	U5	C6	C9
a	^3J(H1′,H2′)^a	0.1	0.1	1.3	4.0	3.3	0.7
	^3J(H2′,H3′)	—	2.8	—	—	—	2.8
	^3J(H3′,H4′)	—	9.7	—	—	—	9.6
	^1J(F2′,C2′)	191.7	194.0	191.2	195.0	193.3	194.0
	^2J(F2′,H2′)	–51.6	–51.6	–51.4	–50.8	–51.7	–51.8
	^3J(F2′,H1′)	13.4	14.1	15.8	14.0	15.6	13.5
b	^3J(H1′,H2′)^a	0.1	0.1	0.9	<1	0.5	0.8
	^3J(H2′,H3′)	3.2	—	—	—	—	—
	^3J(H3′,H4′)	10.2	—	—	—	—	—
	^3J(F2′,H2′)	–51.6	–51.6	–52.1	–51.6	–52.4	–51.7
	^3J(F2′,H1′)	13.3	13.2	14.6	14.0	13.8	14.2

^{a} ^3J(H1′,H2′) couplings in purine nucleotides are smaller than 1 Hz.

3. MEASUREMENT OF SCALAR COUPLING CONSTANTS IN PERDEUTERATED PROTEINS

The now rather routine incorporation of NMR-active isotopes, such as ^{13}C and ^{15}N, into proteins has allowed the development of a standard set of triple-resonance (^{1}H, ^{13}C, ^{15}N) experiments designed for sequential through-bond assignment and structure determination (Bax, 1994). The maximum size of a protein that is amenable to complete structural investigations using heteronuclear NMR spectroscopy has increased, as a consequence, to approximately 30 kDa. One of the major limitations, however, in high-resolution NMR spectroscopy is the increase in the linewidth observed for resonances in larger macromolecules with longer correlation times. In particular, the short transverse ^{13}C relaxation times constitute the principal barrier to the application of triple-resonance NMR techniques to even larger macromolecules uniformly labeled with ^{13}C and ^{15}N.

The problem of short ^{13}C relaxation times can be significantly reduced by deuteration (Crespi *et al.*, 1968; Markley *et al.*, 1968). Because the magnetogyric ratio of ^{2}H is 6.5 smaller than that of ^{1}H, the C_α transverse relaxation in a C_α-D moiety is reduced by a factor of 16 relative to a C_α-H moiety (*see* chapters 2 and 3). This effect is realized because the C_α transverse relaxation time is dominated by the strong dipolar interaction with its attached proton. However, the $^{1}J(C,D)$ coupling, which is on the order of 22 Hz, is often not resolved because of scalar relaxation of the second kind (Abragam, 1961). In fact, one finds that the longitudinal relaxation rates of ^{2}H in biomacromolecules with rotational correlation times on the nanosecond time scale are of the order of a few ten of hertz at high magnetic fields. This leads to a ^{13}C singlet line that is broadened by the scalar relaxation. Utilizing ^{2}H decoupling with an RF field strength stronger than the $^{1}J(^{2}H,^{13}C)$ coupling constant combined with gating the ^{2}H lock channel effectively suppresses the ^{2}H scalar relaxation contribution to the carbon linewidth (Grzesiek *et al.*, 1993a; Kushlan and LeMaster, 1993). The resulting ^{13}C linewidths for deuterated carbons are much narrower than for protonated carbons.

The sensitivity of heteronuclear triple-resonance NMR experiments is improved by deuteration of all but the exchangeable hydrogen atoms usually bound to carbon atoms. Near-complete H^N, ^{15}N, $^{13}C_\alpha$, and $^{13}C_\beta$ assignments are reported for a 64-kDa complex consisting of perdeuterated trp repressor, unlabeled operator DNA, and tryptophan (Yamazaki *et al.*, 1994; Shan *et al.*, 1996).

As a consequence of depletion of ^{1}H spins, distance restraints in a perdeuterated sample can only be derived between exchangeable hydrogen atoms. The limited NOE data set leads to low-resolution protein structures for which determination of global folds is difficult (Venters *et al.*, 1995; Grzesiek *et al.*, 1995). To obtain higher resolution structures, additional structural restraints must be determined, especially for the conformation of sidechains. One approach to this problem is to fractionally label proteins with deuterium. Random fractional deuteration improves the sensitivity of multidimensional NMR experiments designed for the assignment of sidechain proton and carbon resonances (Nietlispach *et al.*, 1996). In addition, this approach increases the global proton density and thus makes NOE contacts between sidechain and backbone proton resonances observable. However, enhancements in resolution are limited because a mixture of isotopomers gives rise to a set of differential isotope shifts (Hansen, 1988) which presents a problem especially for ^{13}C resolution. It is important to maximize the population of one methyl or methylene isotopomer (namely, the fully deuterated isotopomer) because adequate spectral resolution is required in the analysis of larger proteins by NMR. For example, cross-peaks of fractionally deuterated methyl groups in an $^{1}H,^{13}C$ correlation experiment are split into three components arising from a mixture of CH_3, CH_2D, and CHD_2. Each methyl deuteron leads to upfield shifts of both the carbon nucleus (–0.3 ppm) and the remaining methyl protons (–0.02 ppm) (Gardner *et al.*, 1997). Furthermore, methine carbon resonances show an upfield shift (~0.35 ppm) primarily related to the one-bond ^{2}H isotope effect.

Labeling strategies have been developed, in which proteins are highly deuterated in all but methyl positions in Ala, Val, Leu, and Ile residues (Rosen *et al.*, 1996; Metzler *et al.*, 1996; Gardner and Kay, 1997). Type-specific protonation increases local rather than global proton density, thus retaining more of the benefits of perdeuteration. Because of their relatively narrow proton and carbon linewidth, methyl groups are excellent targets for protonation. These amino acids are among the most common in hydrophobic cores and retaining NOE information related to their protonation is especially helpful for structure determination. In addition, conformational averaging in solution can be probed by NMR most efficiently via scalar coupling constants, which in the case of ^{3}J coupling constants can be related to specific torsion angles by Karplus relationships (Karplus, 1959). Homonuclear $J(C,C)$ and heteronuclear $J(C,N)$ as well as $J(H^N,C)$ coupling constants provide a source for local conformational analysis in perdeuterated proteins and will be reviewed in this chapter. The Newman projections around the torsional angles ϕ, ψ, χ_1, and χ_2 in a perdeuterated protein are shown in Fig. 28.

The torsion angles are defined according to the IUPAC-IUB convention (IUPAC-IUB Commission, 1970), where ϕ refers to the torsion angle sequence $C'_{(i-1)}$, N_i, $C_{\alpha i}$, and C'_i and ψ refers to the torsion angle sequence N_i, $C_{\alpha i}$, C'_i, and $N_{(i+1)}$. According to the IUPAC rules, the χ_1 torsion angle is defined by the orientation of the highest-ranking γ-ligand relative to the nitrogen atom. Thus, the definition of χ_1 differs for different residue types, e.g., Thr (N, C_α, C_β, O_γ) and Val (N, C_α, C_β, $C_\gamma^{\text{pro-R}}$). χ_2 is defined as the torsion angle about the C_β-C_γ bond and measured between the highest-ranking δ-ligand and C_α.

Homo- and heteronuclear coupling constants that can be determined in perdeuterated proteins as well as the related experimental methods for the determination of these coupling constants are listed in Table 6. Karplus parametrizations of the mentioned coupling constants are summarized in Table 7.

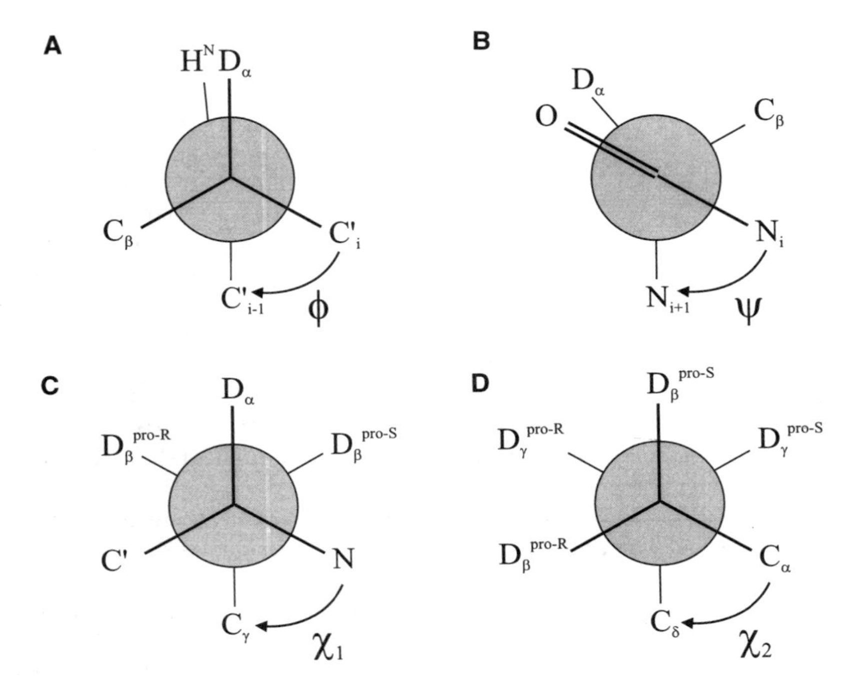

Figure 28. Newman projections around the backbone angles ϕ (A) and ψ (B) and around the sidechain angles χ_1 (C) and χ_2 (D). In the case of the χ_1 angle, the projection for residues Leu, Phe, Tyr, Asn, Glu, Lys, Arg, Trp, His, Met, Asp, and Glu is shown. For Ala, the C_γ must be substituted by a deuteron, for Cys C_γ is substituted by S_γ, and for Ser by O_γ. Additional changes for Val and Ile residues are that C_γ is replaced by C_γ^1 and one deuteron substituent of C_β must be substituted by C_γ^2. Note that for Thr residues again one deuteron substitutent of C_β is replaced by O_γ. The Newman projections around χ_2 refer to aliphatic amino acids.

Applications of various experiments to a 0.6 mM (>95%) deuterated sample of $^{13}C/^{15}N$-enriched calmodulin, complexed with Ca^{2+} (5 mM $CaCl_2$) are shown in the following. Calmodulin is a Ca^{2+}-binding protein of 148 residues that is involved in a wide range of cellular Ca^{2+}-dependent signaling pathways. The crystal (Babu *et al.*, 1985, 1988; Taylor *et al.*, 1991) as well as the solution structures (Ikura *et al.*, 1992) of calcium-bound calmodulin, calcium-free calmodulin, and various complexes with synthetic peptides were solved some years ago.

3.1. Determination of the Backbone Angle ϕ in Perdeuterated Proteins

As shown in Fig. 29, four vicinal coupling constants can be determined around the angle ϕ. The relevant coupling constants in a perdeuterated protein fragment are

Table 6
Scalar Coupling Constants in Perdeuterated Proteins and Related Experiments for Their Determination

Angle	Relevant ^3J coupling constant and related experimental method	Described in section
ϕ	$^3J(H^N,C')$, soft HNCA-COSY	3.1.1
	$^3J(H^N,C_\beta)$, soft HNCA-E.COSY	3.1.2
	$^3J(C',C')$, quantitative HN(CO)CO	3.1.3
	$^3J(C',C_\beta)$,quantititative HN(CO)C	3.3.1
ψ	$^3J(N,N)$,quantititative H(N)N	3.2
	$^3J(N,C_\beta)$, quantititative HNC	3.3.2
χ_1	$^3J(C',C_\gamma)$, quantititative HN(CO)C	3.3.1
	$^3J(N,C_\gamma)$, quantititative HNC	3.3.2
χ_2	$^3J(C_\alpha,C_\delta)$, quantititative HN(CO)CAC$_{ali}$	3.4

Table 7
Coefficients of Karplus Equations, $J = A\cos^2(\phi + \theta) + B\cos(\phi + \theta) + C$ (Hz)[a]

Coupling constant	Karplus parametrization	θ (deg)	Reference
$^3J(H^N,C')$	$A = 4.10$, $B = -1.08$, $C = 0.7$	180	Hu and Bax (1997b)
$^3J(H^N,C_\beta)$	$A = 2.71$, $B = -0.35$, $C = 0.05$	60	Hu and Bax (1997b)
$^3J(C',C')$	A = 1.35, $B = -0.91$, $C = 0.61$	0	Hu and Bax (1997b)
$^3J(C',C_\beta)$	$A = 1.61$, $B = 0.66$, $C = 0.26$	−120	Hu and Bax (1997b)

[a]All listed Karplus coefficients are related to the backbone angle ϕ. Empirical Karplus relations have been derived from experimental ^3J couplings for ubiquitin and ϕ angles from its X-ray structure (Vijay-Kumar *et al.*, 1987) using singular value decomposition. It is not possible to derive Karplus coefficients for ^3J coupling constants related to sidechain angles in a similar manner because the agreement between sidechain angles in crystal structures and solution structures is far worse than between backbone angles. In addition, there is a wide range of sidechain dynamics, namely, rotameric averaging and fluctuations around a given rotameric state, which are difficult to quantify. $^3J(N,C_\beta)$ and $^3J(N,N)$ coupling constants are expected to show Karplus-type dependence on the backbone angle ψ. Both couplings are small and presently no parametrizations are available.

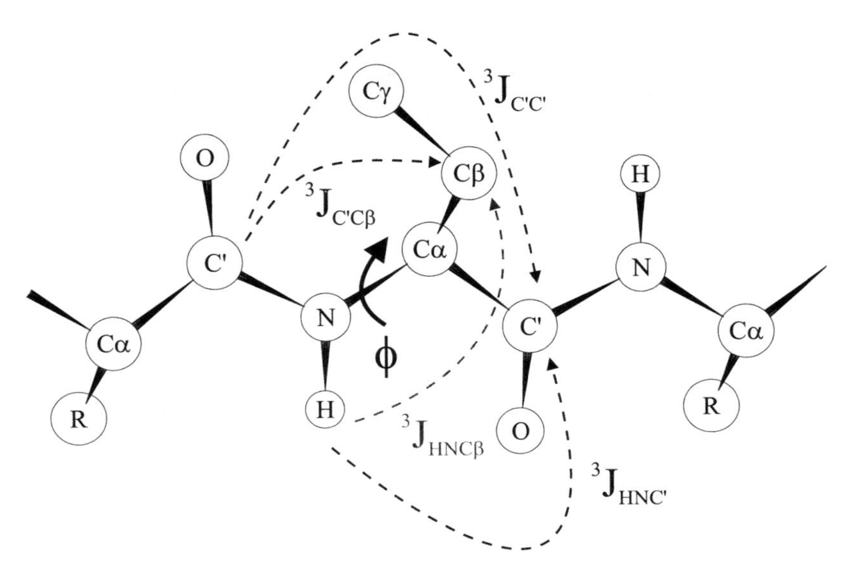

Figure 29. Schematic representation of a peptide backbone showing the dihedral angle ϕ and relevant scalar coupling constants in perdeuterated proteins.

$^3J(H_i^N, C_i')$ $^3J(H_i^N, C_{\beta i})$, $^3J(C_i', C_{i-1}')$, and $^3J(C_{i-1}', C_{\beta i})$. Methods developed for the determination of $^3J(H^N,C')$, $^3J(H^N,C_\beta)$, $^3J(C',C_\beta)$, and $^3J(C',C')$ coupling constants in protonated proteins can be directly applied to deuterated molecules, as the experiments employ an out-and-back-type magnetization transfer and therefore rely only on the ^{15}N nuclei to be protonated.

3.1.1. Measurement of $^3J(H^N,C')$ in a Soft HNCA-COSY Experiment

The soft HNCA-COSY (Fig. 30) (Weisemann *et al.*, 1994; Seip *et al.*, 1994) is derived from the HNCA experiment by introduction of ^2H decoupling and constant-time evolution of the $^{13}C_\alpha$. The HNCA experiment correlates amide ^1H and ^{15}N chemical shifts with the intraresidue $^{13}C_\alpha$ shift making use of the $^1J(N,C_\alpha)$ coupling constant (~11 Hz). The experiment provides additional sequential connectivities by transferring magnetization from ^{15}N to the $^{13}C_\alpha$ of the preceding residue via the smaller $^2J(N,C_\alpha)$ coupling constant (~7 Hz).

Because carbon pulses act selectively on aliphatic carbons in the evolution period t_2 and in the decoupling during acquisition period t_3, C′ remains untouched during the course of the experiment. Thus, the C′ acts as the passive spin in this E.COSY-type experiment with the $^1J(C_\alpha,C')$ of 54 Hz as associated coupling (see Section 1.1.1.4). The soft HNCA-COSY utilizes a CT evolution of the ^{15}N chemical shift during t_1 and of the $^{13}C_\alpha$ chemical shift during t_2. The CT delay for the C_α

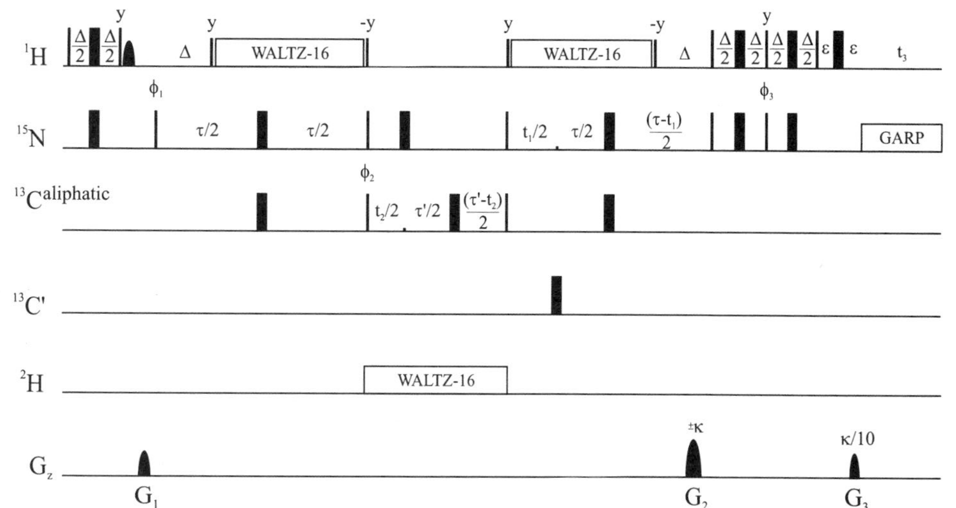

Figure 30. Pulse sequence of the soft HNCA-COSY experiment. In all pulse sequences, narrow and wide bars denote 90° and 180° pulses, respectively, and unless indicated otherwise the phase is x. Phase cycling: $\phi_1 = 2(x)$, $2(-x)$, $\phi_2 = x, -x$, $\phi = -y$, receiver $= x, -x, -x, x$. Quadrature detection in the t_2 dimension is obtained by altering ϕ_2 according to States–TPPI. For each t_1 value, echo and antiecho coherences are obtained by recording data sets where the phase ϕ_3 and the second gradient are inverted. Delay durations: $\Delta = 5.5$ ms, $\tau = 29$ ms, $\tau' = 26$ ms, and $\varepsilon = 1.2$ ms. Proton pulses are applied using a 20.8-kHz RF field with the exception of the 2-ms water-selective 90° flip-back pulse, the 4.2-kHz WALTZ-16 decoupling pulses, and the 4.2-kHz 90° +/− y pulses flanking the decoupling intervals. The ^{15}N pulses are at a field of 8.1 kHz and GARP decoupling during acquisition is applied with a 1-kHz field. All carbon pulses are Gaussian cascades; G3 and G4 pulses had durations of 256 and 409.6 μs, respectively. Every second G4 pulse was time-reversed so as to avoid phase errors. Carbonyl-selective pulses were implemented as phase-modulated pulses. Gradients (sine bell shaped): G1 = (2 ms, 25 G/cm), G2 = (1 ms, 40 G/cm), and G3 = (1 ms, 4.05 G/cm).

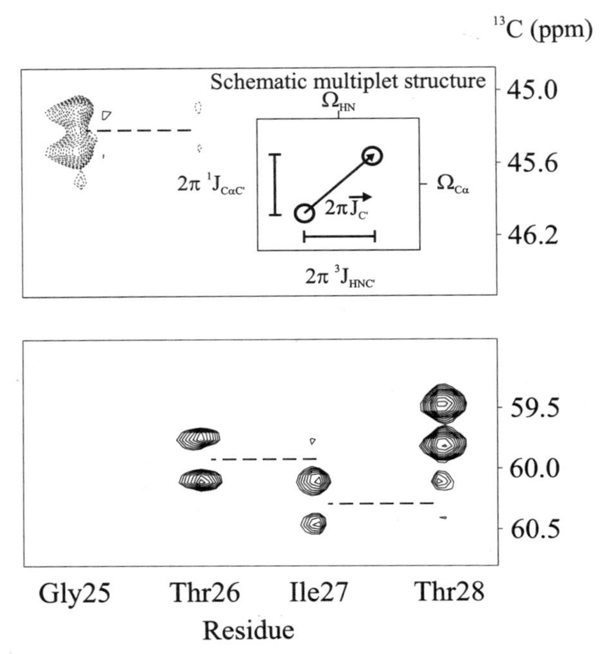

Figure 31. (ω_2) strip plots from the soft HNCA-COSY experiment of perdeuterated calmodulin in H_2O taken at the $^1H(\omega_3)$ and $^{15}N(\omega_1)$ resonance frequencies of Gly-25, Thr-26, Ile-27, and Thr-28. The strong intraresidue cross-peaks are connected with the weaker cross-peaks on the H^N resonance of the following amino acid by broken lines. These sequential interresidual correlations are related to N,C_α transfer via the $^2J(C_{\alpha i},N_{i+1})$ coupling constant. The schematic E.COSY-like cross-peak pattern for the intraresidue correlations is highlighted. The determined $^3J(H^N,C')$ coupling constants and the NMR derived ϕ angles as well as ϕ angles from the refined X-ray structure of calcium-bound calmodulin are shown in Fig. 36 and summarized in Table 8. Because of the different carbon multiplicity, $^{13}C_\alpha$ resonances of glycines, e.g., Gly-25, have phases opposite (dashed contours) to those of other $^{13}C_\alpha$ resonances.

chemical shift evolution is tuned to $1/^1J(C_\alpha,C_\beta)$ so as to remove aliphatic carbon–carbon couplings and to resolve the passive $^1J(C_\alpha,C')$ coupling in ω_2. A significant enhancement in signal-to-noise of this experiment for deuterated proteins compared with protonated proteins is observed because the C_α relaxation is dominated by the dipolar interactions with the directly bound proton. For the majority of residues in larger proteins, relaxation during the CT period for C_α chemical shift evolution greatly attenuates or even eliminates signals from C_α spins with a directly attached proton. Thus, deuteration extends the applicability of this CT-HNCA derived E.COSY-type experiment to larger proteins.

The only modification to the pulse sequence previously applied on protonated proteins is that deuterium decoupling is applied during the t_1 period. Measured

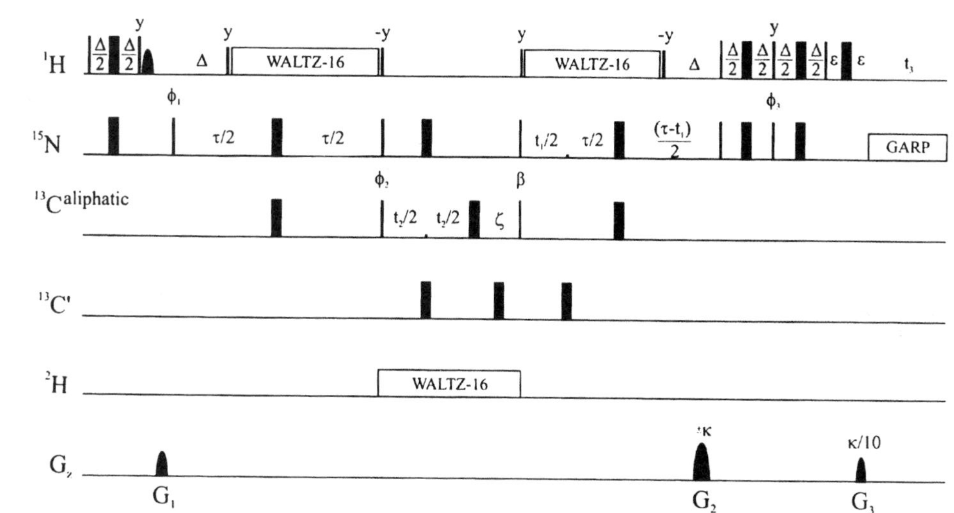

Figure 32. Pulse sequence of the soft HNCA-E.COSY experiment. Phase cycling: $\phi_1 = 2(x), 2(-x)$, $\phi_2 = x, -x$, $\phi_3 = -y$, receiver $= x, -x, -x, x$. Quadrature detection in the t_2 dimension is obtained by altering ϕ_2 according to States–TPPI. For each t_1 value, echo and antiecho coherences are obtained by recording data sets where the phase ϕ_3 and the second gradient are inverted. Delay durations: $\Delta = 5.5$ ms, $\tau = 29$ ms, $\zeta = t_2(0)$, and $\varepsilon = 1.2$ ms. All carbonyl carbon, and the first and last pulses on C_α are Gaussian cascades; G3 pulses had a duration of 256 μs. The 90°(ϕ_2), the 180° refocusing pulse during the C_α evolution period, and the mixing pulse with a flip angle of $\beta = 36°$ are rectangular, with a null in the excitation profile at the ^{13}C' frequency (10.56-kHz RF field).

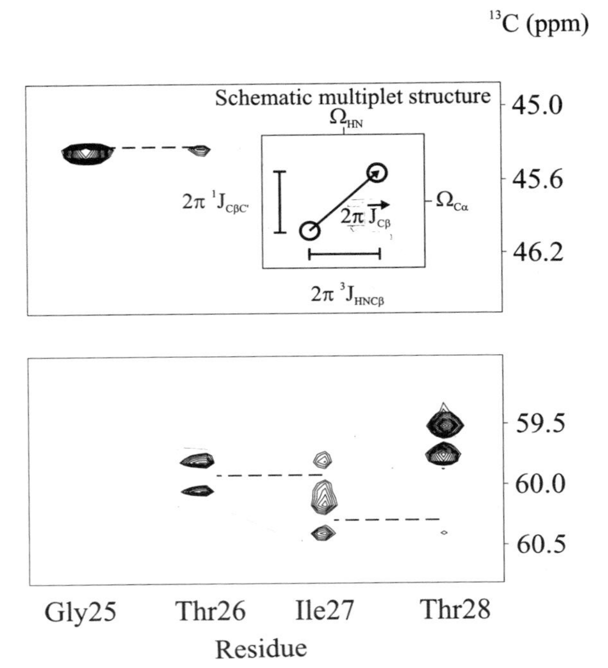

Figure 33. (ω_2) strip plots from the soft HNCA-E.COSY experiment of perdeuterated calmodulin in H_2O taken at the $^1H(\omega_3)$ and $^{15}N(\omega_1)$ resonance frequencies of Gly-25, Thr-26, Ile-27, and Thr-28. The strong intraresidue cross-peaks are connected with the weaker sequential cross-peaks by dashed lines. The schematic E.COSY-like cross-peak pattern for the intraresidue correlations is highlighted. The determined $^3J(H^N,C_\beta)$ coupling constants and the NMR derived ϕ angles as well as ϕ angles from the refined X-ray structure of calcium-bound calmodulin are shown in Fig. 36 and summarized in Table 8.

values of $^3J(H^N,C')$ for the small protein ubiquitin (76 residues) fall in the –0.4 (ϕ ± 120°) to 3.6 Hz (ϕ = 0°) range (Wang and Bax, 1995). Experimental $^3J(H^N,C')$ values for perdeuterated calmodulin are summarized in Table 8 (section 3.1.3).

3.1.2. Measurement of $^3J(H^N,C_\beta)$ in a Soft HNCA-E.COSY Experiment

The pulse sequence for the determination of $^3J(H^N,C_\beta)$ is shown in Fig. 32. In this experiment, the soft HNCA-E.COSY experiment, the C_β is the passive spin. The E.COSY requirement not to touch the spin states is fulfilled by using a small flip angle of $\beta = 36°$. The small β conserves the state of the passive C_β to an extent of $\cos^2(\beta/2)$ while the spin state of the active C_α spin is changed by a factor $\sin^2(\beta/2)$, thus representing a compromise between cancellation of undesired multiplet components and sensitivity. The duration during which the C_α has to be transversal is

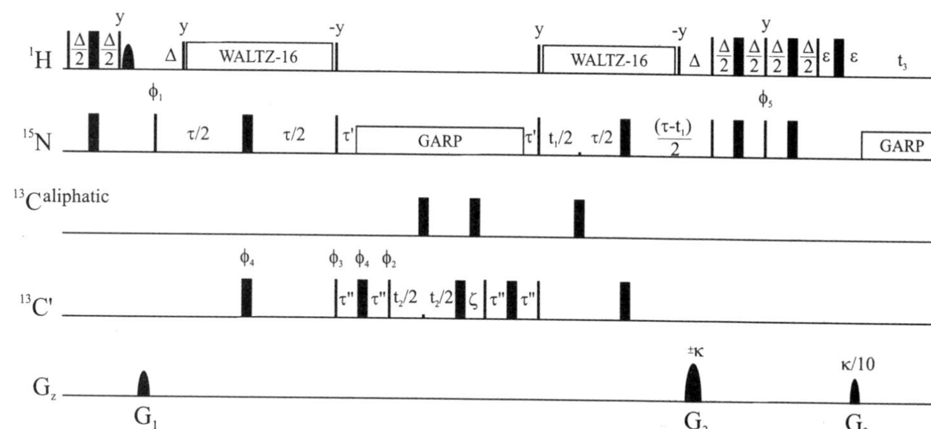

Figure 34. Pulse sequence of the quantitative HN(CO)CO experiment adapted from Hu and Bax (1996). Phase cycling: $\phi_1 = x, -x$, $\phi_2 = 2(x), 2(-x)$, $\phi_3 = 4(x), 4(-x)$, $\phi_4 = x$, $\phi_5 = y$, receiver $= x, -x, x, -x, -x, x, -x, x$. Quadrature detection in the t_2 dimension is obtained by altering ϕ_2, ϕ_3, and ϕ_4 according to States–TPPI. For each t_1 value, echo and antiecho coherences are obtained by recording data sets where the phase ϕ_5 and the second gradient are inverted. Delay durations: $\Delta = 5.5$ ms (or $\Delta' = 2.75$ ms for the observation of additional C′,C$_\gamma$ cross-peaks from sidechain amide protons of asparagines), $\tau = 35$ ms, $\tau' = 32$ ms, $\tau'' = 50$ ms, $\zeta = t_2(0)$, and $\varepsilon = 1.2$ ms. Aliphatic carbon selective pulses were implemented as phase-modulated G3 pulses.

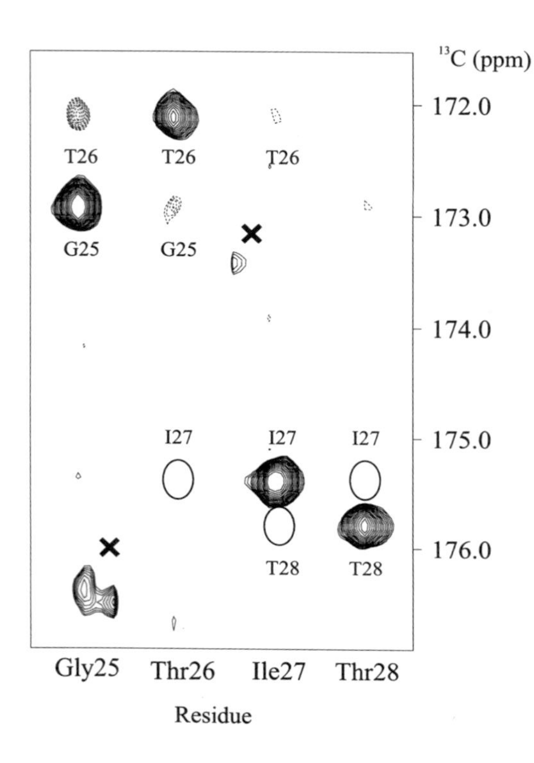

Figure 35. (ω_2) strip plots from the quantitative HN(CO)CO experiment of perdeuterated calmodulin in H_2O taken at the $^1H(\omega_3)$ and $^{15}N(\omega_1)$ resonance frequencies of Asp-24, Gly-25, Thr-26, and Ile-27. Positive and negative peaks are distinguished using solid and dashed contours, respectively; cross-peaks are negative while autocorrelation peaks are positive. Positions of expected C′,C′ correlations, but which are below the noise threshold, are marked by open circles), while correlations from residues with $^1H(\omega_3)$ and $^{15}N(\omega_1)$ chemical shifts in the vicinity of the selected amide strip are marked by X. The determined $^3J(C',C')$ coupling constants and the NMR derived ϕ angles as well as ϕ angles from the refined X-ray structure of calcium-bound calmodulin are shown in Fig. 36 and summarized in Table 8.

determined by the requirement to resolve the passive $^1J(C_\alpha,C_\beta)$ coupling of approximately 35 Hz. Again, for larger proteins such an extended time is possible for deuterated proteins only, in which the relaxation behavior of the C_α spin is significantly improved.

A variant of the HNCA-E.COSY (Wang and Bax, 1996) experiment utilizes a selective excitation pulse, applied to the C_α region. Many of the $^3J(H^N,C_\beta)$ coupling constants are quite small; the measured values for ubiquitin range from -0.2 ($\phi \approx -140°$) to 2.9 ($\phi \approx -60°$) Hz. Experimental $^3J(H^N,C_\beta)$ values for perdeuterated calmodulin are summarized in Table 8 (Section 3.1.3).

3.1.3. Measurement of $^3J(C',C')$ in a Quantitative HN(CO)CO Experiment

Homonuclear $^3J(C',C')$ coupling constants can be determined using quantitative J correlation spectroscopy.

The HN(CO)CO experiment (Fig. 34) (Hu and Bax, 1996) is an out-and-back-type experiment. Coherence is transferred from the amide proton through the nitrogen nuclei to the preceding carbonyl. A COSY-type pulse scheme is then used to generate C′,C′ correlations. Subsequently, carbonyl coherence is dephased with

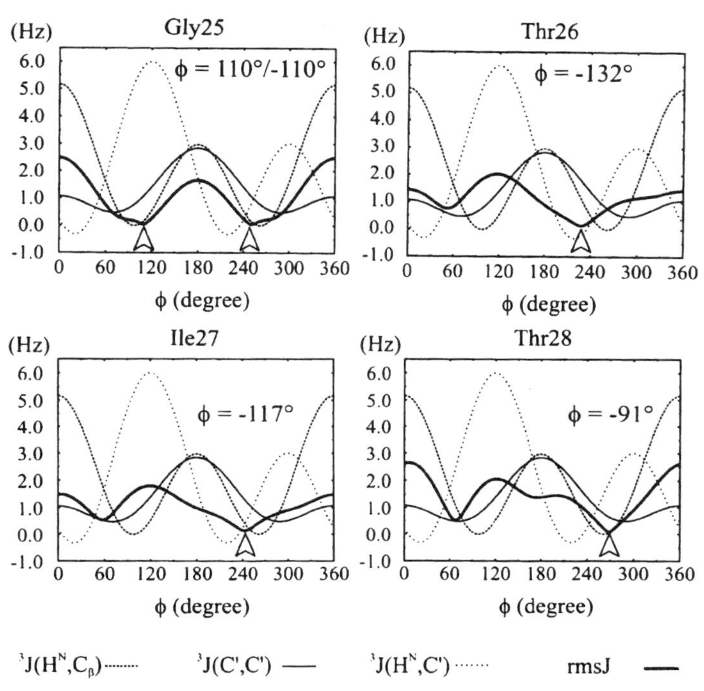

Figure 36. Evaluation of ϕ dihedral angles for residues Gly-25, Thr-26, Ile-27, and Thr-28. The angles were obtained by fitting one ϕ value to the experimental coupling constants $^3J(H^N,C')$, $^3J(H^N,C_\beta)$, and $^3J(C',C')$ listed in Table 8. The dihedral dependence of the ^3J coupling constants is given by a Karplus-type equation, coefficients used for the fit are given in Table 7. The dihedral angle dependence of $^3J(H^N,C')$ on ϕ is represented by a dotted line; of $^3J(H^N,C_\beta)$ on ϕ, by a dashed line; and of $^3J(C',C')$ on ϕ, by a thin solid line. One fixed ϕ value was assumed and the difference between the calculated and the experimental coupling constants is shown as rmsJ curve represented by a thick solid line. Note that for residue Gly-25, because of the absence of a C_β, the $^3J(H^N,C_\beta)$ coupling constant cannot be used to remove the ambiguity around $\phi = \pm 110°$. Because ϕ ranges from 0° to 360°, values for $\phi > 180°$ correspond to $\phi = x\text{-}360°$ in the common representation, where $-180° < \phi < 180°$. Numerical values for ϕ plotted in the figure correspond to the commonly used notation. Arrows indicate the global minima of the rmsJ fits.

respect to the coupling to its adjacent nitrogen, transferred via INEPT to the nitrogen, and, after chemical shift evolution of the ^{15}N nuclei in a CT manner, transferred to the amide proton for detection. The enhancement of this experiment for deuterated proteins compared with protonated ones is only moderate because the C′ relaxation is dominated by the CSA relaxation mechanism and dipolar, CSA cross-correlated relaxation with the neighboring $^{13}C_\alpha$ (Dayie and Wagner, 1997).

Recently, an experiment that yields $^3J(C',C')$ has been demonstrated that utilizes a homonuclear C′,C′-TOCSY transfer step (Grzesiek and Bax, 1997). The

Table 8
Values of $^3J(H^N,C')$, $^3J(H^N,C_\beta)$, and $^3J(C',C')$ Together with the Derived Backbone Angles ϕ Determined for the Amino Acid Residues of Fully Deuterated Calmodulin Shown in Figs. 31, 33, and 35[a]

Residue	$^3J(H^N,C')$ (Hz)	$^3J(H^N,C_\beta)$ (Hz)	$^3J(C',C')$ (Hz)	ϕ (deg)
Gly-25	0.2	—	1.0	±110 (115.4)
Thr-26	1.0	0.0	2.2	−132 (−149.1)
Ile-27	(0.7)	(0.6)	1.3	−117 (−125.4)
Thr-28	0.2	1.9	<1.3	−91 (−93.5)

[a]The extraction ^3J coupling constants from the E.COSY-type experiments was done according to Schwalbe *et al.* (1993b). Because of overlap of the intense intraresidue $C_\alpha C'$ doublet with the weaker $C_\alpha C'$ doublet of the preceding residue in the E.COSY-type experiments, measured couplings in the case of Ile-27 are given in parentheses, reflecting lower precision. The listed ϕ angles of the combined analysis of three different vicinal coupling constants are shown in Fig. 36. In the case of Gly-25, because of the absence of a C_β, the $^3J(H^N,C_\beta)$ coupling constant cannot be used to remove the ambiguity around $\phi = \pm 110°$. For comparison, values for ϕ were extracted for the refined X-ray structure of calcium-bound calmodulin; these values are given in parentheses.

use of homonuclear isotropic mixing instead of pulse-interrupted free precession considerably improves the sensitivity of $^3J(C',C')$ measurements because magnetization transfer is faster by a factor of 2. At the same time, cross-peak amplitudes can be quantified quite accurately using a two-spin approximation. The conditions for achieving maximum cross-peak intensities in both the original and the improved pulse scheme are discussed in greater detail by Grzesiek and Bax (1997).

The experiments rely on the quantification of cross and reference correlation peaks extracted from a COSY-type correlation of two spins B and C (see Section 1.1.1.4). The ratio of cross to reference peak intensity relies only on the constant time in which the coupling J(B,C) evolves, not on the transverse relaxation time or passive couplings of the active nucleus B.

Using the COSY-type HN(CO)CO experiment, it is possible to measure most of the $^3J(C',C')$ values twice because of the symmetry of the experiment, namely, on the H_i^N and on the H_{i+1}^N [for example, see strip plots taken at the $^1H(\omega_3)$ and $^{15}N(\omega_1)$ resonance frequencies of Asp-24 and Gly-25 in Fig. 35]. The pairwise rms difference between those values is on the order of 0.1 Hz.

A reparametrization of the Karplus curve for the $^3J(C',C')$ couplings was performed using the measured values obtained for ubiquitin, a small protein (76 residues) with a relatively short rotational correlation time ($\tau_c \approx 4.1$ ns) (Hu and Bax, 1996). The determined values indicate that this vicinal coupling is useful in restricting the angle ϕ for values around $\pm 180°$ because $^3J(C',C')$ values range from approximately 0.5 ($\phi \pm 120°$) to 3 Hz ($\phi \pm 180°$). Outside β-sheet regions, no sequential connectivities between carbonyl nuclei are observable for most residues in proteins, the reason being the small $^3J(C',C')$ couplings. This absence of cross-peaks, however, may be interpreted in terms of an upper limit for the $^3J(C',C')$ couplings restricting the conformational space of the backbone as well.

3.2. Determination of the Backbone Angle ψ in Perdeuterated Proteins

As shown in Fig. 37, two vicinal coupling constants can be determined around the angle ψ. The relevant coupling constants in a perdeuterated protein fragment are $^3J(N,N)$ and $^3J(N,C_\beta)$.

The interresidue $^3J(N,C_\beta)$ is found to be invariably smaller than 0.4 Hz for all values of the intervening torsion angle ψ. Experimental data have been reported for the small protein ubiquitin. The pulse scheme for the measurement of $^3J(N,C_\beta)$ is of the quantitative J correlation type and is referred to as HNCG (Hu and Bax, 1997a; Konrat *et al.*, 1997). The experiment was proposed for measuring $^3J(N,C_\gamma)$ and, in principle, it is also possible to obtain three- as well as two-bond $J(N,C_\beta)$ couplings from this pulse scheme. Methods for the determination of $J(N,C)$ couplings are discussed in more detail in Section 3.3.2, as the $^3J(N,C_\gamma)$ couplings are significantly larger than $^3J(N,C_\beta)$ and thus can be used in protein structure determination. The methods developed for the determination of $J(N,C_{ali})$ coupling constants in protonated proteins can be directly applied to deuterated molecules, as the experiments are of an out-and-back-type magnetization transfer and rely only on protonated ^{15}N nuclei.

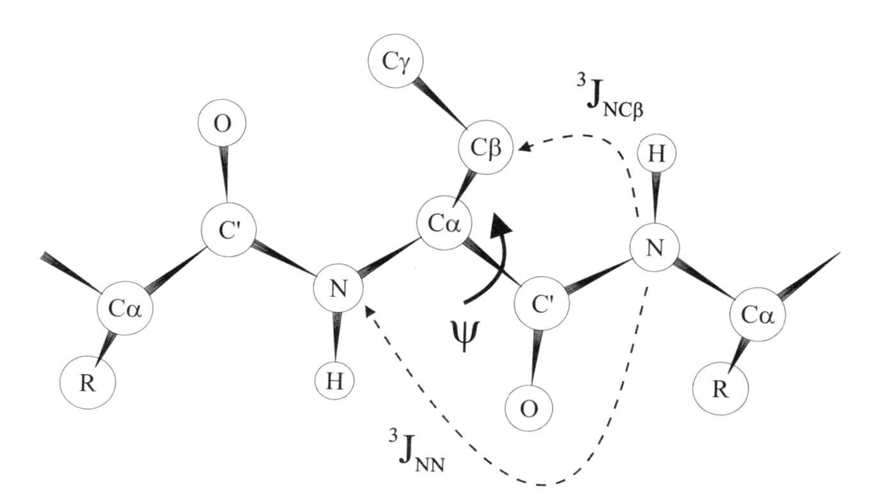

Figure 37. Schematic representation of a peptide backbone showing the dihedral angle ψ and relevant scalar coupling constants in a perdeuterated protein.

The ^3J(N,N) values across the peptide bond are expected to be very small and only recently have some experimental data been reported for a protonated protein (F. Löhr and H. Rüterjans, personal communication). The experiment is similar to the C′,C′ correlation experiment. Coherence transfer between the ^{15}N nuclei can be achieved by either a COSY-type or a homonuclear N,N-TOCSY transfer step. As described for the C′,C′ experiment, the N,N experiment using an isotropic mixing scheme is less sensitive to transverse ^{15}N relaxation. The quantitative H(N)N correlation experiment has been applied to oxidized flavodoxin, and the measured ^3J(N,N) couplings are vanishingly small, ranging from 0.13 to 0.32 Hz.

In conclusion, the utility of both measurable vicinal coupling constants, e.g., $^3J(\mathrm{N,C}_\beta)$ and ^3J(N,N), for the determination of the angle ψ in protein structures is low. Recently, Ottiger and Bax (1997) introduced a method for the determination of the backbone angle ψ based on amide deuterium isotope effects on $^{13}\mathrm{C}_\alpha$ chemical shifts which is applicable to perdeuterated proteins as well.

3.3. Determination of the Sidechain Angle χ_1 in Perdeuterated Proteins

As shown in Fig. 38, two vicinal coupling constants can be determined around the angle χ_1 in a perdeuterated protein. The relevant coupling constants in a perdeuterated protein fragment are $^3J(\mathrm{N,C}_\gamma)$ and $^3J(\mathrm{C',C}_\gamma)$.

3.3.1. Measurement of $^3J(\mathrm{C',C}_\gamma)$ in Quantitative HN(CO)CG Experiments

The HN(CO)CO experiment constitutes a quantitative J experiment (see Section 1.1.1.4) and can also be used for the measurement of vicinal couplings related

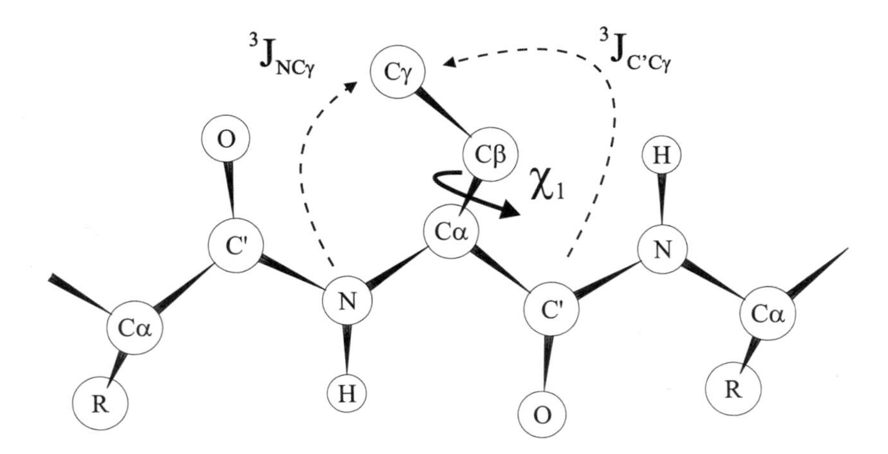

Figure 38. Schematic representation of a peptide backbone showing the dihedral angle χ_1 and relevant scalar coupling constants in a perdeuterated protein.

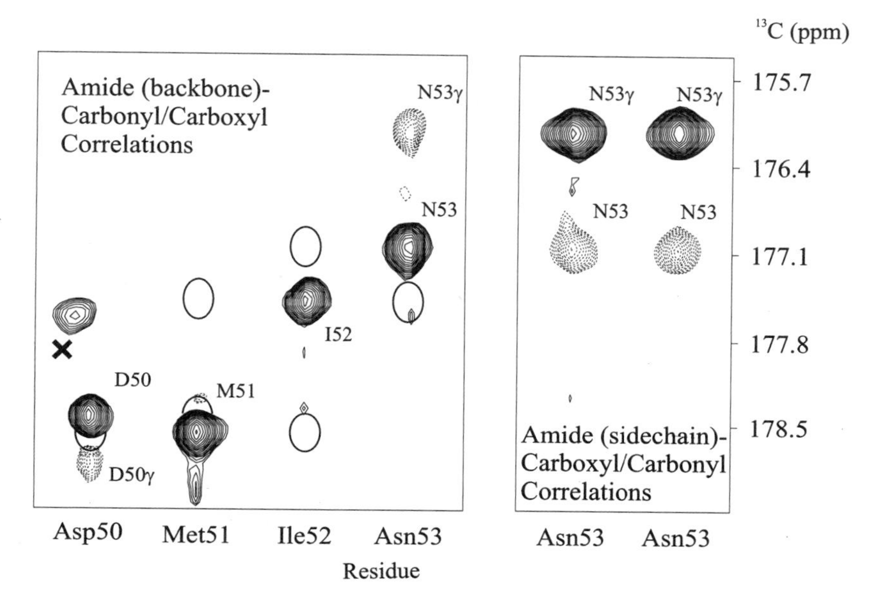

Figure 39. (ω_2) strip plots from the quantitative HN(CO)CG experiment of perdeuterated calmodulin in H_2O taken at the backbone $^1H(\omega_3)$ and $^{15}N(\omega_1)$ resonance frequencies of 49, Asp-50, Met-51, and Ile-52 (left panel) and at the sidechain $^1H(\omega_3)$ and $^{15}N(\omega_1)$ resonance frequencies of Asn-53 (right panel). Positive and negative peaks are distinguished; the diagonal peaks have solid contours while cross-peaks have dashed contours, respectively. Positions of expected C′,C′ correlations, but which are below the noise threshold, are marked by open circles, while correlations from residues with $^1H(\omega_3)$ and $^{15}N(\omega_1)$ chemical shifts in the vicinity of the selected amide strip are marked by X.

to χ_1 for Asn and Asp residues between backbone carbonyl and sidechain carboxyl carbon atoms. Choosing the delay Δ in the HN(CO)CO experiment to be $1/(4^1J(\text{H,N}))$ and thus retaining correlations for NH_2 groups, two additional cross-peaks from sidechain amide protons of asparagines can be observed that yield information in addition to the C',C_γ cross-peak for asparagine residues about the angle χ_1 as shown in Fig. 39 for Asn-53 of calmodulin.

Measured $^3J(C',C_\gamma)$ coupling constants suggest rotameric averaging for amino acid residues Asp-50 and Asn-53 (both located within the third helix of Ca^{2+}-complexed calmodulin), as the values fall in between reported values for the *trans* (4.5 Hz) and *gauche* (0.7 Hz) conformation. The main rotamer is in both cases that with a χ_1 angle close to –60°, with probabilities of 60.5% (Asp-50) and 65.8% (Asn-53).

The analogue of the HN(CO)CG experiment has been applied to different protonated proteins, but can be applied to perdeuterated proteins as well because the pulse schemes are of an out-and-back type. Coherence transfer between $^{13}C'$ and $^{13}C_\gamma$ nuclei in 3D quantitative J correlations can be achieved by either a

Table 9
Values of $^{3}J(C',C_{\gamma})$ Together with Expected Conformations around χ_1 Determined for the Amino Acid Residues of Fully Deuterated Calmodulin Shown in Fig. 39[a]

Residue	$^{3}J(C', C_{\gamma})$ (Hz)	*gauche*$^{-}$ $\chi_1 = -60°$	*gauche*$^{+}$, *trans* $\chi_1 = 60°/180°$
Asp-50	3.0	60.5%	39.5%
Asn-53	3.2	65.8%	34.2%

[a]Populations of *trans* or *gauche*$^{+}$ versus *gauche*$^{-}$ staggered rotameric states were calculated assuming values of $^{3}J(C',C_{\gamma}) = 4.5$ and 0.7 Hz for trans and *gauche* conformations, respectively.

COSY-type (Konrat *et al.*, 1997) or an HMQC (Hu and Bax, 1997c) transfer step, the latter being virtually identical to the regular HN(CO)CA experiment. The HN(CO)CG experiment is more demanding than the HN(CO)CO experiment for two reasons. First, more passive couplings contribute to the cross-peak and the S/N of the cross-peak is therefore lower. Second, 90° excitation pulses and 180° inversion pulses that invert z-magnetization and refocus transverse magnetization at the same time over the complete carbon spectral width, e.g., at 150 MHz, have only recently been developed. Konrat *et al.* (1997) have developed an elegant experiment for the measurement of homonuclear carbon/carbon couplings.

Their pulse scheme allows the measurement of interresidue $^{3}J(C',C_{\beta})$ as well as $^{3}J(C',C')$ in addition to intraresidue $^{3}J(C',C_{\gamma})$ couplings, which are of prime interest. First, coherence originating on the amide proton is transferred through the nitrogen nuclei to the preceding carbonyl. Subsequently, transverse carbonyl coherence is allowed to defocus via long-range homonuclear $^{13}C'$, ^{13}C couplings. The carbon spectrum of proteins contains several well-separated regions. The $^{13}C_{\gamma}$ chemical shifts show a very large dispersion ranging from 180 ppm (δC_{γ} of Asx carboxyl carbons), over 130 ppm (δC_{γ} of Phe, Tyr, Trp, and His aromatic carbons) to 40–20 ppm (δC_{γ} of aliphatic carbons). Thus, care must be taken to invert the complete chemical shift range of the $^{13}C_{\gamma}$ properly. Recently, Böhlen and Bodenhausen (1993) as well as Kupče and Freeman (1995) have described adiabatic pulses for broadband inversion. These pulses are called CHIRP (Böhlen and Bodenhausen, 1993) and WURST (Kupče and Fremann, 1995), respectively. CHIRP pulses of duration 500 μs with a linear frequency sweep of 80 kHz are sufficient to invert the wide $^{13}C_{\gamma}$ chemical shift range while refocusing transverse carbonyl magnetization on a 600-MHz spectrometer as shown in Fig. 41.

During this defocusing period τ, a fraction of magnetization proportional to $\cos(2\pi J(C'C_{\gamma})\tau)\ \Pi_k \cos(2\pi J(C'C_k)\ \tau)$ ($k \neq C',C_{\gamma}$ covers all other J coupled carbons) remains on the carbonyl spin (reference peak) while coherence on C_{γ} (cross-peak) is created with an efficiency proportional to $\sin(2\pi J(C'C_{\gamma})\tau)\ \Pi_k \cos(2\pi J(C'C_k)\tau)$.

After the ^{13}C chemical shift evolution period, the carbon coherences are transferred back the reverse pathway to the amide proton for detection. As found for all quantitative J-correlation experiments, the ratio of cross and reference peak integrals is related to the $^3J(C',C_\gamma)$ coupling by the equation $I^{cross}/I^{ref} = -\tan^2(2\pi J(C'C_\gamma)\tau)$.

For residues Tyr-91, Ile-92 (γ_1), Tyr-93, and Asp-95 of Snase complexed with Ca^{2+} and pdTp, measured $^3J(C',C_\gamma)$ values range from 2.9 to 3.3 Hz, with the exception of the carbonyl/carboxyl correlation of residue Asp-95 for which a coupling of 5.1 Hz was observed (see Fig. 42). All of these couplings correspond to *trans* conformations around χ_1.

Hu and Bax have developed a 3D HMQC-based quantitative J-correlation scheme for the measurement of $^3J(C',C_{ali})$ (Hu and Bax, 1997c). In contrast to the

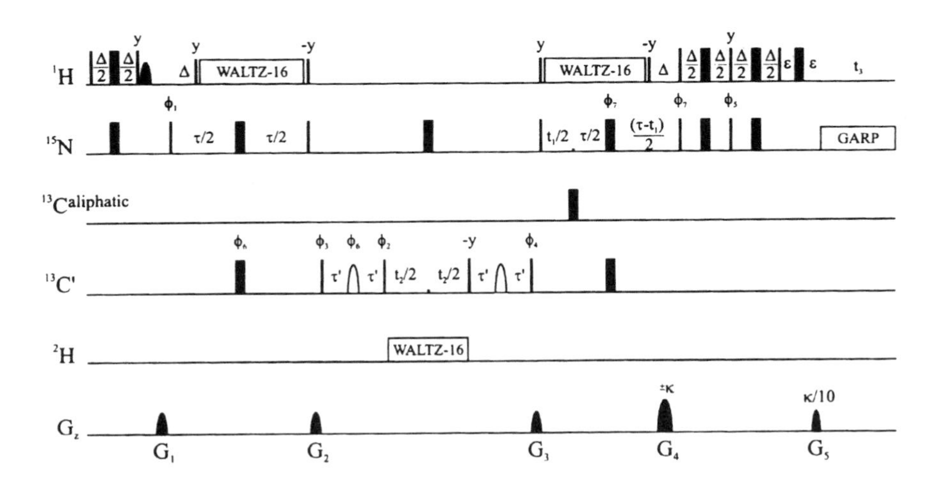

Figure 40. Pulse sequence of the quantitative HN(CO)CG experiment adapted from Konrat *et al.* (1997). Phase cycling: $\phi_1 = x, -x$, $\phi_2 = x$, $\phi_3 = 2(x), 2(-x)$, $\phi_4 = 4(x), 4(-x)$, $\phi_5 = 8(-y), 8(y)$, $\phi_6 = x$, $\phi_7 = 8(x), 8(-x)$, receiver $= x, -x, -x, x, -x, x, x, -x, -x, x, x, -x, x, -x, -x, x$. Quadrature detection in the t_2 dimension is obtained by altering ϕ_2, ϕ_3, and ϕ_6 according to States–TPPI. For each t_1 value, echo and antiecho coherences are obtained by recording data sets where the phase ϕ_5 and the fourth gradient are inverted. Delay durations: $\Delta = 5.5$ ms, $\tau = 25.6$ ms, $\tau' = 37$ ms, and $\varepsilon = 1.2$ ms. Carrier positions: ^{1}H,4.65 ppm; ^{13}C, 100 ppm (equidistant from C′ and aliphatic C_γ regions); ^{15}N, 118.3 ppm; ^{2}H, 4.65 ppm. The excitation center of the ^{13}C pulses of phases ϕ_3 and ϕ_4 is shifted to 173 ppm by phase modulation; the second G4 pulse was time-reversed so as to avoid phase errors. The first and the last carbonyl carbon selective pulses were implemented as phase-modulated G3 pulses. The ^{13}C excitation pulses flanking the t_2 evolution period are applied on resonance with a field strength of 21 kHz. The CHIRP inversion pulses (open half-egg-shaped) are of duration 500 μs with the center of the 80-kHz sweep set at 173 ppm. The amplitude of the CHIRP pulses is apodized at the first and last 30% of each pulse with a sine function. The central plateau was maintained at 10.37 kHz. Aliphatic carbon selective pulses were implemented as phase-modulated G3 pulses.

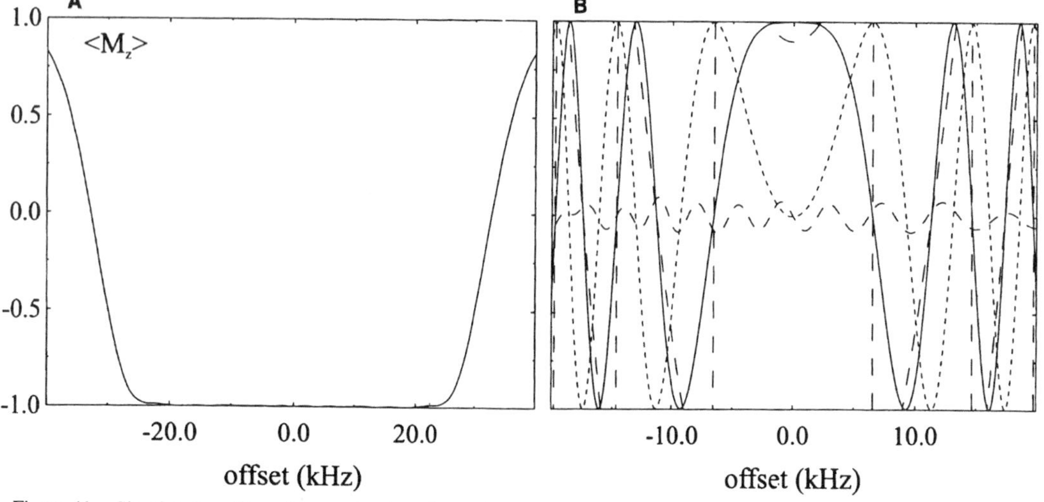

Figure 41. Simulated profiles of magnetization after a single CHIRP pulse with a linear sweep from –40 to +40 kHz in 500 μs with $\gamma B_1/2\pi = 10.37$ kHz with sinusoidal smoothing of the first and last 30%. The resonance offset was stepped from –40 to +40 kHz in increments of 100 Hz. (A) Inversion profile showing the response of initial M_z magnetization after the pulse as a function of offset. The pulse has an inversion bandwidth of ±25.8 kHz (>95% inversion), sufficient to invert the complete $^{13}C_\gamma$ shift range at 600 MHz. (B) Refocusing profile starting from $\langle M_x \rangle$ with a phase x of the CHIRP pulse showing the amount of M_z (dashed line), M_y (dotted line), M_x (solid line), and the phase profile (long dashed line) created after the pulse as a function of offset. The pulse achieves refocusing across a bandwidth roughly equal to ± 2 kHz with phase changes less than 8° over this range centered in the middle of the frequency sweep. Because of the small chemical shift dispersion of carbonyl carbons, the introduced phase errors are small provided that the center of the sweep is set to the middle of the carbonyl region. The profiles were calculated using the program "mule" kindly provided by Dr. W. Bermel.

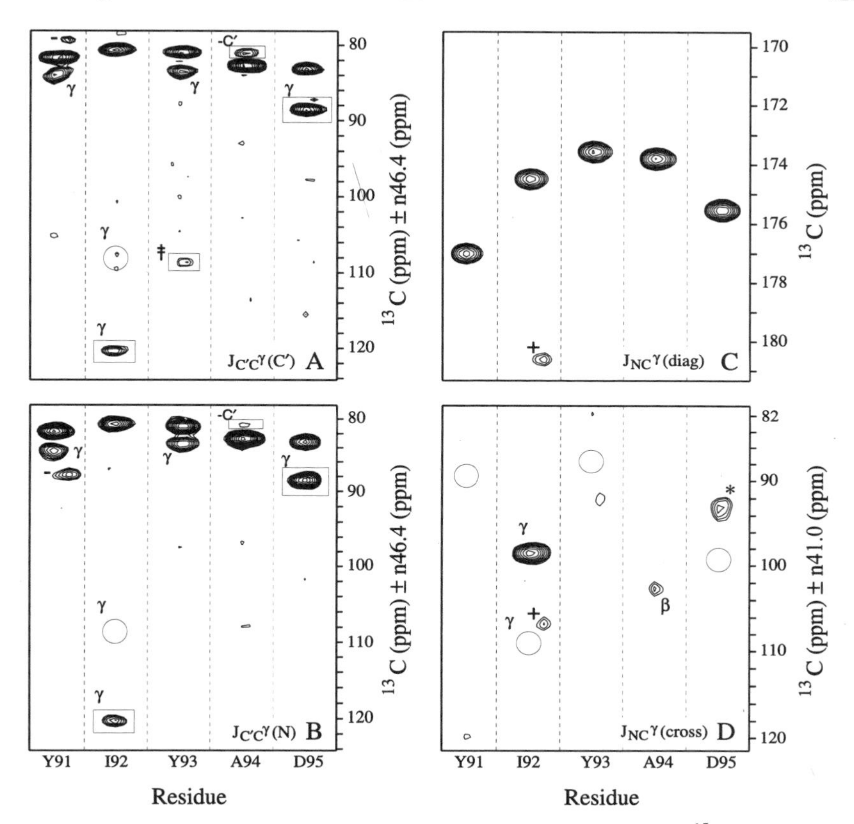

Figure 42. (ω_2) strip plots from the quantitative HN(CO)CG experiment). (A) ^{15}N chemical shift evolution during t_2, and H(N)COCG experiment kindly provided by Konrat *et al.* (1997); (B) ^{13}C′ chemical shift evolution during t_2, applied to a 1.5 mM sample of Snase complexed with Ca^{2+} and pdTp in H_2O taken at the backbone ^{1}H(ω_3) and ^{15}N(ω_1) resonance frequencies from Tyr-91 to Asp-95. Positive and negative peaks are distinguished by placing boxes around peaks of opposite phase. Positions of expected cross-peaks, but which are below the noise threshold, are marked by open circles, while correlations from residues with ^{1}H(ω_3) and ^{15}N(ω_1) chemical shifts in the vicinity of the selected amide strip are marked by +. Correlation peaks to the previous/next carbonyl are indicated by –C′ or +C′, respectively, while cross-peaks labeled with ‡ are not assigned. The value n on the y-axis label corresponds to an integer with $0 \leq n \leq 2$. The majority of the peaks have been aliased twice with $n = 2$ for the C′ and for many of the C_γ correlations because the ^{13}C carrier was centered at 101 ppm and a spectral width (^{13}C) of 5834 Hz was employed.

pulse scheme of Konrat *et al.*, this experiment has been optimized to observe correlations between backbone carbonyl and aliphatic carbons. Coherence is excited on the $^1H^N$ and is transferred through the ^{15}N to the preceding ^{13}C′. After the latter coherence is allowed to defocus via long-range homonuclear ^{13}C′, ^{13}C couplings, a bandselective 90° pulse for aliphatic carbons excites ^{13}C′,$^{13}C_{ali}$ zero-

and double-quantum coherence, giving rise to a C_{ali} cross-peak in the t_2 (^{13}C) chemical shift evolution period. In contrast to the pulse scheme introduced by Konrat *et al.* (1997), the requirements for the RF pulses are less critical. During the multiple-quantum evolution period and the defocusing periods 2τ, the chemical shift evolution of the carbonyl spins is refocused by a bandselective 180° pulse. The multiple-quantum coherence $2C'_xC_{\alpha y}$ does not evolve under the influence of the active scalar coupling, $^1J(C',C_\alpha)$, during the t_2 period. The dephasing effect of $^1J(C',N)$ is refocused by the bandselective 180° pulse as well. Evolution under homonuclear $^3J(C',C')$ scalar coupling is not refocused, but attenuates both spectra, since cross-peak and reference peak intensities have to be recorded in two separate experiments, by the same factor. Note that up to five cross-peaks, resulting from $^1J(C',C_\alpha)$, $^2J(C',C_\alpha)$, $^2J(C',C_\beta)$, $^3J(C',C_\beta)$, and $^3J(C',C_\gamma)$ couplings, can be observed in the HN(CO)CG experiment. Subsequently, magnetization is transferred back the reverse pathway to the amide proton for detection. The duration of the defocusing period τ corresponds to an integral of $(1/^1J(C',C_\alpha))$, such that C′ is in phase with respect to the directly attached C_α when multiple-quantum coherence is excited, giving rise to usually vanishing weak cross-peaks resulting from $^1J(C',C_\alpha)$. This is related to the small variation of $^1J(C',C_\alpha)$ in proteins: 54 ± 1 Hz. The 3D HN(CO)CA reference spectrum is obtained by shortening the defocusing period τ by $(1/2{*}^1J(C',C_\alpha))$, thus only C_α cross-peaks have nonvanishing intensities. Again, the intensity ratio of a C',C_{ali} long-range correlation and the C_α cross-peak in the reference HN(CO)CA experiment is related to the $^3J(C',C_{ali})$ coupling of interest. The largest $^3J(C',C_\gamma)$ coupling constants (~4.0 Hz) measured for the rapidly tumbling protein ubiquitin correspond to conformations where C′ is *trans* with respect to C_γ. However, a wide range of *trans* $^3J(C',C_\gamma)$ coupling constants (2.2–4.0 Hz) similar to the results obtained for Snase complexed with Ca^{2+} and pdTp are reported for ubiquitin and apo-calmodulin.

$^3J(C',C_\gamma)$ coupling constants in aromatic residues can be measured in a 2D spin-echo difference experiment (see also Section 1.1.1.3) (Hu and Bax, 1997b).

In the pulse scheme, antiphase $2C'_yN_z$ coherence is created utilizing two INEPT-type transfer steps. A selective 180° C′ pulse, applied at the midpoint of the following CT spin-echo delay, rephases the effect of $J(C',N)$ and $J(C',C)$ couplings. In the first experiment, a selective 180° C_{arom} pulse at the end of the spin-echo delay refocuses the effect of $J(C',C_{arom})$ as well. In the second experiment, recorded in an interleaved manner with the result of the two experiments stored separately, the 180° C_{arom} pulse is shifted to the midpoint of the spin-echo delay, causing $J(C',C_{arom})$ dephasing occurring during for the full spin-echo period. In the latter experiment the intensity in the 2D ^{15}N, ^1H correlation (I_b) is attenuated by $\cos(2\pi J(C',C_{arom})\tau)$ relative to the first experiment (I_a). As the spin-echo delay τ is known, $J(C',C_{arom})$ can be calculated from $J(C',C_{arom}) = \cos^{-1}(I_b/I_a)/(2\pi\tau)$. Applications of this spin-echo difference experiment to a sample of $^{13}C/^{15}N$-labeled denatured hen lysozyme are shown in Fig. 44. Experiments were performed in

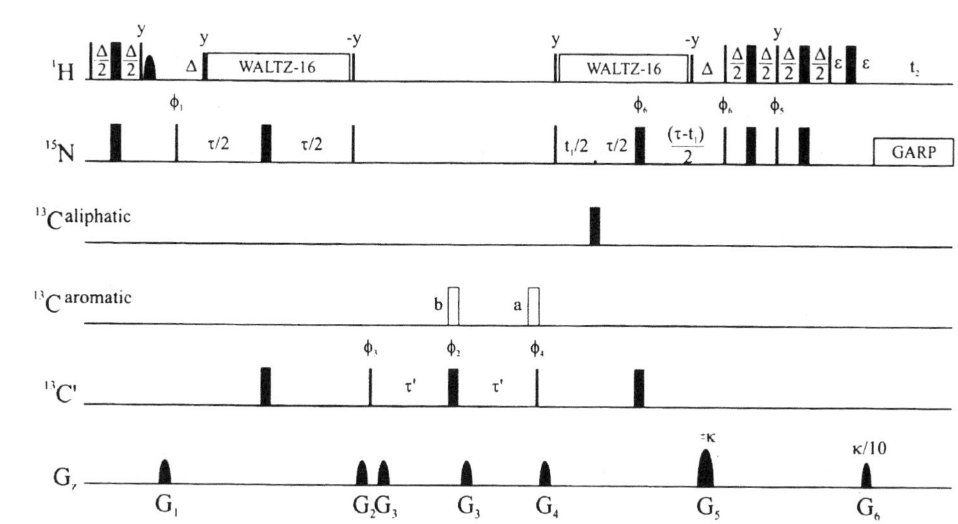

Figure 43. Pulse sequence of the spin-echo difference HN(CO)CG$_{arom}$ experiment adapted from Hu and Bax (1997b). Phase cycling: $\phi_1 = x, -x$, $\phi_2 = 41°$ (Bloch–Siegert shift compensation), $\phi_3 = 2(x), 2(-x)$, $\phi_4 = 4(x), 4(-x)$, $\phi_5 = 8(-y), 8(y)$, $\phi_6 = 8(x), 8(-x)$, receiver $= x, -x, -x, x, -x, x, x, -x, -x, x, x, -x, x, -x, -x, x$. For each t_1 value, echo and antiecho coherences are obtained by recording data sets where the phase ϕ_5 and the fifth gradient are inverted. Delay durations: $\Delta = 5.5$ ms, $\tau = 25.6$ ms, $\tau' = 50$ ms, and $\varepsilon = 1.2$ ms. The reference spectrum is recorded using the ^{13}C$_{arom}$ selective G3 inversion pulse of duration 768 μs in position a, omitting the pulse labeled b, whereas the attenuated spectrum is recorded using pulse b and omitting pulse a. Spectra are recorded in an interleaved manner.

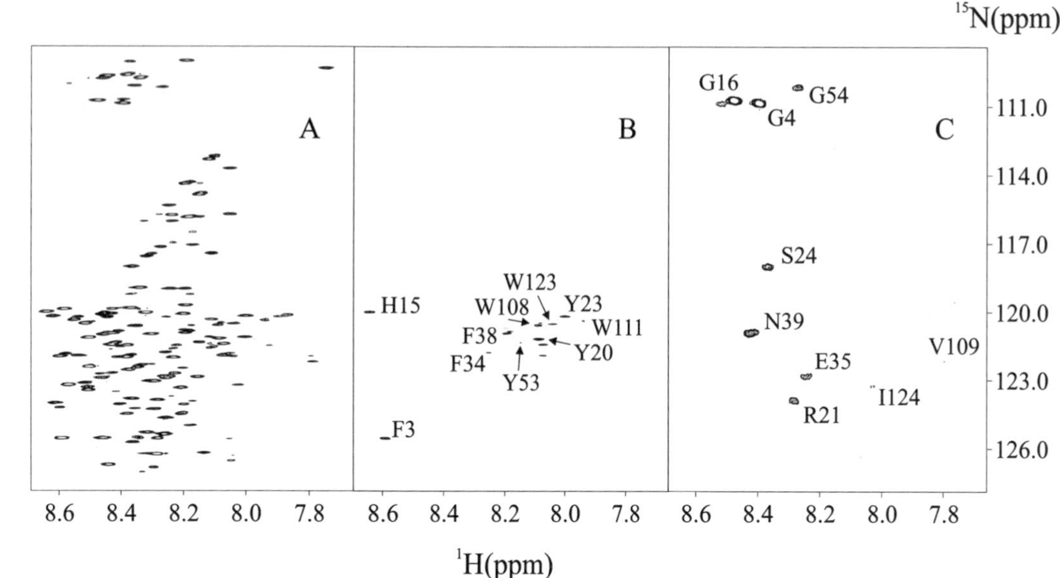

Figure 44. ^{15}N correlation spectra of hen lysozyme denatured in 8 M urea at pH 2. (A) Reference spectrum obtained with the spin-echo difference $HNCG_{arom}$ experiment of Figure 3.18 and the selective $^{13}C_{arom}$ pulse in position a. (B) Difference spectrum from the quantitative $HNCG_{arom}$ experiment (Figure 45), showing ^{1}H, ^{15}N correlations with a significant coupling (> 0.7 Hz) to an aromatic C_{γ}. (C) Difference spectrum from the quantitative $HN(CO)CG_{arom}$ experiment (Fig. 43), showing ^{1}H, ^{15}N correlations for which the preceding backbone carbonyl nuclei has a significant coupling (> 1.0 Hz) to an aromatic C_{γ}. The determined $^{3}J(N,C_{\gamma})$ and $^{3}J(C',C_{\gamma})$ coupling constants and expected conformations for χ_1 of denatured hen lysozyme are summarized in Table 10.

Table 10
Values of $^3J(C',C_\gamma)$ and $^3J(N',C_\gamma)$ Together with Expected Conformations around χ_1 Determined for Aromatic Residues of Denatured Hen Lysozyme[a] Shown in Fig. 44[b]

Residue	$^3J(C',C_\gamma)$ (Hz)	$^3J(N,C_\gamma)$ (Hz)	*gauche*$^+$ ($\chi_1 = 60°$)	*trans* ($\chi_1 = 180°$)	*gauche*$^+$ ($\chi_1 = -60°$)
Phe-3	2.6	0.8	26.7%	20.0%	53.3%
His-15	3.1	1.0	0.0%	30.0%	70.0%
Tyr-20	2.0	1.3	21.7%	45.0%	33.3%
Try-23	2.2	1.0	20.0%	30.0%	40.0%
Phe-34	1.1	1.3	51.7%	45.0%	3.3%
Phe-38	2.3	1.0	26.7%	30.0%	43.3%
Tyr-53	2.2	1.1	25.0%	35.0%	40.0%
Trp-108	1.9	1.6	10.0%	60.0%	30.0%
Trp-111	nd	1.7	—	65.0%	—
Trp-123	2.1	1.4	13.3%	50.0%	36.7%

[a]Schwalbe *et al.* 1997.

[b]Populations of three staggered rotameric states were calculated assuming values of $^3J(C',C_\gamma)$ = 4.0 and 1.0 Hz for *trans* and *gauche* conformations, respectively, or $^3J(N,C_\gamma)$ = 2.4 and 0.4 Hz for *trans* and *gauche* conformations.

90:10 H_2O/D_2O (pH 2.0, 8 M urea) at 293 K. Spectra were obtained by using a protein concentration of about 2.0 mM.

The reported $^3J(C',C_\gamma)$ for aromatic residues in ubiquitin, calmodulin, complexed to a 26-residue unlabeled peptide fragment, and HIV-1 Nef cluster in relatively narrow ranges: 4.0 ± 0.3 Hz for *trans* and ≤1.1 Hz for *gauche* conformations. Measured $^3J(C',C_\gamma)$ and $^3J(N,C_\gamma)$ coupling constants suggest rotameric averaging for all amino acid residues with aromatic sidechains where coupling constants were determined in denatured lysozyme because the values fall in between reported values for the *trans* (4.0 Hz) and *gauche* (≤1.1 Hz) conformation [for $^3J(C',C_\gamma)$] or the *trans* (2.4 Hz) and *gauche* (≤0.5 Hz) conformation [for $^3J(N,C_\gamma)$], respectively. Most residues, e.g., Phe-3, Tyr-20, Tyr-23, Phe-38, and Tyr-53, occupy all three staggered states without significante preferences for a single rotamer. For Trp residues, a conformational equilibrium between two static rotamers is in better agreement with the experimental data. The main rotamer is in all cases that with a χ_1 angle close to 180°, with probabilities of 60.0% (Trp-108), 65.0% (Trp-111), and 50.0% (Trp-123), whereas the rotamer with $\chi_1 = -60°$ is the less favored one. Two sidechain conformations with $\chi_1 = -60°$ (70% probability) and $\chi_1 = 180°$ (30% probability) explain the experimental data for residue His-15 best. Phe-34 occupies two staggered states where $\chi_1 = 180°$ and $\chi_1 = 60°$ are nearly equally distributed.

3.3.2. Measurement of $^3J(N,C_\gamma)$ in a Quantitative HNCG Experiment

Lastly, quantitative *J*-correlation of amide nitrogens with alipathic, aromatic, and carbonyl carbons during a CT delay yields $^3J(N,C)$ coupling constants (Hu and Bax, 1997a,c; Konrat *et al.*, 1997). As opposed to the above class of quantitative correlation experiments, cross and reference peak intensities for correlation of heteronuclear coherence have to be recorded in two separate experiments (see also Section 1.1.1.4 and Fig. 1.5b). 3D versions of the quantitative HNCG experiment allow the measurement of $^3J(N,C_\gamma)$ for all residues with C_γ nuclei. Konrat *et al.* (1997) recorded separate experiments to obtain cross (HNCG) and diagonal peak (HNCO) intensities. Both pulse schemes utilize CHIRP pulses for broadband ^{13}C inversion as described in Section 3.3.1. In the HNCG experiment, coherence originating on the $^1H^N$ is transferred through the ^{15}N nuclei to the $^{13}C_\gamma$. As one- and two-bond $J(C_\alpha,N)$ couplings would significantly attenuate the signal during the lengthy defocusing period τ, selective $^{13}C_\alpha$ pulses are inserted to refocus the effects of these couplings. The transfer function between scalar-coupled ^{15}N and $^{13}C_\gamma$ nuclei is proportional to $\sin(2\pi J(C_\gamma,N)\tau)$. In the case of the reference experiment, the magnetization flow can be described in a manner analogous to the HNCO experiment. As an effect, the diagonal peak transfer amplitudes are proportional to $\cos(2\pi J(C',N)\tau)$. The ratio of the cross-peak volume of a long-range correlation to a $^{13}C_\gamma$ nuclei in the HNCG experiment to the corresponding diagonal peak volume in the reference HNCO experiment is given by $\tan^2(2\pi J(C_\gamma,N)\tau)\cos^2(2\pi J(C',N)\tau)/\sin^2(2\pi J(C',N)\tau)$ allowing a calculation of $^3J(N,C_\gamma)$ provided the $^1J(C',N)$ is well known. Unlike the previously mentioned experiments, accurate estimates of $^3J(N,C_\gamma)$ values require the measurement of volume intensities as the involved $^{13}C'$ and $^{13}C_\gamma$ nuclei in two experiments have significantly different linewidths. Average $^1J(C',N)$ values reported for Snase range from 14 to 16 Hz with $^1J(C',N) = 14.8 \pm 0.5$ Hz for residues within regular secondary structure elements and $^1J(C',N) = 15.6 \pm 0.5$ Hz for residues in random coils (Delaglio *et al.*, 1991). Thus, neglecting the terms related to $^1J(C',N)$ underestimates $^3J(N,C_\gamma)$ by less than 2% provided that values of $^1J(C',N)$ range from 14 to 16 Hz. Corresponding (ω_2) strips from the HNCG experiment and the reference HNCO experiment are shown in Fig. 42 for residues Tyr-91, Ile-92, Tyr-93, Ala-94, and Asp-95 of Snase complexed with Ca^{2+} and pdTp. A strong $^3J(N,C_\gamma)$ coupling of 1.9 Hz is observed for Ile-92(γ_2). Additional cross-peaks, arising from two-bond $^2J(C_\beta,N)$ correlations for Ala-94 or one-bond $^1J(C',N)$ correlations for Asp-95, are also visible.

Hu and colleagues proposed simple 2D spin-echo difference ^1H, ^{15}N HSQC experiments for the measurement of $^3J(N,C_\gamma)$ for residues with aliphatic (Hu and Bax, 1997a) or aromatic carbon nuclei (Hu and Bax, 1997b) which are applicable to perdeuterated proteins as well.

The 2D spin-echo difference ^1H, ^{15}N HSQC experiment for the measurement of $^3J(N,C_\gamma)$ for residues with aromatic carbon nuclei is shown in Fig. 45. Data sets

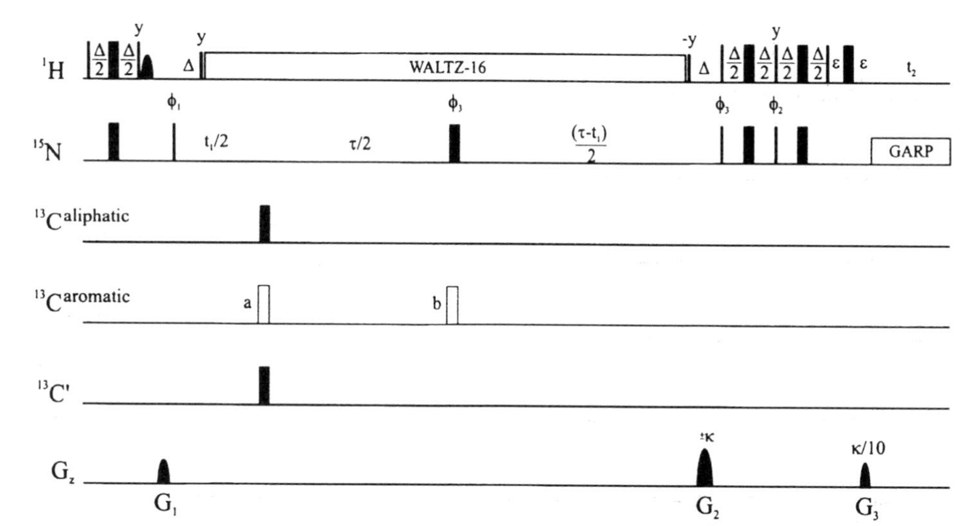

Figure 45. Pulse sequence of the spin-echo difference HNCG$_{arom}$ experiment adapted from Hu and Bax (1997b). Phase cycling: $\phi_1 = x, -x$, $\phi_2 = 2(-y), 2(y)$, $\phi_3 = 2(x), 2(-x)$, receiver $= x, -x, -x, x$. For each t_1 value, echo and antiecho coherences are obtained by recording data sets where the phase ϕ_2 and the second gradient are inverted. Delay durations: $\Delta = 5.5$ ms, $\tau = 128$ ms, and $\varepsilon = 1.2$ ms. The reference spectrum is recorded using the $^{13}C_{arom}$ selective G3 inversion pulse of duration 768 μs in position a, omitting the pulse labeled b, whereas the attenuated spectrum is recorded using pulse b and omitting pulse a. Spectra are recorded in an interleaved manner.

are recorded in an interleaved manner with the result of the two experiments stored separately. In the case of the reference experiment, the scheme is identical to a CT ^{1}H, ^{15}N correlation utilizing refocusing and defocusing periods before and after in-phase N_x coherence evolves during t_1 under composite proton decoupling. In the case of the reference experiment with selective 180° C_{ali} and 180° C_{arom} pulses applied after $t_1/2$, effects of $J(C_\beta,N)$ and $J(C_\gamma,N)$ are refocused. In the second variable experiment either the selective 180° C_{ali} or the 180° C_{arom} pulse is shifted to the midpoint of the spin-echo delay, causing $J(C_{ali},N)$ or $J(C_{arom},N)$ dephasing to be active for the full spin-echo period, respectively. For all schemes ^{15}N dephasing related to $J(C',N)$ and $J(C_\alpha,N)$ is eliminated by selective 180° C′ and 180° C_α pulses applied after $t_1/2$. For the pulse scheme with the selective 180° C_{arom} pulse shifted to the midpoint of the spin-echo delay τ, the intensity in the 2D ^{15}N, ^{1}H correlation (I_b) is attenuated by $\cos(2\pi J(C_{arom},N)\tau)$ relative to the reference experiment (I_a), allowing $^3J(N,C_\gamma)$ for aromatic residues to be calculated straightforwardly. With the selective 180° C_{ali} pulse shifted to the midpoint of the spin-echo delay τ, the net amount of ^{15}N transverse magnetization is attenuated by $\cos(2\pi ^3J(C'_\gamma,N)\tau)\cos(2\pi ^3J(C''_\gamma,N)\tau)\cos(2\pi ^3J(C_{\beta i-1},N_i)\tau)\cos(2\pi ^2J(C_{\beta i},N_i)\tau)$ where the $\cos(2\pi ^3J(C''_\gamma,N)\tau)$ term is only present for Ile and Val residues. For Val and Ile, it is not possible to distinguish whether the difference signal originates primarily from $^3J(C'_\gamma,N)$ or $^3J(C''_\gamma,N)$. Because the interresidue $^3J(C_{\beta i-1},N_i)$ is found to be invariably smaller than 0.4 Hz, one further assumes that either $^3J(C_{\beta i-1},N_i) = {}^2J(C_{\beta i},N_i) = 0$, or $^3J(C_{\beta i-1},N_i) = 0$ and $^2J(C_{\beta i},N_i) = 0.7$ Hz in calculating $^3J(C_\gamma{}',N)$.

Applications of the 2D spin-echo difference ^{1}H,^{15}N HSQC experiments for the measurement of $^3J(N,C_\gamma)$ for residues with aromatic carbon nuclei to a sample of ^{13}C/^{15}N-enriched denatured hen lysozyme are shown in Fig. 44. Reported $^3J(N,C_\gamma)$ values for aromatic residues in ubiquitin, calmodulin, complexed to a 26-residue unlabeled peptide fragment, and HIV-1 Nef cluster in relatively narrow ranges: 2.4 ± 0.2 Hz for *trans* and ≤0.5 Hz for *gauche* conformations suggesting that for these residues χ_1 rotamer averaging is rare. In contrast, the mean experimental $^3J(N,C_\gamma)$ value of 1.2 Hz for aromatic residues in denatured lysozyme is close to the value expected for rotameric averaging.

It has been shown in the literature that $^3J(N,C)$ coupling constants do not exceed 3 Hz.

3.4. Determination of the Sidechain Angle χ_2 in Perdeuterated Proteins

The homonuclear $^3J(C_\alpha,C_\delta)$ coupling constant is the only one that can be determined around the angle χ_2 in perdeuterated proteins. The assignment of χ_2 rotamers derived from $^3J(C_\alpha,C_\delta)$ is of considerable importance because of the limited NOE contacts in perdeuterated sidechains. The HN(CO)CAC$_{ali}$ experiment can be used to improve the structure elucidation of deuterated proteins as the

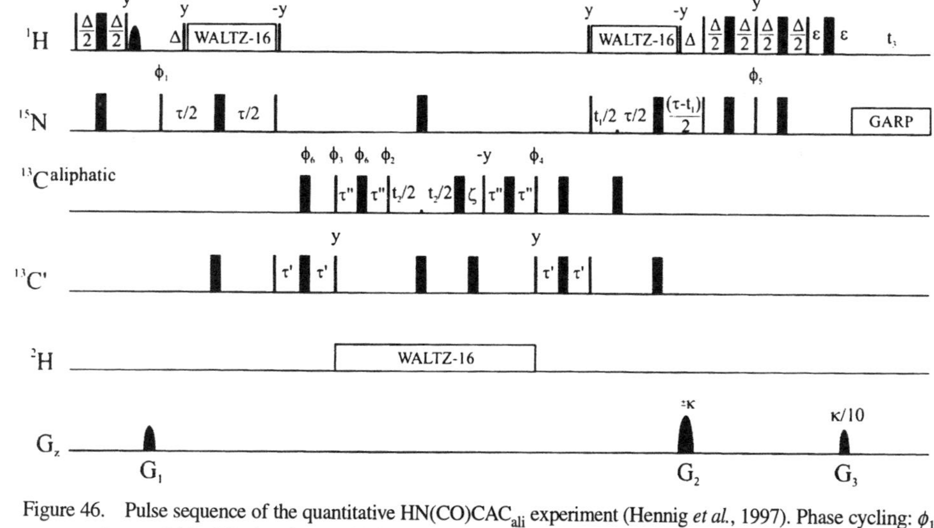

Figure 46. Pulse sequence of the quantitative $HN(CO)CAC_{ali}$ experiment (Hennig *et al.*, 1997). Phase cycling: $\phi_1 = x, -x$, $\phi_2 = 2(y), 2(-y)$, $\phi_3 = 4(x), 4(-x)$, $\phi_4 = 8(x), 8(-x)$, $\phi_5 = -y$, $\phi_6 = x$, receiver = $x, -x, x, -x, -x, x, -x, x, -x, x, -x, x, x, -x, x, -x$. Quadrature detection in the t_2 dimension is obtained by altering ϕ_2, ϕ_3, and ϕ_6 according to States–TPPI. For each t_1 value, echo and antiecho coherences are obtained by recording data sets where the phase ϕ_5 and the second gradient are inverted. Delay durations: $\Delta = 5.5$ ms, $\tau = 35$ ms, $\tau' = 4.55$ ms, $\tau'' = 28.2$ ms, $\zeta = t_2(0)$, and $\varepsilon = 1.2$ ms. Carbonyl-selective pulses were implemented as phase-modulated shaped pulses.

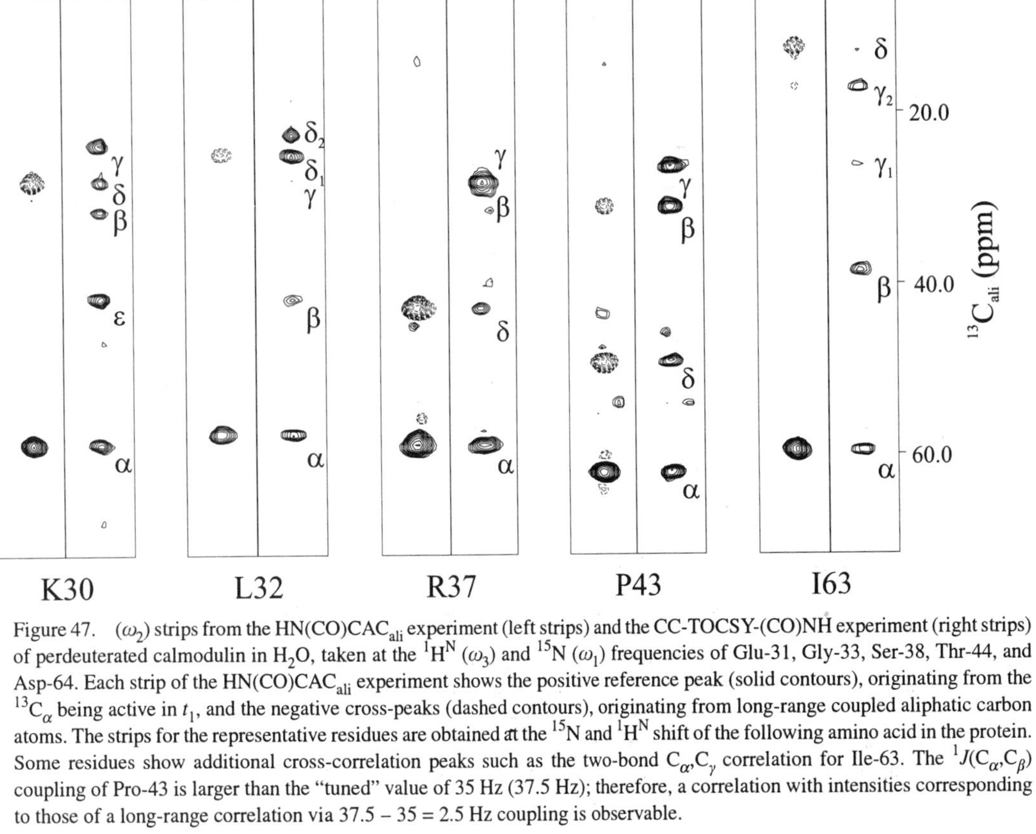

Figure 47. (ω_2) strips from the HN(CO)CAC$_{ali}$ experiment (left strips) and the CC-TOCSY-(CO)NH experiment (right strips) of perdeuterated calmodulin in H_2O, taken at the $^1H^N$ (ω_3) and ^{15}N (ω_1) frequencies of Glu-31, Gly-33, Ser-38, Thr-44, and Asp-64. Each strip of the HN(CO)CAC$_{ali}$ experiment shows the positive reference peak (solid contours), originating from the $^{13}C_\alpha$ being active in t_1, and the negative cross-peaks (dashed contours), originating from long-range coupled aliphatic carbon atoms. The strips for the representative residues are obtained at the ^{15}N and $^1H^N$ shift of the following amino acid in the protein. Some residues show additional cross-correlation peaks such as the two-bond C_α,C_γ correlation for Ile-63. The $^1J(C_\alpha,C_\beta)$ coupling of Pro-43 is larger than the "tuned" value of 35 Hz (37.5 Hz); therefore, a correlation with intensities corresponding to those of a long-range correlation via 37.5 – 35 = 2.5 Hz coupling is observable.

sidechains of Arg, Lys, Leu, Ile, and Pro residues are often involved in intra- and intermolecular interactions (Hennig *et al.*, 1997).

Measurement of $^3J(C_\alpha,C_\delta)$ in a Quantitative HN(CO)CAC$_{ali}$ Experiment

The proposed pulse sequence relies on quantitative correlations of C_α atoms with aliphatic sidechain carbon atoms for the determination of the torsion angle χ_2. The experiment, an out-and-back-type quantitative HN(CO)CAC$_{ali}$ (Fig. 46), creates transverse C_α magnetization that defocuses via long-range homonuclear couplings, i.e., between C_α and C_δ during a period of $2\tau'' = (2/^1J(C,C))$. After chemical shift evolution of C_α (reference peak) and C_δ (cross-peak) coherence during t_2, the two coherences are transferred back via the C_α to the H^N following the reverse pathway. The transfer amplitude for the reference peak is proportional to $\cos^2(2\pi^3J(C_\alpha,C_\delta)\tau'')\ \Pi_k\cos^2(2\pi J(C_\alpha,C_k)\tau'')$ ($k \neq \alpha,\delta$ covers all other J coupled aliphatic carbons, while the transfer amplitude for the cross-peak is proportional to $-\sin^2(2\pi^3J(C_\alpha,C_\delta)\tau'')\ \Pi_k\cos^2(2\pi J(C_\alpha,C_k)\tau'')$. Thus, the ratio of transfer amplitudes for the reference and the cross-peak is given by $-\tan^2(2\pi J(C_\alpha,C_\delta)\tau'')$. As τ'' is known and the line shapes of the reference and the cross-correlation peaks are assumed to be identical in the ^{15}N (ω_1), $^1H^N$ (ω_3), and within the digital resolution of the ^{13}C (ω_2) dimension, the integral ratio of these peaks provides a measure of the magnitude of J(C,C): $I^{cross}/I^{ref} = -\tan^2(2\pi J(C_\alpha,C_\delta)\tau'')$.

Figure 47 shows the results obtained; the carbon resonance frequencies in ω_2 can be inferred from strip plots of the CC-TOCSY-(CO)NH experiment (Farmer and Venters, 1995). The preferred conformation around χ_2 is an antiperiplanar orientation of the two carbon atoms. This is indeed found for most of the residues in calmodulin, where all measured values of $^3J(C_\alpha,C_\delta)$ (2.6–3.8 Hz) fall between the values reported for *trans* (4.0 Hz) and *gauche* (0.9 Hz) aliphatic ^3J(C,C) couplings in proteins (Krivdin and Della, 1991; Bax *et al.*, 1992; Grzesiek *et al.*, 1993; Vuister *et al.*, 1993), with the exception of Arg-37 for which a coupling of

Table 11
Values of $^3J(C_\alpha,C_\delta)$ Together with the Expected Conformations around χ_2 Determined for the Amino Acid Residues of Fully Deuterated Calmodulin Shown in Fig. 47

Residue	$^3J(C_\alpha,C_\delta)$ (Hz)	Conformation
Lys-30	3.8	*Trans*
Leu-32	3.5	*Trans*
Arg-37	5.0	*Trans*
Pro-43	3.7	—
Ile-63	3.3	*Trans*

5.0 Hz is measured (see Table 11). Large values for $^3J(C_\alpha,C_\delta)$ (>3.2 Hz) correspond to *trans* conformations, whereas intermediate couplings are expected if rotamer averaging takes place or if the torsion angles do not correspond to either *trans* or *gauche* rotamer position. The high sensitivity of the HN(CO)CAC$_{ali}$ experiment, in which the C_α atoms are transverse for as long as 100 ms, is related to the extremely long relaxation times of approximately 200 ms for the C_α atoms for perdeuterated calmodulin with an average rotational correlation time of 6.9 ns at 309 K.

4. CROSS-CORRELATED RELAXATION FOR THE MEASUREMENT OF ANGLES BETWEEN BOND VECTORS

This method was briefly introduced in Section 1.1.2. Here, a thorough mathematical description will be given.

4.1. Theoretical Description

4.1.1. Introduction

In high-resolution NMR of isotropic solutions, the magnetic interactions of tensors of rank 2 can only be detected through relaxation. The main source of relaxation is the dipolar interaction between directly bound nuclei. The Hamiltonian describing the dipolar interaction between two spins can be written in the form (Abragam, 1961; Goldman, 1988; Ernst *et al.*, 1989; Cavanagh *et al.*, 1996)

$$H_{kl}^{\mathrm{DD}} = b_{kl}\left\{3\frac{1}{r_{kl}^2}(\mathbf{I}_k\mathbf{r}_{kl})(\mathbf{I}_l\mathbf{r}_{kl}) - \mathbf{I}_k\mathbf{I}_l\right\} = b_{kl}\sum_{q=-2}^{+2} F_{kl}^{(q)}(\theta_{kl}, \phi_{kl})\, A_{kl}^{(q)}(I_k, I_l) \qquad (7)$$

with

$$b_{kl} = -\mu_0\frac{\gamma_k\gamma_l\hbar}{4\pi r_{kl}^3}$$

γ_k and γ_l denote the gyromagnetic ratio of nuclei k and l, $\hbar$ is the Planck constant, and r is the distance between the two nuclei. The angles θ_{kl}, ϕ_{kl} refer to the orientation of the vector $\mathbf{r}_{kl}$ with respect to the applied field. The exact expressions for the second-rank tensor operators $A_{kl}^{(q)}(I_k, I_l)$ and the time-dependent modified spherical harmonics $F_{kl}^{(q)}(\theta_{kl}, \phi_{kl})$ are given in Table 12.

The second main relaxation mechanism is relaxation through chemical shift anisotropy. The contribution to the Hamiltonian related to CSA in the principal axis frame (PAS) of nucleus k and in the laboratory frame (LF) can be written as (Abragam, 1961; Goldman, 1988; Ernst *et al.*, 1989; Cavanagh *et al.*, 1996)

$$H_k^{\mathrm{CSA,PAS}} = \gamma_k \sum_{i=x,y,z} B_i \, \sigma_{ii}^k I_i^k \tag{8a}$$

$$H_k^{\mathrm{CSA,LF}} = b_k \sum_{q=-1}^{1} F_k^{(q)} \, (\theta_k, \phi_k) \, A_k^{(q)}(I_k) \tag{8b}$$

with

$$b_k = \frac{1}{3}\,(\sigma_{\|} - \sigma_{\perp})\gamma_k B_0$$

for axially symmetric CSA tensors. The B_i denote the components of the applied field B_0 in the PAS. In the LF, in analogy to Eq. (7), the CSA Hamiltonian can be separated into the time-dependent orientational functions $F_k^{(q)}\,(\theta_k, \phi_k)$ and the time-independent spin operator terms $A_k^{(q)}(I_k)$ (Goldman, 1988; Ernst *et al.*, 1989; Cavanagh *et al.*, 1996). The expressions for the tensor operators $A_k^{(q)}(I_k)$ and the time-dependent modified spherical harmonics $F_k^{(q)}(\theta_k, \phi_k)$ are summarized in Table 12. The angles θ_k and ϕ_k are the polar coordinates of the main axis of the axially

Table 12
Tensor Operators in the Rotating Frame and Modified Spherical Harmonics for the Dipolar and CSA Interaction[a]

q	Tensor operators for the dipolar interaction $b_{kl} = -\mu_0 \frac{\gamma_k \gamma_l \hbar}{4\pi r_{kl}^3}$ $A_{kl}^{(q)}\,(I_k, I_l)$	Tensor operators for the CSA interaction $b_k = -\frac{1}{3}\,(\sigma_{\|} - \sigma_{\perp})\gamma_k B_0$ $A_k^{(q)}\,(I_k)$	Modified spherical harmonics $F_k^{(q)}(\theta,\phi)$, $F_{kl}^{(q)}(\theta,\phi)$,	Frequency ω_q
−2	$\sqrt{\frac{3}{2}}\, I_k^- I_l^-$	—	$\sqrt{\frac{3}{8}}\,\sin^2\theta \exp(-2i\phi)$	$\omega(I_k) + \omega(I_l)$
−1	$\sqrt{\frac{3}{2}}\, I_{k,z} I_l^-$	—	$\sqrt{\frac{3}{2}}\sin\theta\cos\theta\exp(-i\phi)$	$\omega(I_l)$
−1	$\sqrt{\frac{3}{2}}\, I_k^- I_{l,z}$	$\sqrt{\frac{3}{2}}\, I_k^-$	$\sqrt{\frac{3}{2}}\sin\theta\cos\theta\exp(-i\phi)$	$\omega(I_k)$
0	$2I_{k,z} I_{l,z}$	$2I_{k,z}$	$(3\cos^2\theta - 1)/2$	0
0	$\frac{1}{2}\,(I_k^+ I_l^- + I_k^- I_l^+)$	—	$(3\cos^2\theta - 1)/2$	$\omega(I_k) - \omega(I_l)$
+1	$\sqrt{\frac{3}{2}}\, I_k^+ I_{lz}$	$\sqrt{\frac{3}{2}}\, I_k^+$	$\sqrt{\frac{3}{2}}\sin\theta\cos\theta\exp(+i\phi)$	$\omega(I_k)$
+1	$\sqrt{\frac{3}{2}}\, I_{k,z} I_l^+$	—	$\sqrt{\frac{3}{2}}\sin\theta\cos\theta\exp(+i\phi)$	$\omega(I_l)$
+2	$\sqrt{\frac{3}{2}}\, I_k^+ I_l^+$	—	$\sqrt{\frac{3}{8}}\,\sin^2\theta \exp(+2i\phi)$	$\omega(I_k) + \omega(I_l)$

symmetric tensors in the laboratory frame. The convention we use ensures that the integral

$$\frac{1}{4\pi}\oint F^{(q)}(\theta,\phi)F^{(-q)}(\theta,\phi)d(\cos\theta)d\phi=\frac{1}{5}$$

is independent of q.

For Eq. (8b), we assumed an axially symmetric tensor. However, this is no restriction of generality because each asymmetric CSA tensor can be decomposed into a superposition of two axially symmetric tensors. The bilinear form of the CSA contribution to the Hamiltonian operator in the PAS from Eq. (8) can, after diagonalization of the CSA tensor, be rewritten as (Goldman, 1988)

$$H_k^{\mathrm{CSA,PAS}}=\gamma_k\frac{\sigma_{xx}+\sigma_{yy}+\sigma_{zz}}{3}\mathbf{BI}_k$$

$$+\frac{1}{3}\gamma_k(\sigma_{xx}-\sigma_{zz})\,[2I_{k,x}B_x-I_{k,y}B_y-I_{k,z}B_z]+\frac{1}{3}\gamma_k\,(\sigma_{yy}-\sigma_{zz})\,[2I_{k,y}\,B_y-I_{k,x}\,B_x-I_{k,z}B_z] \tag{9}$$

The first term of Eq. (9) describes the isotropic part of the chemical shift, the second and third terms, the axially symmetric anisotropic contributions. This equation can be used for the description of cross-correlation between arbitrarily anisotropic tensors as an asymmetric anisotropic tensor of rank 2 can be contracted from two axially symmetric tensors lying along orthogonal axes. Using Table 12, $\sigma_{\parallel}-\sigma_{\perp}$ would be $\sigma_{xx}-\sigma_{zz}$ with x as axis of reference from the second term, and $\sigma_{\parallel}-\sigma_{\perp}$ would be $\sigma_{yy}-\sigma_{zz}$ with y as axis of reference from the third term in Eq. (9). For the special case of a symmetric CSA tensor ($\sigma_{xx}=\sigma_{yy}=\sigma_{\perp}$), one finds

$$H_k^{\mathrm{CSA,PAS}}=\gamma_k[\sigma_{\parallel}I_{k,z}B_z+\sigma_{\perp}(I_{k,x}B_x+I_{k,y}B_y)]=$$

$$=\gamma_k\left\{\frac{\sigma_{\parallel}+2\sigma_{\perp}}{3}\mathbf{BI}_k+\frac{\sigma_{\parallel}-\sigma_{\perp}}{3}(2I_{k,z}B_z-I_{k,x}B_x-I_{k,y}B_y)\right\} \tag{10}$$

with

$$\sigma_{\parallel}-\sigma_{\perp}=\sigma_{zz}-1/2(\sigma_{xx}+\sigma_{yy}).$$

4.1.2. Relaxation Superoperator and Spectral Density Functions

The relaxation superoperator in the Liouville– von Neumann equation

$$\frac{d}{dt}\sigma=-i[H_0,\sigma(t)]-\sum_{V,W}\hat{\Gamma}_{V,W}(\sigma(t)-\sigma_0) \tag{11}$$

has the general form

$$\hat{\Gamma}_{VW}\sigma = b_v b_w \sum_{q=-2}^{2} ([A_V^{(-q)}, [A_W^{(q)}, \sigma]] + [A_W^{(q)}, [A_V^{(-q)}, \sigma]]) j_{VW}^q(\omega_q) \tag{12}$$

where the indices V and W refer to the interactions that are the source for relaxation. The term $j_{VW}^q(\omega_q)$ denotes the spectral density function and is obtained by evaluating the correlation function of the modified spherical harmonics

$$j_{VW}^q(\omega_q) = \int_0^\infty d\tau\, \overline{F_V^{(q)}(t) F_W^{(-q)}(t+\tau)} \exp(-\omega_q \tau) \tag{13}$$

The bar indicates time average over t. In the autocorrelation case ($V = W$), both interactions V and W originate from the same source of interaction, e.g., the same pair of nuclei. In the cross-correlation case, the time-dependent spherical harmonics refer to different kinds of tensorial interactions ($V \neq W$).

The cross-correlation case only will be considered in the following. The theory for the description of intramolecular dipolar interaction in coupled multispin systems was developed some time ago (Hubbard, 1958, 1969, 1970; Kuhlmann and Baldeschweiler, 1965; Shimizu, 1969; Grant and Werbelow, 1976; Werbelow and Grant, 1977). Formulas have been derived for the three cases of isotropic, axially symmetric, and generally anisotropic reorientation which are discussed below. A review of this topic was published by Werbelow and Grant (1977).

4.1.2.1. Spherical Top Molecules. The spectral density function for isotropic rotational diffusion has been derived by Hubbard (1958) and by Kuhlmann and Baldeschweiler (1965):

$$\begin{aligned} j_{VW}^q(\omega_q) &= \frac{1}{5}(3\cos^2\theta_{V,W} - 1)/2 \left[\frac{6D}{(6D)^2 + \omega_q^2}\right] \\ &= \frac{1}{10}(3\cos^2\theta_{V,W} - 1)/2 \left[\frac{2\tau_c}{1 + (\omega_q \tau_c)^2}\right] \end{aligned} \tag{14}$$

D denotes the diffusion constant, $1/\tau_c = 6D$. V and W denote the different interactions. $\theta_{V,W}$ describes the projection angle between the principal axes of the two interactions V and W. If V denotes, for example, an N–H and W a C–H dipolar interaction, respectively, as indicated in Fig. 6, θ is the included angle between the two bond vectors N–H and C–H.

4.1.2.2. Asymmetric Top Molecules. The spectral density function for the asymmetric top assumes the form (Shimizu, 1969; Hubbard, 1970; Werbelow and Grant, 1977)

$$j^q_{VW}(\omega_q) = \frac{1}{40} * (12\cos\theta_V\cos\theta_W\sin\theta_V\sin\theta_W\sin\phi_V\sin\phi_W \frac{b_1}{b_1^2+\omega_q^2}$$

$$+ 12\cos\theta_V\cos\theta_W\sin\theta_V\sin\theta_W\cos\phi_V\cos\phi_V \frac{b_2}{b_2^2+\omega_q^2}$$

$$+ 3\sin^2\theta_V\sin^2\theta_W\sin2\phi_V\sin2\phi_W \frac{b_3}{b_3^2+\omega_q^2}$$

$$+ [3\cos^2\left(\frac{\zeta}{2}\right)(\sin^2\theta_V\sin^2\theta_W\cos2\phi_V\cos2\phi_W + \sin^2\left(\frac{\zeta}{2}\right)(3\cos^2\theta_V - 1)\,(3\cos^2\theta_W - 1)$$

$$+ \sqrt{3}\cos\left(\frac{\zeta}{2}\right)\sin\left(\frac{\zeta}{2}\right)((3\cos^2\theta_V - 1)\,\sin^2\theta_W\cos2\phi_W$$

$$+ (3\cos^2\theta_W - 1)\,\sin^2\theta_V\cos2\phi_V)]\,\frac{b_4}{b_4^2+\omega_q^2}$$

$$+ [3\sin^2\left(\frac{\zeta}{2}\right)(\sin^2\theta_V\sin^2\theta_W\cos2\phi_V\cos2\phi_W + \cos^2\left(\frac{\zeta}{2}\right)(3\cos^2\theta_V - 1)\,(3\cos^2\theta_W - 1)$$

$$- \sqrt{3}\cos\left(\frac{\zeta}{2}\right)\sin\left(\frac{\zeta}{2}\right)((3\cos^2\theta_V - 1)\,\sin^2\theta_W\cos2\phi_W$$

$$+ (3\cos^2\theta_W - 1)\,\sin^2\theta_V\cos2\phi_V)]\,\frac{b_5}{b_5^2+\omega_q^2} \qquad (15)$$

where the notation of Woessner (1962 a,b) has been used:

$$D = \tfrac{1}{3}\,(D_{xx} + D_{yy} + D_{zz})$$

$$L^2 = \tfrac{1}{3}\,(D_{xx}D_{yy} + D_{xx}D_{zz} + D_{yy}D_{zz})$$

$$\tan\zeta = \sqrt{3}\left[\frac{D_{xx} - D_{yy}}{2D_{zz} - D_{xx} - D_{yy}}\right]$$

$$b_1 = 4D_{xx} + D_{yy} + D_{zz}$$

$$b_2 = D_{xx} + 4D_{yy} + D_{zz}$$

$$b_3 = D_{xx} + D_{yy} + 4D_{zz}$$

$$b_4 = 6D + 6\sqrt{D^2 - L^2}$$

$$b_5 = 6D - 6\sqrt{D^2 - L^2} \tag{16}$$

4.1.2.3. Axially Symmetric Top Molecules. For the case that $D_{xx} = D_{yy} = D_\perp$, Eq. (15) can be simplified yielding the spectral density function of the symmetric top rotator ($D_{zz} = D_\parallel$) (Schneider, 1964; Hubbard, 1969):

$$j^q_{VW}(\omega_q) = \frac{1}{80}\{(3\cos^2\theta_V - 1)(3\cos^2\theta_W - 1) * J^{q,0}_{VW}$$

$$+ 12\cos\theta_V\cos\theta_W\sin\theta_V\sin\theta_W\cos(\phi_V - \phi_W) * J^{q,1}_{VW}$$

$$+ 3\sin^2\theta_V\sin^2\theta_W\cos(2\phi_V - 2\phi_W) * J^{q,2}_{VW}\} \tag{17}$$

where the reduced spectral density functions ($-2 \le m \le +2$)

$$J^{q,m}_{VW} = \frac{2\tau_{c,m}}{1 + \left(\omega_q \tau_{c,m}\right)^2} \tag{18}$$

have been used. The correlation times $\tau_{c,m}$ can be rewritten as diffusion constants $D_\parallel$ and $D_\perp$ according to

$$1/\tau_{c,m} = 6D_\perp + m^2(D_\parallel - D_\perp) \tag{19}$$

4.1.2.4. Inclusion of Internal Motion Internal motion can be incorporated into the spectral density either by the Lipari and Szabo (1982) approach or by explicit calculation of the motion from, e.g., motional models of molecular dynamics trajectories. In Eq. (13), the spectral densities are Fourier transformations of the motion of the molecule with respect to the external magnetic field. This equation assumed that internal motion is absent. Rewriting the spectral densities as a convolution ($\times$) of a Fourier transformation of the global motion and the Fourier transformation of the local motion, we find for the first term $j^q_{VW}(\omega_q)$ in Eq. (17) for axially symmetric diffusion:

$$j^{q,\text{local motion}}_{VW}(\omega_q) = \frac{1}{20} FT\left\{\overline{P_2[\cos\theta_V(t)]P_2[\cos\theta_W(t+\tau)]}\right\} \times J^{q,0}_{VW}\Big|_{\omega_q} + \cdots \tag{20}$$

The Fourier transformation concerns τ and the average is taken with respect to t. $P_2[\cos\theta]$ denote the Legendre polynomials $(3\cos^2\theta - 1)/2$. If we assume the internal motion to be uncorrelated and fast with respect to the global motion, Eq. (20) can be directly used to compare field-dependent experimental relaxation rates with predicted rates and for the analysis of molecular dynamics trajectories or for the analysis of models.

Application of the Lipari and Szabo approach assumes in addition an exponential decay of the correlation function with the characteristic rate τ_i from time 0 to time τ according to

$$\overline{P_2[\cos\theta_V(t)]\, P_2\,[\cos\theta_W(t+\tau)]}$$

$$= \exp\left(-\frac{\tau}{\tau_i}\right)\{\overline{P_2[\cos\theta_V(t)]\, P_2\,[\cos\theta_W(t)]} - \overline{P_2[\cos\theta_V(t)]\, P_2\,[\cos\theta_W\,(t+\infty)]}\}$$

$$+\overline{P_2[\cos\theta_V(t)]\, P_2\,[\cos\theta_W(t+\infty)]}$$

$$= \exp\left(-\frac{\tau}{\tau_i}\right)\{[1-(S_{VW}^{q,0})^2] + (S_{VW}^{q,0})^2\}\, \overline{P_2[\cos\theta_V(t)]\, P_2\,[\cos\theta_W(t)]} \tag{21}$$

where $(S_{VW}^{q,0})^2$ is the order parameter for the respective interaction and τ_i corresponds to the internal correlation time (Brüschweiler and Case, 1994; Brutscher *et al.*, 1998).

4.2. Application of Cross-Correlated Relaxation to High-Resolution NMR as a Tool to Obtain Structural Information

Dalvit and Bodenhausen (1988) introduced a triple quantum filtered NOESY where a system of three nuclei—H_α, $H_\beta^{\text{pro-R}}$, and $H_\beta^{\text{pro-S}}$—in a protein is investigated to yield structural information about the sidechain conformation. The complementary experiment is the triple quantum filtered ROESY, reported by Brüschweiler *et al.* (1989). The drawback of the two experiments is their inherently low sensitivity, which is related to the large distances of the spins involved. Furthermore, the experiments can only be carried out if the involved nuclei are scalar coupled. Quantitative interpretation of the spectra with respect to the size of the involved angle $\theta_{kl,km}$ is very difficult because the multiplets are affected by scalar couplings and relaxation which are difficult to disentangle.

Vold *et al.* (1977; Vold and Vold, 1978) have shown that in dilute solutions of trisubstituted benzenes such as 1,2,3- trichlorobenzene, cross-correlation rates can also be used to determine the motional anisotropy of a molecule. Because the rate R_1 for each individual line is a function of the corresponding spectral density for the transition between the respective spin states (Freeman *et al.*, 1970), D_{xx}, D_{yy} and D_{zz} can be estimated after calculation of the transition probabilities for an AB_2 spin system—assuming a nonspherical reorientational process.

Another experiment characterizing hindered or unhindered rotation of sidechains in a protein (χ_1) has been introduced by Ernst and Ernst (1994). Sign changes in the cross-correlation rate are interpreted as a function of the motional model of the sidechain.

However, none of these approaches provided structural information in a simple way.

4.3. Cross-Correlated Relaxation of Zero and Double Quantum Coherences

In the following, we consider two spin pairs, an N–H^N vector and a C^α–H^α vector in a protein, for which we excite double and zero quantum coherences between N and C^α. The equation of motion for the different components of double quantum and zero quantum coherences $\sigma^{DQ/ZQ}_{\mu,\mu'}$ under the influence of the scalar-coupled, directly bound, protons has the general form

$$\left(\sigma^{DQ/ZQ}_{\mu,\mu'}\right)^{\bullet} = \left[-\Gamma^{DQ/ZQ} - i\Omega^{DQ/ZQ}\right]\left(\sigma^{DQ/ZQ}_{\mu,\mu'}\right) \tag{22}$$

The terms $\sigma^{DQ/ZQ}_{\mu,\mu'}$ summarize the double and zero quantum coherences, $C^+N^+H^\mu_C H^{\mu'}_N$, $C^-N^-H^\mu_C H^{\mu'}_N$, $C^+N^-H^\mu_C H^{\mu'}_N$, and $C^-N^+H^\mu_C H^{\mu'}_N$. $H^\mu_C H^{\mu'}_N$ $(\mu,\mu' = \alpha,\beta)$ stands for the spin polarization operators of the nitrogen and carbon bound protons, respectively. The isotropic chemical shift $\Omega^{DQ/ZQ}$ (the index DQ refers to coherences C^+N^+, ZQ to coherences C^+N^-) for the four resonance lines $\alpha\alpha$, $\alpha\beta$, $\beta\alpha$, and $\beta\beta$ of the doublet of doublets is given by

$$\Omega^{DQ/ZQ} = \left[(\Omega_C \pm \Omega_N)\mathbf{1} + \pi\begin{pmatrix} \pm J_{NH} + J_{CH} & 0 & 0 & 0 \\ 0 & \mp J_{NH} + J_{CH} & 0 & 0 \\ 0 & 0 & \pm J_{NH} - J_{CH} & 0 \\ 0 & 0 & 0 & \mp J_{NH} - J_{CH} \end{pmatrix}\right] \tag{23}$$

Ω_C and Ω_N are the carbon and nitrogen chemical shifts and π^1J_{NH} and π^1J_{CH} the relative chemical shifts of the different multiplet lines related to the scalar coupling. The relaxation matrix $\Gamma^{DQ/ZQ}$ assumes the form

$$\Gamma^{DQ/ZQ} = \begin{pmatrix} \Gamma^a + \Gamma_1 + \Gamma^{DQ/ZQ}_{\alpha\alpha} & -\Gamma_{T1}(H^N) & -\Gamma_{T1}(H^C) & -W_2 \\ -\Gamma_{T1}(H^N) & \Gamma^a + \Gamma_1 + \Gamma^{DQ/ZQ}_{\alpha\beta} & -W_0 & -\Gamma_{T1}(H^C) \\ -\Gamma_{T1}(H^C) & -W_0 & \Gamma^a + \Gamma_1 + \Gamma^{DQ/ZQ}_{\beta\alpha} & -\Gamma_{T1}(H^N) \\ -W_2 & -\Gamma_{T1}(H^C) & -\Gamma_{T1}(H^N) & \Gamma^a + \Gamma_1 + \Gamma^{DQ/ZQ}_{\beta\beta} \end{pmatrix}$$

$$\Gamma_1 = \Gamma_{T1}(H^N) + \Gamma_{T1}(H^C) \tag{24}$$

In the relaxation matrix ΓDQ/ZQ, the terms $\Gamma^{DQ/ZQ}_{\mu\mu'}$ denote the different cross-relaxation rates related to the heteronuclear dipolar interaction, including the dipole dipole cross-relaxation rate $\Gamma^c_{NH,CH}$, the CSA dipole cross-relaxation rates

$\Gamma^{c}_{C,NH}$, $\Gamma^{c}_{N,NH}$, $\Gamma^{c}_{C,CH}$ and $\Gamma^{c}_{N,CH}$, and the secular part of the relaxation of double quantum and zero quantum transitions related to NOE between the two protons H^N and H^C. The nonsecular part of the latter mechanism is also described in the off-diagonal elements W_2 and W_0. These are as well included in the expressions $\Gamma^{DQ/ZQ}_{\mu\mu'}$. Γ^{a} contains the contributions related to autocorrelated relaxation and external relaxation of C,N-DQ/ZQ coherences. $\Gamma_{T1}(H^N)$ and $\Gamma_{T1}(H^C)$ denote the contributions related to T_1 relaxation of the proton directly attached to the carbon and nitrogen, respectively. The influence of nonsecular contributions in the relaxation matrix $\Gamma^{DQ/ZQ}$ on the angle-dependent dipole dipole cross-relaxation rate is discussed in detail later in this section. Secular contributions in $\Gamma^{DQ/ZQ}$ are discussed next.

4.3.1. Dipole Dipole Cross-Correlated Relaxation

The relaxation superoperator in Eq. (12) which describes double and zero quantum coherences contains contributions from autocorrelated relaxation ($V = W$)

$$\left[\hat{\Gamma}^{a}_{NH,NH} + \hat{\Gamma}^{a}_{CH,CH}\right](\sigma^{DQ/ZQ}_{\mu,\mu'}) = b^2_{NH} \sum_q \left[A^{(-q)}_{NH}, [A^{(q)}_{NH}, \sigma^{DQ/ZQ}_{\mu,\mu'}]\right] j^q_{NH,NH}(\omega_q)$$
$$+ b^2_{CH} \sum_q \left[A^{(-q)}_{CH}, [A^{(q)}_{CH}, \sigma^{DQ/ZQ}_{\mu,\mu'}]\right] j^q_{CH,CH}(\omega_q) \qquad (25)$$

and from cross-correlated relaxation ($V \neq W$)

$$\left[\hat{\Gamma}^{c}_{NH,CH} + \hat{\Gamma}^{c}_{CH,NH}\right](\sigma^{DQ/ZQ}_{\mu,\mu'}) = b_{NH} b_{CH} \sum_q [A^{(-q)}_{NH}, [A^{(q)}_{CH}, \sigma^{DQ/ZQ}_{\mu,\mu'}]] j^q_{NH,CH}(\omega_q)$$
$$+ b_{NH} b_{CH} \sum_q \left[A^{(-q)}_{CH}, [A^{(q)}_{NH}, \sigma^{DQ/ZQ}_{\mu,\mu'}]\right] j^q_{CH,NH}(\omega_q) \qquad (26)$$

in which $V = NH$ and $W = CH$. In the autocorrelation case, the relaxation superoperator contains second-rank tensor operators stemming from only one interaction, whereas in the cross-correlated case, double commutators containing tensor operators from two distinct interactions have to be evaluated. In the secular approximation, only double commutators with a Larmor frequency of 0 contribute (*vide infra*). The terms b_{NH} and b_{CH} are given in Eq. (7). The expressions for the different relaxation rates for auto- and cross-correlated relaxation will be illustrated in the following. As an example, the double commutator from Eqs. (25) and (26) is applied to double quantum coherences $C^+N^+H^{\mu}_{C}H^{\mu'}_{N}$. The complementary expressions $C^-N^-H^{\mu}_{C}H^{\mu'}_{N}$ and zero quantum coherences behave accordingly. The four terms $C^+N^+H^{\mu}_{C}H^{\mu'}_{N}$ ($\mu,\mu' = \alpha,\beta$) corresponding to the four multiplet lines of the doublet of doublets are subjected to the different double commutators for auto- and cross-correlated relaxation. A graphical representation of the respective rates is given in Fig. 48. It should be noted that in all figures and calculations given in this chapter we consider γ_N to be positive and consequently also ${}^1J_{NH}$ and Ω_N. While the appearance of the multiplets is not changed e.g. the labelling of the four submultiplet lines does.

In the autocorrelated case, as well as in the cross-correlated case, the *single line operators* $C^+N^+H_C^{\mu}H_N^{\mu'}$ are eigenoperators with respect to the $j^0(0)$ part. For the autocorrelated case, all lines relax equally fast, whereas in the cross-correlated case, for the two pairs of lines $\alpha\alpha$ and $\beta\beta$, as well as $\alpha\beta$ and $\beta\alpha$ the rates have the same absolute value but opposite sign. The right-hand side of Fig. 48 shows a graphical representation of the respective rates. Altogether, the multiplet is governed by a superposition of the rates of Γ^a and $\Gamma^c_{NH,CH}$ which is indicated by the sum of the two rates at the bottom right-hand side of Fig. 48. Finally, the dipole dipole cross-correlated relaxation rate for each multiplet line of the doublet of doublets according to Eq. (26), with the spectral density function given in Eq. (14) for a spherical top molecule, can be written as

$$\Gamma^c_{NH,CH} = \frac{\gamma_H\gamma_N}{(r_{NH})^3}\frac{\gamma_H\gamma_C}{(r_{CH})^3}\left(\frac{\mu_0}{4\pi}\hbar\right)^2\frac{1}{5}(3\cos^2\theta_{NH,CH} - 1)\tau_c \tag{27}$$

$\theta_{NH,CH}$ denotes the projection angle between the NH and the CH vector.

4.3.2. Dipole CSA Cross-Correlated Relaxation

Up to now, only dipole (NH)–dipole (CH) cross-correlation rates have been considered. The cross-correlation rate to the CSA tensors of the respective heteroatoms, e.g., dipole (NH)–CSA (N) or dipole (NH)–CSA (C), influences also the intensity distribution on the four resonance lines. The quantitative evaluation of the individual effects is described in the following. In the same manner as in the case of dipole dipole cross-correlated relaxation, the rates for the CSA dipole cross-correlated relaxation, together with the corresponding double commutators are given for one example double commutator and are summarized in Fig. 49. Note that the assignment of the order of the spin polarization states of the multiplet components differs between DQ and ZQ coherences.

For the double quantum operator, the sums of the rates of the cross-correlated spectral densities for CSA (^{13}C) and CSA (^{15}N) with dipole NH and CH, respectively, are observed. Accordingly, for the respective zero quantum coherences, the difference of the rates will be observed, which allows determination of each rate individually, as shown later. According to Eq. (26), together with Eqs. (8)–(10) and (14), the general form for the dipole CSA cross-correlated relaxation rate can be written as

$$\Gamma^c_{N,CH} = -\frac{2}{15}\gamma_N B_0 \tau_c \hbar \frac{\mu_0}{4\pi}\frac{\gamma_H\gamma_C}{(r_{CH})^3}\{(\sigma_{xx} - \sigma_{zz})(3\cos^2\theta_{CH,\sigma_{xx}} - 1) + (\sigma_{yy} - \sigma_{zz})(3\cos^2\theta_{CH,\sigma_{yy}} - 1)\} \tag{28a}$$

Auto Correlation

$$\left(C^+N^+H_C^\alpha H_N^\alpha\right)^\bullet \rightarrow b_{NH}^2 j_{NH,NH}^0\left[2H_z^N N_z,\left[2H_z^N N_z, C^+N^+H_C^\alpha H_N^\alpha\right]\right]+b_{CH}^2 j_{CH,CH}^0\left[2H_z^C C_z,\left[2H_z^C C_z, C^+N^+H_C^\alpha H_N^\alpha\right]\right]=$$

$$=+1/4\left(b_{NH}^2 j_{NH,NH}^0+b_{CH}^2 j_{CH,CH}^0\right)C^+N^+H_C^\alpha H_N^\alpha=+\left[\Gamma_{NH,NH}^a+\Gamma_{CH,CH}^a\right]\left(C^+N^+H_C^\alpha H_N^\alpha\right)$$

$$\left(C^+N^+H_C^\alpha H_N^\beta\right)^\bullet \rightarrow b_{NH}^2 j_{NH,NH}^0\left[2H_z^N N_z,\left[2H_z^N N_z, C^+N^+H_C^\alpha H_N^\beta\right]\right]+b_{CH}^2 j_{CH,CH}^0\left[2H_z^C C_z,\left[2H_z^C C_z, C^+N^+H_C^\alpha H_N^\beta\right]\right]=$$

$$=+1/4\left(b_{NH}^2 j_{NH,NH}^0+b_{CH}^2 j_{CH,CH}^0\right)C^+N^+H_C^\alpha H_N^\beta=+\left[\Gamma_{NH,NH}^a+\Gamma_{CH,CH}^a\right]\left(C^+N^+H_C^\alpha H_N^\beta\right)$$

$$\left(C^+N^+H_C^\beta H_N^\alpha\right)^\bullet \rightarrow b_{NH}^2 j_{NH,NH}^0\left[2H_z^N N_z,\left[2H_z^N N_z, C^+N^+H_C^\beta H_N^\alpha\right]\right]+b_{CH}^2 j_{CH,CH}^0\left[2H_z^C C_z,\left[2H_z^C C_z, C^+N^+H_C^\beta H_N^\alpha\right]\right]=$$

$$=+1/4\left(b_{NH}^2 j_{NH,NH}^0+b_{CH}^2 j_{CH,CH}^0\right)C^+N^+H_C^\beta H_N^\alpha=+\left[\Gamma_{NH,NH}^a+\Gamma_{CH,CH}^a\right]\left(C^+N^+H_C^\beta H_N^\alpha\right)$$

$$\left(C^+N^+H_C^\beta H_N^\beta\right)^\bullet \rightarrow b_{NH}^2 j_{NH,NH}^0\left[2H_z^N N_z,\left[2H_z^N N_z, C^+N^+H_C^\beta H_N^\beta\right]\right]+b_{CH}^2 j_{CH,CH}^0\left[2H_z^C C_z,\left[2H_z^C C_z, C^+N^+H_C^\beta H_N^\beta\right]\right]=$$

$$=+1/4\left(b_{NH}^2 j_{NH,NH}^0+b_{CH}^2 j_{CH,CH}^0\right)C^+N^+H_C^\beta H_N^\beta=+\left[\Gamma_{NH,NH}^a+\Gamma_{CH,CH}^a\right]\left(C^+N^+H_C^\beta H_N^\beta\right)$$

Rates

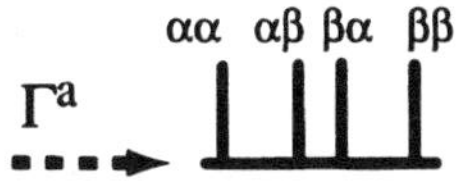

αα αβ βα ββ

Γ^a

Cross Correlation

$$\left(C^{+}N^{+}H_{C}^{\alpha}H_{N}^{\alpha}\right)^{\bullet} \rightarrow b_{NH}b_{CH}j_{NH,CH}^{0}\left\{\left[2H_{z}^{N}N_{z},\left[2H_{z}^{C}C_{z},C^{+}N^{+}H_{C}^{\alpha}H_{N}^{\alpha}\right]\right]+\left[2H_{z}^{C}C_{z},\left[2H_{z}^{N}N_{z},C^{+}N^{+}H_{C}^{\alpha}H_{N}^{\alpha}\right]\right]\right\} =$$

$$= +1/2b_{NH}b_{CH}j_{NH,CH}^{0}C^{+}N^{+}H_{C}^{\alpha}H_{N}^{\alpha} = +\Gamma_{NH,CH}^{c}\left(C^{+}N^{+}H_{C}^{\alpha}H_{N}^{\alpha}\right)$$

$$\left(C^{+}N^{+}H_{C}^{\alpha}H_{N}^{\beta}\right)^{\bullet} \rightarrow b_{NH}b_{CH}j_{NH,CH}^{0}\left\{\left[2H_{z}^{N}N_{z},\left[2H_{z}^{C}C_{z},C^{+}N^{+}H_{C}^{\alpha}H_{N}^{\beta}\right]\right]+\left[2H_{z}^{C}C_{z},\left[2H_{z}^{N}N_{z},C^{+}N^{+}H_{C}^{\alpha}H_{N}^{\beta}\right]\right]\right\} =$$

$$= -1/2b_{NH}b_{CH}j_{NH,CH}^{0}C^{+}N^{+}H_{C}^{\alpha}H_{N}^{\beta} = -\Gamma_{NH,CH}^{c}\left(C^{+}N^{+}H_{C}^{\alpha}H_{N}^{\beta}\right)$$

$$\left(C^{+}N^{+}H_{C}^{\beta}H_{N}^{\alpha}\right)^{\bullet} \rightarrow b_{NH}b_{CH}j_{NH,CH}^{0}\left\{\left[2H_{z}^{N}N_{z},\left[2H_{z}^{C}C_{z},C^{+}N^{+}H_{C}^{\beta}H_{N}^{\alpha}\right]\right]+\left[2H_{z}^{C}C_{z},\left[2H_{z}^{N}N_{z},C^{+}N^{+}H_{C}^{\beta}H_{N}^{\alpha}\right]\right]\right\} =$$

$$= -1/2b_{NH}b_{CH}j_{NH,CH}^{0}C^{+}N^{+}H_{C}^{\beta}H_{N}^{\alpha} = -\Gamma_{NH,CH}^{c}\left(C^{+}N^{+}H_{C}^{\beta}H_{N}^{\alpha}\right)$$

$$\left(C^{+}N^{+}H_{C}^{\beta}H_{N}^{\beta}\right)^{\bullet} \rightarrow b_{NH}b_{CH}j_{NH,CH}^{0}\left\{\left[2H_{z}^{N}N_{z},\left[2H_{z}^{C}C_{z},C^{+}N^{+}H_{C}^{\beta}H_{N}^{\beta}\right]\right]+\left[2H_{z}^{C}C_{z},\left[2H_{z}^{N}N_{z},C^{+}N^{+}H_{C}^{\beta}H_{N}^{\beta}\right]\right]\right\} =$$

$$= +1/2b_{NH}b_{CH}j_{NH,CH}^{0}C^{+}N^{+}H_{C}^{\beta}H_{N}^{\beta} = +\Gamma_{NH,CH}^{c}\left(C^{+}N^{+}H_{C}^{\beta}H_{N}^{\beta}\right)$$

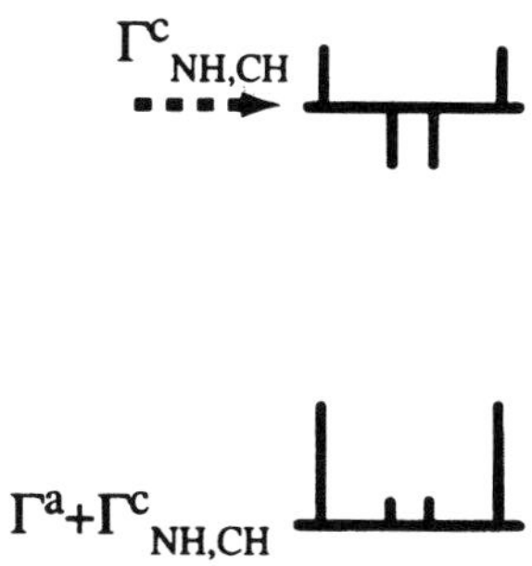

Figure 48. Double commutators for dipole dipole auto- and cross-correlated relaxation and respective rates in a graphical representation. The identification of the submultiplet lines assume $\gamma_N > 0$, ${}^{1}J_{NH} > 0$ and $\Omega_N > 0$ as explained in the text.

where $\theta_{\mathrm{CH},\sigma_{xx}}$ and $\theta_{\mathrm{CH}\sigma_{yy}}$ denote the angle between the CH vector and the two principal components of the nitrogen CSA tensor. An equivalent formulation is

$$\Gamma^{c}_{\mathrm{N,CH}} = -\frac{2}{15}\gamma_{\mathrm{N}} B_0 \tau_c \hbar \frac{\mu_0}{4\pi}\frac{\gamma_{\mathrm{H}}\gamma_{\mathrm{C}}}{(r_{\mathrm{CH}})^3}\{(\sigma_{\parallel}-\sigma_{\perp})(3\cos^2\theta_{\mathrm{CH},\sigma_{\parallel}}-1)$$

$$+\frac{3}{4}(\sigma_{xx}-\sigma_{yy})(\sin^2\theta_{\mathrm{CH},\sigma_{\parallel}}\cos 2\phi_{\mathrm{CH},\sigma_{xx}})\} \tag{28b}$$

In the following, this tensor is assumed to be axially symmetric. Equation (28b) simplifies accordingly, setting $\sigma_{xx}=\sigma_{yy}=\sigma_{\perp}$ and $\sigma_{zz}=\sigma_{\parallel}$:

Dipole–CSA cross correlated relaxation

$$b_C b_{NH} j^0_{C.NH}(0)\left\{\left[2H^N_z N_z,\left[2C_z, C^+N^+H^\alpha_C H^\beta_N\right]\right]+\left[2C_z,\left[2H^N_z N_z, C^+N^+H^\alpha_C H^\beta_N\right]\right]\right\}=$$

$$=-b_C b_{NH} j^0_{C.NH}(0)C^+N^+H^\alpha_C H^\beta_N$$

$$=-\Gamma^c_{C.NH}\left(C^+N^+H^\alpha_C H^\beta_N\right)$$

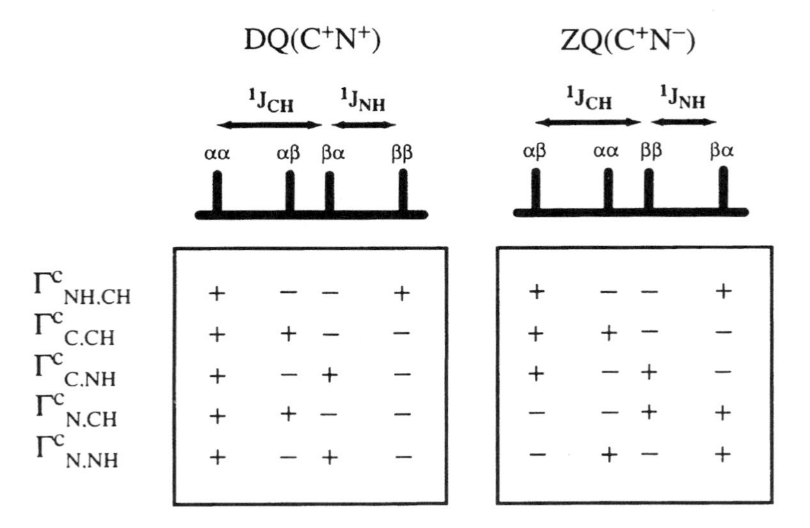

Figure 49. Schematic representation of the dipole CSA cross-correlated relaxation rates of C,N DQ and ZQ coherences. The double commutator serves as an example. The sign of the rates for the respective submultiplet line are given in the boxes. Again γ_{N}, ${}^1J_{\mathrm{NH}}$ and Ω_{N} are assumed to be positive.

$$\Gamma^{c}_{\mathrm{N,CH}} = -\frac{2}{15}\gamma_{\mathrm{N}} B_0 \tau_c \hbar \frac{\mu_0}{4\pi}\frac{\gamma_{\mathrm{H}}\gamma_{\mathrm{C}}}{(r_{\mathrm{CH}})^3}\{(\sigma_{\parallel}-\sigma_{\perp})\,(3\cos^2\theta_{\mathrm{CH},\sigma_{\parallel}}-1)\} \tag{28c}$$

4.3.3. NOE Effects on the Dipole Dipole Cross-Correlated Relaxation Rates

The NOE between the two protons H^N and H^C can, in principle, lead to an additional contribution to the measured dipole dipole cross-correlated relaxation rate. A quantification of the size of this effect is given in this subsection. The NOE between the two protons H^N and H^C comes from the autocorrelated dipolar relaxation from the $j^{(0)}(\omega_k-\omega_l)$ (W_0) and the $j^{(2)}(\omega_k+\omega_l)$ (W_2) term. The respective double commutators are

W_0:

$$[-\frac{1}{2}(H_{\mathrm{C}}^{+}H_{\mathrm{N}}^{-}+H_{\mathrm{C}}^{-}H_{\mathrm{N}}^{+}),[-\frac{1}{2}(H_{\mathrm{C}}^{+}H_{\mathrm{N}}^{-}+H_{\mathrm{C}}^{-}H_{\mathrm{N}}^{+}),C^{+}N^{+}H_{\mathrm{C}}^{\alpha}H_{\mathrm{N}}^{\beta}]]$$

$$=\frac{1}{2}(H_{\mathrm{C}}^{\alpha}H_{\mathrm{N}}^{\beta}-H_{\mathrm{C}}^{\beta}H_{\mathrm{N}}^{\alpha})C^{+}N^{+}$$

W_2:

$$[\sqrt{\tfrac{3}{2}}\,H_{\mathrm{C}}^{+}H_{\mathrm{N}}^{+},[\sqrt{\tfrac{3}{2}}\,H_{\mathrm{C}}^{-}H_{\mathrm{N}}^{-},H_{\mathrm{N}}^{\alpha}H_{\mathrm{C}}^{\alpha}C^{+}N^{+}]]+[\sqrt{\tfrac{3}{2}}\,H_{\mathrm{C}}^{-}H_{\mathrm{N}}^{-},[\sqrt{\tfrac{3}{2}}\,H_{\mathrm{C}}^{+}H_{\mathrm{N}}^{+},H_{\mathrm{C}}^{\alpha}H_{\mathrm{N}}^{\alpha}C^{+}N^{+}]]$$

$$=3(H_{\mathrm{C}}^{\alpha}H_{\mathrm{N}}^{\alpha}-H_{\mathrm{C}}^{\beta}H_{\mathrm{N}}^{\beta})C^{+}N^{+} \tag{29}$$

The NOE contributes a secular and a nonsecular term. We will see in the following that the nonsecular term can be safely ignored. The secular term, however, adds to the linewidth of the $\alpha\alpha$, $\beta\beta$ line the rate W_2 and to the $\alpha\beta$, $\beta\alpha$ line the rate W_0. This is taken into account in the simulations carried out for the evaluation of the dipole dipole cross-correlation rate $\Gamma^{c}_{\mathrm{NH,CH}}$ in the following section. The cross-correlation rate $\Gamma^{\mathrm{NOE}}_{H^C,H^N} = W_2 - W_0$ (Ernst *et al.*, 1994; Cavanagh *et al.*, 1996) is given by

$$\Gamma^{\mathrm{NOE}}_{\mathrm{H^C,H^N}} = \frac{1}{10}\left(\frac{\mu_0}{4\pi}\right)^2\left(\frac{\hbar\gamma_{\mathrm{H}}^2}{r^3_{\mathrm{H^C},H^N}}\right)^2\tau_c\left[-1+\frac{6}{1+4(\omega_{\mathrm{H}}\tau_c)^2}\right] \tag{30}$$

To get an impression of the size of the effect, Fig. 50 shows the cross-correlation rate $\Gamma^{\mathrm{NOE}}_{\mathrm{H^C,H^N}}$ between the two protons H_k^{α} and H_{k+1}^{N} in a protein as a function of the backbone angle ψ. The overall correlation time τ_c was assumed to be 6.4 ns.

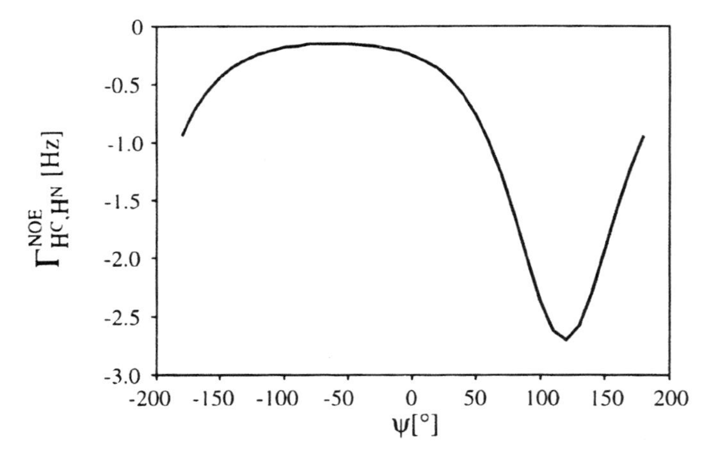

Figure 50. NOE cross-correlation rate $\Gamma^{NOE}_{HC,HN}$ between the protons H^{α}_{k} and H^{N}_{k+1} in a protein as a function of the peptide backbone angle ψ (τ_c = 6.4 ns) according to Eq. (30) ψ is correlated with the distance between the two protons by the relation given in Fig. 56b.

4.4. Practical Extraction Procedure

The relative signs of the relaxation rates of the individual lines $\alpha\alpha$, $\alpha\beta$, $\beta\alpha$, and $\beta\beta$ can be written—as stated above—in the secular approximation as follows for the DQ spectrum

$$\Gamma^{DQ}_{\alpha\alpha} = +\Gamma^{a} + \Gamma^{c}_{NH,CH} + \Gamma^{c}_{N,NH} + \Gamma^{c}_{C,NH} + \Gamma^{c}_{N,CH} + \Gamma^{c}_{C,CH} + W_2$$

$$\Gamma^{DQ}_{\alpha\beta} = +\Gamma^{a} - \Gamma^{c}_{NH,CH} - \Gamma^{c}_{N,NH} - \Gamma^{c}_{C,NH} + \Gamma^{c}_{N,CH} + \Gamma^{c}_{C,CH} + W_0$$

$$\Gamma^{DQ}_{\beta\alpha} = +\Gamma^{a} - \Gamma^{c}_{NH,CH} + \Gamma^{c}_{N,NH} + \Gamma^{c}_{C,NH} - \Gamma^{c}_{N,CH} - \Gamma^{c}_{C,CH} + W_0$$

$$\Gamma^{DQ}_{\beta\beta} = +\Gamma^{a} + \Gamma^{c}_{NH,CH} - \Gamma^{c}_{N,NH} - \Gamma^{c}_{C,NH} - \Gamma^{c}_{N,CH} - \Gamma^{c}_{C,CH} + W_2 \qquad (31a)$$

and the ZQ spectrum

$$\Gamma^{ZQ}_{\alpha\alpha} = +\Gamma^{a} + \Gamma^{c}_{NH,CH} - \Gamma^{c}_{N,NH} + \Gamma^{c}_{C,NH} - \Gamma^{c}_{N,CH} + \Gamma^{c}_{C,CH} + W_2$$

$$\Gamma^{ZQ}_{\alpha\beta} = +\Gamma^{a} - \Gamma^{c}_{NH,CH} + \Gamma^{c}_{N,NH} - \Gamma^{c}_{C,NH} - \Gamma^{c}_{N,CH} + \Gamma^{c}_{C,CH} + W_0$$

$$\Gamma^{ZQ}_{\beta\alpha} = +\Gamma^{a} - \Gamma^{c}_{NH,CH} - \Gamma^{c}_{N,NH} + \Gamma^{c}_{C,NH} + \Gamma^{c}_{N,CH} - \Gamma^{c}_{C,CH} + W_0$$

$$\Gamma^{ZQ}_{\beta\beta} = + \Gamma^{a} + \Gamma^{c}_{NH,CH} + \Gamma^{c}_{N,NH} - \Gamma^{c}_{C,NH} + \Gamma^{c}_{N,CH} - \Gamma^{c}_{C,CH} + W_2 \qquad (31b)$$

The relaxation rate of a signal is reflected in the linewidth at half height. In the experiment described below, double and zero quantum coherences are evolved in a *constant-time* manner during time T. The relaxation rate of each multiplet component is directly reflected in the intensity of the signal by $I_{\mu\nu} \propto \exp(-\Gamma_{\mu\nu}T)$. Correspondingly, the cross-correlated relaxation rates can be extracted from the multiplet intensities according to

$$\Gamma^{c,DQ}_{NH,CH} = \frac{1}{4T} * \ln\left(\frac{I^{DQ}(\alpha\beta) * I^{DQ}(\beta\alpha)}{I^{DQ}(\alpha\alpha) * I^{DQ}(\beta\beta)}\right) - \frac{1}{2}(W_2 - W_0)$$

$$\Gamma^{c,ZQ}_{NH,CH} = \frac{1}{4T} * \ln\left(\frac{I^{ZQ}(\alpha\alpha) * I^{ZQ}(\beta\beta)}{I^{ZQ}(\alpha\beta) * I^{ZQ}(\beta\alpha)}\right) - \frac{1}{2}(W_2 - W_0) \qquad (32)$$

Note that this implies taking always the product of the intensities of the inner lines in the nominator and the product of the intensities of the outer lines in the denominator of the logarithm.

The dipole dipole cross-correlated relaxation rates on the left-hand side of Eq. (32) are obtained with simulations with the program WTEST (Madi and Ernst, 1989). The bases for these simulations are Eqs. (25) and (26). To obtain the angular information about the included projection angle between the bond vectors $N–H^N$ and $C^{\alpha}–H^{\alpha}$, Eq. (27), which describes the angular dependence of the dipole dipole cross-correlation rate, is combined with Eq. (32).

The four cross-correlated relaxation rates $\Gamma^{c}_{N,NH}$, $\Gamma^{c}_{C,NH}$, $\Gamma^{c}_{N,CH}$, and $\Gamma^{c}_{C,CH}$ can be extracted from Eq. (31) in a similar way as the dipole dipole cross-correlated relaxation rates $\Gamma^{c}_{NH,CH}$ were obtained. The single dipole (NH)–CSA cross-correlation rates are given by

$$\Gamma^{c}_{N,NH} = \frac{1}{8T} * \ln\left(\frac{I^{DQ}(\alpha\beta) * I^{DQ}(\beta\beta)}{I^{DQ}(\alpha\alpha) * I^{DQ}(\beta\alpha)}\right) * \left(\frac{I^{ZQ}(\alpha\beta) * I^{ZQ}(\beta\beta)}{I^{ZQ}(\alpha\alpha) * I^{ZQ}(\beta\alpha)}\right)$$

$$\Gamma^{c}_{C,NH} = \frac{1}{8T} * \ln\left(\frac{I^{DQ}(\alpha\beta) * I^{DQ}(\beta\beta)}{I^{DQ}(\alpha\alpha) * I^{DQ}(\beta\alpha)}\right) * \left(\frac{I^{ZQ}(\alpha\alpha) * I^{ZQ}(\beta\alpha)}{I^{ZQ}(\alpha\beta) * I^{ZQ}(\beta\beta)}\right) \qquad (33)$$

and similarly for the dipole (CH)–CSA cross-correlated relaxation rate

$$\Gamma^{c}_{C,CH} = \frac{1}{8T} * \ln\left(\frac{I^{DQ}(\beta\beta) * I^{DQ}(\beta\alpha)}{I^{DQ}(\alpha\alpha) * I^{DQ}(\alpha\beta)}\right) * \left(\frac{I^{ZQ}(\beta\beta) * I^{ZQ}(\beta\alpha)}{I^{ZQ}(\alpha\alpha) * I^{ZQ}(\alpha\beta)}\right)$$

$$\Gamma^{c}_{N,CH} = \frac{1}{8T} * \ln\left(\frac{I^{DQ}(\beta\beta) * I^{DQ}(\beta\alpha)}{I^{DQ}(\alpha\alpha) * I^{DQ}(\alpha\beta)}\right) * \left(\frac{I^{ZQ}(\alpha\alpha) * I^{ZQ}(\alpha\beta)}{I^{ZQ}(\beta\beta) * I^{ZQ}(\beta\alpha)}\right) \qquad (34)$$

The orientation and size of the CSA tensor are known from solid-state NMR studies both for amides (Hartzell *et al.*, 1987; Lumsden *et al.*, 1994) and for aliphatic carbons (Janes *et al.*, 1983) and can be used for these studies in high-resolution NMR. A review of investigations of CSA tensors of all kinds of heteronuclei that have been determined by means of solid-state NMR is given by Duncan (1990).

As average values for the main components of the ^{15}N CSA tensor for a peptide, one finds in the literature (Duncan, 1990): $\sigma_{11} = 223 \pm 7$ ppm, $\sigma_{22} = 79 \pm 8$ ppm, $\sigma_{33} = 55 \pm 9$ ppm, and therefore $\Delta\sigma = \sigma_{\parallel} - \sigma_{\perp} = 156$ ppm. The orientation of the ^{15}N CSA tensor is indicated in Fig. 51. The ^{13}C CSA tensors for aliphatic carbons only show small anisotropy values. One finds the following values for L-threonine (Janes *et al.*, 1983): $\sigma_{11} = 69.0 \pm 0.4$ ppm, $\sigma_{22} = 58.9 \pm 0.4$ ppm, and $\sigma_{33} = 52.6 \pm 0.3$ ppm. Therefore, DQ and ZQ spectra should show the same rates with respect to the scalar $^1J_{CH}$ coupling and one would expect no change of the cross-correlation rate by variation of the angle between the main axis of the C^α CSA tensor and the NH vector.

4.5. Constant-Time versus Real-Time Evolution

Up to now, cross-correlation between two different interactions was considered to take place during a constant time period. Under real-time evolution, the rates are no longer reflected in a simple way in the intensities except that the line shape of each multiplet component is known including all small long-range coupling constants. Therefore, the cross-correlated relaxation rates must be extracted differently. If we consider a multiplet with four lines with an arbitrary but constant line shape $L(\omega)$ for all multiplet components on top of the Lorentzian line shape as described by Eq. (31), the Fourier transformation including an apodization function $w(t)$ will yield the following line shape $F_{\mu\nu}(\omega)$ for the multiplet line $I_{\mu\nu}$:

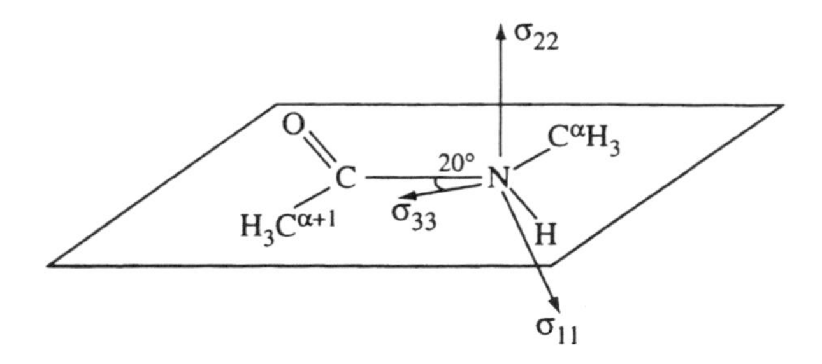

Figure 51. Orientation of the ^{15}N CSA tensor in a peptide according to Lumsden *et al.* (1994). σ_{11} and σ_{33} are oriented in the peptide plane, where σ_{33} is rotated about ca. 20° relative to the C′N bond. σ_{22} is oriented orthogonal to the peptide plane. For an amide nitrogen, σ_{11} is the most shielded component of the CSA tensors.

$$F_{\mu\nu}(\omega) = L(\omega) \times FT[\exp(-\Gamma_{\mu\nu}t)] \times FT[w(t)] \tag{35}$$

As described before, the desired cross-correlated relaxation rate can be extracted from

$$\Gamma^{c}_{\mathrm{NH,CH}} + \frac{1}{2}(W_2 - W_0) = \frac{1}{4}(\Gamma_{\alpha\alpha} + \Gamma_{\beta\beta} - \Gamma_{\alpha\beta} - \Gamma_{\beta\alpha}) \tag{36}$$

The difference of the rates of the multiplet components, e.g., $\Gamma_{\alpha\beta} - \Gamma_{\alpha\alpha}$, can be obtained by fitting the line shape $F_{\alpha\alpha}(\omega)$ to the line shape $F_{\alpha\beta}(\omega)$ by convolution ($\times$) of the $\alpha\alpha$ multiplet component with a Lorentzian with a trial linewidth $(\Gamma_{\alpha\beta} - \Gamma_{\alpha\alpha})^{\mathrm{trial}}$. The best fit for $(\Gamma_{\alpha\beta} - \Gamma_{\alpha\alpha})^{\mathrm{trial}}$ and similarly $(\Gamma_{\beta\alpha} - \Gamma_{\beta\beta})^{\mathrm{trial}}$ yields the desired cross-correlated relaxation rate. An example of the application of this technique is given by Reif *et al.* (1998).

4.6. Nonsecular Terms in the DQ/ZQ Relaxation Matrix

To be able to evaluate properly the effect of cross-correlated relaxation, the involved scalar couplings must be resolved. However, the resolved couplings also ensure that the relaxation matrix in Eq. (24) is faithfully evaluated taking into account only the secular components. Therefore, spectral density functions of nonzero frequency $j^q(\omega \neq 0)$, proton T_1 relaxation and proton proton NOE that introduce nonsecular terms in the DQ/ZQ relaxation matrix $\Gamma^{\mathrm{DQ/ZQ}}$ can be ignored. These terms lead to coherence transfer between the multiplet components. We show here the influence of these terms that eventually limit the applicability of the method to molecules that are very large by NMR standards.

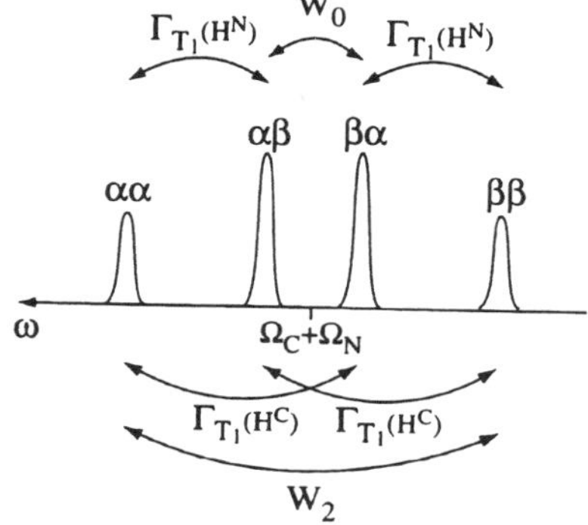

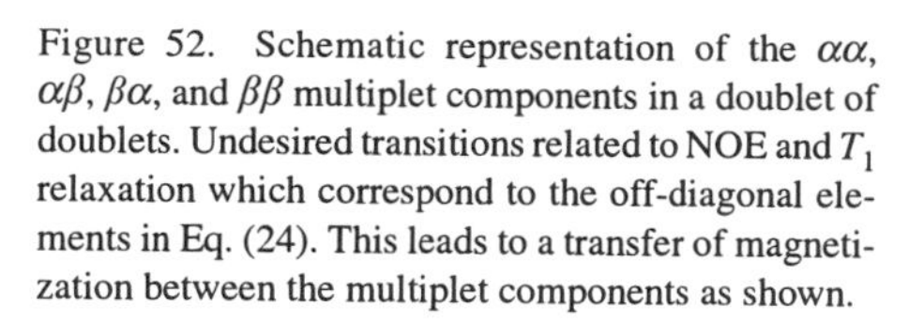
Figure 52. Schematic representation of the $\alpha\alpha$, $\alpha\beta$, $\beta\alpha$, and $\beta\beta$ multiplet components in a doublet of doublets. Undesired transitions related to NOE and T_1 relaxation which correspond to the off-diagonal elements in Eq. (24). This leads to a transfer of magnetization between the multiplet components as shown.

4.6.1. Contributions to the Cross-Correlated Relaxation Rate $\Gamma^{c}_{NH,CH}$ from Spectral Density Functions of Higher Order

In Eq. (26), only double commutators have been considered so far with Larmor frequency $\omega = 0$. We show here that additional contributions from other spectral density functions either do not exist or contribute only nonsecular relaxation terms. Except for the $A^{(\pm1)}_{NH}$, $A^{(\mp1)}_{CH}$ terms with the C and N operators being longitudinal $\sqrt{\frac{3}{2}}\, N_z H^{\pm}_{N}$ and $\sqrt{\frac{3}{2}}\, C_z H^{\pm}_{N}$, there are no further contributions related to incompatible Larmor frequency in the rotating frame. Evaluation of one of the four possible permutations of the double commutator yields

$$b_{NH} b_{CH} [\sqrt{\tfrac{3}{2}}\, H^{+}_{N} N_z, [\sqrt{\tfrac{3}{2}}\, H^{-}_{C} C_z, C^{+}N^{+}]] * j^{1}_{NH,CH}(\omega_H) = \Gamma^{c,q=1}_{NH,CH}\,(H^{-}_{C} H^{+}_{N} C^{+} N^{+}) \quad (37)$$

This is a nonsecular term, provided the chemical shifts of the two involved protons are different. We show for this example that this relaxation channel does not have any effect on the cross-correlation rate $\Gamma^{c}_{NH,CH}$. The subset of operators connected in the Liouville–von Neumann differential equation by the double commutator of Eq. (37) is given in Eq. (38). The differential equation describes a transition between the lines of $C^{+}N^{+}$ and $C^{+}N^{+} H^{+}_{N} H^{-}_{C}$

$$\frac{d}{dt}\begin{pmatrix} C^{+}N^{+} \\ C^{+}N^{+}H^{+}_{N}H^{-}_{C} \end{pmatrix} = \begin{pmatrix} 0 & \Gamma \\ \Gamma & i(\Omega_{H^N} - \Omega_{H^C}) \end{pmatrix} \begin{pmatrix} C^{+}N^{+} \\ C^{+}N^{+}H^{+}_{N}H^{-}_{C} \end{pmatrix} \quad (38)$$

The Γ terms in the matrix expression of Eq. (39) are obtained after evaluation of the double commutator from Eq. (33). The relative chemical shift of $C^{+}N^{+}$ and $C^{+}N^{+} H^{+}_{N} H^{-}_{C}$ is given by 0 and $\Delta\Omega = \Omega_{HN} - \Omega_{HC}$, respectively. The eigenvalues λ_1, λ_2 of the matrix are

$$\lambda_{1/2} = i\frac{\Delta\Omega}{2} + \sqrt{-\frac{(\Delta\Omega)^2}{4} + \Gamma^2} \overset{\Delta\Omega >> \Gamma}{\approx} i\,\frac{\Delta\Omega}{2}\left[1 \pm \left(1 - \frac{2\Gamma^2}{\Delta\Omega^2}\right)\right]$$

$$\lambda_1 \approx i\Delta\Omega$$

$$\lambda_2 \approx i\frac{\Gamma^2}{\Delta\Omega} \quad (39)$$

It is obvious that for $\Gamma < \Delta\Omega/2$, both eigenvalues are purely imaginary. This is fulfilled for H^C and H^N because of their large difference in chemical shifts. Therefore, as long as this inequality holds, there is no contribution to the linewidths of the doublet of doublets of $C^{+}N^{+}$, but only a contribution to the relative chemical shift originating from double commutators of higher order.

4.6.2. Proton T_1 Relaxation

In addition, scalar relaxation of the second kind (Abragam, 1961; Harbison, 1993; Norwood, 1993) related to longitudinal T_1 relaxation through the interaction of the two involved protons with other nuclei also contributes nonsecular elements to the relaxation matrix Γ^{DQ} of Eq. (24). T_1 relaxation equilibrates the intensities of the α line and β multiplet line of a given doublet which can lead to an underestimation of the angle-dependent cross-correlation rate (Fig. 53). This will be discussed next.

To evaluate the contribution related to T_1 relaxation, it is sufficient to consider here a resonance line that is split by a scalar coupling in an α- and a β-multiplet component. Their respective relaxation rates are Γ_α, Γ_β and their relative chemical shifts $\pm\pi J$. The rates Γ_α and Γ_β contain all of the secular contributions. Furthermore, intensity is transferred from the α line to the β line by means of T_1 relaxation leading to exchange of coherence between the α line and β line with the rate $1/T_1$. The system can therefore be described as

$$\begin{pmatrix} I^\alpha \\ I^\beta \end{pmatrix}^{\bullet} = -\Gamma \begin{pmatrix} I^\alpha \\ I^\beta \end{pmatrix} = \begin{pmatrix} -\Gamma_\alpha - \frac{1}{T_1} + i\pi J & \frac{1}{T_1} \\ \frac{1}{T_1} & -\Gamma_\beta - \frac{1}{T_1} - i\pi J \end{pmatrix} \begin{pmatrix} I^\alpha \\ I^\beta \end{pmatrix} \tag{40}$$

Γ denotes the rate matrix. The eigenvalues λ_1 and λ_2 of the matrix Γ are

$$\lambda_{1/2} = -\left(\Gamma_\Sigma + \frac{1}{T_1}\right) \pm \sqrt{(i\pi J - \Gamma_\Delta)^2 + \left(\frac{1}{T_1}\right)^2} \tag{41}$$

with $\Gamma_\Sigma = 1/2(\Gamma_\alpha + \Gamma_\beta)$ and $\Gamma_\Delta = 1/2(\Gamma_\alpha - \Gamma_\beta)$. After separation of the real and imaginary parts, one obtains for the root expression $(\lambda_1 - \lambda_2)/2$:

$$(\lambda_1 - \lambda_2)/2 = \sqrt{(i\pi J - \Gamma_\Delta)^2 + \left(\frac{1}{T_1}\right)^2}$$

$$= \frac{1}{\sqrt{2}} \sqrt{\sqrt{(\pi J)^4 - 2(\pi J)^2\left(\frac{1}{T_1^2} - \Gamma_\Delta^2\right) + \left(\frac{1}{T_1^2} + \Gamma_\Delta^2\right)^2} + \left(\frac{1}{T_1^2} + \Gamma_\Delta^2 - (\pi J)^2\right)}$$

$$- \frac{i}{\sqrt{2}} \sqrt{\sqrt{(\pi J)^4 - 2(\pi J)^2\left(\frac{1}{T_1^2} - \Gamma_\Delta^2\right) + \left(\frac{1}{T_1^2} + \Gamma_\Delta^2\right)^2} - \left(\frac{1}{T_1^2} + \Gamma_\Delta^2 - (\pi J)^2\right)}$$

$$= \frac{1}{2}\Gamma^c - \frac{i}{2}\Delta\Omega^{\alpha,\beta} \tag{42}$$

The real part contains the effective relaxation rate of lines α and β, the imaginary part the influence on the relative line position. The effects are illustrated in Fig. 54

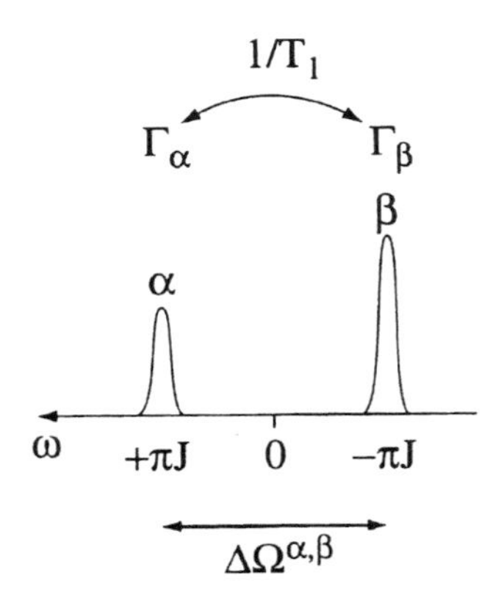

Figure 53. Schematic representation of a signal that is split by a scalar coupling of the size $2\pi J$ into a doublet. T_1 relaxation leads to an averaging of signal intensity on the α and the β line and therefore to a downscaling of the cross-correlated relaxation rate $\Gamma_c = \mathrm{Re}(\lambda_1 - \lambda_2)$ as derived from Eq. (42).

for assumed cross-correlation rates of 20 Hz (a,c) and 5 Hz (b). The splitting of the two lines is related to the NH coupling, which was assumed to be 90 Hz. Figure 54 shows that the cross-correlated relaxation rate is influenced by T_1 relaxation only for rates $R_1 = 1/T_1$ on the order of π^1J_{HN}. This is independent of the assumed cross-correlation rate of 20 or 5 Hz. As the rate $R_1 = 1/T_1$ increases, the difference of the chemical shifts of the two signals of the doublet becomes smaller. Expansion of the root expression in Eq. (42) according to Taylor yields the eigenvalues

$$\lambda_1 = -\Gamma_\alpha - \frac{1}{T_1} \pm i\sqrt{(\pi J)^2 - \frac{1}{T_1^2}}$$
$$\lambda_2 = -\Gamma_\beta - \frac{1}{T_1} \pm i\sqrt{(\pi J)^2 - \frac{1}{T_1^2}} \tag{43}$$

Therefore, the dipole dipole cross-correlated relaxation rate is independent of T_1 relaxation, as long as $\pi J >> 1/T_1$.

4.7. Experimental Implementation

The backbone angle ψ in proteins (Fig. 55) is relatively difficult to access by means of conventional NMR spectroscopic parameters. Both the scalar $^3J(\mathrm{H}^{\mathrm{N}}_{k+1}, \mathrm{C}^{\alpha}_{k})$ coupling (Wang and Bax, 1995) as well as distance measurements between the protons H^{α}_{k} and $\mathrm{H}^{\mathrm{N}}_{k+1}$ (Wüthrich, 1986) (Fig. 56b) are too inaccurate to define the angle properly. A different approach consists in the measurement of the relative displacements of the ^{1}H and ^{15}N resonance frequencies in 1:1 mixtures of D_2O and H_2O (Ottiger and Bax, 1997). It turns out that the solvent-induced chemical shift is a function of the backbone geometry around ψ. Yang *et al.* (1997) also suggested measuring the backbone angle ψ in a protein based on cross-correlated relaxation between the H^{α}–C^{α} dipolar and the C′ chemical shift anisotropy relaxation mechanism.

Measurement of the parameter of cross-correlated relaxation rate of double and zero quantum coherences involving the two dipolar vectors $C_k^{\alpha} - H_k^{\alpha}$ and $N_{k+1} - H_{k+1}^{N}$ allows a quite accurate determination of the backbone angle ψ. This is discussed next.

Correlation of θ and ψ, as shown in Fig. 56, reveals that regions of different secondary structure elements, such as α-helices and β-sheets, can be differentiated. The $3\cos^2\theta - 1$ dependence of the cross-correlation is also not degenerate for the two secondary structure elements. The solid line $\cos(\theta) = 0.163 + 0.819*\cos(\psi - 119°)$ can be obtained by means of geometrical considerations. Bond lengths of $NH^N = 1.03$ Å, $C^{\alpha}H^{\alpha} = 1.09$ Å, $NC^{\alpha} = 1.47$ Å, $NC' = 1.33$ Å, and $C'C^{\alpha} = 1.52$ Å, as well as tetrahedral symmetry and the planarity of the peptide backbone have been assumed.

The pulse sequence, shown in Fig. 57, is essentially a HNCOCA-like experiment (Bax and Ikura, 1991). A correlation between the two pairs of atoms $C_k^{\alpha} - H_k^{\alpha}$ and $N_{k+1} - H_{k+1}^{N}$ is achieved by excitation of double and zero quantum coherences during t_2. Starting from the proton H_{k+1}^{N}, magnetization is transferred to

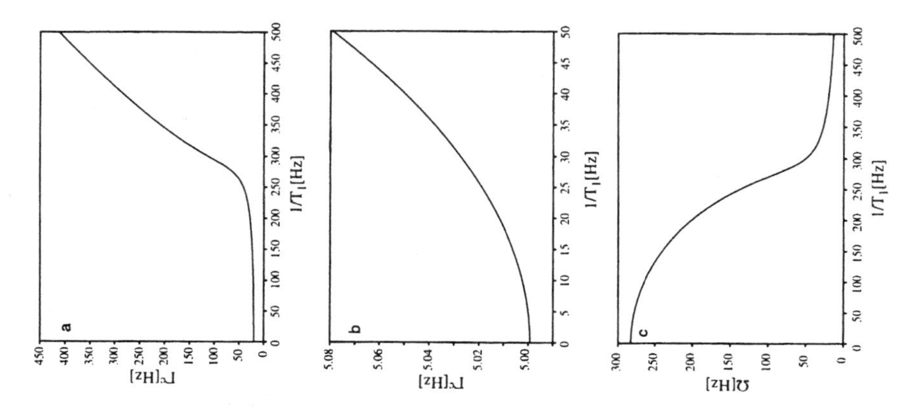

Figure 54. Cross-correlated relaxation rate Γ^c of a doublet resonance line (a,b) and the relative chemical shift of one signal of the doublet $\Delta\Omega^{\alpha,\beta}/2$ (c) as a function of the proton T_1 time. The simulation is based on cross-correlation rates of 20 Hz (a,c) and 5 Hz (b). The two resonance lines of the signal are split by a scalar coupling of 90 Hz. The difference of real and imaginary parts of the eigenvalues $(\lambda_1 - \lambda_2)$ as obtained from Eq. (42) is shown. The real part corresponds to the cross-correlated relaxation rate, the imaginary part to the difference of chemical shift of the α and β component of the resonance line. The effects of dipole dipole cross-correlation are averaged out by the T_1 relaxation when the rate $1/T_1$ is on the order of the scalar coupling. In this case, however, the doublet is not revealed anymore because of the decrease of α- and β-frequency difference $\Delta\Omega^{\alpha,\beta}$ with increasing rate $1/T_1$ (c).

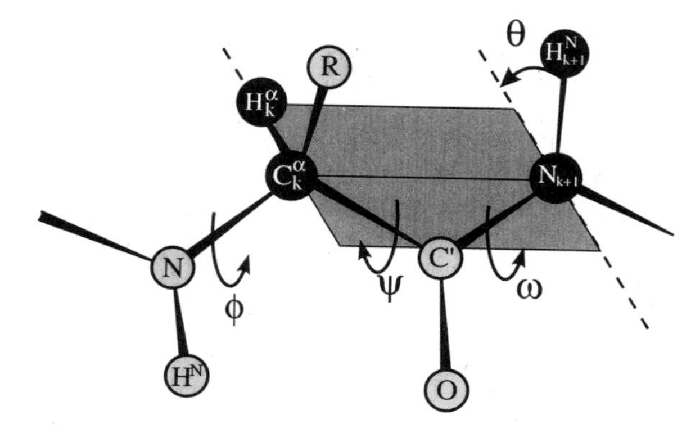

Figure 55. Schematic representation of the peptide backbone with backbone angles ϕ, ψ, ω, and θ of amino acids k and k+1 in a protein. Assuming a planar backbone geometry, the measured angle θ can be correlated with the backbone angle ψ (see text). The plane that is spanned by the atoms H_k^α, C_k^α, and N_{k+1} is highlighted.

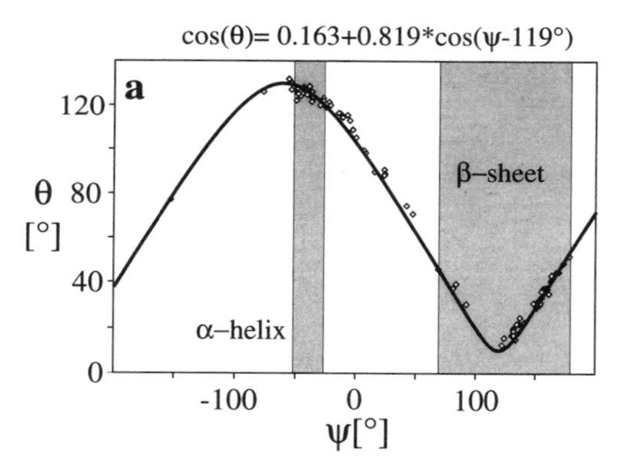

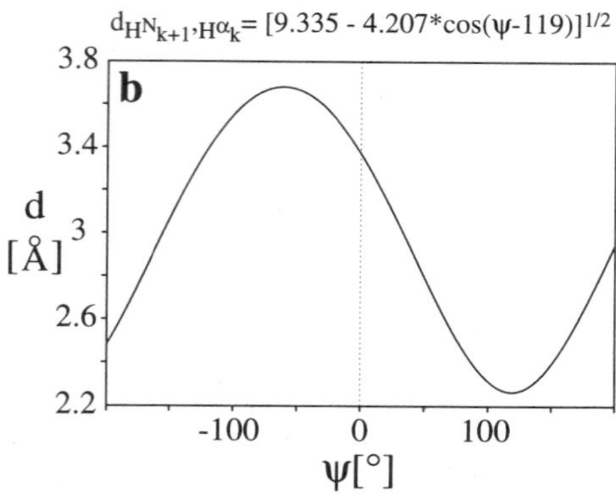

Figure 56. (a) Correlation of the angles θ and ψ in the protein backbone from Fig. 55. Actually assumed values of pairs of θ and ψ are depicted as diamonds for the protein rhodniin. (b) Distance between the H_{k+1}^N and the H_k^α proton as a function of the backbone angle ψ in a protein.

the nitrogen N_{k+1}. After a further INEPT transfer over C'_k, magnetization is finally located on the carbon C^{α}_k. Through application of two simultaneous 90° pulses on ^{15}N and $^{13}C^{\alpha}$, DQ and ZQ coherences are excited for a constant time τ''. The delay τ'' is optimized with respect to the $C^{\alpha}C^{\beta}$ coupling.

Proton magnetization is not decoupled during t_2. The experiment is symmetric around t_2. In the backtransfer step to the detected proton H^N_{k+1}, a COS-CT (coherence order selective coherence transfer) element is employed for optimal sensitivity.

The experiment has been applied to the protein rhodniin, a thrombin inhibitor. The protein consists of 103 amino acids and folds into two domains that are connected via a flexible linker. Each domain contains a triple-stranded β-sheet and an α-helical region. Experimentally obtained spectra are shown in Fig. 58. The doublet of doublet lines are not completely symmetric with respect to the resonance frequency. This is related to a different X-CSA contribution which can be observed on the multiplet lines correlated with the $^1J_{HX}$ coupling (X = C^{α} or N).

The cross-correlated relaxation rates $\Gamma^c_{NH,CH}$ can be extracted from the intensities of the lines $\alpha\alpha$, $\alpha\beta$, $\beta\alpha$, and $\beta\beta$ with the procedure described above. All extracted rates have been collected in Fig. 59. The ψ angles on the x-axis have been taken from an X-PLOR structure calculated using NOE and dihedral restraints. The experimental as well as the theoretical cross-correlation rates $\Gamma^c_{NH,CH}$ are drawn on the y-axis.

Dipole (NH)–dipole (CH) cross-correlated relaxation rates tend to assume positive values for β-sheet regions, whereas they are around zero for α-helices. Coils and turns show negative rates. The theoretical angle dependent cross-correlation rate is drawn with a solid line in Figures 59 and 61. The curve was simulated with the progam WTEST (Madi and Ernst, 1989) assuming an overall correlation time of the molecule of τ_c = 6.0 ns. This value is consistent with ^{15}N relaxation measurements (Hennig, 1996) that yield a value of 6.4 ns for the second domain and a value of 5.4 ns for the first domain in rhodniin.

Figure 60 shows the experimental cross-correlation rates $\Gamma^c_{NH,CH}$ together with the respective ψ angles as obtained from structure calculations using deviations from the theoretical rate curve as restraints in the X-PLOR protocol as additional restraints. If the trajectories contain more than one possible ψ value, all possible values were taken into account and are indicated in the figure by a double-sided arrow. After the refinement procedure, the rmsd value for the backbone dropped from 1.4 Å to 1.2 Å, which indicates that the measured cross-correlation rates are valuable for structure refinement.

Similar to a Karplus relation, the $3\cos^2\theta - 1$ function of the cross-correlated relaxation rate can also assume different ψ values for a single cross-correlation rate. Interpretation of the cross-correlated relaxation between the $(C^{\alpha}-H^{\alpha})_k$ vector and the N_{k+1} CSA tensor can help to work around these ambiguities (Fig. 61).

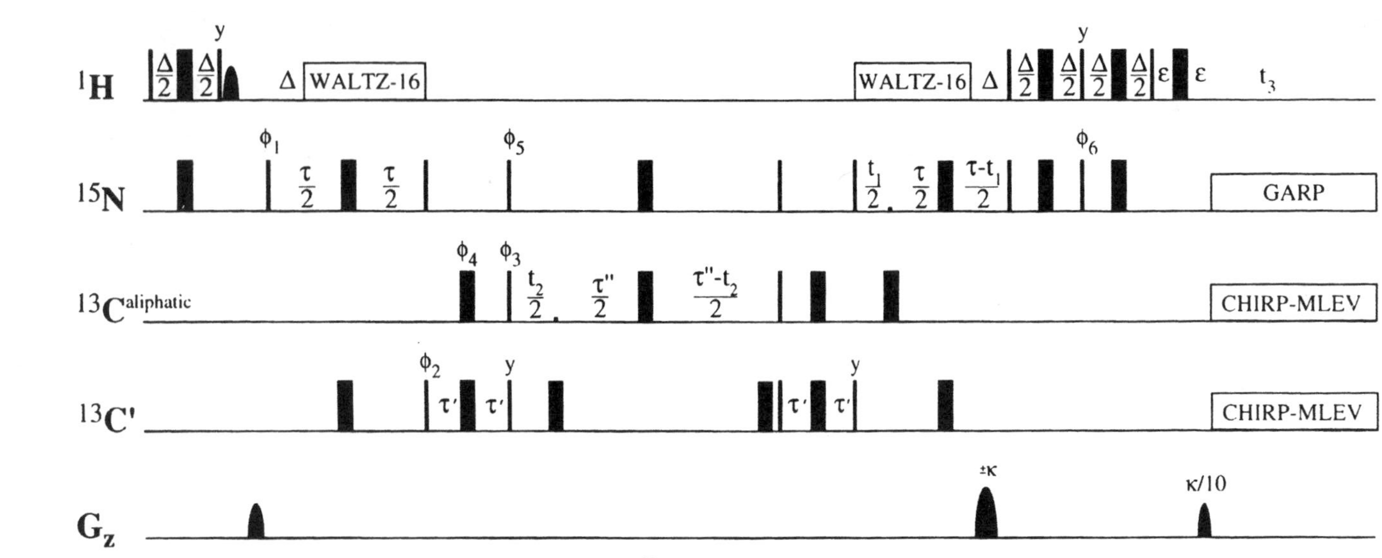

Figure 57. HNCOCA-like pulse sequence for the measurement of N–H^N, C^α-H^α projection angles. DQ and ZQ coherences evolve chemical shift during t_1. Delays are as follows: $\Delta = 5$ ms, $2\tau' = 9$ ms, $\tau'' = 26$ ms, $\varepsilon = 1.2$ ms. G3 and G4 Gaussian cascades (Emsley and Bodenhausen, 1989) have been used as selective 90° and 180° pulses on C^α and C′ resonances. If not otherwise indicated, RF pulses have phase *x*. $\phi_1 = x, -x$, $\phi_2 = 2(x), 2(-x)$, $\phi_3 = 4(x), 4(-x)$, $\phi_5 = 8(x), 8(-x)$, $\phi_{rec} = \phi_1 + \phi_2 + \phi_3 + \phi_5$. Quadrature in t_2 is achieved by variation of phases ϕ_3 and ϕ_4 in States–TPPI manner. Echo–antiecho coherences are selected during t_1 by inversion of phases $\phi_6 = -y$ together with the sign of the second gradient (Palmer *et al.*, 1991; Kay *et al.*, 1992; Schleucher *et al.*, 1993; Sattler *et al.*, 1995). The phases ϕ_3 and ϕ_5 are shifted by 90° in subsequent FIDS and stored separately to be able to differentiate DQ and ZQ coherences during t_2. Aliphatic and carbonyl resonances are decoupled during acquisition using MLEV expanded CHIRP pulses (Fu and Bodenhausen, 1995).

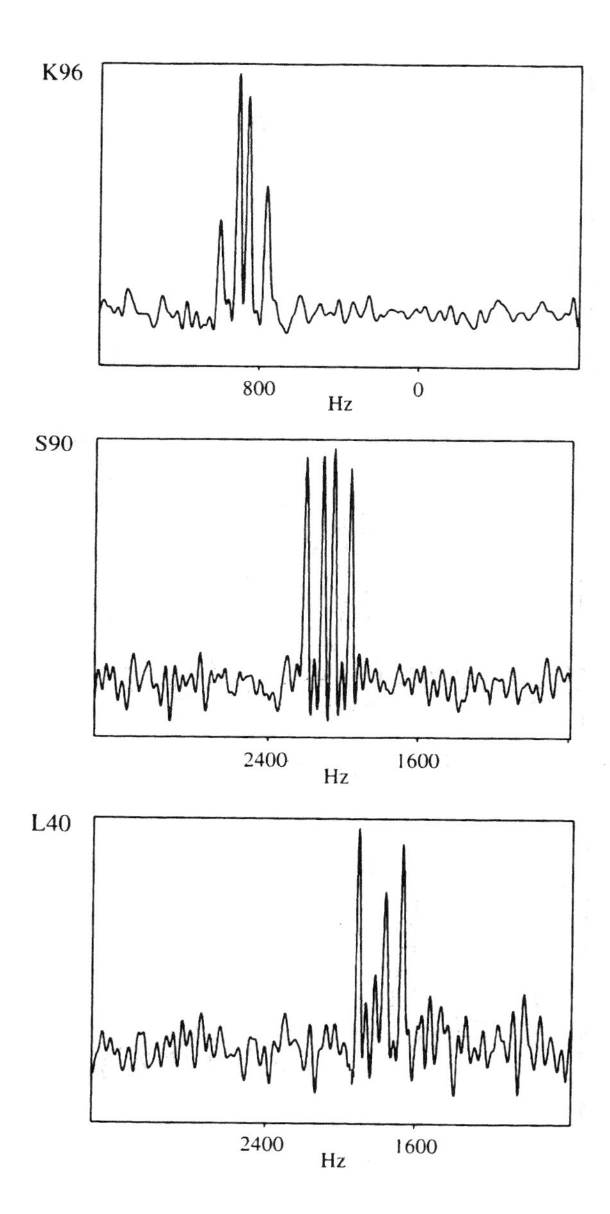

Figure 58. 1D strips from the 3D DQ/ZQ HNCOCA for the residues K96, S90, and L40 in rhodniin corresponding to backbone angles ψ_{k-1}.

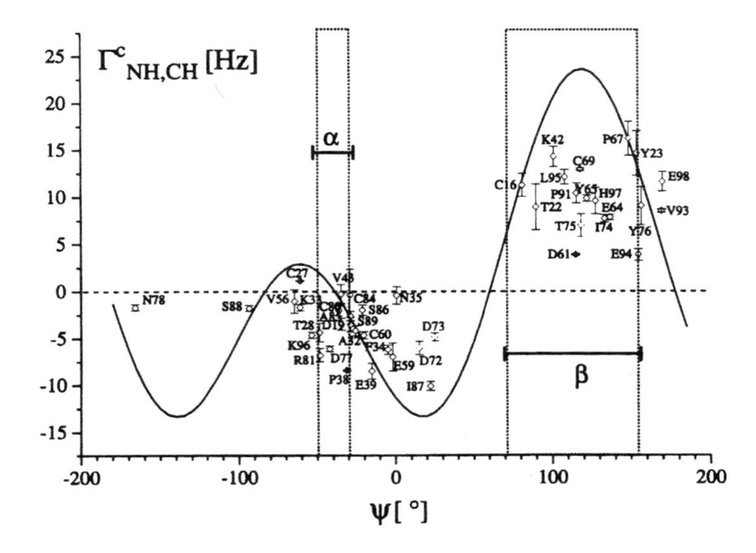

Figure 59. Graphical representation of the extracted dipole $(NH)_{k+1}$–dipole$(CH)_k$ cross-correlated relaxation rates as a function of the backbone angle ψ_k as found from structure calculations (before refinement). The solid line corresponds to the theoretical dipole dipole cross-correlation rate obtained with WTEST for a correlation time $\tau_c = 6.0$ ns.

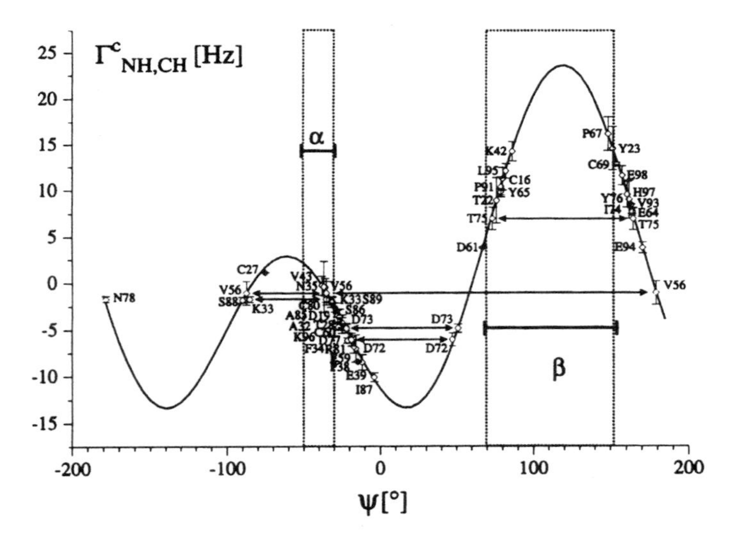

Figure 60. Graphical representation of the extracted dipole $(NH)_{k+1}$–dipole $(CH)_k$ cross-correlated relaxation rates as a function of the backbone angle ψ_k as found from structure calculations (after refinement).

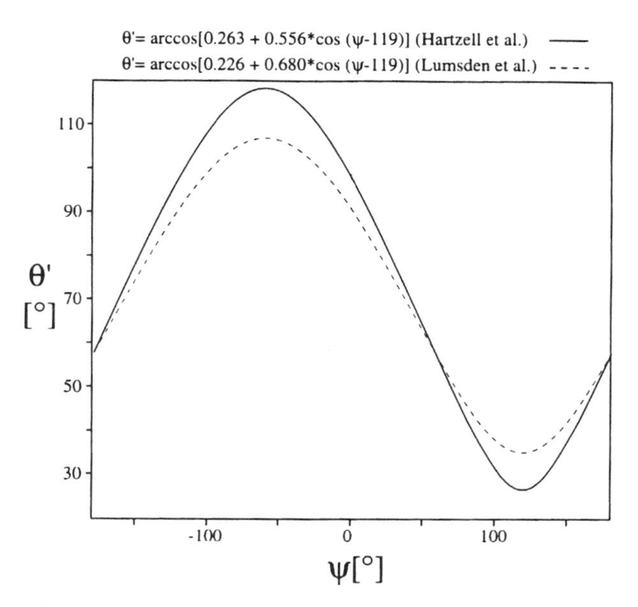

Figure 61. $\theta'[(C^{\alpha}H^{\alpha})_k,\mathrm{CSA(N)}_{k+1}]$ as a function of the backbone angle ψ_k. The orientation of the CSA tensor is taken from solid-state NMR spectra (Hartzell *et al.*, 1987; Lumsden *et al.*, 1994).

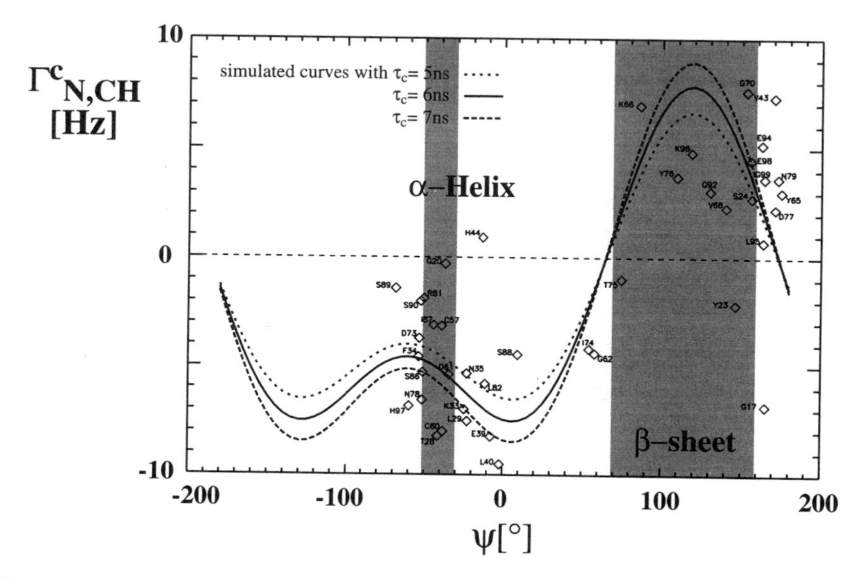

Figure 62. Graphical representation of the extracted dipole $(C^{\alpha}H^{\alpha})_k$–CSA $(N)_{k+1}$ cross-correlated relaxation rate $\Gamma^c_{N,CH}$ as a function of the backbone angle ψ_{k+1}.

Figure 62 shows the dipole $(C^{\alpha}H^{\alpha})_k$–CSA $(N)_{k+1}$ cross-correlated relaxation rate $\Gamma^{c}_{N,CH}$ as a function of the backbone angle ψ_{k+1} as found in structures calculated with X-PLOR using only NOE and dihedral restraints.

In principle, the cross-correlated relaxation rate $\Gamma^{c}_{N,NH}$ between the ^{15}N CSA tensor and the N–H^N vector as well as the cross-correlation rate $\Gamma^{c}_{C,CH}$ between the $^{13}C^{\alpha}$ CSA tensor and the $C^{\alpha}H^{\alpha}$ vector can be obtained. However, these rates can be measured more reliably by means of ^{15}N and $^{13}C^{\alpha}$ relaxation measurements.

REFERENCES

Aboul-ela, F., and Varani, G., 1995, *Curr. Opin. Biotechnol.* **6:**89.

Abragam, A., 1961, in *Principles of Nuclear Magnetism*, Oxford University Press (Clarendon), London.

Agrofoglio, L. A., Jacquinet, J.-C., and Lancelot, G., 1997, *Tetrahedron* **38:**1411.

Allain, F. H.-T., and Varani, G., 1997, *J. Mol. Biol.* **267:**338.

Altona, C., and Sundaralingam, M., 1972, *J. Am. Chem. Soc.* **94:**8205.

Altona, C., Geise, H. J., and Romers, C., 1968, *Tetrahedron* **24:**13.

Babu, Y. S., Sack, J. S., Greenhough, T. J., Bugg, C. E., Means, A. R., and Cook, W. J., 1985, *Nature* **315:**37.

Babu, Y. S., Bugg, C. E., and Cook, W. J., 1988, *J. Mol. Biol.* **204:**191.

Bastiaan, E. W., MacLean, C., van Zijl, P. C. M., and Bothner-By, A. A., 1987, *Annu. Rep. NMR Spectrosc.* 35.

Batey, R. T., Inada, M., Kujawinski, E., Puglisi, J. D., and Williamson, J. R., 1992, *Nucleic Acids Res.* **20:**4515.

Batey, R. T., Battiste, J. L., and Williamson, J. R., 1995, *Methods Enzymol.* **261:**300.

Bax, A., 1994, *Curr. Opin. Struct. Biol.* **4:**738.

Bax, A., and Ikura, M., 1991, *J. Biomol. NMR* **1:**99.

Bax, A., Max, D., and Zax, D., 1992, *J. Am. Chem. Soc.* **114:**6923.

Bermel, W., Wagner, K., and Griesinger, C., 1989, *J. Magn. Reson.* **83:**223.

Blake, P. R., Summers, M. F., Adams, M. W. W., Park, J.-B., Zhou, Z. H., and Bax, A., 1992, *J. Biomol. NMR* **2:**527.

Bock, K., and Pederson, C., 1977, *Acta Chim. Scand. Ser. B* **31:**354.

Böhlen, J. M., and Bodenhausen, G., 1993, *J. Magn. Reson. Ser. A* **102:**293.

Brüschweiler, R., and Case, D. A., 1994, *Prog. NMR Spectrosc.* **26:**27.

Brüschweiler, R., Griesinger, C., and Ernst, R. R., 1989, *J. Am. Chem. Soc.* **111:**8034.

Brutscher, B., Skrynnikov, N. R., Bremi, T., Brüschweiler, R., and Ernst, R. R., 1998, *J. Magn. Reson.* in press.

Bystrov, V. F., 1976, *Prog. NMR Spectrosc.* **10:**41.

Carlomagno, T., Schwalbe, H., Rexroth, A., Sørensen, O. W., and Griesinger, C., *J. Magn. Reson.* in press.

Cavanagh, J., Fairbrother, W. J., Palmer, A. G., III, and Skelton, N. J., 1996, in *Protein NMR Spectroscopy*, Academic Press, San Diego.

Chu, W.-C., and Horowitz, J., 1992, *FEBS Lett.* **295:**159.

Chu, W.-C., Feiz, V., Derrick, W. B., and Horowitz, J., 1992, *J. Mol. Biol.* **227:**1164.

Crespi, H. L., Rosenberg, R. M., and Katz, J. J., 1968, *Science* **161:**795.

Cyr, N., and Perlin, A. S., 1979, *Can. J. Chem.* **57:**2504.

Cyr, N., Hamer, G. K., and Perlin, A. S., 1978, *Can. J. Chem.* **56:**297.

Dabbagh, G., Weli, D. P., and Tycko, R., 1994, *Macromolecules* **27:**6183.

Dalvit, C., and Bodenhausen, G., 1988, *J. Am. Chem. Soc.* **110:**7924.
Davies, D. B., Rajani, P., MacCoss, M., and Danyluk, S. S., 1985, *Magn. Reson. Chem.* **23:**72.
Dayie, K. T., and Wagner, G., 1997, *J. Am. Chem. Soc.* **119:**7797.
Delaglio, F., Torchia, D. A., and Bax, A., 1991, *J. Biomol. NMR* **1:**439.
Duncan, T. M., 1990, in *A Compilation of Chemical Shift Anisotropies*, The Farragut Press, Chicago.
Eaton, H. L., Fesik, S. W., Glaser, S. J., and Drobny, G. P., 1990, *J. Magn. Reson.* **90:**425.
Eggenberger, U., Schmidt, P., Sattler, M., Glaser, S. J., and Griesinger, C., 1992, *J. Magn. Reson.* **100:**604.
Emsley, L., and Bodenhausen, G., 1989, *J. Magn. Reson.* **82:**211.
Emsley, L., and Bodenhausen, G., 1990, *Chem. Phys. Lett.* **165:**469.
Ernst, M., and Ernst, R. R., 1994, *J. Magn. Reson. Ser. A* **110:**202.
Ernst, R. R., Bodenhausen, G., and Wokaun, A., 1989, in *Principles of Nuclear Magnetic Resonance in One and Two Dimensions*, Oxford University Press (Claredon), London.
Farmer, B. T., II, and Venters, R. A., 1995, *J. Am. Chem. Soc.* **117:**4187.
Farmer, B. T., II, Müller, L., Nikonowicz, E. P., and Pardi, A., 1993, *J. Am. Chem. Soc.* **115:**11040.
Farmer, B. T., II, Müller, L., Nikonowicz, E. P., and Pardi, A., 1994, *J. Biomol. NMR* **4:**129.
Feng, X., Lee, Y. K., Sandstrøm, D., Edén, M., Maisel, H., Sebald, A., and Levitt, M. H., 1996, *Chem. Phys. Lett.* **257:**314.
Feng, X., Verdegem, P. J. E., Lee, Y. K., Sandstrøm, D., Edén, M., Bovee-Geurts, P., de Grip, J. W., Lugtenburg, J., de Groot, H. J. M., and Levitt, M. H., 1997, *J. Am. Chem. Soc.* **119:**6853.
Freeman, R., Wittekoek, S., and Ernst, R. R., 1970, *J. Chem. Phys.* **52:**1529.
Fu, R., and Bodenhausen, G., 1995, *Chem. Phys. Lett.* **245:**415.
Gardner, K. H., and Kay, L. E., 1997, *J. Am. Chem. Soc.* **119:**7599.
Gardner, K. H., Rosen , M. K., and Kay, L. E., 1997, *Biochemistry* **36:**1389.
Glaser, S. J., and Quant, J. J., 1996, *Adv. Magn. Opt. Reson.* **19:**60.
Glaser, S. J., Schwalbe, H., Marino, J. P., and Griesinger, C., 1996, *J. Magn. Reson. Ser. B* **112:**160.
Goldman, M., 1988, in *Quantum Description of High-Resolution NMR in Liquids*, Oxford University Press (Clarendon), London.
Grant, D. M., and Werbelow, L. G., 1976, *J. Magn. Reson.* **21:**369.
Griesinger, C., and Eggenberger, U., 1992, *J. Magn. Reson.* **75:**426.
Griesinger, C., Sørensen, O. W., and Ernst, R. R., 1985, *J. Am. Chem. Soc.* **107:**6394.
Griesinger, C., Sørensen, O. W., and Ernst, R. R., 1986, *J. Chem. Phys.* **85:**6837.
Griesinger, C., Sørensen, O. W., and Ernst, R. R., 1987, *J. Magn. Reson.* **75:**474.
Grzesiek, S., and Bax, A., 1997, *J. Biomol. NMR* **9:**207.
Grzesiek, S., Anglister, J., Ren, H., and Bax, A., 1993a, *J. Am. Chem. Soc.* **115:**4369.
Grzesiek, S., Vuister, G. W., and Bax, A., 1993b, *J. Biomol. NMR.* **3:**487.
Grzesiek, S., Wingfield, P., Stahl, S., Kaufman, J. D., and Bax, A., 1995, *J. Am. Chem. Soc.* **117:**9594.
Haasnoot, C. A. G., deLeeuw, F. A. A. M., deLeeuw, H. P. M., and Altona, C., 1981, *Org. Magn. Reson.* **15:**43.
Hansen, P. E., 1988, *Prog. NMR Spectrosc.* **20:**207.
Harbison, G. S., 1993, *J. Am. Chem. Soc.* **115:**3026.
Hartzell, C. J., Whitfield, M., Oas, T. G., and Drobny, G. P., 1987, *J. Am. Chem. Soc.* **109:**5966.
Hennig, M., 1996, *Diplomarbeit*, Frankfurt/M.
Hennig, M., Ott, D., Schulte, P., Löwe, R., Krebs, J., Vorherr, T., Bermel, W., Schwalbe, H., and Griesinger, C., 1997, *J. Am. Chem. Soc.* **119:**5055.
Hester, R. K., Ackermann, J. L., Neff, B. L., and Waugh, J. S., 1976, *Phys. Rev. Lett.* **36:**1081.
Heus, H. A., Wijmenga, S. S., van de Ven, F. J. M., and Hilbers, C. W., 1994, *J. Am. Chem. Soc.* **116:**4983.
Hines, J. V., Laundry, S. M., Varani, G., and Tinoco, I., Jr., 1993, *J. Am. Chem. Soc.* **115:**11002.
Hines, J. V., Laundry, S. M., Varani, G., and Tinoco, I., Jr., 1994, *J. Am. Chem. Soc.* **116:**5823.
Hong, M., Gross, J. D., and Griffin, R. G., 1997, *J. Phys. Chem. B* **101:**5869.

Hu, J.-S., and Bax, A., 1996, *J. Am. Chem. Soc.* **118:**8170.
Hu, J.-S., and Bax, A., 1997a, *J. Biomol. NMR* **9:**323.
Hu, J.-S., and Bax, A., 1997c, *J. Am. Chem. Soc.* **119:**6360.
Hu, J.-S., Grzesiek, S., and Bax, A., 1997b, *J. Am. Chem. Soc.* **119:**1803.
Hubbard, P. S., 1958, *Phys. Rev.* **109:**1153.
Hubbard, P. S., 1969, *J. Chem. Phys.* **51:**1647.
Hubbard, P. S., 1970, *J. Chem. Phys.* **52:**563.
Ikura, M., Clore, G. M., Gronenborn, A. M., Zhu, G., Klee, C. B., and Bax, A., 1992, *Science* **256:**632.
Ippel, J. H.; Wijmenga, S. S., de Jong, R., Heus, H. A., Hilbers, C. W., de Vroom, E., van der IUPAC-IUB Commission on Biochemical Nomenclature, 1970, *J. Mol. Biol.* **52:**1.
Janes, N., Ganapathy, S., and Oldfield, E., 1983, *J. Magn. Reson.* **54:**111.
Kainosho, M., 1997, *Nature Struct. Biol.* **4:**858, and references therein.
Kalinowski, H.-O., Berger, S., and Braun, S., 1994, in *NMR-Spektroskopie von Nichtmetallen*, Vol. IV, *^{19}F-NMR Spektroskopie*, Thieme Verlag, Stuttgart.
Karplus, M., 1959, *J. Chem. Phys.* **30:**11.
Karplus, M., 1963, *J. Am. Chem. Soc.* **85:**2870.
Kay, L. E., Keifer, O., and Saarinen, T., 1992, *J. Am. Chem. Soc.* **114:**10663.
Kessler, H., Anders, U., and Gemmecker, G., 1988, *J. Magn. Reson.* **78:**582.
Kilpatrick, J. E., Pitzer, K. S., and Spitzer, R., 1947, *J. Am. Chem. Soc.* **69:**2483.
Konrat, R., Muhandiram, D. R., Farrow, N. A., and Kay, L. E., 1997, *J. Biomol. NMR* **9:**409.
Krivdin, L. B., and Della, E. W., 1991, *Prog. NMR Spectrosc.* **23:**301.
Kuhlmann, K. F., and Baldeschweiler, J. D., 1965, *J. Chem. Phys.* **43:**572.
Kung, H. C., Wang, K. Y., Goljer, I., and Bolton, P. H., 1995, *J. Magn. Reson. Ser. B* **109:**323.
Kupče, E., and Freeman, R., 1995, *J. Magn. Reson. Ser. A* **115:**273.
Kurz, M., Schmieder, P., and Kessler, H., 1991, *Angew. Chem.* **103:**1341.
Kushlan, D. M., and LeMaster, D. M., 1993, *J. Biomol. NMR* **3:**701.
Lankhorst, P. P., Haagnoot, C. A. G., Erkelens, C., and Altona, C., 1984, *J. Biomol. Struct. Dyn.* **1:**1387.
Legault, P., Jucker, F. M., and Pardi, A., 1995, *FEBS Lett.* **362:**156.
Lemieux, R. U., Nagabhushan, T. L., and Paul, B., 1972, *Can. J. Chem.* **50:**773.
Levitt, M. H., Freeman, R., and Frenkiel, T., 1982, *J. Magn. Reson.* **47:**328.
Linder, M., Höhener, A., and Ernst, R. R., 1980, *J. Chem. Phys.* **73:**4959.
Lipari, G., and Szabo, A. J., 1982, *J. Am. Chem. Soc.* **104:**4546.
Lohman, J. A. B., and MacLean, C., 1978a, *Chem. Phys.* **35:**269.
Lohman, J. A. B., and MacLean, C., 1978b, *Chem. Phys.* **43:**144.
Lohman, J. A. B., and MacLean, C., 1979, *Mol. Phys.* **38:**1255.
Lumsden, M. D., Wasylishen, R. W., Eichele, K., Schindler. M., Penner, G. H., Power, W. P., and Curtis, R. D., 1994, *J. Am. Chem. Soc.* **116:**1403.
Madi, Z., and Ernst, R. R., 1989, *WTEST program package*, Eidgenössische Technische Hochschule Zürich.
Marino, J. P., Schwalbe, H., Anklin, C., Bermel, W., Crothers, D. M., and Griesinger, C., 1994a, *J. Am. Chem. Soc.* **116:**6472.
Marino, J. P., Schwalbe, H., Anklin, C., Bermel, W., Crothers, D. M., and Griesinger, C., 1994b, *J. Biomol. NMR* **5:**87.
Marino, J. P., Schwalbe, H., Glaser, S. J., and Griesinger, C., 1996, *J. Am. Chem. Soc.* **118:**4388.
Marino, J. P., Diener, J. L., Moore, P. B., and Griesinger, C., 1997, *J. Am. Chem. Soc.* **119:**7361.
Marino, J. P., Schwalbe, H., and Griesinger, C., *Acc. Chem. Rev.*, in press.
Markley, J. L., Potter, I., and Jardetzky, O., 1968, *Science* **161:**1249.
Metzler, W. J., Wittekind, M., Goldfarb, V., Mueller, L., and Farmer, B. T., II, 1996, *J. Am. Chem. Soc.* **118:**6800.
Michnicka, M. J., Harper, J. W., and King, G. C., 1993, *Biochemistry* **32:**395.

Montelione, G. T., Winkler, M. E., Rauenbühler, P., and Wagner, G., 1989, *J. Magn. Reson.* **82:**198.
Mooren, M. M. W., Wijmenga, S. S., van der Marel, G. A., van Boom, J. H., and Hilbers, C. W., 1994, *Nucleic Acids Res.* **22:**2658.
Nietlispach, D., Clowes, R. T., Broadhurst, R. W., Ito, Y., Keeler, J., Kelly, M., Ashurst, J., Oschkinat, H., Domaille, P. J., and Laue, E. D., 1996, *J. Am. Chem. Soc.* **118:**407.
Nikonowicz, E. P., and Pardi, A., 1992, *Nature* **335:**184.
Nikonowicz, E. P., Sirr, A., Legault, P., Jucker, F. M., Baer, L. M., and Pardi, A., 1992, *Nucleic Acids Res.* **20:**4507.
Norwood, T. J., 1993, *J. Magn. Reson. Ser. A* **104:**106.
Olsen, H. B., Ludvigsen, S., and Sørensen, O. W., 1993, *J. Magn. Reson. Ser. A* **104:**226.
Ottiger, M., and Bax, A., 1997, *J. Am. Chem. Soc.* **119:**8070.
Otting, G., and Wüthrich, K., 1989, *J. Magn. Reson.* **85:**586.
Otting, G., and Wüthrich, K., 1990, *Q. Rev. Biophys.* **23:**39.
Otting, G., Senn, H., Wagner, K., and Wüthrich, K., 1986, *J. Magn. Reson.* **70:**500.
Palmer , A. G., III, Cavanagh, J., Wright, P. E., and Rance, M., 1991, *J. Magn. Reson.* **93:**151.
Parisot, D., Malet-Martino, M. C., Martino, R., and Crashnier, P., 1991, *Appl. Environ. Microbiol.* **57:**3605.
Plavec, J., and Chattopadhyaya, J., 1995, *Tetrahedron Lett.* **36:**1949.
Quant, S., Wechselberger, R. W., Wolter, M. A., Wörner, K.-H., Schell, P., Engels, J. W., Griesinger, C., and Schwalbe, H., 1994, *Tetrahedron Lett.* **35:**6649.
Ramos, A., Gubser, C. C., and Varani, G., 1997, *Curr. Opin. Struct. Biol.* **7:**317.
Rastinejad, F., and Lu, P., 1993, *J. Mol. Biol.* **232:**105.
Reif, B., Wittmann, V., Schwalbe, H., Griesinger, C., Wörner, K., Jahn-Hofmann, K., Engels, J. W., and Bermel, W., 1997a, *Helv. Chim. Acta* **80:**1952.
Reif, B., Hennig, M., and Griesinger, C., 1997b, *Science* **276:**1230.
Reif, B., Steinhagen, H., Junker, B., Reggelin, M., and Griesinger, C., 1998, *Angew. Chem. Int. Ed. Engl.* **37:**1903.
Rexroth, A., Schmidt, P., Szalma, S., Geppert, T., Schwalbe, H., and Griesinger, C., 1995, *J. Am. Chem. Soc.* **117:**10389.
Rosen, M. K., Gardner, K. H., Willis, R. C., Parris, W. E., Pawson, T., and Kay, L. E., 1996, *J. Mol. Biol.* **263:**627.
Sänger, W., 1988, in *Principles of Nucleic Acids Structure*, 2nd ed., Springer, Berlin.
Sattler, M., Schwalbe, H., and Griesinger, C., 1992, *J. Am. Chem. Soc.* **114:**1126.
Sattler, M., Schwendinger, M. G., Schleucher, J., and Griesinger, C., 1995, *J. Biomol. NMR* **5:**11.
Schleucher, J., Schwörer, B., Zirngibl, C., Koch, U., Weber, W., Egert, E., Thauer, R. K., and Griesinger, C., 1992, *FEBS* **31:**440.
Schleucher, J., Sattler, M., and Griesinger, C., 1993, *Angew. Chem. Int. Ed. Engl.* **32:**1489.
Schmidt-Rohr, K., 1996a, *Macromolecules* **29:**3975.
Schmidt-Rohr, K., 1996b, *J. Am. Chem. Soc.* **118:**7601.
Schmieder, P., Kurz, M., and Kessler, H., 1991, *J. Biomol. NMR* **1:**403.
Schmieder, P., Ippel, J. H., van der Elst, H., van der Marel, G. A., Boom, J. H., Altona, C., and Kessler, H., 1992, *Nucleic Acids Res.* **20:**4747.
Schneider, H., 1964, *Z. Naturforsch. A* **19:**510.
Schwalbe, H., Samstag, W., Engels, J. W., Bermel, W., and Griesinger, C., 1993a, *J. Biomol. NMR* **3:**479.
Schwalbe, H., Rexroth, A., Eggenberger, U., Geppert, T., and Griesinger, C., 1993b, *J. Am. Chem. Soc.* **115:**7878.
Schwalbe, H., Marino, J. P., King, G. C., Wechselberger, R., Bermel, W., and Griesinger, C., 1994, *J. Biomol. NMR* **4:**631.
Schwalbe, H., Schmidt, P., and Griesinger, C., 1995a, in *Encyclopedia of NMR* (D. M. Grant and R. K. Harris, eds.), Wiley, New York, 1489.

Schwalbe, H., Marino, J. P., Glaser, S. J., and Griesinger, C., 1995b, *J. Am. Chem. Soc.* **117:**7251.
Schwalbe, H., Fiebig, K. M., Buck, M., Jones, J. A., Grimshow, S. B., Spencer, A., Glaser, S. J., Smith, L. J., and Dobson, C. M., 1997, *Biochemistry*, **36:**8977.
Schwarcz, J. A., Cyr, N., and Perlin, A. S., 1975, *Can. J. Chem.* **53:**1872.
Seip, S., Balbach, J., and Kessler, H., 1994, *J. Magn. Reson. Ser. B* **104:**172.
Shan, X., Gardner, K. H., Muhandiram, D. R., Rao, N. S., Arrowsmith, C. H., and Kay, L. E., 1996, *J. Am. Chem. Soc.* **118:**6570.
Shimizu, H., 1969, *J. Chem. Phys.* **37:**765.
Simorre, J. S., Zimmerman, G. R., Pardi, A., Farmer, B. T., II, and Müller, L., 1995, *J. Biomol. NMR* **6:**427.
Simorre, J. S., Zimmerman, G. R., Müller, L., and Pardi, A., 1996, *J. Biomol. NMR* **7:**153.
Sklenar, V., Peterson, R. D., Rejante, M. R., and Feigon, J., 1993, *J. Biomol. NMR* **3:**721.
Sklenar, V., Peterson, R. D., Rejante, M. R., and Feigon, J., 1994, *J. Biomol. NMR* **4:**117.
Sklenar, V., Dieckman, T., Butcher, S. E., and Feigon, J., 1996, *J. Biomol. NMR* **7:**83.
Smith, D. E., Su, J.-Y., and Jucker, F. M., 1997, *J. Biomol. NMR* **10:**245.
Tate, S., Ono, A., and Kainosho, M., 1993, *J. Am. Chem. Soc.* **116:**5977.
Taylor, D. A., Sack, J. S., Maune, J. F., Beckingham, K., and Quiocho, F. A., 1991, *J. Biol. Chem.* **266:**21375.
Tjandra, N., and Bax, A., 1997a, *J. Magn. Reson.* **124:**512.
Tjandra, N., Grzesiek, S., and Bax, A., 1996, *J. Am. Chem. Soc.* **118:**6264.
Tjandra, N., and Bax, A., 1997b, *Science* **278:**1111.
Tjandra, N., Omichinski, J. B., Gronenborn, A. M., Clore, G. M., and Bax, A., 1997, *Nature Struct. Biol.* **4:**732.
Tolman, J. R., and Prestegard, J. H., 1996, *J. Magn. Reson. Ser. B* **112:**245.
Tolman, J. R., Flanagan, J. M., Kennedy, M. A., and Prestegard, J. H., 1995, *Proc. Natl. Acad. Sci. USA* **92:**9279.
van de Ven, F. J. M., and Hilbers, C. W., 1988, *Eur. J. Biochem.* **178:**1.
Varani, G., Aboul-ela, F., Allain, F., and Gubser, C. C., 1995, *J. Biomol. NMR* **5:**315.
Varani, G., Aboul-ela, F., and Allain, F.H.-T., 1996, *Prog. NMR Spectrosc.* **29:**51.
Venters, R. A., Metzler, W. J., Spicer, L. D., Mueller, L., and Farmer, B. T., II, 1995, *J. Am. Chem. Soc.* **117:**9592.
Vijay-Kumar, S., Bugg, C. E., and Cook, W. J., 1987, *J. Mol. Biol.* **194:**531.
Vold, R. L., and Vold, R. R., 1978, *Prog. NMR Spectrosc.* **12:**79.
Vold, R. L., Vold, R. R., and Canet, D., 1977, *J. Chem. Phys.* **66:**1202.
Vuister, G. W., Wang, A. C., and Bax, A., 1993a, *J. Am. Chem. Soc.* **115:**5334.
Vuister, G. W., Yamazaki, T., Torchia, D. A., and Bax, A., 1993b, *J. Biomol. NMR* **3:**297.
Vuister, G. W., Grzesiek, S., Delaglio, S., Wang, A. C., Tschudin, R., and Bax, A., 1994, *Methods Enzymol.* **239:**79, and references therein.
Wang, A. C., and Bax, A., 1995, *J. Am. Chem. Soc.* **117:**1810.
Wang, A. C., and Bax, A., 1996, *J. Am. Chem. Soc.* **118:**2483.
Weisemann, R., Rüterjans, H., Schwalbe, H., Schleucher, J., Bermel, W., and Griesinger, C., 1994, *J. Biomol. NMR* **4:**231.
Werbelow, L. G., and Grant, 1977, D. M., *Adv. Magn. Reson.* **9:**189.
Wijmenga, S. S., Heus, H. A., van de Ven, F. J. M., and Hilbers, C. W., 1994, in *NMR on Biological Macromolecules* (C. I. Stassinopoulou, ed.), NATO ASI Series, **87:**307, Springer, Berlin.
Wijmenga, S. S., Heus, H. A., Leeuw, H. A., Hoppe, H., van der Graf, M., and Hilbers, C. W., 1995, *J. Biomol. NMR* **5:**82.
Woessner, D. E., 1962a, *J. Chem. Phys.* **36:**1.
Woessner, D. E., 1962b, *J. Chem. Phys.* **37:**647.
Wüthrich, K., 1986, in *NMR of Proteins and Nucleic Acids*, Wiley, New York.

Yamazaki, T., Lee, W., Arrowsmith, C. H., Muhandiram, D. R., and Kay, L. E., 1994, *J. Am. Chem. Soc.* **116:**11655.
Yang, D., Konrat, R., and Kay, L. E., 1997, *J. Am. Chem. Soc.* **119:**11938.
Zhu, G., Live, D., and Bax, A., 1994, *J. Am. Chem. Soc.* **116:**8370.
Zimmer, D. P., and Crothers, D. M., 1995, *Proc. Natl. Acad. Sci. USA* **92:**3091.
Zimmer, D. P., Marino, J. P., and Griesinger, C., 1996, *Magn. Reson. Chem.* **34:**S177.
Zuiderweg, E. R. P., and Fesik, S. W., 1991, *J. Magn. Reson.* **93:**653.

Contents of Previous Volumes

VOLUME 3

VOLUME 4

VOLUME 6

VOLUME 7

VOLUME 9

VOLUME 10

Index